T0324758

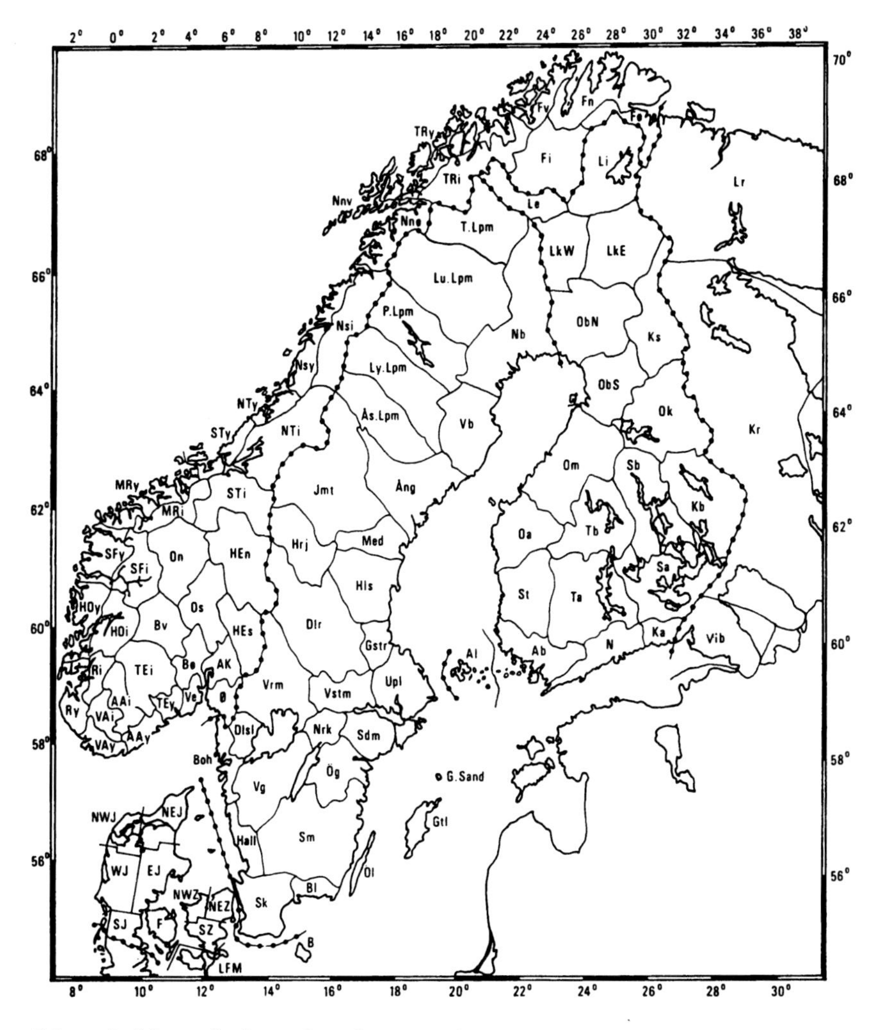

List of abbreviations for the provinces used throughout the text, on the map and in the following tables.

DENMARK

SJ	South Jutland	LFM	Lolland, Falster, Møn
EJ	East Jutland	SZ	South Zealand
WJ	West Jutland	NWZ	North West Zealand
NWJ	North West Jutland	NEZ	North East Zealand
NEJ	North East Jutland	B	Bornholm
F	Funen		

SWEDEN

Sk.	Skåne	Vrm.	Värmland
Bl.	Blekinge	Dlr.	Dalarna
Hall.	Halland	Gstr.	Gästrikland
Sm.	Småland	Hls.	Hälsingland
Öl.	Öland	Med.	Medelpad
Gtl.	Gotland	Hrj.	Härjedalen
G. Sand.	Gotska Sandön	Jmt.	Jämtland
Ög.	Östergötland	Ång.	Ångermanland
Vg.	Västergötland	Vb.	Västerbotten
Boh.	Bohuslän	Nb.	Norrbotten
Dlsl.	Dalsland	Ås. Lpm.	Åsele Lappmark
Nrk.	Närke	Ly. Lpm.	Lycksele Lappmark
Sdm.	Södermanland	P. Lpm.	Pite Lappmark
Upl.	Uppland	Lu. Lpm.	Lule Lappmark
Vstm.	Västmanland	T. Lpm.	Torne Lappmark

NORWAY

Ø	Østfold	HO	Hordaland
AK	Akershus	SF	Sogn og Fjordane
HE	Hedmark	MR	Møre og Romsdal
O	Opland	ST	Sør-Trøndelag
B	Buskerud	NT	Nord-Trøndelag
VE	Vestfold	Ns	southern Nordland
TE	Telemark	Nn	northern Nordland
AA	Aust-Agder	TR	Troms
VA	Vest-Agder	F	Finnmark
R	Rogaland		

n northern s southern ø eastern v western y outer i inner

FINLAND

Al	Alandia	Kb	Karelia borealis
Ab	Regio aboensis	Om	Ostrobottnia media
N	Nylandia	Ok	Ostrobottnia kajanensis
Ka	Karelia australis	ObS	Ostrobottnia borealis, S part
St	Satakunta	ObN	Ostrobottnia borealis, N part
Ta	Tavastia australis	Ks	Kuusamo
Sa	Savonia australis	LkW	Lapponia kemensis, W part
Oa	Ostrobottnia australis	LkE	Lapponia kemensis, E part
Tb	Tavastia borealis	Li	Lapponia inarensis
Sb	Savonia borealis	Le	Lapponia enontekiensis

USSR

Vib Regio Viburgensis Kr Karelia rossica Lr Lapponia rossica

FAUNA ENTOMOLOGICA SCANDINAVICA
Volume 7, part 2 1981

The Auchenorrhyncha (Homoptera) of Fennoscandia and Denmark

Part 2: The Families Cicadidae, Cercopidae, Membracidae, and Cicadellidae (excl. Deltocephalinae)

by

F. Ossiannilsson

SCANDINAVIAN SCIENCE PRESS LTD.
Klampenborg . Denmark

Fauna entomologica scandinavica
is edited by »Societas entomologica scandinavica«

World list abbreviation
Fauna ent. scand.

Printed by
Vinderup Bogtrykkeri A/S
7830 Vinderup, Denmark

ISBN 87-87491-36-2
ISSN 0106-8377

Family Cicadidae
Cicadas

Middle-sized or large species. Fore wings usually extended in length, often transparent, in repose held roof-like over the back. Wing veins usually strong, well-marked. In the fore wing they start from a closed so-called basal cell. First antennal segment concealed in a pit. Pronotum above with oblique furrows. Scutellum large, convex, caudally with an x-shaped swelling. Fore femora ventrally with 2–3 strong spines. Tymbal apparatus of males often with accessory structures absent in other Auchenorrhyncha, e.g. the so-called opercula, a pair of ventral horizontal plates directed caudad arising from the metathoracal epimera. These are much varying in size and shape, sometimes covering most of the abdominal venter, sometimes rudimentary. In the females a tymbal apparatus appears to be absent. A tympanal organ is present in both sexes. Larvae subterranean with fore legs modified for digging (Text-fig. 734). Their development lasts from 2 to several years. In Denmark and Fennoscandia only one genus with one species.

Genus *Cicadetta* Kolenati, 1857

Cicadetta Kolenati, 1857: 417 (as subgenus of *Cicada*).
  Type-species: *Cicada montana* Scopoli, 1772, by monotypy.
Melampsalta Kolenati, 1857: 425 (as subgenus of *Cicada*).
  Type-species: *Cicada musiva* Germar, 1830, by monotypy.

Head short, triangular. Pronotum at least as broad as head, fore border somewhat convex, hind border straight. The two longitudinal veins of corium arise close to each other or have a common stem from the narrow basal cell. Abdomen long, conical. Opercula present or missing. Fore femora beneath with three spines.

89. ***Cicadetta montana*** (Scopoli, 1772)
  Plate-fig. 37, text-fig. 734.

Cicada montana Scopoli, 1772: 109.
(Swedish: bergsångstrit, Finnish: vuorilaulukaskas).

Black, with a fine yellowish pubescence. A short longitudinal streak on pronotum and lateral borders of the x-shaped swelling on mesonotum often orange-coloured. Sometimes with a pair of orange-coloured longitudinal stripes on mesonotum. Hind margins of abdominal terga orange, abdominal venter and legs more or less largely yellowish or orange. Wings transparent, veins fuscous or partly reddish, costal border of fore wing pale orange-coloured, base of fore and hind wings bright orange. Opercula of male small, black with margins more or less broadly whitish. Length of body 16–20 mm, length with wings 23–28 mm.

Distribution. Not found in Denmark. – Rare in Sweden, found in Vg.: Kinnekulle by Gyllenhal and Bjerkander and in Skaraborg near Skara by Hackwitz. Boheman found it in Bohuslän, Lars Brundin in Sm.: Gårdsby. Several times captured in Ög.: by an anonymous collector according to a specimen in the collection of the former Swedish Plant Protection Institute, and by Holmgren (specimen in the Swedish State Museum of Natural History); Lohmander found it near the mansion of Stjärnorp; Ander recorded it from the vicinity of Söderköping. Notini captured a female in Sdm.: Botkyrka, Ahlby; S. Stjärnfelt found a specimen in Sdm.: Huddinge, Ågesta 15.VI.-1948; Gonnert took it in the same locality and in Huddinge: Stuvsta. Lundevall found *C. montana* in Upl.: Uppsala 27.VII.1937, and Lindroth captured a larva in Upl.: Lovö in May, 1946; Coulianos found a larval exuvium in Upl.: Värmdö, Boo, Sågsjön 16.VIII.-1959, and a female in Värmdö, Boo 5.VIII.1965; L.-Å. Jansson took a specimen in Värmdö, Velamsund 20.VI.1970. In June, 1948, Bo and Anders Tjeder heard something probably being the song of *C. montana* in Sk., Kivik, Bredarör. – Very rare in Norway, found in Bø: Modum 1832 (Esmark); AK: Etterstad (Siebke); Baerum, Kolsås 13.VI.-1875 (Sølsberg); Oslo, Grefsenåsen (N. G. Moe); TEy: Versvik near Porsgrunn 1937 and 1938 (K. Fægri and Gunvor Knaben). – Rare also in East Fennoscandia, found in Ab: Karislojo (Forsius, Rolf and Harry Krogerus, Håkan Lindberg), Lojo (Forsius, Frey, R. and H. Krogerus, K. Rein), Lojo, Storön (Håkan and P. H. Lindberg; Ta: Jockis (E. Bonsdorff); Sa: Rantasalmi (Westerlund). – Kr: Sortavala (Tiensuu), Kuujärvi (Platonoff), Kirjavalahti (Kanervo). – Austria, Belgium, Bulgaria, Bohemia, Moravia, Slovakia, France, German D. R. and F. R., England, Greece, Hungary, Italy, Poland, Romania, Yugoslavia, Spain, Switzerland, n., m., and s. Russia, Anatolia, Palestine, Azerbaijan, Georgia, Kazakhstan, Moldavia, Ukraine, Tadzhikistan, Siberia, China.

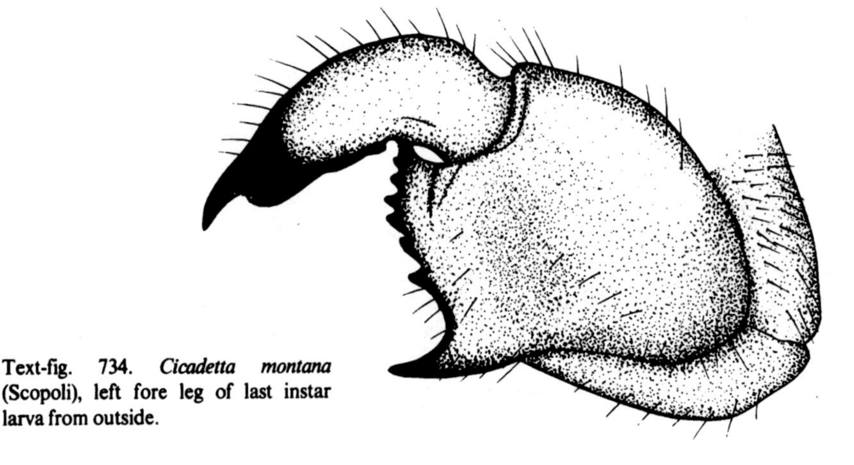

Text-fig. 734. *Cicadetta montana* (Scopoli), left fore leg of last instar larva from outside.

Biology. Hibernation takes place in the larval stage (Müller, 1957). – Mainly on *Quercus* and *Prunus spinosa* in the brushwood belt of dryish slopes (Schwoerbel, 1957). – "An warmen, trockenen Hängen. In den Nordostalpen mit Vorliebe an *Pinus silvestris*. Die Larven überwintern im Boden. – Die letzte Häutung findet meist an Grashalmen statt. Die Imagines leben in den Baumkronen, von denen sie bei Wind bisweilen abgeschüttelt werden" (Wagner & Franz, 1961). "Larven mehrjährig" (Schiemenz, 1969). "Stenotope Art der Trockenrasen" (Schiemenz, l.c.). – On warm, sun-exposed slopes of dry hills and on open south-exposed birch-wood hillsides (H. Krogerus in litt.). – Adults in June–August.

Family Cercopidae
Froghoppers

Body elongate or short and broad, structure robust. Frontoclypeus strongly swollen, partly visible from above as a rectangular, triangular, or transversely oval field enclosed in the vertex. Anterior surface of clypeus usually distinctly transversely furrowed. Pronotum frontally straight or convex, caudally more or less broadly W-shaped. Fore wings leathery, longer than abdomen, in resting position held roof-like over abdomen or together strongly vaulted. Fore border of hind wings basally with an accessory coupling apparatus consisting of a triangular lobe equipped with a number of strong hooks. Hind tibiae on outside with two strong spines, and apically with two transverse rows of spines (Text-fig. 8). First and second tarsal segments each with a transverse row of spines. In the larva, the terga and pleura of abdominal segments III–IX are curved beneath the abdomen, concealing the true sternum. – The larval development takes place in the well-known froth-lumps called "cuckoo spit", see p. 27. Our species are phlegmatic and slow in their movements but they can leap vigorously.

Key to genera of Cercopidae

1 Fore border of head without a groove 2

– Fore border of head with a distinct transverse groove either laterally above the antennae (in the vertex part) or medially (in the frontoclypeus part) 3

2 (1) Dorsum hairless. Pronotum frontally convex, angular
Peuceptyelus J. Sahlberg (p.228)

– Dorsum with a fine pilosity. Fore border of pronotum evenly arched, not angular .... *Lepyronia* Amyot & Serville (p.226)

3 (1) Pronotum medially with a distinct longitudinal carina. Distance between ocelli equals half distance between one ocellus and the compound eye of the same side. Large species
Aphrophora Germar (p.238)

- Pronotum medially without a carina but with a shallow longitudinal groove. Distance between ocelli approximately equal to distance between the ocellus and the compound eye of the same side. Smaller species 4

4 (3) Face in frontal aspect rhombic. Frontoclypeus (median) part of fore border of head longer than one of the (lateral) parts belonging to vertex. Outline of body seen from above with wings in resting position ovoid .... *Philaenus* Stål (p.248)

- Face in frontal aspect approximately triangular. Frontoclypeus part of fore border of head at most as long as one of the parts belonging to vertex. Dorsal outline of body with wings elongate, almost parallel-sided .... *Neophilaenus* Haupt (p.230)

Genus *Lepyronia* Amyot & Serville, 1843

Lepyronia Amyot & Serville, 1843: 567.
Type-species: *Cicada coleoptrata* Linné, 1758, by subsequent designation.

Body short and broad. Fore wings and dorsum of fore body finely punctured with a dense recumbent pilosity. Fore border of head without a groove. Dorsal plate of frontoclypeus broadly inversely cordiform. Rostrum short. Pronotum medially little longer than vertex, with a longitudinal furrow. Fore wings strongly convex, their veins almost obsolete. One species in Denmark and Fennoscandia.

90. ***Lepyronia coleoptrata*** (Linné, 1758)
Plate-figs. 38, 39, text-figs. 735–740.

Cicada coleoptrata Linné, 1758: 437.
Cercopis angulata Fabricius, 1794: 53.

Venter largely black, dorsum of fore body and fore wings greyish yellow to brownish yellow. Anterior part of dorsum especially in the male with more or less prevalent, diffusely delimited dark markings. Fore wings each with a fuscous L-shaped figure, the angle of which touches the costal border, a short longitudinal fuscous streak near wing basis, and often a fuscous spot in clavus at its commissural border. Fore wing veins and numerous transverse veinlets partially dark. Usually the light colour is predominant but in some individuals, especially males, the dark markings may extend very much, leaving only three light spots on each fore wing. Male pygofer as in Text-figs. 735, 736,

Text-figs. 735–740. *Lepyronia coleoptrata* (Linné). – 735: male pygofer from the left; 736: male pygofer in ventral aspect; 737: right genital style from below; 738: aedeagus in ventral aspect; 739: aedeagus from the left; 740: caudal part of female abdominal venter (pilosity not considered). Scale: 0.5 mm for 740, 0.1 mm for the rest.

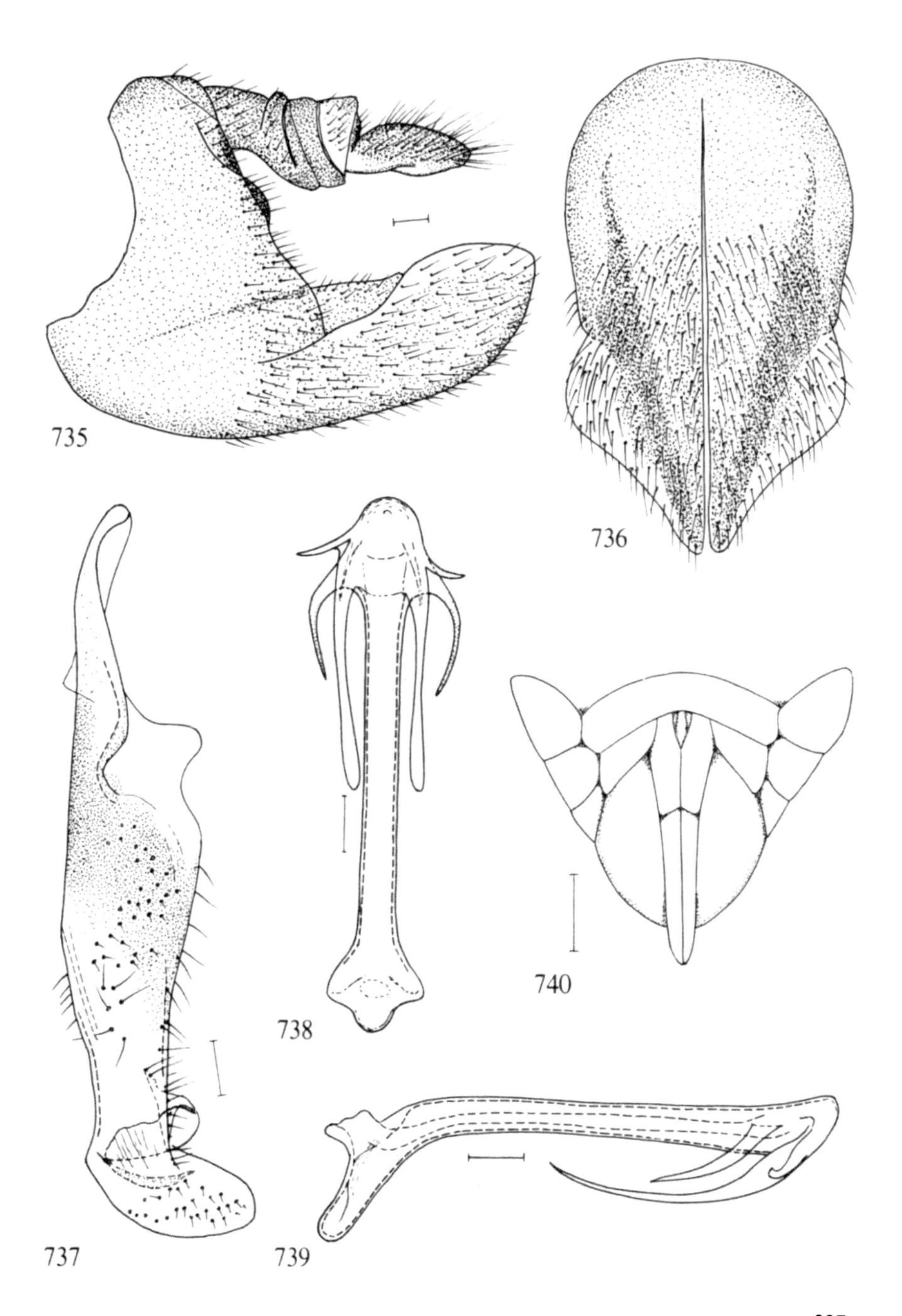
735
736
737
738
739
740

genital style as in Text-fig. 737, aedeagus as in Text-figs. 738, 739, ventral aspect of posterior part of female abdomen as in Text-fig. 740. Overall length of males 5.5–6 mm, of females 6.5–8 mm. – Larva (Plate-fig. 39) with sharply delimited jet-black markings on head and thorax. Fore border of head with a broad black transverse stripe. Abdomen with a distinct light middle stripe.

Distribution. Denmark: scarce, found in SJ, SZ, NEZ, and B. – Sweden: very local, locally abundant, found in Sk., Hall., Öl., Ög., Vg., Nrk., Sdm., Upl., Vstm., Gstr. – Appears to be sporadic in Norway but locally abundant in the vicinity of Oslo, found in AK, Bø, Bv, and VAy. – Comparatively common in the southern part of East Fennoscandia, found in Ab, N, Ka, St, Ta; Kr. – Europe (not Great Britain), Algeria, Iraq, Syria, Anatolia, Altai, Armenia, Azerbaijan, Georgia, Kazakhstan, Moldavia, Ukraine, Kirghizia, Uzbekistan, e. and m. Siberia, Mongolia.

Biology. Ossiannilsson (1946c) mentions *Galium mollugo, Hypericum perforatum, Rumex acetosella,* and *Cirsium arvense* as hostplants. Ossiannilsson (1950) found the larvae in their froth-lumps on *Salix pentandra, cinerea* and *repens, Populus tremula, Betula pubescens,* and *Corylus avellana,* mostly on adventitious shoots or on twigs of current and last year's growth; further on *Vaccinium myrtillus* and *uliginosum, Carex* sp., one or two undetermined grasses; *Polygonum viviparum; Anthriscus silvestris; Chamaenerium angustifolium; Trifolium pratense* and *hybridum; Vicia hirsuta; Lotus corniculatus; Lathyrus montanus; Rubus saxatilis; Potentilla anserina* and *argentea; Alchemilla* sp.; *Geum rivale; Filipendula ulmaria* and *hexapetala; Hypericum maculatum; Ranunculus acris; Plantago major; Galium mollugo* and *boreale; Taraxacum* sp.; *Artemisia vulgaris; Cirsium palustre; Equisetum arvense.* Apparently the species is polyphytophagous. It hibernates in the egg stage (Müller, 1957) or in the larval stages (Kuntze, 1937). Univoltine, adults in June–September.

Genus *Peuceptyelus* J. Sahlberg, 1871

Peuceptyelus J. Sahlberg, 1871: 84.
Type-species: *Cercopis coriacea* Fallén, 1826, by monotypy.

Body short and broad. Fore wings and dorsum of fore body strongly punctured, without pilosity. Vertex and pronotum with an indistinct median carina. Fore border of head without a groove. Space between ocelli (in our species) approximately as long as their distance from hind border of vertex and half as long as their distance from eyes. Rostrum long, reaching far beyond hind coxae. Pronotum about twice as long as vertex. Scutellum with a central depression. Veins of fore wings distinct, partly prominent. One species in Europe.

Text-figs. 741–747. *Peuceptyelus coriaceus* (Fallén). – 741: male pygofer from the left; 742: male pygofer from below; 743: genital style in ventral aspect; 744: genital style from outside; 745: aedeagus from the right; 746: aedeagus in ventral aspect; 747: ventral aspect of posterior abdominal segments in female. Scale: 0.1 mm.

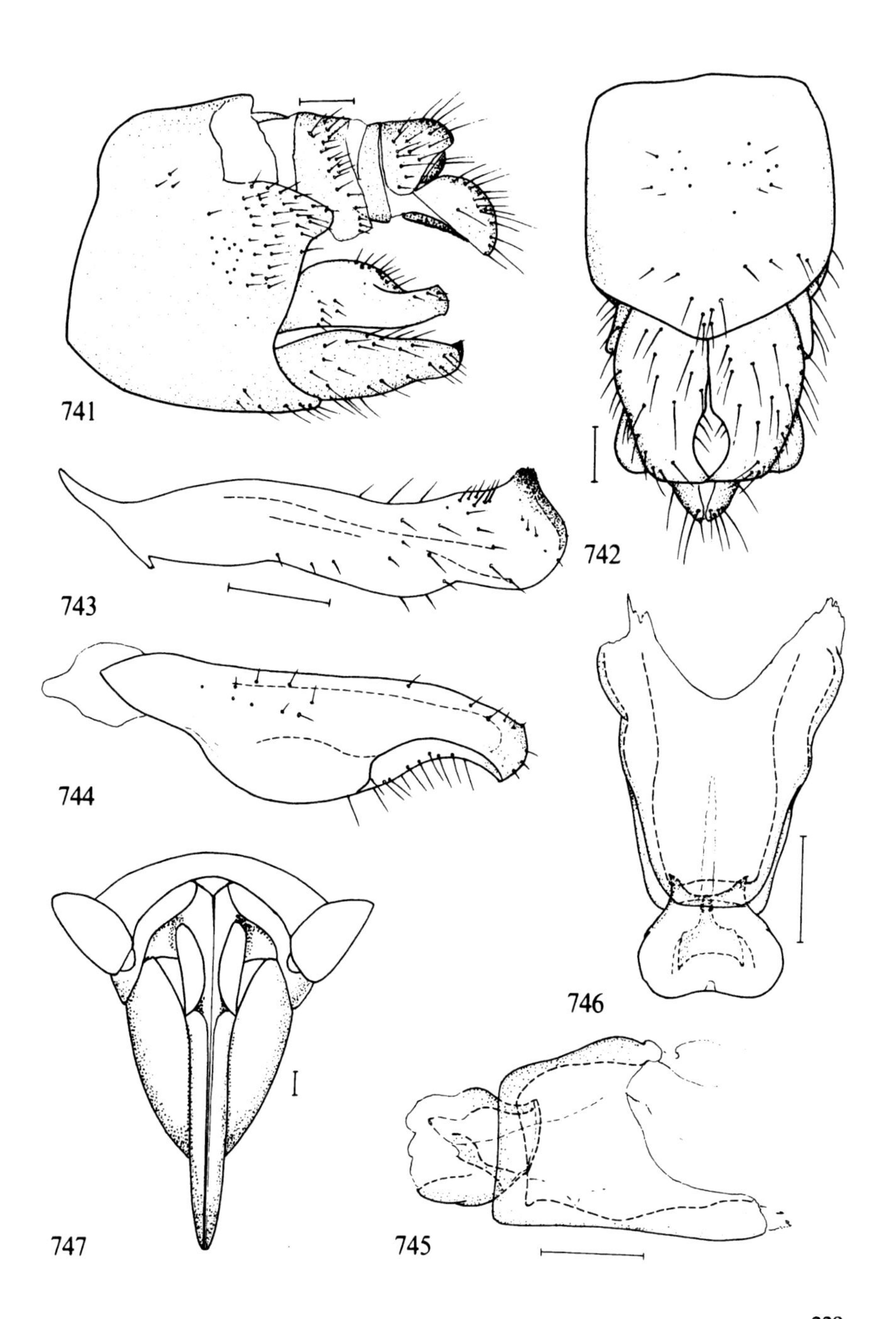
741
742
743
744
745
746
747

91. ***Peuceptyelus coriaceus*** (Fallén, 1826)
Plate-fig. 40, text-figs. 741–747.

Cercopis coriacea Fallén, 1826: 14.

Greyish yellow with ivory-white and fuscous markings, usually more distinct in males than in females. Frontoclypeus strongly swollen, with transverse rows of usually dark punctuation on either side of a pale median band, often with a dark spot near anteclypeus. Antennal cavities black or fuscous. Vertex and pronotum with irregular fuscous mottlings varying in extension, pronotum in females often largely light. Venter and legs with a fine pilosity. Thoracal venter light with more or less extended black or fuscous markings, abdomen largely black or fuscous with light segmental borders, terminal segments light. Legs light yellowish, anterior femora each with a fuscous spot near basis and a black or fuscous band near apex, median femora with a fuscous band near apex. Anterior and median tibiae with apices and a band near basis dark, anterior and median tarsi apically fuscous. Fore wings with 2–3 more or less incomplete transverse fuscous bands or rows of spots; distal half of median cell black or fuscous; an indistinct light transverse band consisting of locally ivory-white veins is present immediately basally of the median fuscous band. Apical veins bordered with fuscous. Male pygofer as in Text-figs. 741, 742, genital style as in Text-figs. 743, 744, aedeagus as in Text-figs. 745, 746. Ventral aspect of apical part of female abdomen as in Text-fig. 747. Overall length of males 6.45–7 mm, of females 7–8 mm.

Distribution. Not found in Denmark, Sweden and Norway. Scarce in East Fennoscandia, found in Ab, N, St, Ta, Sa, Kb, Vib, Kr. – East Poland, Latvia, Estonia, Altai Mountains, Belorussia, N. Russia, w. Siberia, Ural Mountains, Kurile Islands.

Biology. On *Picea abies* (Kontkanen, 1938). On *Abies excelsa;* hibernation takes place in the adult stage (Lindberg, 1947). Adults in April, May, June, August, September, October, December (Lindberg, l.c.). On spruce in moist *Oxalis* – *Myrtillus* spruce woods (Linnavuori, 1952a). On spruces 2-3 m in height, or on the lower branches of larger spruce trees (Vilbaste, 1971).

Genus *Neophilaenus* Haupt, 1935

Neophilaenus Haupt, 1935:155.
Type-species: *Cicada lineata* Linné, 1758, by original designation.

Body elongate, almost parallel-sided, finely and densely punctured and finely pubescent. Veins of fore wings thin, more or less obsolescent. Head in some species medially protruding. In Denmark and Fennoscandia four species, all of them living on grasses, sedges and rushes.

Key to species of *Neophilaenus*

1 Frontoclypeus strongly convex, little pilose, shining 2

- Frontoclypeus less convex, distinctly hairy, less shining 4

2 (1) Dorsum of fore body and fore wings entirely fuscous or black
92. *exclamationis* (Thunberg), f. *lindbergi* Metcalf

- Fore wings not entirely dark 3

3 (2) Fore body above and fore wings greyish brown to black-brown. Fore wing along fore border with an !-shaped light marking .... 92. *exclamationis* (Thunberg), f. *typica*

- Dorsum of fore body and fore wings yellowish grey, marking as above but less distinct .... 92. *exclamationis* (Thunberg), f. *diluta* (J. Sahlberg)

4 (1) Greyish yellow to greyish brown, often with a reddish tinge. Fore wing along costal border with two light spots. Dorsum often with a dark median longitudinal stripe .... 93. *campestris* (Fallén)

- Marking of fore wing not as above 5

5 (4) Dorsum of fore body and fore wings unicolorous, black
94. *lineatus* (Linné), f. *aterrima* (J. Sahlberg)

- Not as above 6

6 (5) Dorsum of fore body and fore wings largely fuscous. Costal border of fore wing narrowly pale to near apex where the light stripe is widening into a patch
94. *lineatus* (Linné), f. *pulchella* (J. Sahlberg)

- Dorsum greyish yellow or straw-coloured, fore wing with a black stripe along basal 2/3 of the light fore border 7

7 (6) Male genital style as in Text-figs. 760, 763, 765 .... 94. *lineatus* (Linné)

- Male genital style as in Text-fig. 770 .... 95. *minor* (Kirschbaum)

92. ***Neophilaenus exclamationis*** (Thunberg, 1784)
Plate-fig. 57, text-figs. 748-752.

Cicada exclamationis Thunberg, 1784: 24.
Philaenus dilutus J. Sahlberg, 1871: 94.
Philaenus campestris Jensen-Haarup, 1920: 80, (nec Fallén, 1805).
Philaenus exclamationis nigerrimus Lindberg, 1923a: 40.
Philaenus exclamationis lindbergi Metcalf, 1955: 264 (n.n. for *Philaenus exclamationis nigerrimus* Lindberg (nec Strobl, 1900)).

Above blackish brown or greyish brown. Proximal half of costal border of fore wing yellowish white. On the costal border between middle and apex there is an usually triangular light spot sharply bordered with dark colour on its basal as well as on its apical side. In light individuals this dark delimitation remains as two dark spots. At the hind margin of the fore wing distally of the apex of clavus there is a small light spot and apically of this spot the wing border is blackish brown even in light specimens. Venter largely light, abdomen blackish brown. In the brownish yellow or dirty yellowish f. *diluta* (J. Sahlberg) the dark pigmentation remains only in an oblique streak and a spot at the fore border of the fore wing near its apex and in the sharply black-brown hind margin of the wing apex. In forma *lindbergi* Metcalf, the dorsum of fore body and fore wings are entirely dark brown or black. Pygofer of male as in Text-figs. 748, 749, genital style as in Text-fig. 750, aedeagus as in Text-figs. 751, 752. Overall length of f. *typica* 3.5–5 mm, of f. *diluta* 4.2–5.0 mm.

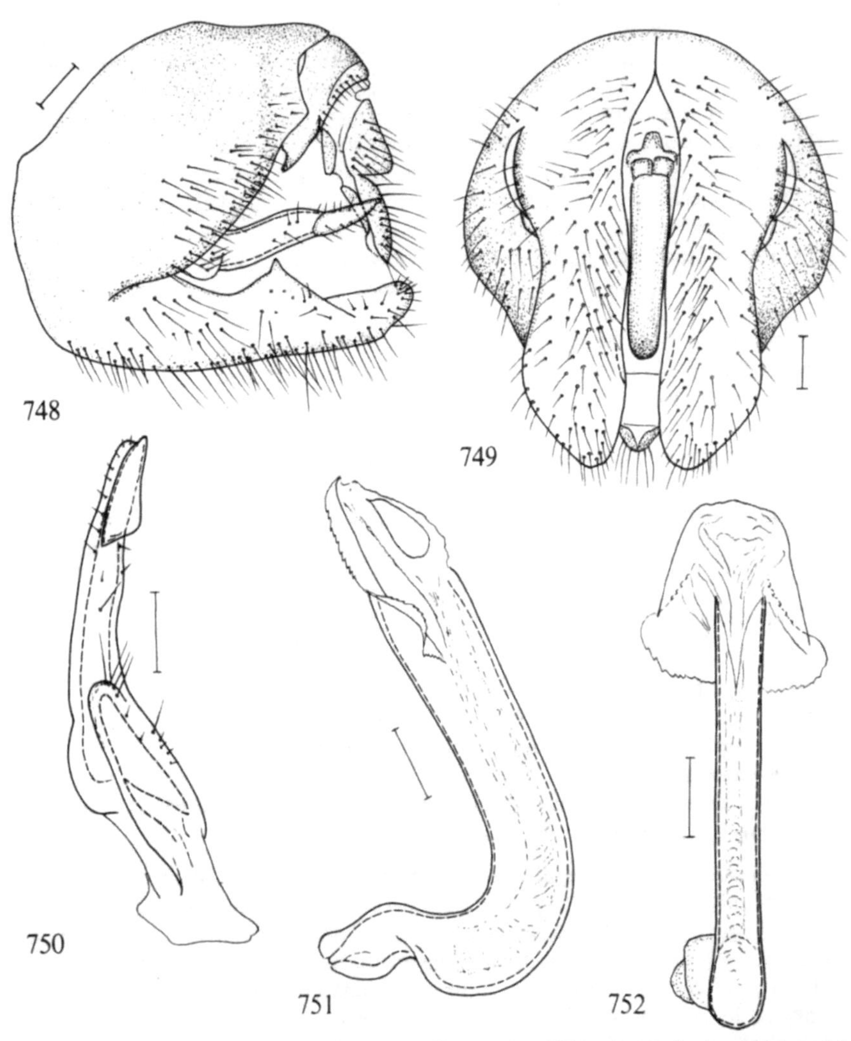

Text-figs. 748–752. *Neophilaenus exclamationis* (Thunberg). – 748: male pygofer in ventral aspect; 749: male pygofer from the left, aedeagus not considered; 750: right genital style from outside; 751: aedeagus from the left; 752: aedeagus in ventral aspect. Scale: 0.1 mm.

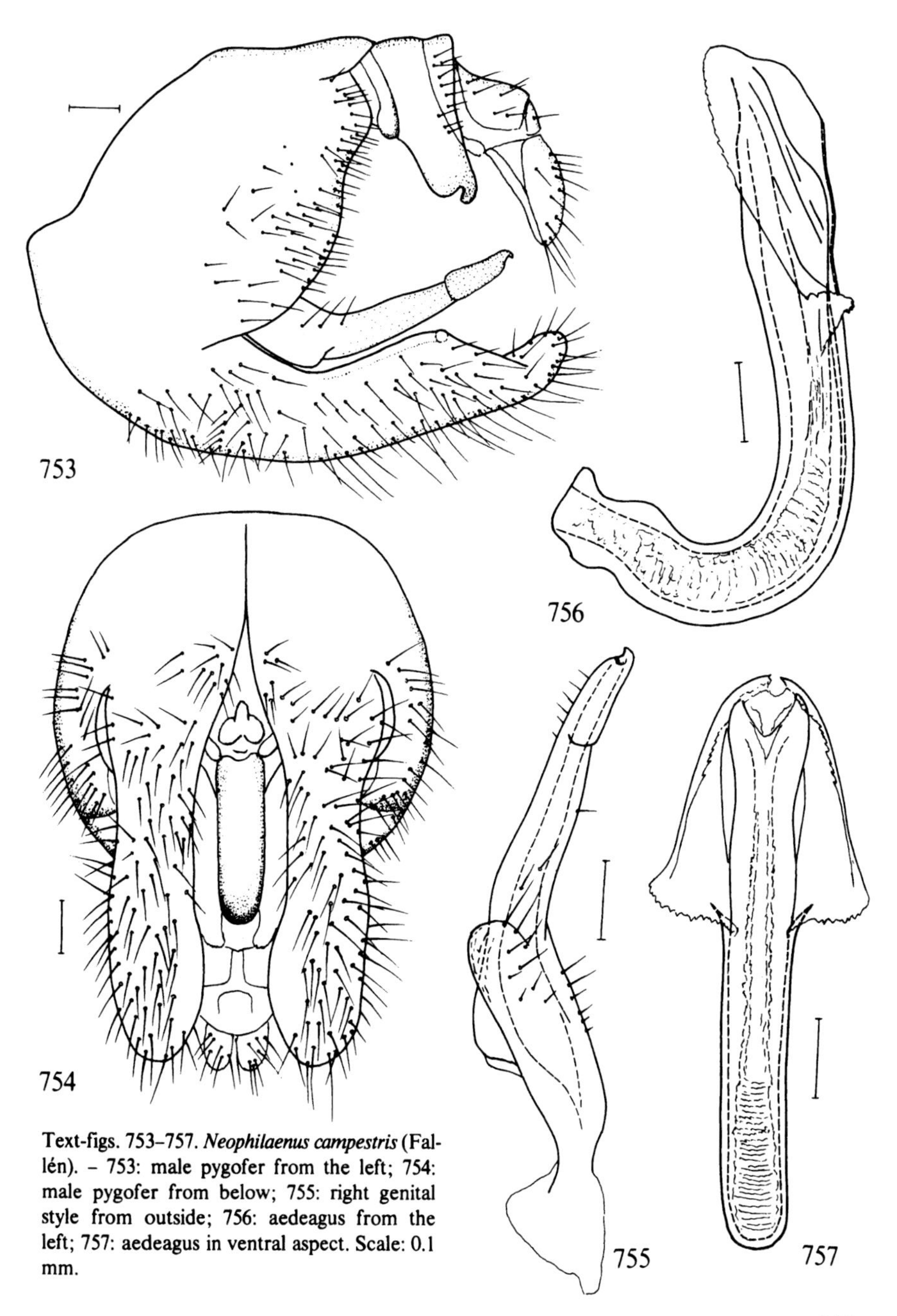

Text-figs. 753–757. *Neophilaenus campestris* (Fallén). – 753: male pygofer from the left; 754: male pygofer from below; 755: right genital style from outside; 756: aedeagus from the left; 757: aedeagus in ventral aspect. Scale: 0.1 mm.

Distribution. Fairly common in Denmark. – Very common in Sweden (Sk.- Lu. Lpm.) and East Fennoscandia; common in the east af Norway, more sparsely distributed in the coastal districts. F. *diluta* has so far not been found in Denmark. In Sweden it is fairly common in Vb. and Nb. Its distribution in East Fennoscandia is little studied; found in Al and Ab; in Norway recorded from TRi. F. *lindbergi* has been found in Norway, AK: Oslo, and in Finland, Ab: Pargas. – *Neophilaenus exclamationis* is widespread in Europe, also found in Algeria and Kazakhstan.

Biology. Adults in June–October, in "Binnendünen, Sandfeldern, Heiden, Hochmooren, Wäldern" (Kuntze, 1937). "Im Grase an trocknen Waldrändern im August und September" (Kontkanen, 1938). In the dryish meadow zone and the rocky islets and shores of seashores; in moist sloping meadows and cultivated fields (Linnavuori, 1952a). Hibernation in the egg stage (Müller, 1957). "Eurytope Art der Trockenrasen, auf xerothermen Biotopen und mesophilen Wiesen" (Schiemenz, 1969). Univoltine.

93. ***Neophilaenus campestris*** (Fallén, 1805)
Plate-fig. 45, text-figs. 753–757.

Cercopis campestris Fallén, 1805: 252.
Philaenus albipennis Jensen-Haarup, 1920: 80; nec Fabricius, 1798.

Above greyish yellow, greyish brown or yellowish brown with a reddish tinge, often with a fuscous longitudinal stripe on head and notum. Fore wings with two whitish spots at fore border and often with another one in clavus. A small light spot is present apically of the claval apex. Apically of this spot the wing border is fuscous. Wing veins often darker than the wing surface between veins. Venter largely light. Male pygofer as in Text-figs. 753, 754, genital style as in Text-fig. 755, aedeagus as in Text-figs. 756, 757. Overall length 3.5–6 mm.

Distribution. Not particularly common in Denmark (EJ, WJ, NEJ, B). – Scarce in Sweden, only found in Sk. – Not found in Norway. – In East Fennoscandia only recorded from Vib. – Widespread in Central and South Europe, including England, Ireland, Estonia, Latvia, and S. Russia, also in N. Africa, Cyprus, Anatolia, Israel, Jordan, Azerbaijan and Georgia.

Biology. In "Stranddünen, Sandfeldern und besonnten Hängen" (Kuntze, 1937). "Auf den Trockenrasen regelmässig dominante Art" (Schwoerbel, 1957). "Bewohnt Trockenrasen in warmen Lagen" (Wagner & Franz, 1961). "Stenotope Art der Trockenrasen" (Schiemenz, 1969). Univoltine, hibernation in the egg stage (Schiemenz, l.c.).

Text-figs. 758–762. *Neophilaenus lineatus* (Linné). – 758: male pygofer from the left; 759: male pygofer from below; 760: right genital style from outside; 761: aedeagus from the left; 762: aedeagus in ventral aspect. Scale: 0.1 mm.

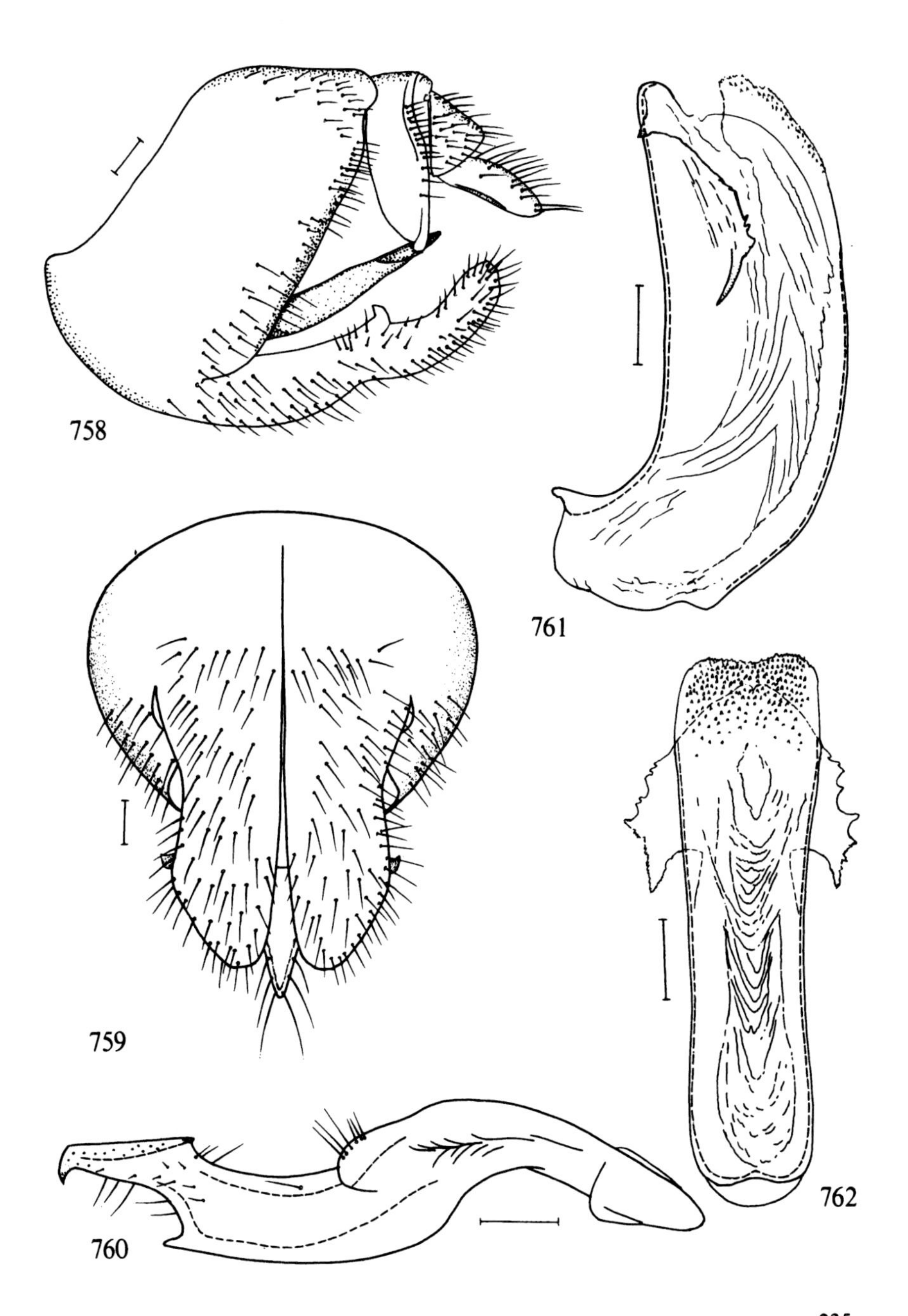
758
761
759
762
760

94. ***Neophilaenus lineatus*** (Linné, 1758)
Plate-fig. 56, text-figs. 758–767.

Cicada lineata Linné, 1758: 438.
Philaenus aterrimus J. Sahlberg, 1871: 92.
Philaenus pulchellus J. Sahlberg, 1871: 93.
Philaenus lineatus pallida Haupt, 1917: 238.
Philaenus lineatus danicus W. Wagner, 1935b: 167.
Neophilaenus lineatus, f. *fulva* Gyllensvärd, 1965: 228.

The typical form is straw-coloured to yellowish white. Upper part of face and an inversely V-shaped marking on its lower part fuscous. Fore wing with costal border broadly whitish. Behind this there is an apically broader black-brown or dark brown longitudinal stripe along basal 2/3 of the wing. Also the apical margin of the fore wing may be fuscous to a varying extent. Fore wing veins brownish. A short but sharply marked black-brown streak is present on the hind border of the fore wing just distally of the claval apex. Venter largely dirty yellow, abdominal dorsum often blackish. In f. *pulchella* (J. Sahlberg) the entire upper side except the broadly pale costal border of the fore wing is black-brown or dark brown. In f. *aterrima* (J. Sahlberg), the upper side is entirely black. – Populations on *Elymus* and *Ammophila* in dunes make the impression of belonging to a separate taxon. On an average, individuals in such populations are a little paler and larger than "typical" *lineatus*. This is f. *pallida* (Haupt) (=f. *fulva* Gyllensvärd). W. Wagner (1952, 1955) regarded *pallida* as a separate species, but distinct morphological characters making separation of "*N. pallidus*" from *N. lineatus* possible have not been found. In the extensive Norwegian dune material of Ardö (see below), several males are less than 5 mm in length. Therefore I regard f. *pallida* as an ecological form of *N. lineatus*. See also Kristensen (1965) and Le Quesne (1965). – In populations of f. *pulchella*, transition forms between the latter and f. *typica* are usually present. – Male pygofer as in Text-figs. 758, 759, genital style as in Figs. 760, 763, 765, aedeagus as in Text-figs. 761, 762, 764, 766, venter of posterior part of female abdomen as in Text-fig. 767. Overall length of males 4.6–5.6 mm, of females 5.4–6.8 mm.

Distribution. Common and widespread in Denmark, Sweden and Norway (so far not recorded from AAi, STy, STi, NTi, Nsi, Nn, TR, F). – Common and widespread also in East Fennoscandia (except Lk, Le, Li, Lr). Forma *aterrima* was originally described from Öl. Wagner (1935b) mentions this variety from SJ, Holgersen (1944a) recorded it from VAy and Ry. F. *pulchella* was described from Hall., and I have seen specimens also from Sk., Bl., Sm., Öl., and Dlr. A. Løken found a specimen in Norway, HOy: Fjell. – F. *pallida* is probably a constant inhabitant of *Elymus*-dunes. Trolle (in litt.) records it from EJ, WJ, NWJ, NEJ, and B; I have seen specimens from Sk., Hall., Gtl., and G. Sand. Ardö collected great number of specimens in Norway, VE: Larvik, Oddane, Sand, in a special investigation of the dune fauna. – Range of the species: holarctic.

Biology. On grasses. "Fast in allen Biotopen" (Kuntze, 1937). "Auf *Carex*-

Weissmooren und seggenreichen Moorwiesen" (Kontkanen, 1938). In the drier meadow area of seashores; in drier and wet peaty meadows, in bogs and marshes, in rich swampy woods (Linnavuori 1952a). Belongs to the "Cariceto canescentis – Agrostidetum caninae, Subass. v. *Carex inflata*" (Marchand, 1953). In "Hochmooren, abundant bei Wasserzügen, im *Eriophorum vaginatum* Hochmoor- und Zwischenmoor-Bultgesellschaft, im *Calluna*-Hochmoor-Bultgesellschaft, und im Reitgras-Fichtenwald" (Schiemenz, 1975). "In many types of vegetation, particularly wet meadows on shores, treeless fens, pine bogs, spruce-birch swamps, and grass-herb forests. Also frequent on grasses growing in pastures and leys and along ditches.

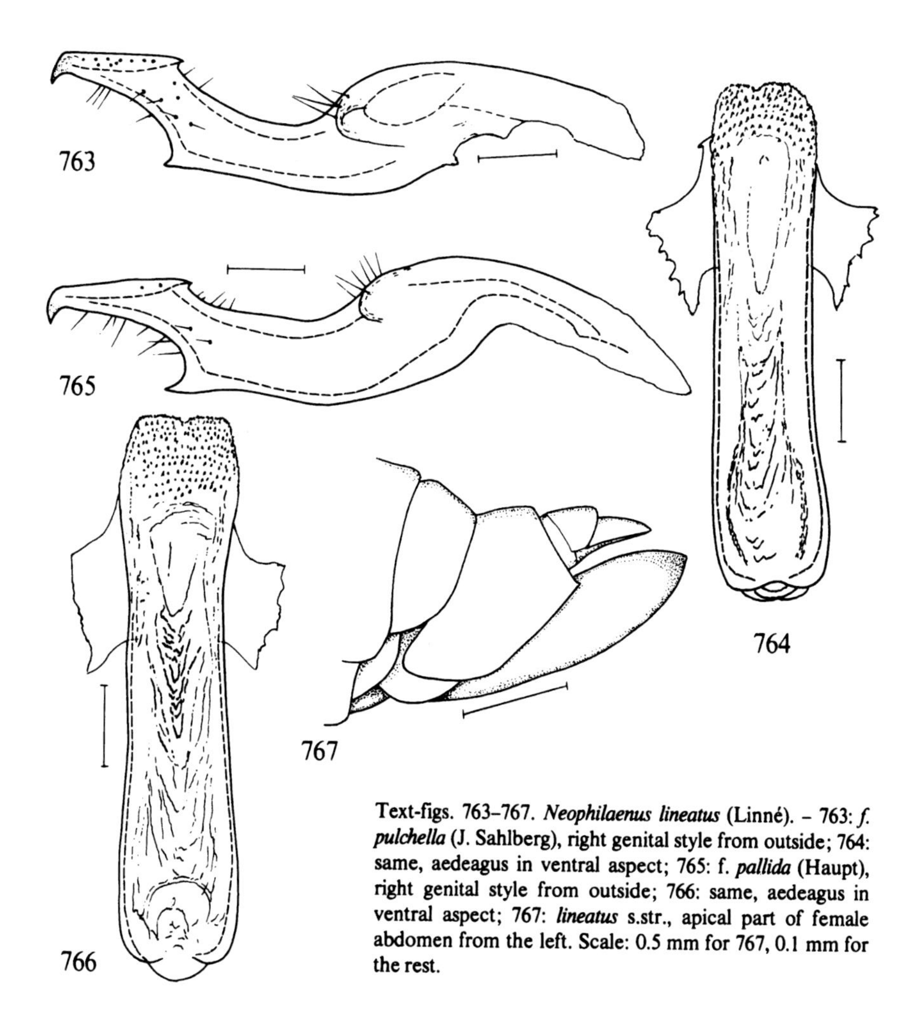

Text-figs. 763–767. *Neophilaenus lineatus* (Linné). – 763: *f. pulchella* (J. Sahlberg), right genital style from outside; 764: same, aedeagus in ventral aspect; 765: f. *pallida* (Haupt), right genital style from outside; 766: same, aedeagus in ventral aspect; 767: *lineatus* s.str., apical part of female abdomen from the left. Scale: 0.5 mm for 767, 0.1 mm for the rest.

Reared for several weeks on *Festuca rubra, Deschampsia caespitosa, D. flexuosa, Alopecurus pratensis, Phalaris arundinacea, Carex nigra* and *C. magellanica* – – – and *Calamagrostis epigeios* and *Festuca arundinacea*" (Raatikainen & Vasarainen, 1976). Hibernation in the egg stage (Müller, 1957, Remane, 1958). – F. *aterrima* and its allies (incl. *danicus* and *pulchellus*) are "Bewohner von Mooren im Tiefland und in höheren Lagen der Mittelgebirge" (Wagner & Franz, 1961). According to Le Quesne (1965) the dark forms of *lineatus* are associated with *Molinia.*

95. ***Neophilaenus minor*** (Kirschbaum, 1868)
Text-figs. 768–772.

Ptyelus minor Kirschbaum, 1868b: 65.

Resembling *lineatus,* differing by the smaller body size and by details in the male genitalia. Male pygofer as in Text-figs. 768, 769, genital style as in Text-fig. 770, aedeagus as in Text-fig. 771, apex of female abdomen as in Text-fig. 772. Overall length; ♂ 4.0–4.5 mm, ♀ 4.1–5.05 mm.

Distribution. Not found in Denmark, Sweden and Norway. Recorded from East Fennoscandia, Sa: Joutseno, by Linnavuori (1953b). – Albania, Austria, Belgium, Bulgaria, Bohemia, Moravia, Slovakia, France, German D.R. and F.R., Hungary, Italy, Netherlands, Poland, Portugal, Ukraine, Yugoslavia, Kazakhstan, Kirghizia.

Biology. In "Binnendünen, Sandfeldern, besonnten Hängen und Wäldern" (Kuntze, 1937). "Auf sandigen Grasstellen, besonders auf *Weingaertneria canescens* Bernh. Juli bis September." (Wagner, 1939). "Für die Corynephoreten typisch" (Marchand, 1953). "Lebt wahrscheinlich monophag an *Corynephorus canescens*" (Wagner & Franz, 1961). "Hauptart der Silbergras-Sandtrockenrasen" (Schiemenz, 1969a). Hibernation in the egg stage (Remane, 1958). Univoltine (Schiemenz, 1969b).

Genus *Aphrophora* Germar, 1821

Aphrophora Germar, 1821: 48.
Type-species: *Cercopis alni* Fallén, 1805, by subsequent designation under the Plenary Powers of the International Commission on Zoological Nomenclature.

Body moderately elongate. Dorsum coarsely punctured. Dorsal plate of frontoclypeus at least twice as broad as long. Vertex and pronotum with a median longitudinal carina. Veins of fore wings distinct. In Denmark and Fennoscandia four species.

Key to species of *Aphrophora*

1 Surface of fore wings hairless. Lateral borders of scutellum elevated .... 96. *corticea* Germar
– Fore wings finely pilose. Surface of scutellum plain 2

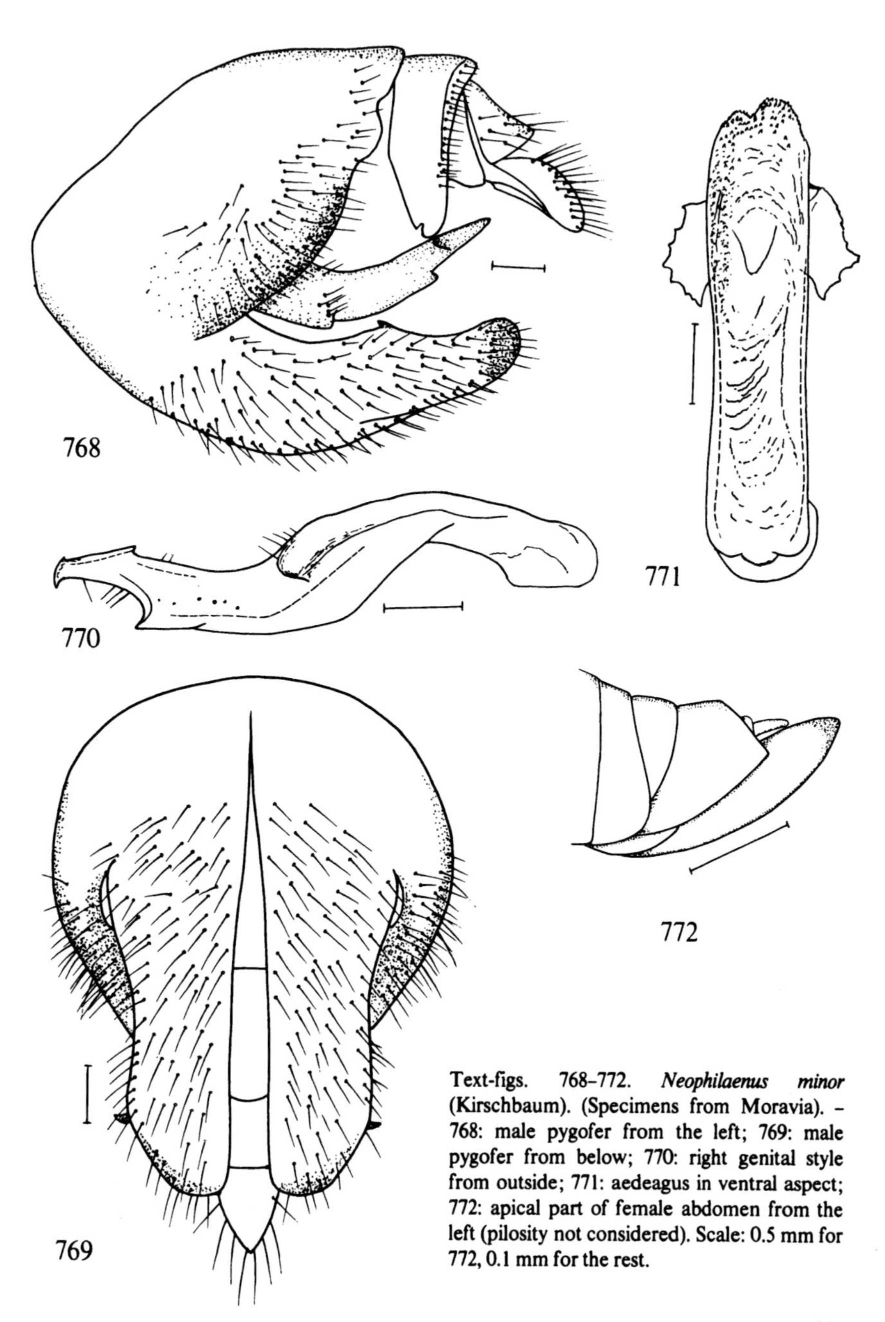

Text-figs. 768–772. *Neophilaenus minor* (Kirschbaum). (Specimens from Moravia). – 768: male pygofer from the left; 769: male pygofer from below; 770: right genital style from outside; 771: aedeagus in ventral aspect; 772: apical part of female abdomen from the left (pilosity not considered). Scale: 0.5 mm for 772, 0.1 mm for the rest.

2 (1) Fore wings at costal border with two light spots, one larger proximally of middle and one smaller half-way between middle and apex of costal margin. These markings may be indistinct only in very dark specimens .... 97. *alni* (Fallén)
– Fore wings unicolorous, light or with diffuse markings 3
3 (2) Fore wings almost unicolorous. Pilosity of fore wings very short, hairs usually not longer than diameter of punctures. In lateral aspect, the part of ovipositor projecting beyond apex of pygofer is at most 1½ times as long as wide (Text-fig. 791) .... 98. *salicina* (Goeze)
– Fore wings with an oblong yellowish spot along basal 2/5 of costal border, and with an indistinct oblique dark transverse band distally of this spot. These markings are more distinct in live specimens. Hairs on fore wings distinctly longer than diameter of punctures. In lateral aspect, the part of ovipositor projecting beyond apex of pygofer is nearly twice as long as broad (Text-fig. 798) .... 99. *costalis* Matsumura

96. ***Aphrophora corticea*** Germar, 1821
Text-figs. 773–778.

Aphrophora corticea Germar, 1821: 50.

Elongate, yellowish brown. Head above finely pubescent, brownish mottled. Frontoclypeus broad, very convex, punctuation and transverse furrows more or less obsolete. Pronotum, scutellum and fore wings (except costal border) hairless, strongly punctured. Pronotum largely ivory white. Scutellum transversely furrowed with elevated lateral borders and a fine whitish median streak. Fore wings mottled with ivory, yellowish brown, rusty brown and warm brown, with an indistinct pattern including two transverse bands of more or less confluent light spots, one of them proximally, the other distally of middle. Venter darker or lighter brownish, legs yellowish, femora and tibiae transversely dark-banded. Pygofer of male as in Text-figs. 773, 774, genital style as in Text-fig. 775, aedeagus as in Text-figs. 776–778. Overall length 5.75–11 mm.

Distribution. Not found in Denmark, nor in East Fennoscandia. Rare in Sweden, found in Sk., Hall., and Gtl. According to Holgersen (1944a) one male was collected in Norway, Ø: Kirkøya, Hvaler, in July, 1925, by Munster. – Albania, Belgium, Bohemia, Moravia, Slovakia, France, German D.R. and F.R., Italy, Poland, Portugal, Spain, Switzerland, Yugoslavia, Anatolia, Altai Mountains, m. Siberia.

Biology. Hibernation takes place in the egg stage (Müller, 1957). "Imagines auf *Picea, Abies, Pinus,* Larven auf Heidelbeere und Erdbeere" (Wagner & Franz, 1961). "Monovoltiner Eiüberwinterer, als Larven weitgehend oligophag bodennah in der mehr oder weniger beschatteten Krautschicht – – – in trockneren, lichten

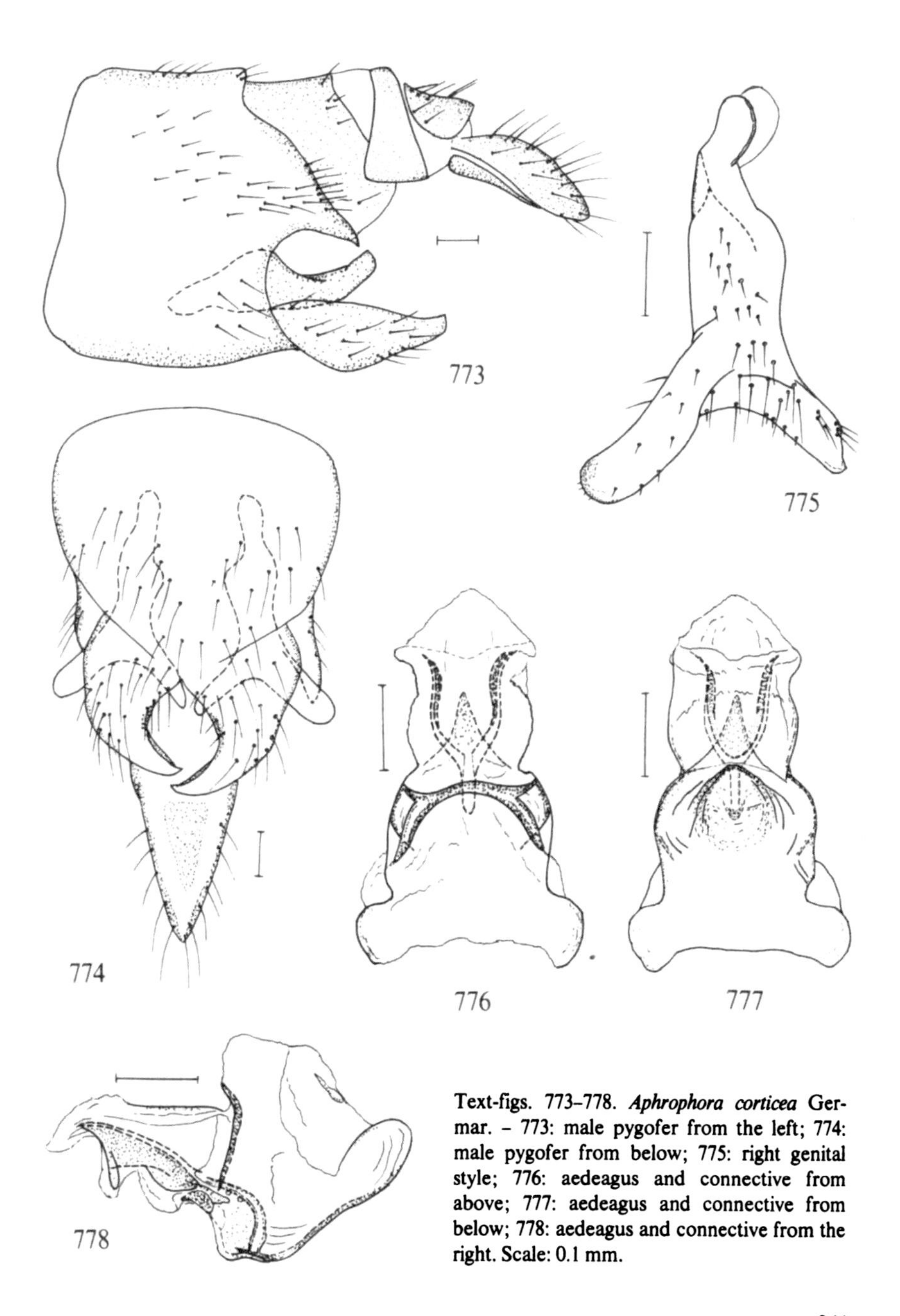

Text-figs. 773–778. *Aphrophora corticea* Germar. – 773: male pygofer from the left; 774: male pygofer from below; 775: right genital style; 776: aedeagus and connective from above; 777: aedeagus and connective from below; 778: aedeagus and connective from the right. Scale: 0.1 mm.

Kiefernwäldern" (Müller, 1978). – In July, 1965, I collected 52 adult specimens of *A. corticea* on pines in Gtl., Väskinde, Nors. The dominant plant under these pines was *Arctostaphylos Uva-ursi* (L.) Spreng. which is therefore a probable hostplant. – Adults in June–August.

97. ***Aphrophora alni*** (Fallén, 1805)
Plate-figs. 41, 42, text-figs. 779–784.

Cicada cincta Thunberg, 1784: 23 [Nomen oblitum].
Cercopis alni Fallén, 1805: 240.
Aphrophora alni fuscata Haupt, 1925: 11.
Aphrophora alni umbrina Linnavuori, 1950: 182.

Less oblongate than our other *Aphrophora* species, greyish yellow. Dorsally and ventrally finely pubescent. Frontoclypeus and dorsum strongly punctured. Dorsal plate of frontoclypeus often with a pair of dark spots. Fore wing often with a light patch near basis, usually with an oblique whitish transverse band extending from the costal border somewhat proximally of middle towards scutellum. Inside clavus this band is usually less distinct. It is missing in the dark-brown f. *fuscata* Haupt. Further there is an oblong whitish patch extending along the costal border from a point somewhat distally of middle to another half-way to the wing apex. These light areas are usually broadly dark-bordered. Fore wing veins at intervals darker than surrounding surfaces. Thus the two claval veins are black-brown on a short section near middle. In f. *umbrina* Linnavuori the fore wings are dark brown with an indistinct lighter spot replacing the proximal transverse band. Male pygofer as in Text-figs. 779, 780, genital style as in Text-fig. 781, aedeagus as in Text-figs. 782, 783, 784. Overall length 6–8 mm. Larvae (Plate-fig. 42) pale or flesh-coloured with coalescent brownish markings, abdomen not lighter than fore part of body. Abdominal terga each with a transverse row of small dark points or very short streaks. Hind tibiae (in older stages) at least twice as long as hind tarsi.

Distribution. Common and widespread in Denmark. – Common in Sweden up to Ång. – Common and widespread in Norway up to NTi. – In East Fennoscandia up to Om, also in Vib and Kr. – Widespread in Europe, also in Algeria, Morocco, north, central and eastern Asia.

Biology. Ossiannilsson (1946c) mentioned *Hypericum* and *Erigeron* as hostplants. Ossiannilsson (1950b) found larvae of *Aphrophora alni* in their froth-lumps on *Polygonum viviparum, Geum rivale, Filipendula ulmaria, Trifolium hybridum* and *medium, Lotus corniculatus, Galium verum, Ranunculus flammula, Hieracium umbellatum, Sonchus ar-*

Text-figs. 779–784. *Aphrophora alni* (Fallén). – 779: male pygofer from the left; 780: male pygofer from below; 781: right genital style from below; 782: aedeagus and connective in dorsal view; 783: aedeagus and connective in ventral aspect; 784: aedeagus and connective from the left. Scale: 0.1 mm.

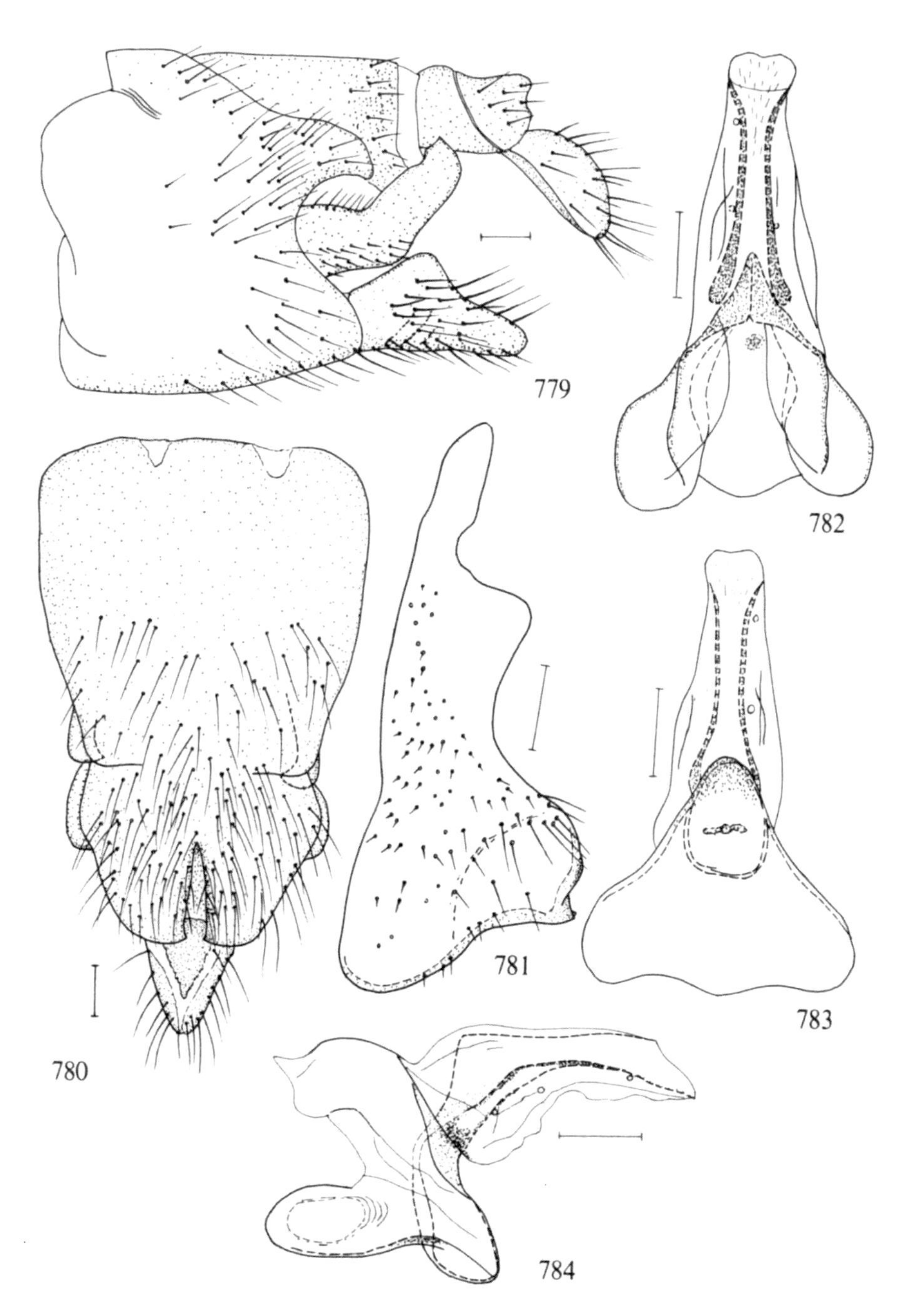
779
780
781
782
783
784

vensis, *Carduus crispus*, and *Cirsium palustre*, and on adventitious shoots of *Salix caprea*, *Betula pubescens*, and *Alnus glutinosa*. To this list of hostplants *Viola canina*, *Potentilla reptans*, *P. anserina* and *Trifolium repens* can now be added. "As a rule, the nymphs of

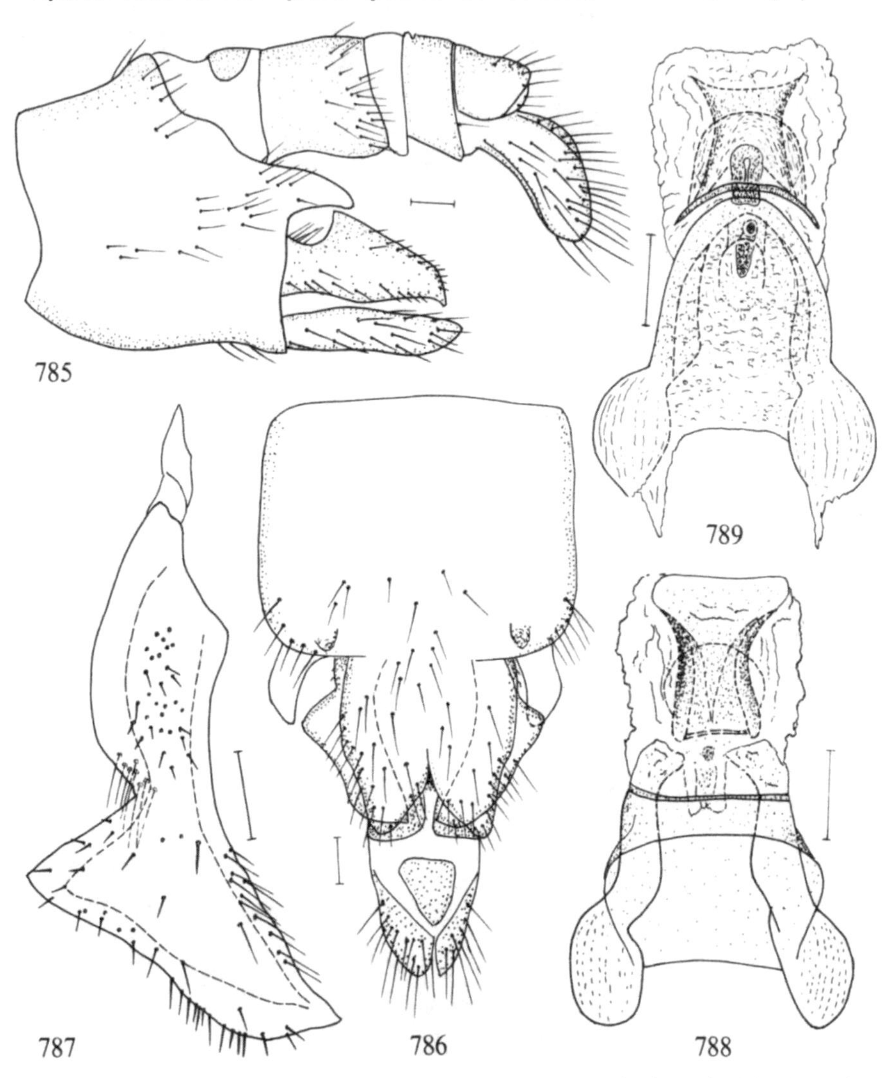

Text-figs. 785–789. ***Aphrophora salicina*** **(Goeze). – 785: male pygofer from the left; 786: male pygofer from below; 787: right genital style from below; 788: aedeagus and connective from above; 789: aedeagus and connective from below. Scale: 0.1 mm.**

this species are found near the ground, very often at the very base of the stems of the host-plant. Its large froth-lumps are therefore less conspicuous than those of *Philaenus*" (Ossiannilsson, 1950b). In Sweden adults have been found in June-September. They frequently visit foliiferous trees and bushes. According to Nuorteva (1951), *Betula* spp., *Salix phylicifolia,* and *S. aurita* are preferred hostplants of adult *Aphrophora alni* in Finland. Univoltine. Hibernation takes place in the egg stage (Müller, 1957). Ubiquitous (Schwoerbel, 1957).

98. ***Aphrophora salicina*** (Goeze, 1778)
Text-figs. 785–791.

Cicada spumaria salicis De Geer, 1773: 180 [Invalid].
Cicada salicina Goeze, 1778: 153.

Elongate, greyish yellow to brownish yellow, dorsally and ventrally finely pubescent. Frontoclypeus and dorsal surface strongly punctured. Dorsum of fore body and fore wings almost unicolorous; a small dark spot is present in the apex of clavus. Abdomen largely black. Male pygofer as in Text-figs. 785, 786, genital style as in Text-fig. 787, aedeagus as in Text-figs. 788–790, apical part of female abdomen as in Text-fig. 791. Overall length of males 8.7–10.2 mm, of females 9–10.8 mm.

Distribution. Common and widespread in Denmark. – Common in the south of Sweden, found in Sk., Hall., Sm., Öl., Gtl., and Vrm. – Holgersen (1944a) recorded *A. salicina* from AK, Os, Bø, and TEy but these records have not been revised (at that time "*salicina*" included *costalis*.). – In East Fennoscandia *salicina* is absent. – Widespread in Central and Southern Europe, including England, Ireland, Estonia, Latvia and n., m., and s. Russia; also in Morocco, Altai, Armenia, Azerbaijan, Kazakhstan, Kirghizia, Moldavia, m. and w. Siberia, Uzbekistan, Mongolia, and Maritime Territory.

Biology. On *Salix* spp., but exact information on the species of hostplants is scarce in the literature. Schwoerbel (1957) gives *Salix aurita* as a hostplant. I collected *A. salicina*

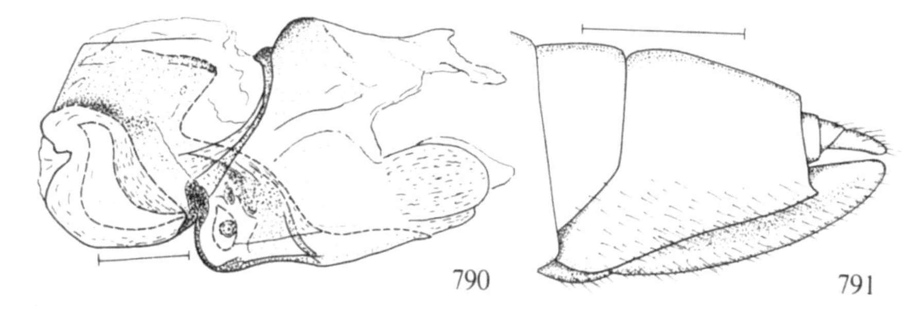

Text-figs. 790–791. *Aphrophora salicina* (Goeze). – 790: aedeagus and connective from the right; 791: apical part of female abdomen from the left. Scale: 0.1 mm for 790, 1 mm for 791.

on *Salix viminalis* in Scania and on *S. cinerea* in Öl. Univoltine, adults in June-September. Hibernation in the egg stage (Müller, 1957).

99. ***Aphrophora costalis*** Matsumura, 1903
Plate-figs. 43, 44, text-figs. 792–798.

Aphrophora costalis Matsumura, 1903: 36.
Aphrophora salicis forneri Haupt, 1919: 165.
Aphrophora maculata Edwards, 1920: 53 (nec Capanni, 1894).

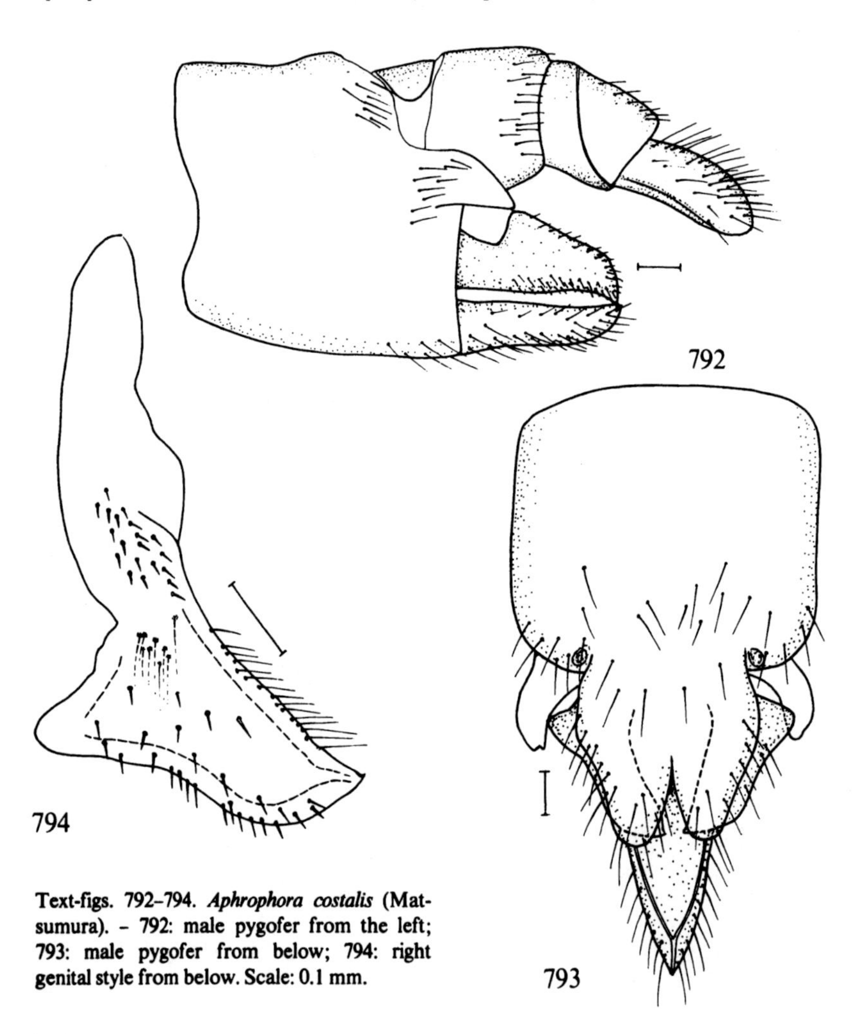

Text-figs. 792–794. *Aphrophora costalis* (Matsumura). – 792: male pygofer from the left; 793: male pygofer from below; 794: right genital style from below. Scale: 0.1 mm.

Very like *A. salicina,* differing by the presence of an indistinct colour pattern consisting of an elongate light yellow patch along basal 2/5 of the costal border, and an indistinct dark transverse band distally of this patch. The pubescence of fore wings and thoracal venter is distinctly longer in *costalis*. Male pygofer as in Text-figs. 792, 793, genital style as in Text-fig. 794, aedeagus as in Text-figs. 795–797, apical part of female abdomen as in Text-fig. 798. Overall length of males 9–10 mm, of females 9.5–11 mm. Larvae with fore part of body brownish, often considerably darker than abdomen, with or without

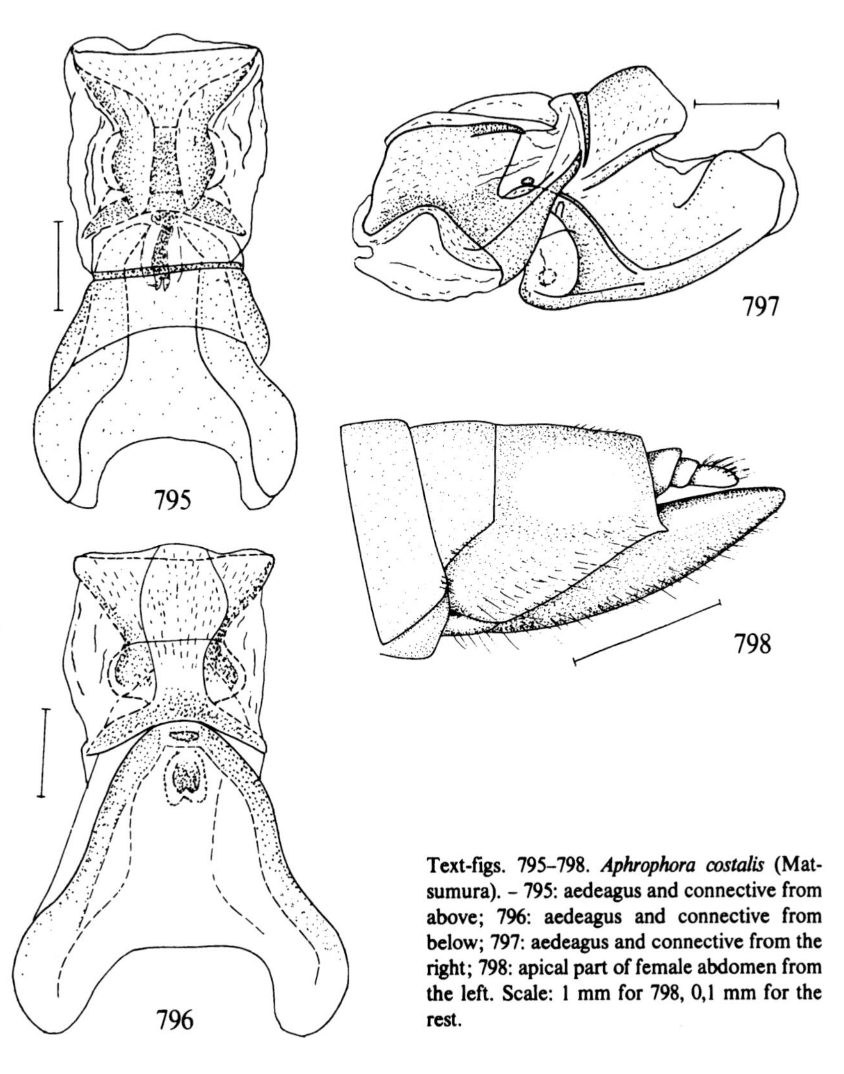

Text-figs. 795–798. *Aphrophora costalis* (Matsumura). – 795: aedeagus and connective from above; 796: aedeagus and connective from below; 797: aedeagus and connective from the right; 798: apical part of female abdomen from the left. Scale: 1 mm for 798, 0,1 mm for the rest.

small lighter spots (Plate-fig. 44). Abdominal terga each with a transverse row of dark points as in *alni,* in addition usually a scattered irregular punctuation, more dense on the posterior segments. Hind tibiae as a rule shorter than twice the length of hind tarsi (in older stages).

Distribution. Not found in Denmark. - Common in Central Sweden, found in Vg., Upl., Vstm., Vrm., Gstr. - Concerning Norway, see above under *salicina.* - Fairly common in Southern and Central East Fennoscandia (Al, Ab, N, Ka, St, Ta, Tb, Sb, Kb; Vib). - Austria, Bohemia, Moravia, Slovakia, France, German D.R. and F. R., England, Italy, Poland, Estonia, Altai, m. Siberia, Manchuria, Korean Peninsula, Japan.

Biology. On *Salix* spp. (e.g. *cinerea, pentandra, purpurea*). Univoltine, adults in June–September.

Genus *Philaenus* Stål, 1864

Philaenus Stål, 1864: 66.
Type-species: *Cicada spumaria* Linné, 1758, by subsequent designation.

Body oval, finely and densely punctured and with a fine pubescence. Head anteriorly rather obtuse-angled. Veins of fore wings thin, fairly indistinct. In Denmark and Fennoscandia one species.

100. ***Philaenus spumarius*** (Linné, 1758)
Plate-fig. 48, text-figs. 799–818.

Cicada spumaria Linné, 1758: 437.
Cicada leucophthalma Linné, 1758: 437.
Cicada lateralis Linné, 1758: 437.
Cicada leucocephala Linné, 1758: 437.
Cicada spumaria graminis De Geer, 1773: 163.
Cercopis gibba Fabricius, 1775: 689.
Cicada albomaculata Schrank, 1776: 76.
Cicada quadrimaculata Schrank, 1776: 77.
Cicada trilineata Schrank, 1776: 75.
Cercopis fasciata Fabricius, 1787: 275.

Text-figs. 799–813. *Philaenus spumarius* (Linné), dorsal aspect (schematic). -799: *f. typica;* 800: *f. populella* Metcalf; 801: *f. vittata* (Fabricius); 802: *f. trilineata* (Schrank); 803: *f. praeusta* (Fabricius); 804: *f. fasciata* (Fabricius); 805: *f. gibba* (Fabricius); 806: *f. leucocephala* (Linné); 807: *f. marginella* (Fabricius); 808: *f. lateralis* (Linné); 809: *f. quadrimaculata* (Schrank); 810: *f. albomaculata* (Schrank); 811: *f. leucophthalma* (Linné); 812, 813: unnamed colour forms.

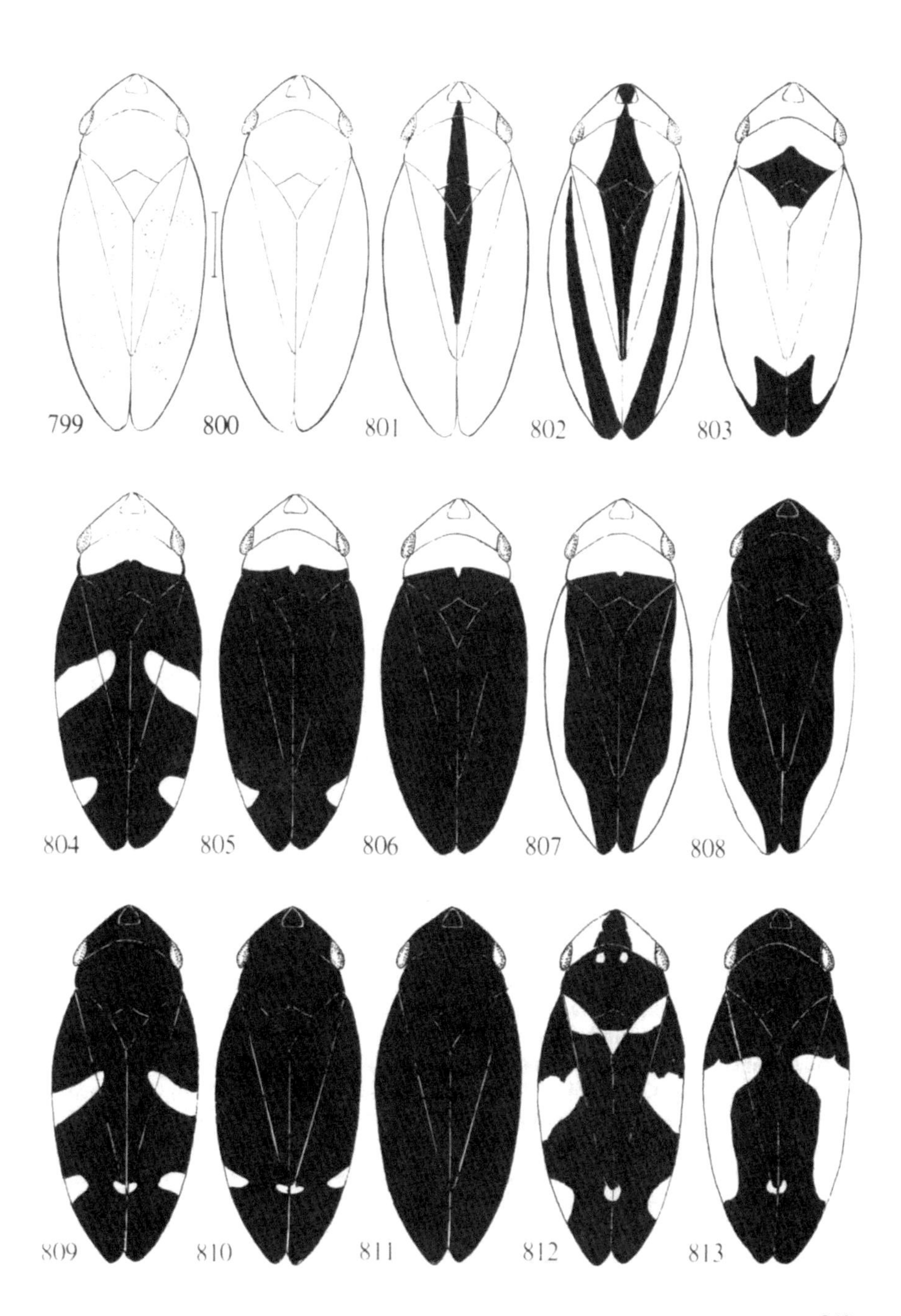
799
800
801
802
803
804
805
806
807
808
809
810
811
812
813

Cercopis marginella Fabricius, 1794: 52.
Cercopis praeusta Fabricius, 1794: 53.
Cercopis vittata Fabricius, 1794: 53.
Cercopis xanthocephala Schrank, 1801: 57.
Cicada oenotherae Scopoli, 1763: 114.
Cercopis spumaria ustulata Fallén, 1826: 19.
Cercopis spumaria pallida Zetterstedt, 1828: 515.
Philaenus leucophthalmus, var. *populellus* Metcalf, 1962: 303.

Danish: almindelig skumcikade. Swedish: vanlig spottstrit. Norwegian: vanlig skumsikade. Finnish: sylkikaskas. British: meadow spittlebug.

Extremely varying in colour, especially in the female sex. The typical form (Plate-fig. 48, text-fig. 799) is above greyish yellow with indistinct markings, among which especially two pale, often dark-bordered spots near the costal border may be noticed. But the species varies from unicolorous yellowish white to unicolorous black with numerous more or less characteristic intermediate forms. Most of these were originally described as species. The list of synonyms given above is far from complete. The colour variation has recently been studied, also from the genetic point of view, by Halkka & al. (1962–1976). An idea of this variation can be gained by studying the series of simplified pictures, Text-figs. 799–813. Male pygofer as in Text-figs. 814, 815, genital style as in Text-fig. 816, aedeagus as in Text-figs. 817, 818. – According to Wagner (1955a), there exists a geographic variation in the shape of the appendages of the aedeagus, these decreasing in length with increasing latitude. Also altitude appears to affect the structure of this organ. Overall length of males 5.3–6 mm, of females 5.4–6.9 mm. The larvae are light green or pale yellow without dark markings.

Distribution. Very common in Denmark and Sweden, found in all provinces except G. Sand. – Very common in Norway, found in all districts except VAi, NTi, TRy, Fv, Fn, and Fø. – Very common in East Fennoscandia, so far not found in LkE, LkW, and Le. – Widespread in the Palaearctic region, also in the Nearctic region (introduced).

Biology. Ubiquitous, polyphytophagous, especially on herbaceous plants in meadows and cultivated fields. Halkka & al. (1967) listed 166 plant species (among these also grasses and pteridophytes) as hostplants in Finland, but the total number of hostplants established in the world exceeds 1000. According to Ossiannilsson (1950b), also adventitious shoots and shoots of the current year's growth of a number of woody phanerogams can serve as breeding plants. Univoltine (Kontkanen, 1954). Hibernates in the egg stage (De Geer, 1741; Müller, 1957; Remane, 1958). In the autumn the eggs are deposited in cracks in treetrunk bark, on herbs etc. (De Geer, 1741), on dead parts of plants or in litter (Halkka & al., 1967). They "are laid in masses held together by a hardened frothy cement. – – – The number of eggs per mass varies from 1 to 30 and averages about 7" (Weaver & King, 1954). In the experiments of Halkka & al. (1967) in Finland, the mean number of eggs per group was 2. "It may be inferred that in southern Finland the females lay an average of perhaps 30 or more eggs. – – – In the field experiments performed at Tikkurila, the first eggs hatched at the end of May. The nymphs usually crept to the nearest dicotyledonous plants and began forming spittle. They remained within the spittle for about 5 or 6 weeks, but cold weather reduced the speed of development considerably. - - - In the southernmost provinces of Finland and along

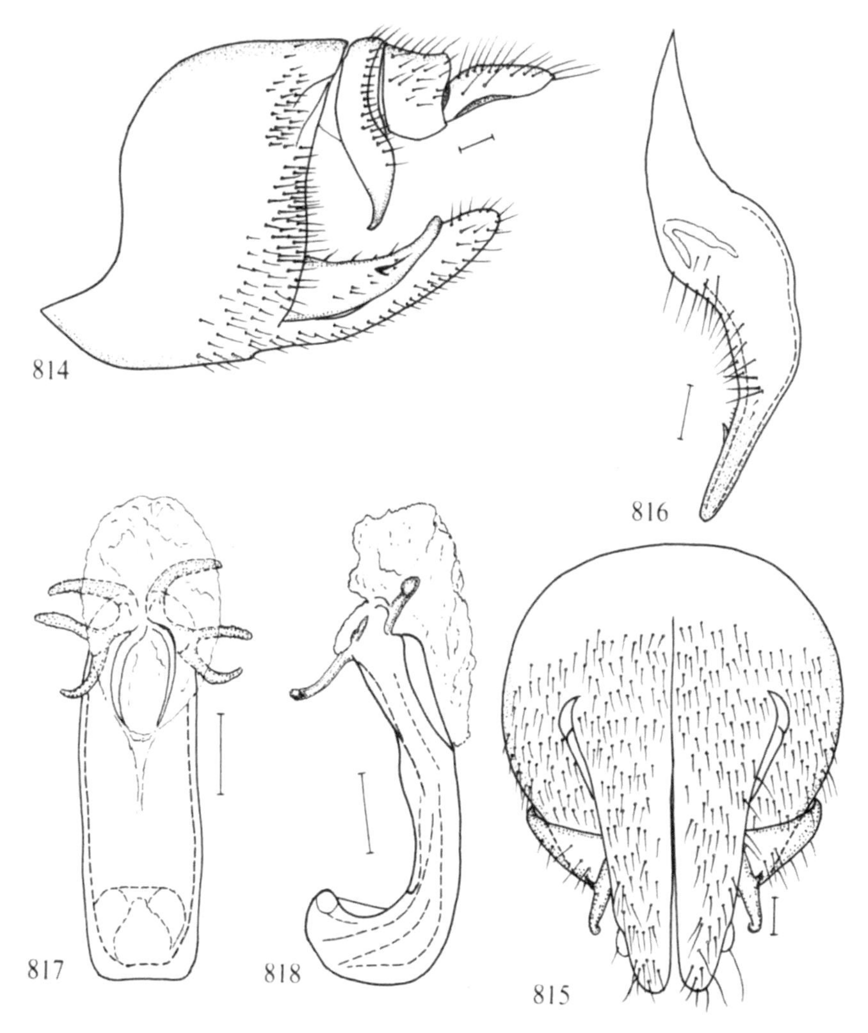

Text-figs. 814–818. *Philaenus spumarius* (Linné). – 814: male pygofer from the left; 815: male pygofer from below; 816: right genital style from below; 817: aedeagus in ventral aspect; 818: aedeagus from the left. Scale: 0.1 mm.

the coast of the Gulf of Bothnia, the first adults appear at the very end of June in favourable summers" (Halkka & al., 1967). In Sweden adults occur in June–October; in the south of Sweden (Bl.) one adult male has been found as early as 12. VI. Males emerge earlier then females (Halkka & al., l.c.). Adults are capable of flying but they are not strong fliers. The direction of migration from one field to another is determined largely by wind (Weaver & King, 1954). These authors "have observed meadow spittlebugs traveling for more than 100 feet in a single flight. Marked adults have traveled as much as 300 feet from a release point within 24 hours".

Economic importance, see p. 28. The importance of *Philaenus* as a virus vector is considered to be negligible in spite of its ability to transmit e.g. peach yellows virus and the virus of Pierce's disease of vines having been demonstrated in the laboratory.

Family Membracidae
Treehoppers

Face directed downwards, almost horizontal. Postclypeus projecting over anteclypeus, often entirely concealing the latter. Vertex and anterior part of pronotum vertical. Coronal suture distinct even in adults. Pronotum caudally with a prolongation extending over meso- and metanotum and often also abdomen. Pronotum often with lateral horn-like processes. In many exotic species the pronotum carries superstructures with a more or less monstrous appearance. Fore wings membranous, basally leathery with strong veins. Tibiae short, with a triangular cross-section, without spines. Larvae with pronotum strongly elevated, gibbous, last abdominal segment slender, approximately as long as rest of abdomen (in our species).

Key to genera of Membracidae

1 Pronotum with a pair of lateral horn-like processes. Caudal prolongation of pronotum long, almost extending to apex of abdomen, sinuate .... *Centrotus* Fabricius (p.252)

– Lateral processes on pronotum absent. Caudal prolongation shorter, straight .... *Gargara* Amyot & Serville (p.254)

Genus *Centrotus* Fabricius, 1803

Centrotus Fabricius, 1803: 16.

Type-species: *Cicada cornuta* Linné, 1758, by subsequent designation.

Body robust. Ridges over antennae broad, sharp-edged. Postclypeus inflated, forming a drooping cone. Pronotum strongly convex, on each side with a triangular pointed process and caudally with a slender, weakly curved, dorsally sharply carinate process

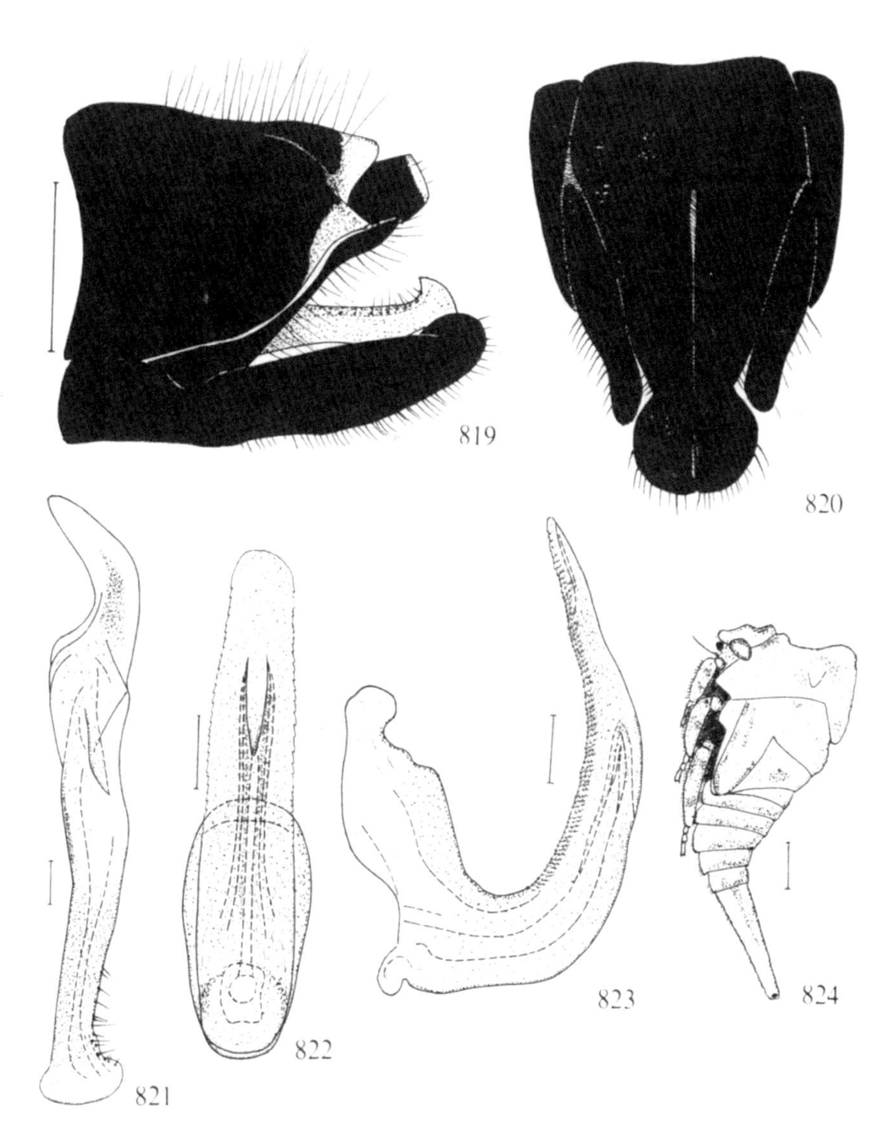

Text-figs. 819–824. *Centrotus cornutus* (Linné). – 819: male genital segment from the left; 820: male genital segment from below; 821: left genital style from outside; 822: aedeagus in ventral aspect; 823: aedeagus from the left; 824: 5th instar larva from the left. Scale: 0.5 mm for 819 and 820, 1 mm for 824, 0.1 mm for the rest.

extending almost to apex of abdomen. Fore wings longer than abdomen, and in repose roof-like covering it. Hind tibiae short, equipped with fine setigerous tubercles. One species in Denmark and Fennoscandia.

101. ***Centrotus cornutus*** (Linné, 1758)
Plate-fig. 47, text-figs. 819–824.

Cicada cornuta Linné, 1758: 435.
Danish: skjoldcikade, torncikade, horncikade. Swedish: hornstrit. Norwegian: hornsikade, tornsikade. Finnish: sarvikaskas.

Black, coarsely punctured with a depressed whitish or yellowish pubescence. In the female, the margins of the lateral and caudal processes are yellowish brown. Fore wings transparent with strong brownish veins and a fuscous spot on the hind border a little distally of the claval apex. Apices of femora, tibiae (at least on outside) and tarsi yellowish brown. Margins of abdominal segments narrowly yellowish brown. Male pygofer as in Text-figs. 819, 820, genital style as in Text-fig. 821, aedeagus as in Text-figs. 822, 823. Overall length 7–9.5 mm. Larva (Text-fig. 824) brownish, often mottled, abdomen apically extended into a tube-like process.

Distribution. Somewhat local but widespread in Denmark. – Fairly common but rarely abundant in Sweden up to Hls. (missing in G. Sand.), also found in Jmt. and Ång. – Norway: found in Ø, AK, HEs, Os, Bø, VE, TEy, AAy, VAy, VAi, Ry, and Ri. – Fairly common in southern and central East Fennoscandia, recorded from Al, Ab, N, St, Ta, Sa, Oa, Sb, Kb; Vib, Kr. – Widespread in Europe, also in Anatolia, Armenia, Azerbaijan, Georgia, Kazakhstan, m. and w. Siberia.

Biology. "An Laubhölzern (besonders Eiche)" (Kuntze, 1937). On leafy trees, especially *Populus tremula* (Ossiannilsson, 1946c). In leafy woods "an allen Stellen häufig, besonders auf Brombeere, aber auch auf anderen Sträuchern" (Schwoerbel, 1957). "In den Ostalpen heliophil und ganz vorwiegend auf Kalk- und kalkhaltigen Gesteinen. Bevorzugt als Futterpflanze im Gebiet deutlich *Cynanchum vincetoxicum*" (Wagner & Franz, 1961). Trolle (in litt.) found *Centrotus* repeatedly on nettles (*Urtica*). Adults in May–October. Hibernates in the larval stage (Müller, 1957). Semivoltine (Müller, 1978).

Genus *Gargara* Amyot & Serville, 1843

Gargara Amyot & Serville, 1843: 537.
Type-species: *Membracis genistae* Fabricius, 1775, by monotypy.

Body short and broad, robust. Pronotum anteriorly strongly convex, without lateral processes, caudally with a spine-like carinate prolongation extending to apex of clavus or a little longer. One species in Europe.

102. ***Gargara genistae*** (Fabricius, 1775)
Plate-fig. 46, text-figs. 825–827.

Membracis genistae Fabricius, 1775: 677.

Body black (♂) or dark reddish brown (♀), densely punctured, with a semi-depressed yellowish or whitish pubescence. Venter of thorax with a dense depressed whitish pubescence. Antennae, rostrum, tibiae, and tarsi brownish yellow, femora largely black. Fore wings basally black or brownish, for the rest brownish yellow with light brownish or yellowish hairy veins, with a dark spot in apex of clavus and another near costal border 2/3 from wing basis (on proximal end of subapical cells). Abdomen black, segmental borders narrowly yellowish. Male genital style as in Text-fig. 825, aedeagus as in Text-figs. 826, 827. Overall length 3–4 mm.

Distribution. Rare in Denmark, only found in SJ: Haderslev and Kidskelund Plantage 21.VIII.1955 (O. Bøggild). – Not in Sweden, Norway, and East Fennoscandia. –

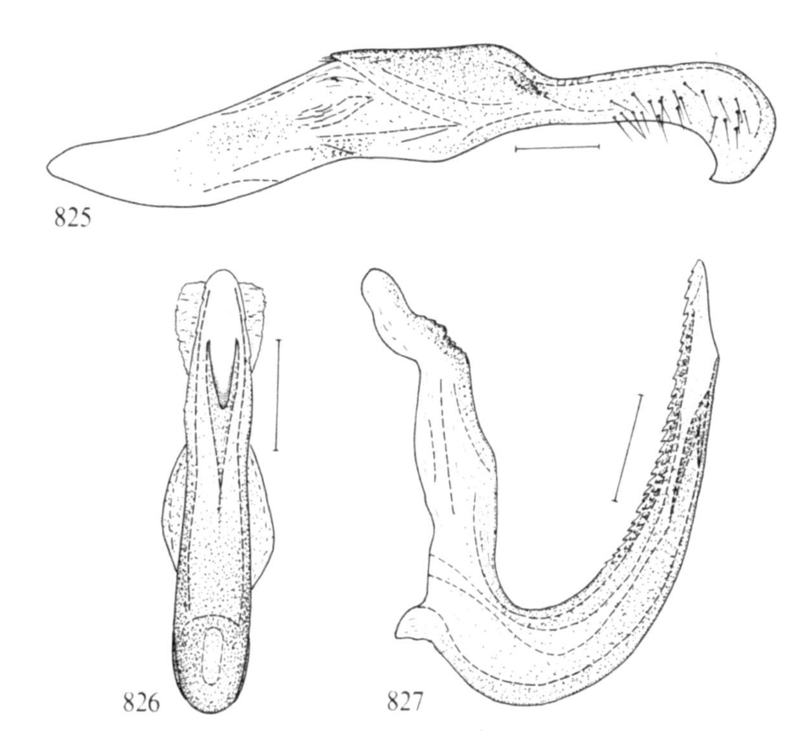

Text-figs. 825–827. *Gargara genistae* (Fabricius). – 825: left genital style from above; 826: aedeagus in ventral aspect; 827: aedeagus from the left. Scale: 0.1 mm.

Widespread in m., s., and e. Europe, including England, also in Algeria, Anatolia, Iran, Altai Mts., Georgia, Kazakhstan, Uzbekistan, m. and w. Siberia, China, Japan, Korean Peninsula, Maritime Territory, Ethiopian, Nearctic, and Oriental regions.

Biology. On *Sarothamnus scoparius* (Kuntze, 1937). "In Mitteleuropa vorwiegend auf *Sarothamnus scoparius*, in Süddeutschland auf *Ononis, Genista, Medicago*" (Wagner & Franz, 1961). "Eier überwintern, schlüpfen im Juni. Larvalzeit etwa ein Monat. Larven und ad. von Ameisen gepflegt, träge" (Müller, 1956). Univoltine. "In xerophilen und mesophilen Biotopen" (Schiemenz, 1969b).

Family Cicadellidae
Leafhoppers

Clypeus usually more or less enlarged, fused with frons. Position of ocelli varying. Pronotum usually broad, rarely with processes. Hind tibiae long, carinate, often more or less flattened, usually with longitudinal rows of mobile setae on the keels. Oviposition usually into living plant tissues.

Key to subfamilies of Cicadellidae

1 Lateral carina of hind tibia strongly widened into a thin and broad plate, in our genus armed with a number of teeth on its outside. Ocelli dorsal. Pronotum in our genus with two large lobiform processes .... **Ledrinae** (p.261)
– Lateral carina of hind tibia not especially broad. Pronotum without processes 2
2 (1) Apex of hind femur without setae. Genae elevated, well delimited from adjacent lateral parts of face (Text-fig. 828) ...... **Ulopinae** (p.257)
– Apex of hind femur with a group of setae (Text-fig. 9). Genae not sharply delimited from adjacent lateral parts of face 3
3 (2) Longitudinal veins of corium distinct even basally. Fore wing with transverse veins also proximally of apical part (Text-fig. 16) 4
– Longitudinal veins of corium basally indistinct. Fore wing with transverse veins in apical part only (Text-fig. 18) ...... **Typhlocybinae** (p.393)
4 (3) Frontal and epicranial sutures marked by broad carinae (Text-fig. 833) .... **Megophthalminae** (p.258)
– Frontal sutures not carinate 5

5 (4) Frontoclypeus inflated, its upper part visible from above as a part of the upper side of the head .... **Cicadellinae** (p.383)
– Frontoclypeus not partly included in the upper side of the head 6
6 (5) Face and vertex medially with a strong longitudinal carina. Pronotum on each side with two longitudinal carinae .... **Dorycephalinae** (p.358)
– No median carina on vertex, nor on the face 7
7 (6) Ocelli situated on the face 8
– Ocelli situated on the fore border of the head or immediately over or under it 11
8 (7) Muscle traces on frontoclypeus appearing as two reniform or semilunar spots (Text-fig. 885) .... **Macropsinae** (p.263)
– Muscle traces on frontoclypeus different 9
9 (8) Well-marked ridges present above antennae, extending from compound eyes to frontoclypeus. Epicranial suture weak but distinct, semicircular .... **Iassinae** (p.352)
– Antennal ridges more or less distinct, not extending from eyes to frontoclypeus 10
10 (9) Frontal sutures distinct. Antennal ridges directly running into the clypeal sutures. Always macropterous .... **Idiocerinae** (p.306)
– Frontal sutures absent. Antennal ridges weakly marked .... **Agalliinae** (p.298)
11 (7) 9th abdominal sternum in males laterally fused with the pygofer .... **Aphrodinae** (p.361)
– 9th abdominal sternum in the male laterally touching but not fused with the pygofer, forming the so-called genital valve .... **Deltocephalinae** (part 3)

SUBFAMILY ULOPINAE

Small leafhoppers. A distinct subgenal suture demarcates each of the genae from the corresponding maxillar plate (Text-fig. 828). Frontoclypeus convex but flattened towards the sharp border of vertex. Antennal ridges sharp. Hind tibiae with a quadrangular cross-section, with weak setae but otherwise unarmed. In Denmark and Fennoscandia one genus.

Genus *Ulopa* Fallén, 1814

Ulopa Fallén, 1814: 19.
Type-species: *Cicada obtecta* Fallén, 1806, by subsequent designation.

Body short and broad. Vertex flat, anteriorly sharp-edged, on each side with a shallow depression. Pronotum more than twice as broad as long, frontally faintly convex, caudally somewhat concave, on each side with a shallow depression. In Denmark and Fennoscandia one wing-dimorphous species.

103. ***Ulopa reticulata*** (Fabricius, 1794)
Plate-fig. 55, text-figs. 828–832.

Cercopis reticulata Fabricius, 1794: 57.
Cicada obtecta Fallén, 1806b: 114.

In the brachypterous form hind wings are absent. Fore wings of this form a little longer than abdomen, leathery, strongly vaulted and coarsely punctured like the remaining part of dorsum as well as the face. Head and pronotum with a median carina. Reddish brown to greyish brown. Depressions of head and prothorax partly blackish, lateral corners of scutellum black. Fore wings with elevated dark veins and two oblique light transverse bands within which the veins are ivory white. Claval suture absent. Abdomen ventrally often blackish. In the macropterous form hind wings are present, the fore wings are not vaulted, their distal half is transparent and there is a distinct claval suture. Male pygofer as in Text-fig. 829, genital style as in Text-fig. 830, aedeagus as in Text-figs. 831, 832. Overall length of f. brach. 3–4 mm. The larvae are very like the adults in structure and colour.

Distribution. Common in Denmark. - Common in Sweden up to Hls. - Common in the south of Norway, found in Ø, AK, HEs, TEi, AAy, VAy, VAi, Ry, Ri, HOy. - Fairly common in s. and c. East Fennoscandia, recorded from Al, Ab, N, Ka, St, Ta, Sa, Tb, Kb; Kr. - Austria, Belgium, Bohemia, Moravia, France, German D.R. and F.R., England, Scotland, Ireland, Italy, Netherlands, Poland, Portugal, Spain, Switzerland, Estonia, Latvia, n. and m. Russia, Yugoslavia; Morocco.

Biology. On *Calluna vulgaris.* Larvae in April–August; adults can be found at any time of the year. Hibernates as imago (Müller, 1957; Remane, 1958; Schiemenz, 1969). Univoltine (Schwoerbel, 1957). Under *Erica* and *Calluna* (Le Quesne, 1965). Usually brachypterous. I have seen only one macropterous specimen collected by Lohmander in Ög., Gusum, Stickkärr 11.VIII.1932. The statement of Jensen-Haarup (1920) that *Ulopa reticulata* is usually macropterous in Denmark is erroneous (Trolle, in litt.).

SUBFAMILY MEGOPHTHALMINAE

Small leafhoppers. Head traversed by high carinae, i.a. marking the position of epistomal and frontal sutures. Vertex strongly concave. By an angular carina it is divided into three parts, one median and two lateral ones, the latter enclosing the ocelli at a considerable distance from the compound eyes. Venation of fore wings normal. Hind tibiae on distal half with a few small spines and setae. Body coarsely punctured. In Denmark and Fennoscandia one genus.

Genus *Megophthalmus* Curtis, 1833

Megophthalmus Curtis, 1833: 193.
Type-species: *Megophthalmus bipunctatus* Curtis, 1833, by monotypy.
Paropia Germar, 1833: 181.
Type-species: *Cicada scanica* Fallén, 1806.

Vertex short, its median field frontally obtuse-angled. Antennal ridges high. Fore wings with strong veins. In Denmark and Fennoscandia only one species.

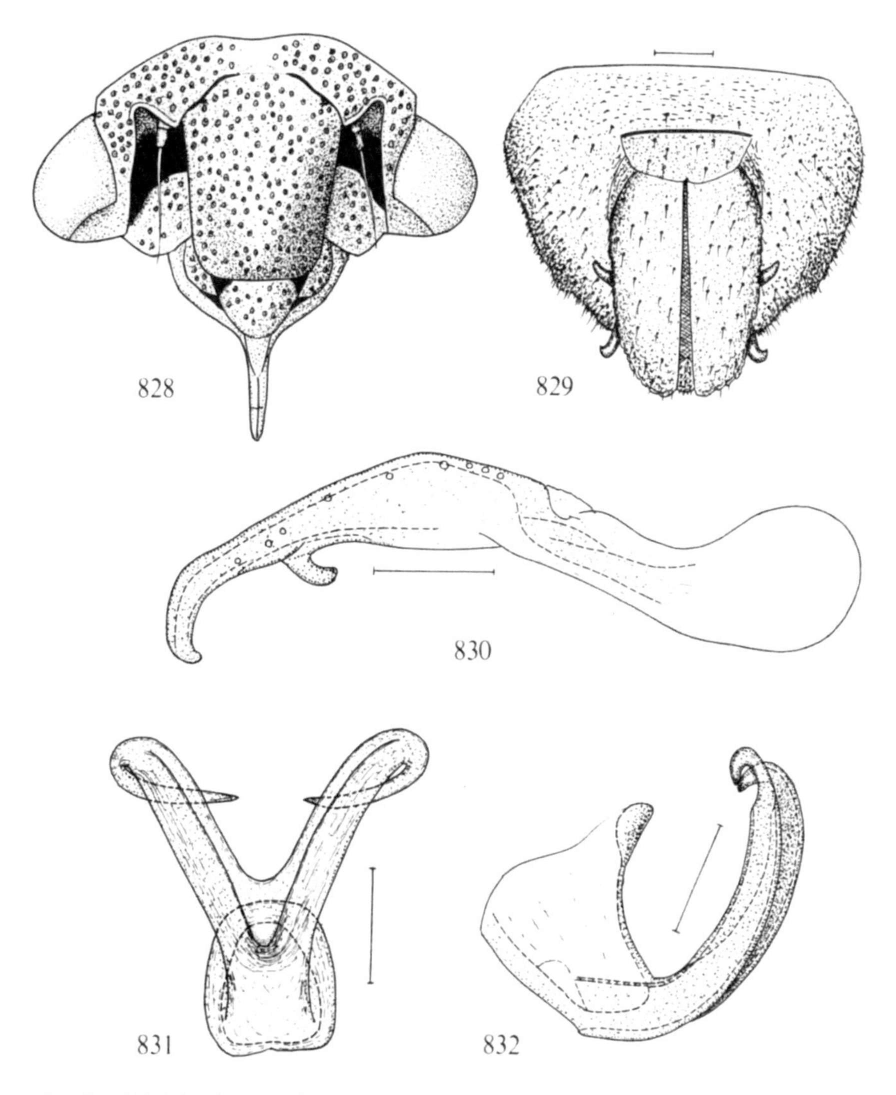

Text-figs. 828–832. *Ulopa reticulata* (Fabricius). – 828: face; 829: male genital segment from below; 830: left genital style from below; 831: aedeagus in ventral aspect; 832: aedeagus from the left. Scale: 0.1 mm.

104. ***Megophthalmus scanicus*** (Fallén, 1806)
Plate-fig. 54, text-figs. 833–837.

Cicada scanica Fallén, 1806b: 113.
Megophthalmus bipunctatus Curtis, 1833: 194.
Paropia scanica Sahlberg, 1871: 375.

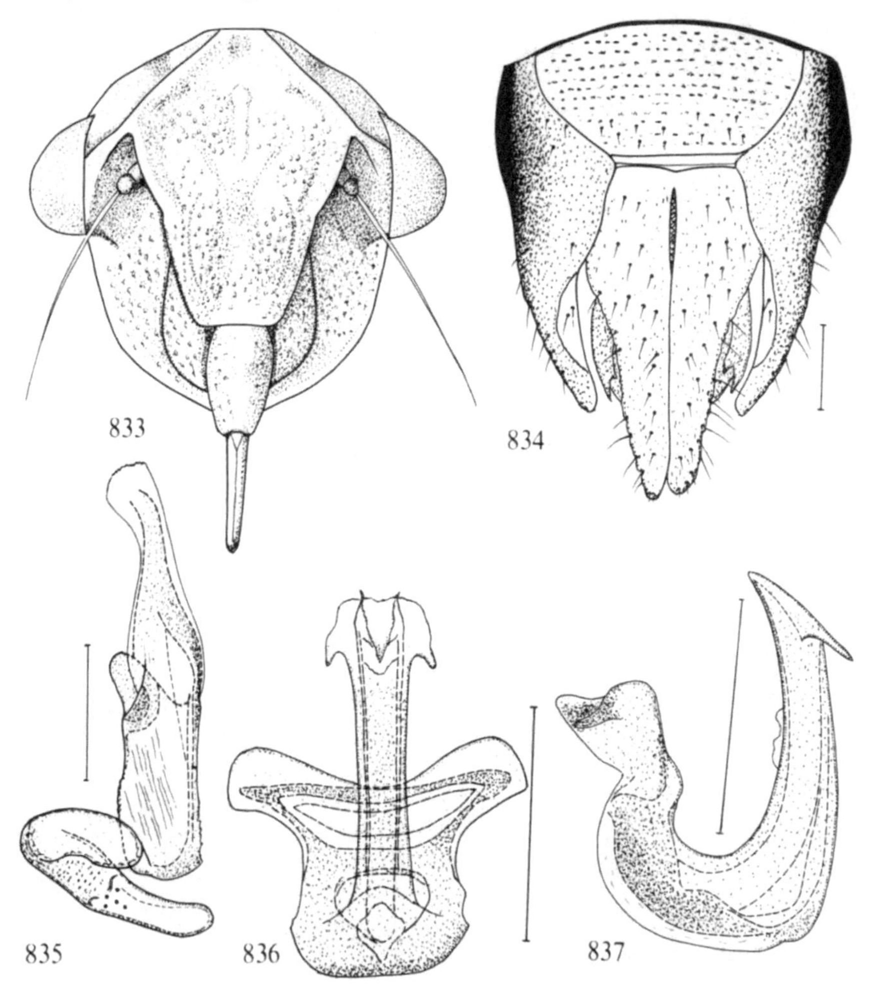

Text-figs. 833–837. *Megophthalmus scanicus* (Fallén). – 833: face of female; 834: apex of male abdomen from below; 835: left genital style from below; 836: aedeagus in ventral aspect; 837: aedeagus from the left. Scale: 0.1 mm.

Body short and broad. Body and also base of fore wings in both sexes coarsely punctured. Fore wings considerably longer than abdomen. Male (Plate-fig. 54) largely black. Borders and carinae of the head, vertex except three blackish spots, borders of pronotum, a pair of elevated oblique carinae on scutellum, antennae, thoracal pleura (partly), legs, borders of abdominal segments, and pygofer yellowish white. Fore wings basally yellowish, leathery, towards apex somewhat thinner, whitish. Female dirty yellowish white with diffusely limited, often indistinct, fuscous patches, fore wings with partly dark veins, commissural border dark between the points where claval veins and claval suture reach the border. Male pygofer as in Text-fig. 834, genital style as in Text-fig. 835, aedeagus as in Text-figs. 836, 837. Overall length of males 2.5–3 mm, of females 3.3–4 mm. - The larvae are yellowish white, on anterior part of body with three black longitudinal bands, the median one being double, and on abdomen with four longitudinal rows of black spots.

Distribution. Common and widespread in Denmark. - Common in the south of Sweden up to Upland (not yet found in G. Sand.), also in Dlr. and Ång. - "Known only from the coastal districts in Southern and Western Norway" (Holgersen, 1944a) (Ø, AAy, VAy, Ry, Ri, HOy, SFy). - Moderately common in southern and central East Fennoscandia: Al, Ab, N, Ka, St, Ta, Sa, Tb, Kb; Kr. - Widespread in Europe, also found in Algeria, Tunisia, Syria, Azerbaijan, Georgia.

Biology. "Auf Wiesen im allgemeinen. - - - Die ♀♀ scheinen zu überwintern" (Kuntze, 1937). In the "Molinio-Arrhenatheretea", auf mesophilen bis hygrophilen Graslandtypen (Marchand, 1953). "Bewohnt in den Nordostalpen - - - ebenso Trockenrasen wie Sumpfränder, steigt ganz vereinzelt bis zur alpinen Waldgrenze empor" (Wagner & Franz, 1961). "Auf feuchten Wiesen, im Auenwald, und in Waldsteppen" (Okáli, 1960). Univoltine, hibernation in the egg stage (Remane, 1958, Müller, 1978).

SUBFAMILY LEDRINAE

Head strongly depressed and flattened, ventrally concave. Anteclypeus narrow, often pear-shaped. Antennal ridges indistinct. Ocelli dorsal. Pronotum collar-like with or without lateral lobiform processes. Fore wings often with many supernumerary veinlets. Hind tibiae strongly flattened, with or without spines.

Genus *Ledra* Fabricius, 1803

Ledra Fabricius, 1803 VII: 13.

Type-species: *Cicada aurita* Linné, 1758, by subsequent designation.

Fore margin of head obtuse-angled. Vertex with an elevated longitudinal median carina and two shorter lateral keels. Pronotum caudally on each side with an ear-like process. Fore wings leathery with many supernumerary transverse veins. In Europe one species.

105. ***Ledra aurita*** (Linné, 1758)
Plate-fig. 49, text-figs. 838–842.

Cicada aurita Linné, 1758: 435.
Danish: Ørecikade; Swedish: öronstrit.

Greyish green or greyish yellow, mottled with brownish. Dorsum of fore body with reddish tubercles and blackish dots. Fore wings with an indistinct light transverse band proximally of middle. Venter light, finely pilose, legs with rather long pilosity. Hind tibiae on outside with 5–6 coarse saw-teeth. Male pygofer as in Text-fig. 838, genital style as in Text-fig. 841, aedeagus as in Text-figs. 839, 840. Overall length of male about 13 mm, of female 15–18 mm. – Larvae much flattened, leaf-thin, in colour similar to adults, dorsally with several rows of small tubercles (Text-fig. 842).

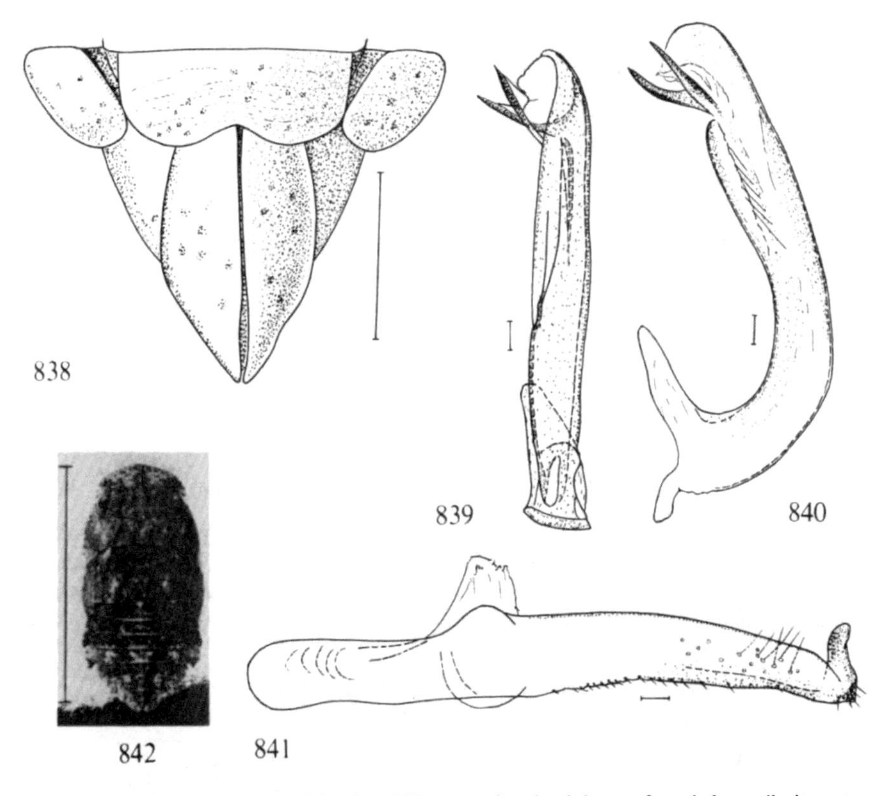

Text-figs. 838–842. *Ledra aurita* (Linné). – 838: apex of male abdomen from below, pilosity not considered; 839: aedeagus in ventral aspect; 840: aedeagus from the left; 841: left genital style from above; 842: larva. Scale: 1 mm for 838, 10 mm for 842, 0.1 mm for the rest.

Distribution. Scarce in Denmark (EJ, F, LFM, SZ). - Rare in the south of Sweden: Sk.: Brunnby, Kullen (Boheman, E. Olofsson, Lindroth, Ossiannilsson), Laröd (Ringdahl), Skäralid (Ringdahl, Kemner), Brösarp (Ingvar Svensson), Sandhammaren (Bengt Wickholm); Bl. (Afzelius), Lyckeby (Boheman), Marielund, Nättraby (Lundblad); Sm.: Hornsö near point Korstalden about 22 km west of Timmernabben (L. Rytterfalk); Öl.: Högsrum, Halltorp (A. Jansson, H. Andersson and R. Danielsson); Hall., Vippentorpet (Mortonson); Vg.: Kinnekulle (Gyllenhal); Ög.: Törnevalla, Alvestad (Håkan Lindberg, Ossiannilsson. - Norway: Ø: Hvaler (T. Helliesen), Kirkøya (Hvaler) (Munster). - Not found in East Fennoscandia. - Albania, Austria, Belgium, Bulgaria, Bohemia, Moravia, Slovakia, France, German D. R. and F. R., England, Hungary, Italy, Netherlands, Poland, Portugal, Romania, Switzerland, Yugoslavia, s. Russia, Ukraine, Moldavia, Azerbaijan, Georgia, Anatolia, China.

Biology. On *Quercus*, sometimes also on other deciduous trees and bushes *(Alnus glutinosa, Corylus)*. "In den Ostalpen nur in tieferen Lagen, an Eichen, nirgends häufig" (Wagner & Franz, 1961). Hibernates as larva (Müller, 1957). Development probably takes more than one year (Schwoerbel, 1957). Adults and larvae are very sluggish but they can leap vigorously. Lindroth caught some specimens (males only) in light traps. Adults 30.VI.-2.X.; small larvae were found in July and August, large larvae in June and July.

SUBFAMILY MACROPSINAE

Vertex very short. Muscle traces on frontoclypeus reniform. Head usually with a strongly punctured or wrinkled surface structure. Ocelli ventral. Side margins of pronotum quite short. Fore wings in resting position held roof-like over abdomen, their apical part with a narrow membranous edge, "apical appendage", outside the marginal vein. Hind tibiae with several rows of strong spines and fine setae. The pigmentation of the face has been found to be useful for specific separation. This necessitates a special terminology illustrated in Text-figs. 843, 885, and 942.

Key to genera of Macropsinae

1 Striations of pronotum more or less parallel to its posterior margin, at least caudally .... *Oncopsis* Burmeister (p.264)

\- Striations of pronotum running obliquely from middle of its front margin towards lateral angles (less distinctly in *Hephathus*) (Text-figs. 894, 908) 2

2 (1) Vertex medially invisible from above, being covered by frontal part of pronotum (Text-figs. 894, 895) .... *Pediopsis* Burmeister (p.281)

\- Vertex as seen from above short but visible, not covered by pronotum (Text-figs. 908, 909) 3

3 (2) Striations of pronotum less distinctly oblique. Fore wings at most 2.8 mm long. Face in lateral aspect uniformly curved. Male pygofer without an appendage, or appendage quite short and bud-like .... *Hephathus* Ribaut (p.297)

– Striations of pronotum strongly oblique. Fore wings at least 2.9 mm long. Face in lateral aspect more strongly curved in its upper part (Text-fig. 909). Male pygofer with a long pointed appendage emanating from ventral margin (Text-fig. 933) .... *Macropsis* Lewis (p.281)

Genus *Oncopsis* Burmeister, 1838

Oncopsis Burmeister, 1838a: (10).

Type-species: *Cicada flavicollis* Linné, 1761, by subsequent designation.

Body structure robust. Head in dorsal aspect short, obtuse-angled, medially somewhat shorter than by the eyes. Frontoclypeus in males much broadened in its lower part, reducing the lora to narrow lists. Face more convex in females than in males. Thyridia of vertex always distinct. Pronotum frontally convex, obtuse-angled, caudally concave, obtuse-angled or arched, surface finely transversely wrinkled. Scutellum large. Fore wings not punctured, with stron veins. Hind tibiae with several rows of strong spines. In Denmark and Fennoscandia seven species, all of them living on trees and bushes.

Key to species of *Oncopsis*

1 Males 2

– Females 8

2 (1) Ventral outline of aedeagus S-shaped (Text-figs. 851, 859) 3

– Ventral outline of aedeagus constituting a simple curve (Text-figs. 869, 881, 891) 5

3 (2) Appendages of anal collar each consisting of two short and stout branches of approximately the same length, distance between their apices roughly equalling their length (Text-fig. 862) .... 108. *subangulata* (J. Sahlberg)

– Appendages of anal collar longer and thinner, different in length 4

4 (3) Lower margin of genital style in lateral aspect concave in apical third (Text-figs. 849, 850). Shorter branch of anal collar appendage straight or faintly curved outwards (Text-figs. 846–848). Interocular band on face in strongly pigmented individuals reaching compound eyes. On *Betula* 106. *flavicollis* (Linné)

– Lower margin of genital style in apical third straight or faintly convex (Text-fig. 857). Shorter branch of anal collar appendage straight or faintly curved inwards. Interocular band usually not extending laterad of interocular spots. On *Carpinus* .... 107. *carpini* (J. Sahlberg)

5 (2) Branches of anal collar appendages equal in length (Text-fig. 879) .... 111. *planiscuta* (Thomson)

Branches of anal collar appendages differing in length 6

6 (5) Longer branch of anal collar appendage strongly retrorse (Text-fig. 873) .... 110. *appendiculata* W. Wagner

– Longer branch of anal collar appendage moderately curved 7

7 (6) Large species, >5 mm. Genital style broad, its apical margin concave (Text-fig. 889) .... 112. *alni* (Schrank)

– Small species, <5 mm. Genital style narrow (Text-fig. 868) 109. *tristis* (Zetterstedt)

8 (1) Caudal margin of 7th abdominal sternum almost straight, medially with a slight incision (Text-fig. 883). Black markings of face normally complete, with interocular band, median band, discoidal band and discoidal spot, all forming a coherent figure. On *Alnus incana* .... 111. *planiscuta* (Thomson)

– Caudal margin of 7th abdominal sternum convex 9

9 (8) Caudal margin of 7th abdominal sternum with a deep incision between angular corners (Text-figs. 853, 854). Frontoclypeus in lateral aspect usually strongly convex 10

– Incision on caudal margin of 7th abdominal sternum shallow, arched (Text-figs. 871, 892). Frontoclypeus less strongly convex 12

10 (9) Muscle traces on frontoclypeus and major part of surface between them rust-coloured, a cordiform, laterally more or less narrowly black-edged rust-coloured figure occupying centre of face. Ovipositor caudally extending considerably beyond apex of pygofer (Text-fig. 866) ...... 108. *subangulata* (J. Sahlberg)

– Markings of face different or absent 11

11 (10) Fore wings of strongly pigmented specimens with black-brown or black markings. Ventral margin of saw-case usually somewhat arched, its projecting apex short (Text-fig. 855). On *Betula* .... 106. *flavicollis* (Linné)

– Pigmentation of fore wings brownish, not black. Ventral margin of saw-case straight, projecting apex a little longer (Text-fig. 860) On *Carpinus* .... 107. *carpini* (J. Sahlberg)

12 (9) Large species, >5 mm. On *Alnus glutinosa* .... 112. *alni* (Schrank)

– Smaller species, <5 mm. On *Betula* 13

13 (12) Face with three pairs of black spots. Genital segment in-

cluding ovipositor considerably longer than pre-genital part of abdomen (Text-fig. 877) .... 110. *appendiculata* (W. Wagner)

– Dark markings of face rarely consisting of three pairs of spots. Genital segment including ovipositor little longer than pre-genital part of abdomen (Text-fig. 872) .... 109. *tristis* (Zetterstedt)

106. ***Oncopsis flavicollis*** (Linné, 1761)
Plate-figs. 59, 60, text-figs. 843–855.

Cicada flavicollis Linné, 1761: 242.
Cicada triangularis Fabricius, 1794: 46.
Jassus fruticola Fallén, 1806: 120.
Jassus fruticola, var. *obscurus* Zetterstedt, 1828: 543.
Jassus fruticola, var. *pallens* Zetterstedt, 1828: 543.
Jassus nigritulus Zetterstedt, 1828: 544.
Jassus fruticola, var. *capucinus* Zetterstedt, 1838: 303.
Bythoscopus dubius Fieber, 1868: 457.
Oncopsis flavicollis, var. *hyalina* W. Wagner, 1944: 131.
Oncopsis flavicollis, var. *luteomaculata* W. Wagner, 1944: 131.

Face usually strongly convex, especially in females (Text-figs. 843–845). Extremely varying in colour. The male is yellowish with more or less extended black markings. Face often with a continuous figure consisting of the discoidal spots, discoidal band, median band, and interocular band, but this figure is as often more or less reduced, and the face may also be devoid of dark pigmentation. Pronotum greyish, anteriorly with some black spots often extending over the entire or almost entire surface. Scutellum even in light males usually black with two small light spots. Fore wings greyish, semi-transparent or transparent with black or fuscous dark-bordered veins. The hind margin of the fore wing is whitish in two streaks, one proximally and the other one distally of the points where the claval veins reach the commissural border. The dark pigment bordering the veins may form spots, e.g. one at the distal end of each claval vein, and one around each of the transverse veins between M and Cu. The fore wing apex may also be dark. Venter black-spotted, abdomen black with yellow segmental borders, legs with black longitudinal streaks.

The female is still more varying in colour than the male. The body is whitish yellow, sulphur yellow or yellowish brown with or without black spots. One form resembles the male but is lighter, scutellum largely yellowish brown (f. *obscura* (Zetterstedt)). In three common varieties the upper side is unicolourous: black-brown in f. *nigritula* (Zetterstedt), yellow in f. *dubia* (Fieber), yellowish brown in f. *pallens* (Zetterstedt). In f. *typica* the anterior part of dorsum and the proximal part of the fore wing are yellow while the rest of the latter is rust-coloured. F. *capucina* (Zetterstedt) resembles f. *typica* but the distal part of the fore wing is black-brown. In f. *triangularis* (Fabricius) the upper side is black-brown or black except basis of clavus which is yellow or whitish. In f.

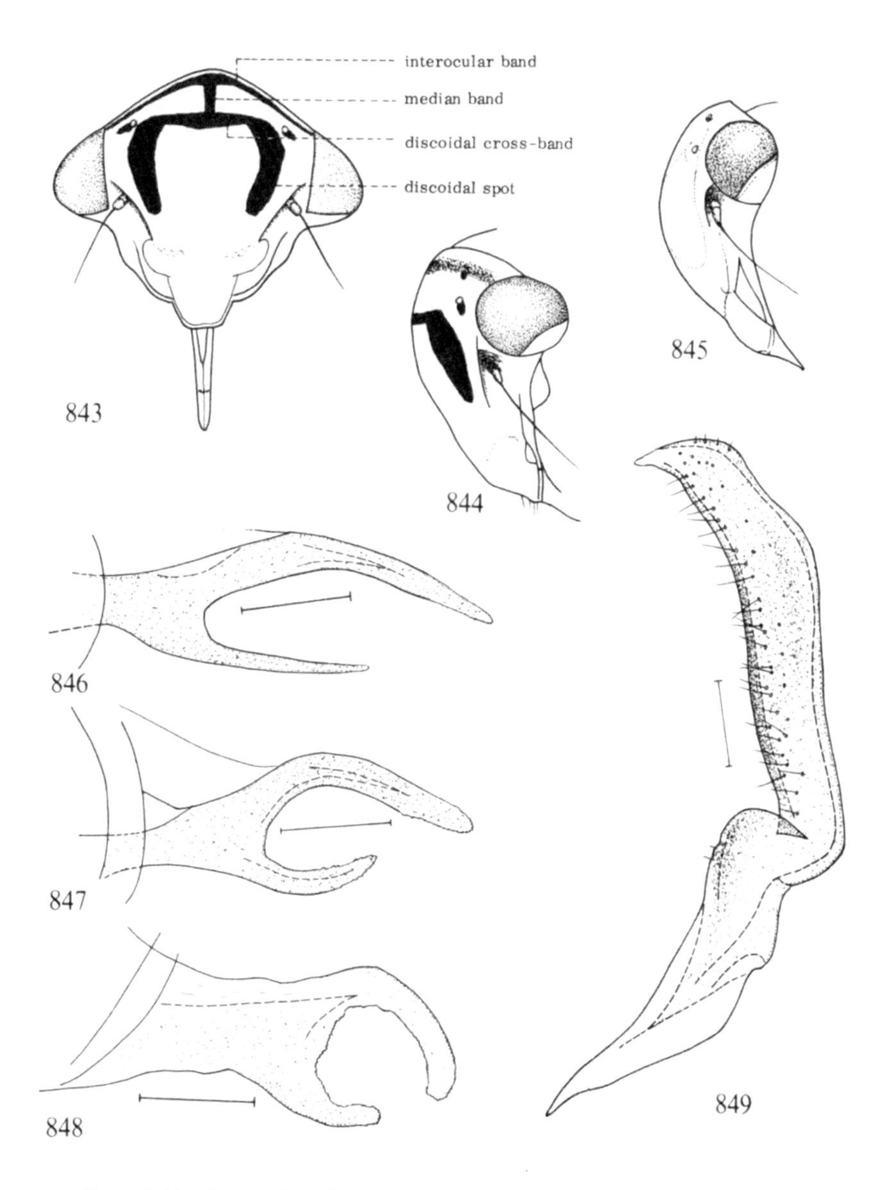

Text-figs. 843–849. *Oncopsis flavicollis* (Linné). – 843: face of male; 844: head of male from the left; 845: head of female from the left; 846–848: left anal collar appendages of three different males; 849: left genital style from the left. Scale: 0.1 mm.

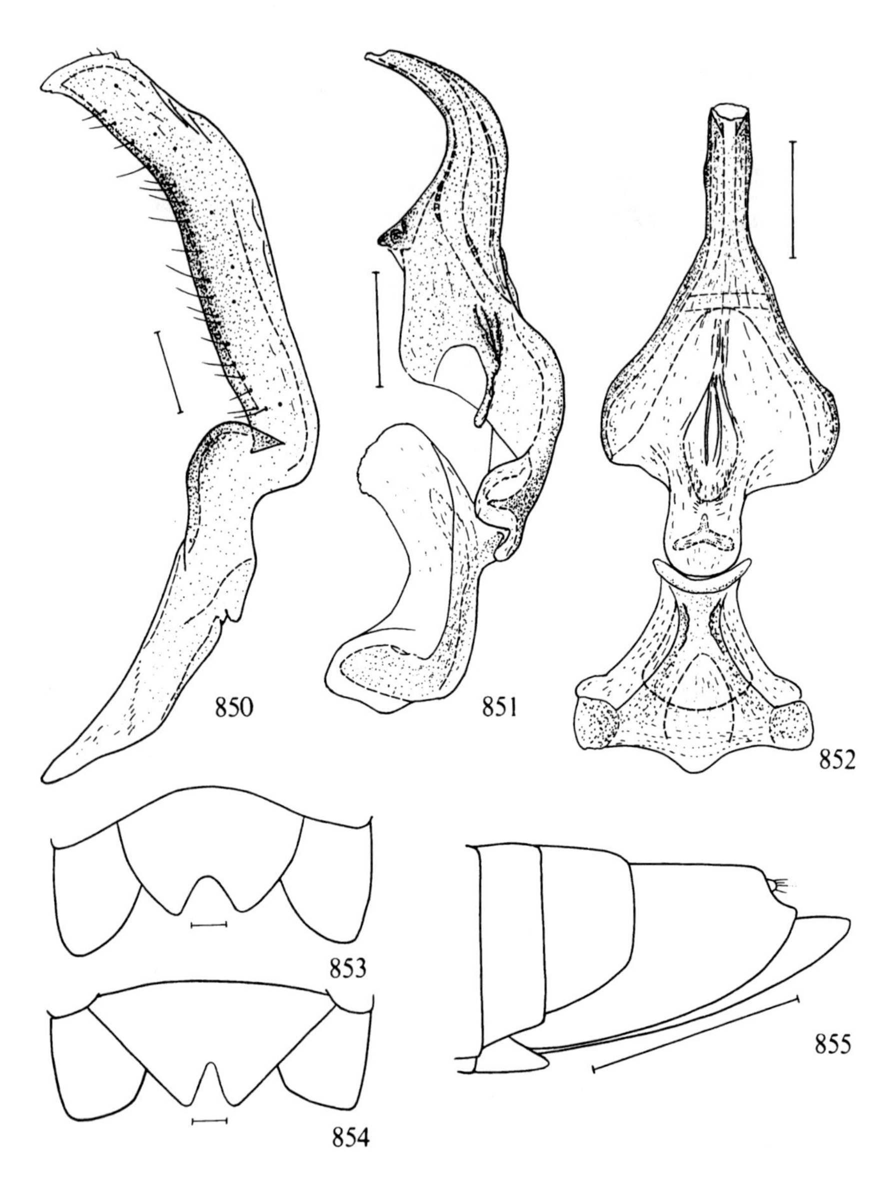

Text-figs. 850–855. *Oncopsis flavicollis* (Linné). – 850: left genital style of a second male, from the left; 851: aedeagus and connective from the left; 852: aedeagus and connective in ventral aspect; 853, 854: 7th abdominal sterna of two different females, from below; 855: posterior segments of female abdomen from the left. Scale: 1 mm for 855, 0.1 mm for the rest.

luteomaculata W. Wagner the fore body is yellow, corium brown with one or two yellow spots. F. *hyalina* W. Wagner is yellow with hyaline fore wings. – Shape of appendages of male anal collar much varying, see Text-figs. 846–848, male genital style as in Text-figs. 849, 850, aedeagus and connective as in Text-figs. 851, 852. 7th abdominal sternum in females with an angular incision, see Text-figs. 853, 854, side view of abdominal apex in female as in Text-fig. 855. Overall length 4.4–5.4 mm.

Last instar larva brownish, dorsally sometimes with a pale median longitudinal band extending from vertex to apex of abdomen, and with a lateral light longitudinal band on each side of abdomen, dorsal pilosity very short, inconspicuous. Head inflated, vertex medially about twice as long as near eyes, muscle traces on face reniform, moderately large.

Distribution. Common and widespread in Denmark, Norway, Sweden, and East Fennoscandia. – Widespread in Europe, found also in Algeria, Tunisia, Anatolia, Syria, Altai Mts., Armenia, Georgia, Kazakhstan, Uzbekistan, Kirghizia, Mongolia, Siberia, Sakhalin, Kamchatka, Maritime Territory.

Biology. On *Betula verrucosa* and *B. pubescens*. Univoltine. Hibernation in the egg stage (Müller, 1957). "An *Betula*, vielleicht ausschliesslich *B. alba*" (Wagner & Franz, 1961). "Especially on moist sites with young *Betula pubescens* and *B. verrucosa*, where the species reproduces on birch and also occurs in the field layer. – – – It is an able flier – – –." (Raatikainen & Vasarainen, 1976). "Eggs are laid in the dormant buds of both *Betula pendula* and *B. pubescens*. – – – Eggs hatch very early, immediately following bud burst – – –" (Claridge & Reynolds, 1972; for more detailed information on oviposition, see this paper). There is a chromosome polymorphism in *O. flavicollis*, and a close correlation between this polymorphism and populations on *Betula pendula* and *B. pubescens* has been established (John & Claridge, 1974). "Females of *O. flavicollis* have the ability to discriminate between, and to oviposit differentially on, *B. pendula* and *B. pubescens*" (Claridge & al., 1977). – Adults in May–September.

107. ***Oncopsis carpini*** (J. Sahlberg, 1871)
Text-fig. 856–860.

Pediopsis carpini J. Sahlberg, 1871: 123.
Oncopsis carpinicola Edwards, 1920: 54.

Resembling *flavicollis*, differing by characters mentioned in the key. Generally paler and more weakly pigmented. Claval commissure in males often entirely light. The median band is normally discernible even if the discoidal band is missing. Anal collar appendage of male as in Text-fig 856, genital style as in Text-fig. 857, aedeagus as in Text-figs. 858, 859, apical part of female abdomen as in Text-fig. 860. Overall length 4–5.2 mm.

Distribution. Fairly common in Denmark. – In Sweden found only in Sk. and Bl. but distribution in this country has not been studied thoroughly. – Not found in Norway

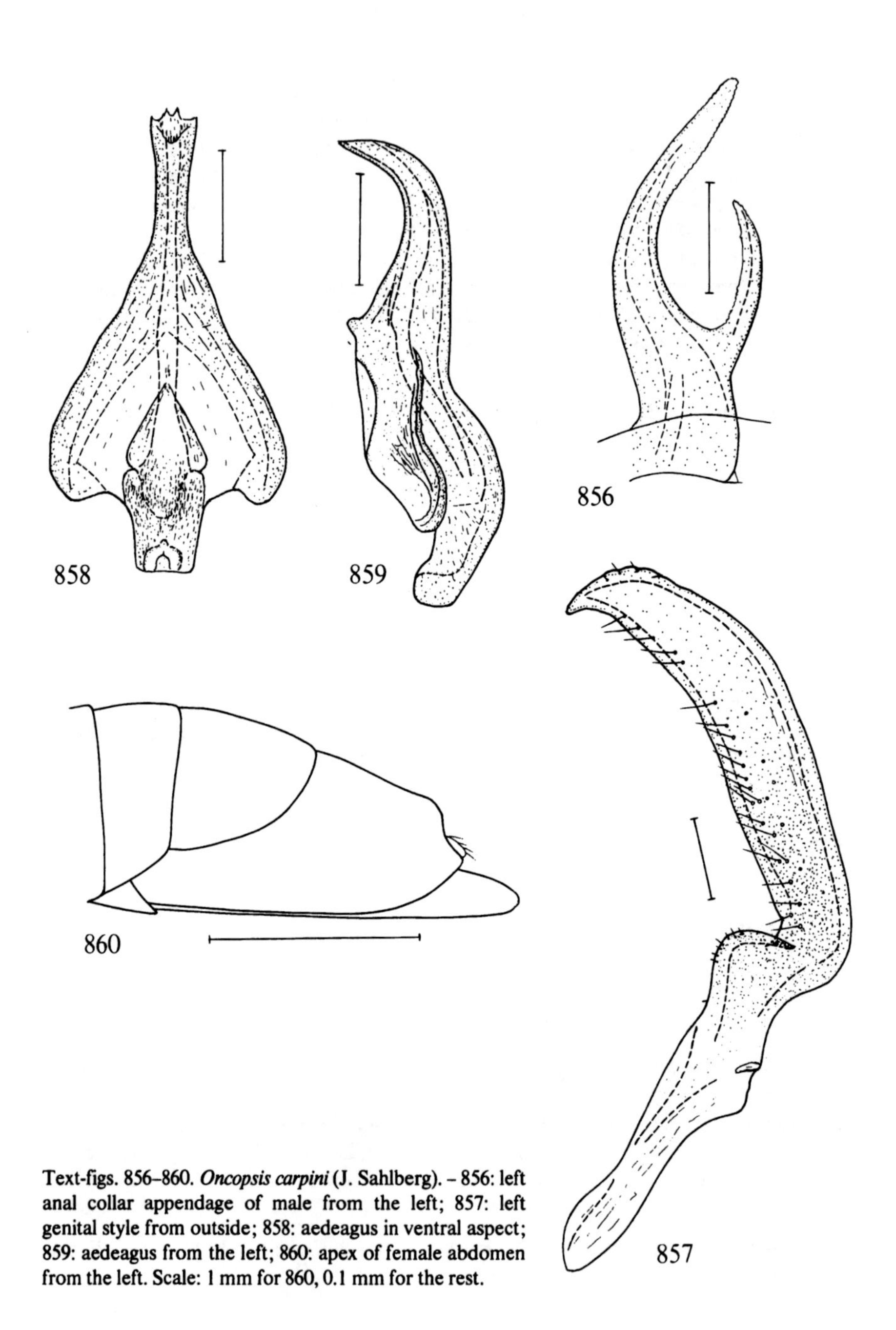

Text-figs. 856–860. *Oncopsis carpini* (J. Sahlberg). – 856: left anal collar appendage of male from the left; 857: left genital style from outside; 858: aedeagus in ventral aspect; 859: aedeagus from the left; 860: apex of female abdomen from the left. Scale: 1 mm for 860, 0.1 mm for the rest.

and East Fennoscandia. - Belgium, France, German D.R. and F.R., England, Netherlands, Poland, Bohemia, Moravia, Slovakia, Yugoslavia.

Biology. On *Carpinus betulus*. "Eggs were inserted in the new apical stem growth, above leaf bases and at the bases of dormant buds. The eggs were found to descend into the shoot away from the point of insertion at a bud base" (Claridge & Reynolds, 1972). "Females collected from hornbeam were caged separately on hornbeam and hazel and also given a choice between hornbeam and hazel. A total of forty-one eggs were later dissected from hornbeam stems and none from hazel" (Claridge & al., 1977). Adults in June–August.

108. ***Oncopsis subangulata*** (J. Sahlberg, 1871)
Text-figs. 861–866.

Pediopsis subangulatus J. Sahlberg, 1871: 125.
Oncopsis fortior W. Wagner, 1944: 129.

Resembling *flavicollis*, body a little larger. In the male, all that remains of the discoidal spots is a narrow black stripe along the upper border of the muscle traces. A narrow discoidal band often present, median band absent. Fore wings of both sexes hyaline with dark veins. Male genitalia as in Text-figs. 861–865. The anal collar appendices are much varying in length and in the angle between their branches. Overall length of males 4.7–5 mm., of females 4.9–5.5 mm. Apical part of female abdomen as in Text-fig. 866. - Last instar larva largely dark brownish or black, dorsal pilosity slightly more distinct than in *flavicollis*. Shape of head as in *flavicollis*, muscle traces on face larger, usually shining black. Pronotum caudally more coarsely transversely wrinkled than in *flavicollis*.

Distribution. Denmark: so far recorded from EJ: Spentrup 14.VII.1968 (Trolle) and NEZ: Rudehegn 18.VI.1916 (C. C. R. Larsen). - Fairly common and widespread in Sweden (Sk.–Ög. and Vg., Dlsl., Upl., Vrm., Dlr., Hls., Med., Ång., Lu. Lpm.). - Norway: so far established in On, TEi, Ry, Ri, HOy, HOi, SFi, MRi, Nnø, and TRi. - Moderately common in southern and central East Fennoscandia (Al, Ab, N, St, Ta, Oa, Tb, Sb, Om; Kr). - Austria, Bohemia, Moravia, Slovakia, France, German D.R. and F.R., England, Italy, Netherlands, Poland, n. Russia.

Biology. "Auf *Betula alba*" (Wagner & Franz, 1961). Oviposition as in *flavicollis* (Claridge & Reynolds, 1972). "We have found both *O. flavicollis* and *O. subangulata* breeding regularly on *B. pubescens* as well as on *B. pendula*" (Claridge & Reynolds, l.c.). "*O. subangulata* appears to oviposit exclusively in birch" (Claridge & al., 1977). Adults in June–September.

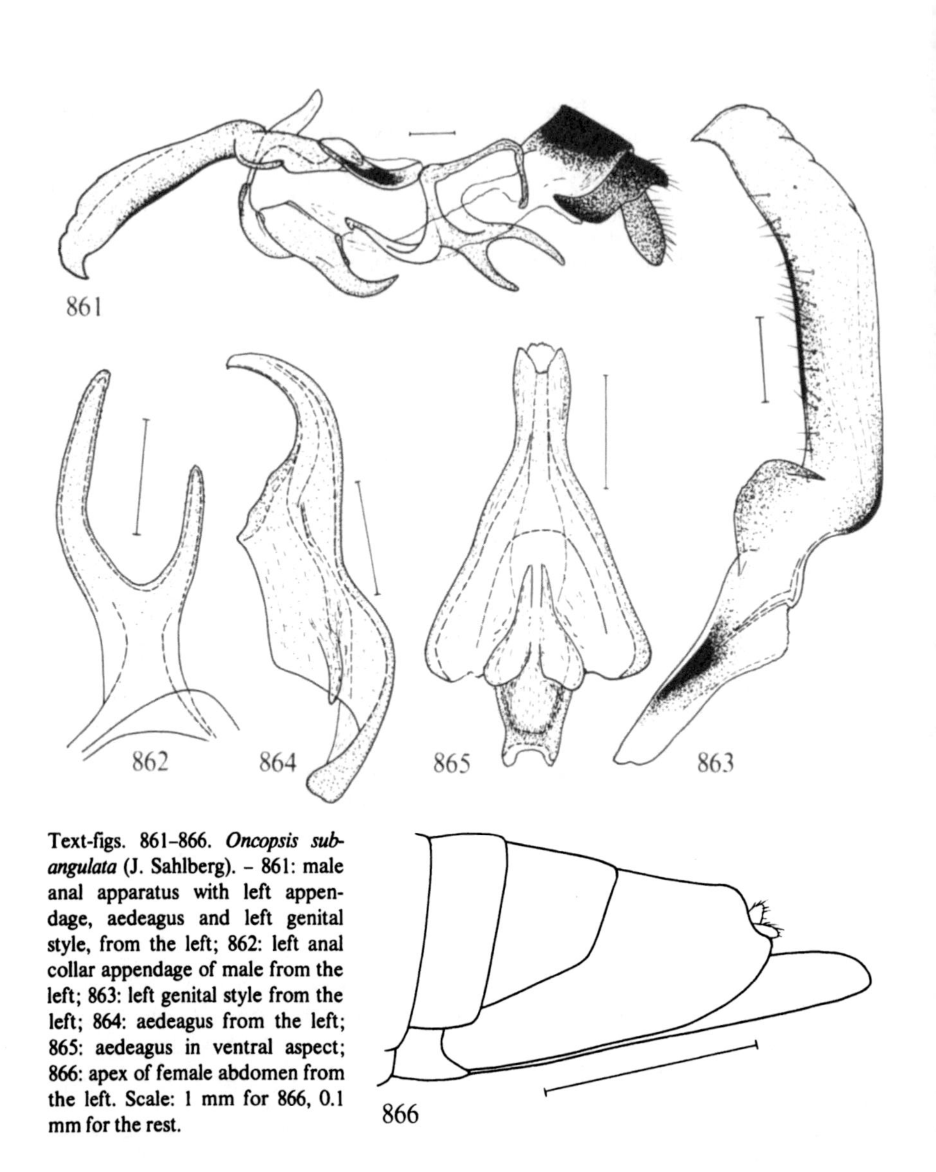

Text-figs. 861–866. *Oncopsis subangulata* (J. Sahlberg). – 861: male anal apparatus with left appendage, aedeagus and left genital style, from the left; 862: left anal collar appendage of male from the left; 863: left genital style from the left; 864: aedeagus from the left; 865: aedeagus in ventral aspect; 866: apex of female abdomen from the left. Scale: 1 mm for 866, 0.1 mm for the rest.

109. ***Oncopsis tristis*** (Zetterstedt, 1838)
Text-figs. 867–872.

Jassus fruticola, var. *tristis* Zetterstedt, 1838: 303.
Bythoscopus rufusculus Fieber, 1868: 456.
Pediopsis brevicauda Thomson, 1870: 318.

Face less convex than in typical *flavicollis*. Male usually yellowish with black markings. Discoidal spots, discoidal band and interocular band often present but as often partly reduced or absent. Pronotum greyish or yellowish brown with black transverse wrinkles and some black spots at its fore border. Scutellum tawny or reddish brown with fuscous or black markings: one spot at each lateral corner, often also a median

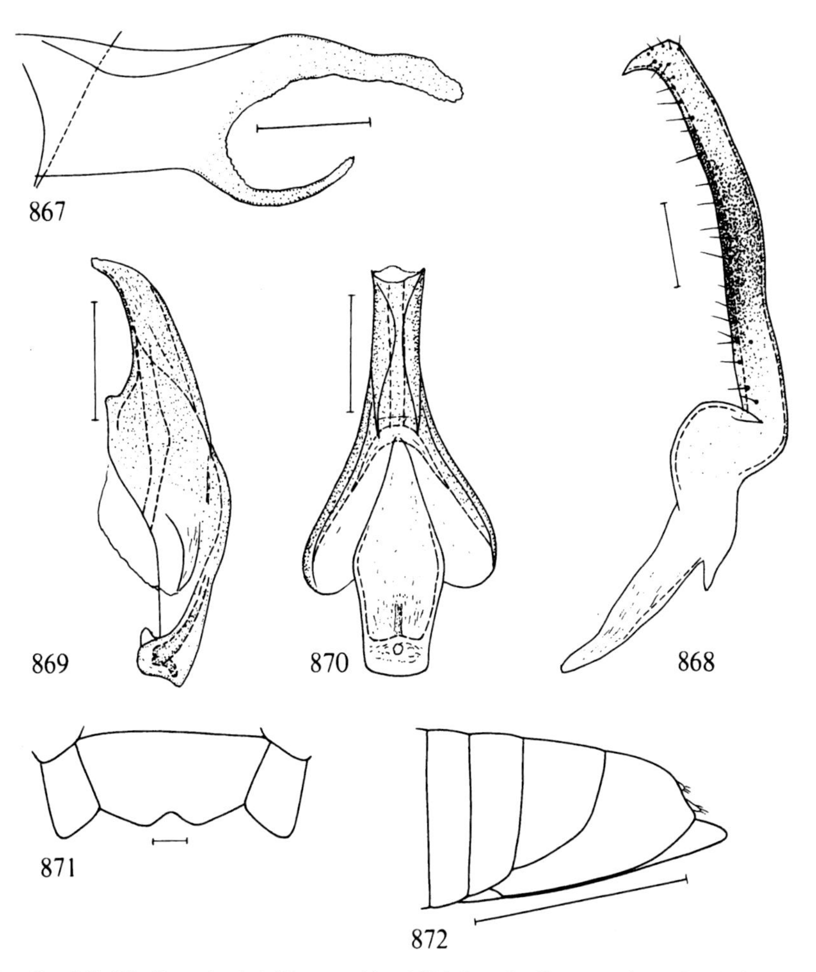

Text-figs. 867–872. *Oncopsis tristis* (Zetterstedt). – 867: left anal collar appendage of male from the left; 868: left genital style from outside; 869: aedeagus from the left; 870: aedeagus in ventral aspect; 871: ventral aspect of 7th abdominal segment of female; 872: apex of female abdomen from the left. Scale: 1 mm for 872, 0.1 mm for the rest.

stripe and a pair of roundish spots on each side of the latter. Fore wings greyish white, transparent, with black veins. Claval commissure whitish proximally and distally of the points where the claval veins reach it; a dark spot is present between these points, another near claval apex, and at third spot of varying size surrounds transverse veins distally, delimiting radial and median cells. Apex of fore wing also dark. Venter dark spotted, abdomen black with more or less broadly yellow segmental margins, legs with dark streaks. In the corresponding female form the dark markings are brownish and less extended. Another female form is unicolorous yellowish brown without markings. Male anal collar appendage as in Text-fig. 867, genital style as in Text-fig. 868, aedeagus as in Text-figs. 869–870; ventral aspect of 7th abdominal segment of female as in Text-fig. 871, apical part of female abdomen as in Text-fig. 872. Length of saw-case <1.3 mm. Length of males 3.7–4.3 mm, of females 3.9–4.6 mm. – Last instar larva brownish, often with three broad light longitudinal bands, dorsal pilosity inconspicuous. Head little inflated, vertex medially only 1½ times as long as near eyes.

Distribution. Common in Denmark. – Common and widespread in Sweden (Sk.–T. Lpm.). – Spread over most of southern Norway, found also in several localities in the north (Nnø, Nnv, TRi). – Common and widespread in East Fennoscandia. – Widespread in Europe, also found in Altai Mts., Kazakhstan, Sakhalin, Siberia, Mongolia, Japan, and Oriental region.

Biology. On *Betula verrucosa* and *pubescens*. Adult males appear earlier than females but they also die earlier (Nuorteva, 1951). "The eggs of *O. tristis* are laid inside the inner bud scales of dormant buds of birches, both *Betula pendula* and *B. pubescens*. – – – First instar nymphs hatch soon after bud burst, as the bud scales open" (Claridge & Reynolds, 1972). Univoltine. Adults in June–September (–November).

110. ***Oncopsis appendiculata*** W. Wagner, 1944
Text-figs. 873–877.

Oncopsis appendiculata W. Wagner, 1944: 130.

Face in lateral aspect as in *tristis*. Resembling *tristis*, on an average a little larger. Discoidal spots at most three times as large as interocular spots, ocellar spots always present, discoidal band absent. Anterior part of body in females without yellow pigment, fore wings transparent with dark veins. Male anal collar appendage as in Text-fig. 873, genital style as in Text-fig. 874, aedeagus as in Text-figs. 875, 876, apical part of female abdomen as in Text-fig. 877. Length of saw-case > 1.4 mm. Overall length of males 4.1–4.7 mm, of females 4.5–4.7 mm.

Distribution. Not yet found in Denmark, nor in Norway. – Rare in Sweden, found in Sk., Sandhammaren (Lohmander); Bl., Förkärla (Gyllensvärd), Kyrkhult, Emmedal (Ossiannilsson); Hall., Halmstad (Ossiannilsson); Sm., Långsjö (Lohmander); Sdm.,

Botkyrka, Ahlby (Ossiannilsson); Upl., Ekerö (Ossiannilsson), Nysätra, vicinity of Bunkhus (Ossiannilsson); Nrk., Ö. Mark (Anton Jansson). - Very rare in East Fennoscandia, found in Sa, Joutseno by Thuneberg. - Austria, France, German D.R. and F.R., Netherlands, Poland, Latvia.

Biology. On *Betula verrucosa* in sun-exposed sites, apparently thermophilous. Adults 23.VI.-16.VIII.

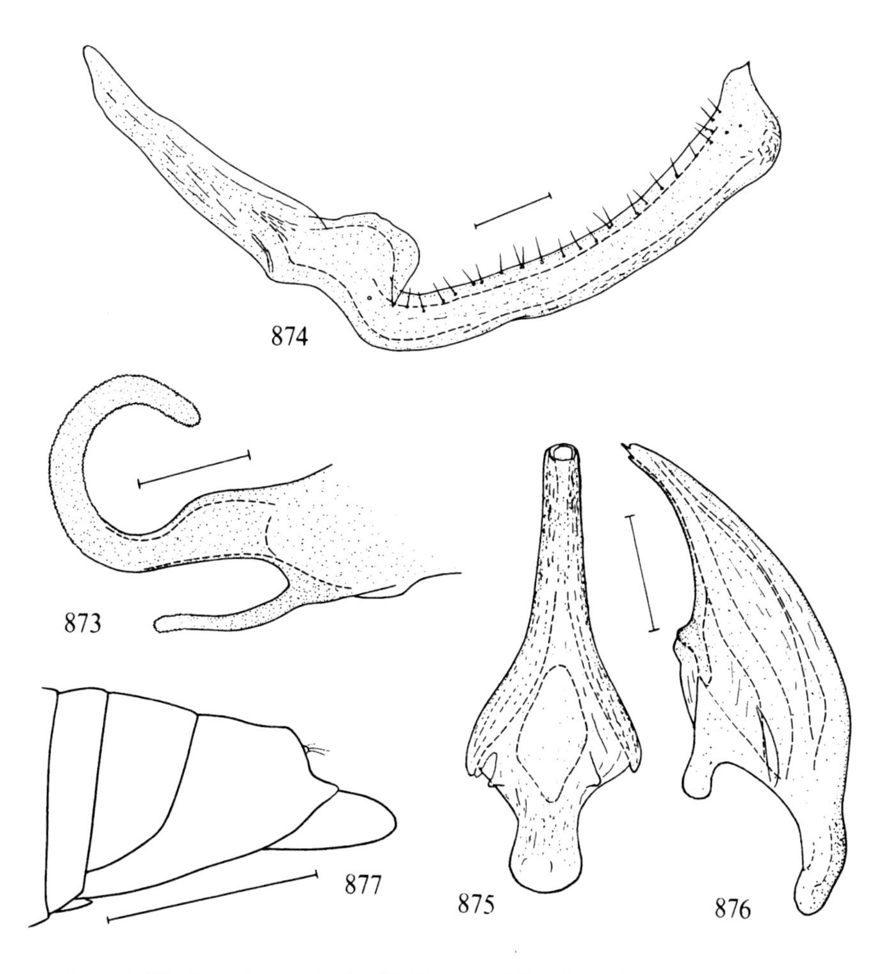

Text-figs. 873–877. *Oncopsis appendiculata* W. Wagner. - 873: right anal collar appendage of male from the right; 874: left genital style from outside; 875: aedeagus in ventral aspect; 876: aedeagus from the left; 877: apex of female abdomen from the left. Scale: 1 mm for 877, 0.1 mm for the rest.

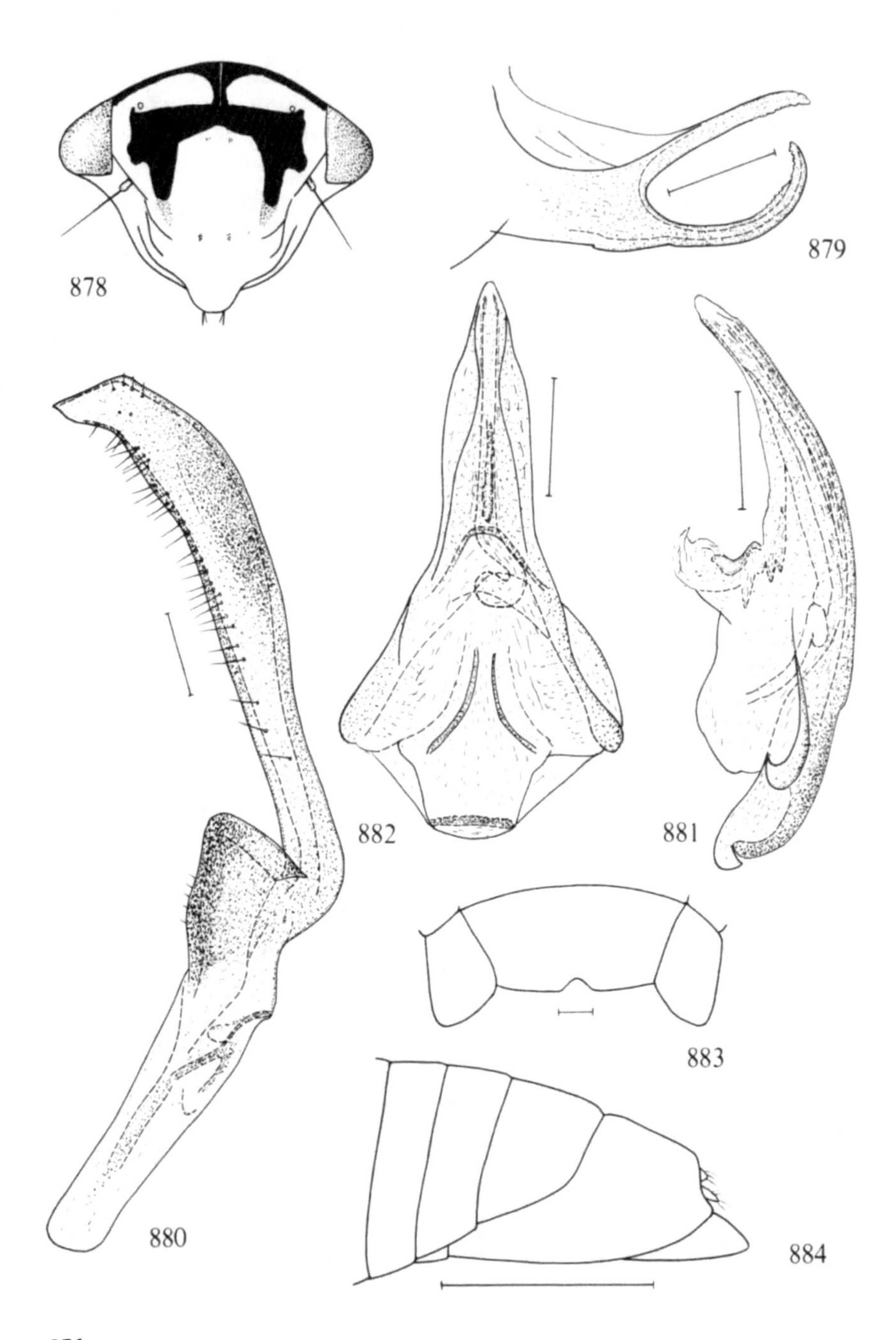
878
879
882
881
880
883
884

111. ***Oncopsis planiscuta*** (Thomson, 1870)
Text-figs. 878–884.

Pediopsis planiscuta Thomson, 1870: 318.

Colour and pigmentation fairly invariable. Venter yellow with black spots, dorsum of fore-body greyish to yellowish brown with black markings. Face (Text-fig. 878) with discoidal spots and band, interocular band, and median band, the latter sometimes indistinct. Pronotum greyish brown (♂) or yellowish brown, anteriorly with yellow spots and brown or black muscle traces. Scutellum yellowish brown with a triangular black spot in each lateral corner, often with an interrupted black longitudinal median streak and an additional pair of small black spots. Fore wings transparent with black (♂) or brown (♀) veins. Often with a marked pattern resembling that of male *flavicollis*. Commissural border whitish except along the section between claval veins and at apex of clavus. Abdomen above black with yellow segmental borders, below entirely yellow or with dark transverse stripes. Legs blackish. Male anal collar appendage as in Text-fig. 879, genital style as in Text-fig. 880, aedeagus as in Text-figs. 881, 882, ventral aspect of 7th abdominal segment in female as in Text-fig. 883, apical part of female abdomen as in Text-fig. 884. Overall length 4–5 mm. - Last instar larva brownish, dorsum and face covered with many short, but conspicuous, depressed or semi-erect, flattened white hairs about uniform in length. Vertex medially little longer than near eyes.

Distribution. Not found in Denmark. - Common, often abundant on its hostplant in the north of Sweden (Dlr., Hls., Hrj., Jmt., Ång., Vb., Nb., Ly.Lpm. - T.Lpm.). - Scarce in Norway, found in Bv, TEi, AAi, Ry, HOi, MRy, Nsy, Nsi, Nnø, TRi. - Rare and sporadic in central and northern East Fennoscandia, found in St, Ta, Sb, ObN, and Ks. - Kurile Is., n. Russia, e. Siberia, Maritime Territory.

Biology. On *Alnus incana*, adults in July–September.

112. ***Oncopsis alni*** (Schrank, 1801)
Text-figs. 885–893.

Cicada fenestrata Schrank, 1776: 74 (nec Fabricius, 1775).
Cicada schrankii Gmelin, 1789: 2109 (n.n.) (nomen oblitum).
Cicada alni Schrank, 1801: 50.

Male resembling the male of *planiscuta* but body larger, dark markings more varying especially on the face where the black spots may be strongly reduced. Female sometimes resembling the male but usually largely reddish to yellowish brown, ventrally yellowish

---

Text-figs. 878–884. *Oncopsis planiscuta* (Thomson). - 878: face of male; 879: left anal collar appendage of male from outside; 880: left genital style from outside; 881: aedeagus from the left; 882: aedeagus in ventral aspect; 883: ventral aspect of 7th abdominal segment of female; 884: apex of female abdomen from the left. Scale: 1 mm for 884, 0.1 mm for the rest.

white, and without or with slight black markings. In a common colour variety the proximal half of clavus is whitish. Commissural border as in *planiscuta*. Anal collar appendage in male as in Text-fig. 888, genital style as in Text-fig. 889, aedeagus as in Text-figs. 890, 891, ventral aspect of 7th abdominal sternum in female as in Text-fig. 892, apical part of female abdomen as in Text-fig. 893. Overall length 5.2–6 mm. – Last instar larva

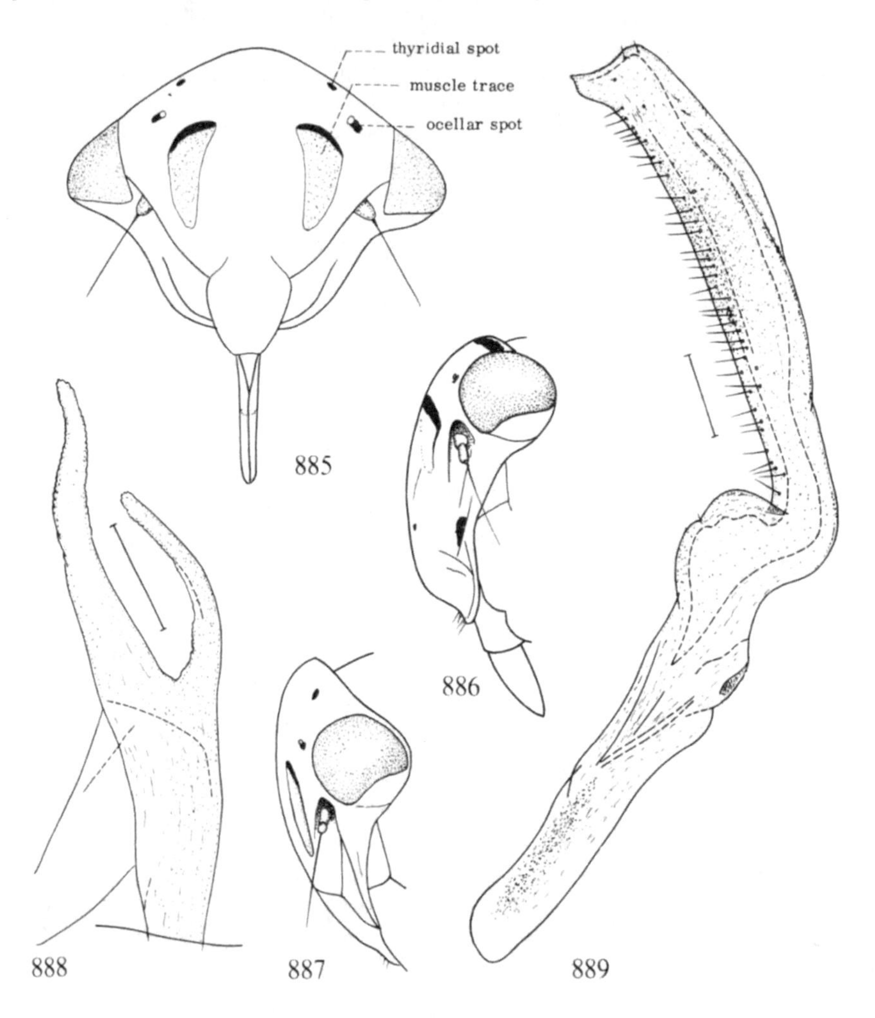

Text-figs. 885–889. *Oncopsis alni* (Schrank). – 885: face of female; 886: head and rostrum of male from the left; 887: head of female from the left; 888: left anal collar appendage of male from outside; 889: left genital style from outside. Scale: 0.1 mm.

brownish, black pigmentation much varying in extension, dorsal hairs depressed, whitish, much varying in length but usually shorter than in *planiscuta*. Shape of head as in *planiscuta*.

Distribution. Fairly common in Denmark. - Common in Sweden up to Hls., also found in Ång. - Common and widespread in Norway (AK-TRi). - Common in southern and central East Fennoscandia (Al, Ab, N, St, Ta, Sa, Oa; Vib.).

Biology. On *Alnus glutinosa* and *incana*. Hibernation in the egg stage (Müller, 1957). "Steigt in den Alpen bis in subalpine Lagen empor" (Wagner & Franz, 1961). "Eggs of *O. alni* are laid in new stem growth, usually in close proximity to overwintering dormant buds. - - - The precise site of oviposition in this species is more variable than in any of the others so far investigeted" (Claridge & Reynolds, 1972).

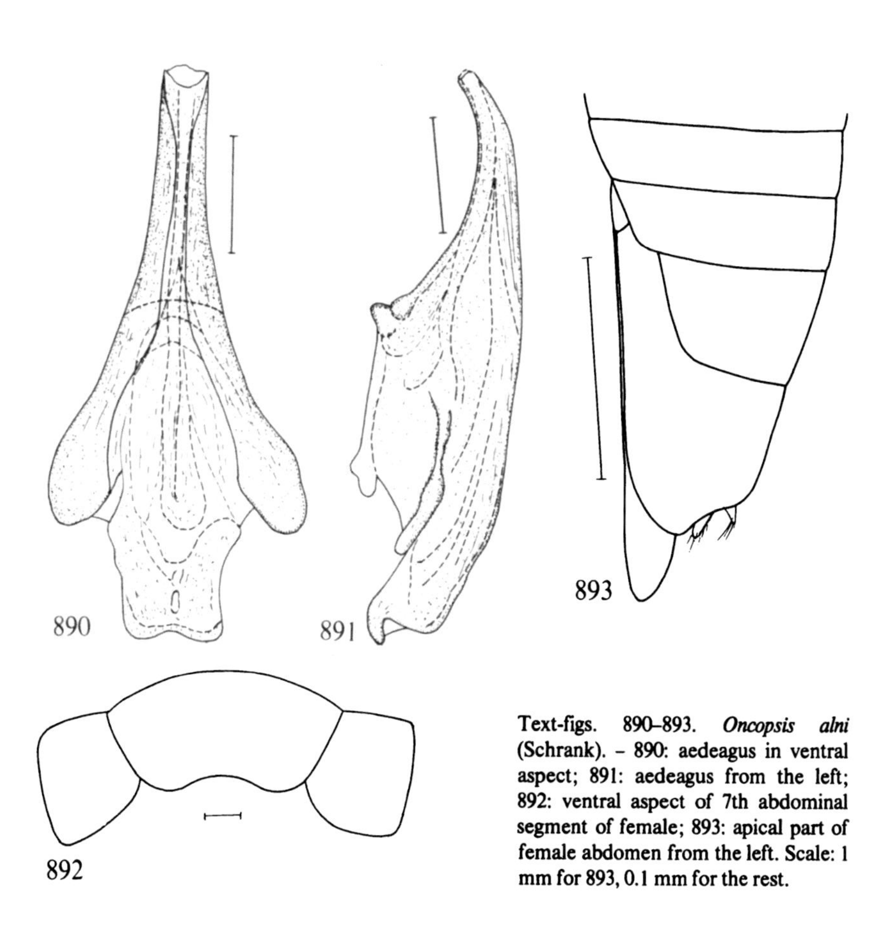

Text-figs. 890-893. *Oncopsis alni* (Schrank). - 890: aedeagus in ventral aspect; 891: aedeagus from the left; 892: ventral aspect of 7th abdominal segment of female; 893: apical part of female abdomen from the left. Scale: 1 mm for 893, 0.1 mm for the rest.

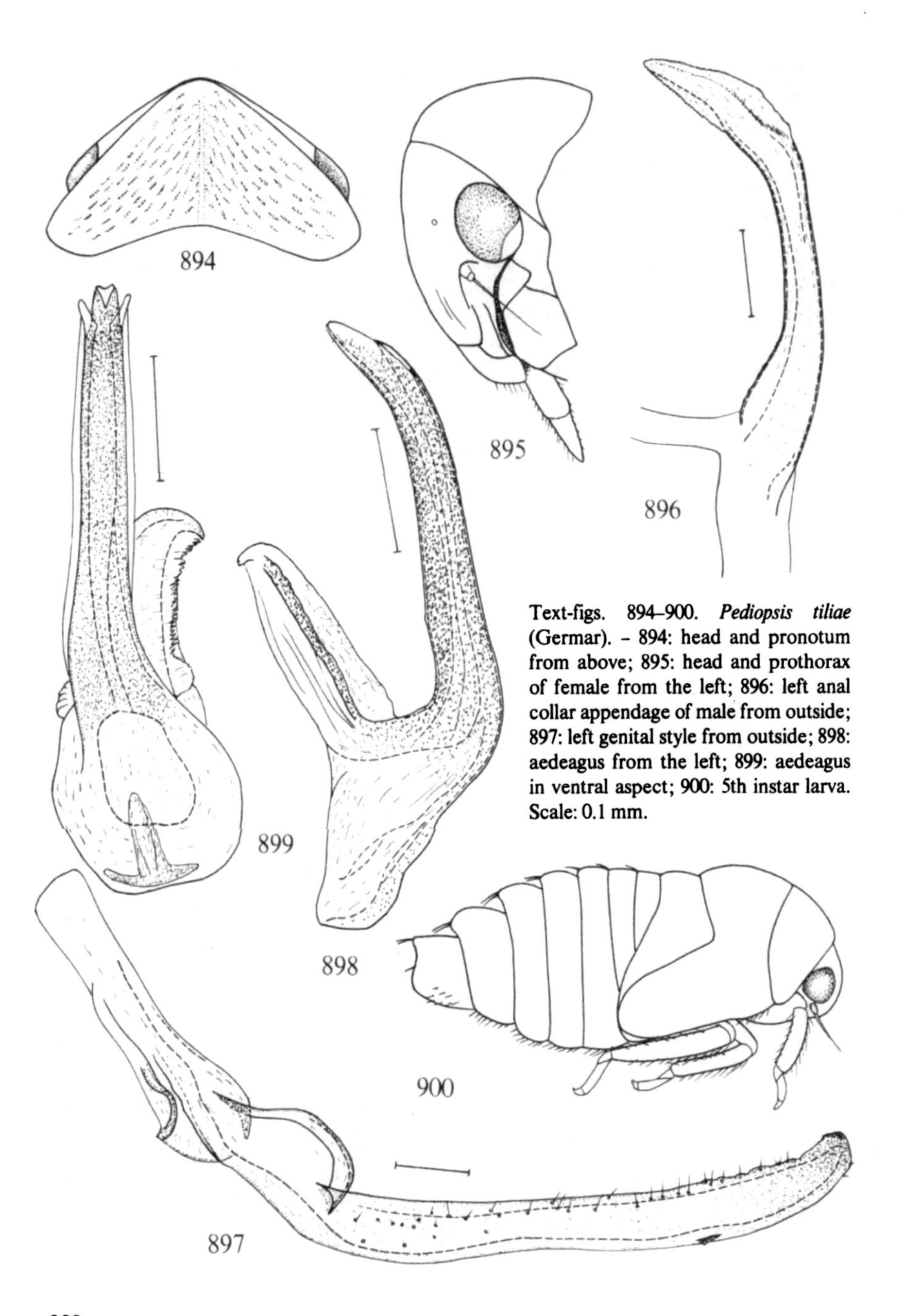

Text-figs. 894–900. *Pediopsis tiliae* (Germar). – 894: head and pronotum from above; 895: head and prothorax of female from the left; 896: left anal collar appendage of male from outside; 897: left genital style from outside; 898: aedeagus from the left; 899: aedeagus in ventral aspect; 900: 5th instar larva. Scale: 0.1 mm.

Genus *Pediopsis* Burmeister, 1838

Pediopsis Burmeister, 1838a: (11).
Type-species: *Jassus tiliae* Germar, 1831, by subsequent designation.

As *Macropsis,* differing by characters given in the key. Pronotum broader than head. Basis of anteclypeus in males dilated. In Europe one species.

113. ***Pediopsis tiliae*** (Germar, 1831)
Plate-fig. 61, text-figs. 894–900.

Jassus tiliae Germar, 1831: pl. 14.

Large and broad. Pronotum covering head like a sort of hood (Text-fig. 895). Head and pronotum yellow or greenish yellow, scutellum and fore wings black-brown (♂) or yellowish brown (♀), the latter often with a pair of indistintly marked darker transverse bands. Venter and legs brownish. Male anal collar appendage as in Text-fig. 896, genital style as in Text-fig. 897, aedeagus as in Text-figs. 898, 899. Overall length of male 5.1–5.5 mm, of female 5.4–5.9 mm, width of thorax about 2 mm. Larva (Text-fig. 900) light brownish or more or less brownish mottled.

Distribution. Scarce in Denmark (SJ, LFM, SZ, NEZ). – Fairly common in Sweden (Sk., Bl., Hall., Sm., Gtl., Ög., Vg., Boh., Sdm., Upl., Vrm.). – Rare in Norway, so far only found in Bø: Tofteholmen 10.VII. 1953 by Holgersen. – Rare also in East Fennoscandia, recorded from Ab: Pargas, Runsala, Ispois, and Lojo. – Austria, Belgium, Bohemia, Moravia, Slovakia, France, German D. R. and F. R., England, Hungary, Italy, Netherlands, Poland, Romania, Switzerland, Latvia, n. and m. Russia, Moldavia, Ukraine, n. Yugoslavia, also in Algeria, Siberia, Manchuria, Korean Peninsula, and Japan.

Biology. On *Tilia cordata.* Adults in June–September.

Genus *Macropsis* Lewis, 1834

Macropsis Lewis, 1834: 49.
Type-species: *Iassus prasinus* Boheman, 1852, by subsequent designation under the Plenary Powers of the International Commission on Zoological Nomenclature.

Body structure robust. Head as seen from above short, obtuse-angled, longer near eyes than medially. Pronotum at least twice as broad as long, surface with a dense sculpture consisting of fine wrinkles directed obliquely forwards and towards the median line. Scutellum large. Fore wings in repose held roof-like over abdomen. Hind tibiae with several rows of strong spines. In Denmark and Fennoscandia 9 species.

Key to species of *Macropsis*

1 Face in lateral aspect moderately convex (Text-fig. 909). Body narrow, fairly parallel-sided. Aedeagus in lateral aspect broader (Text-figs. 904, 913, 920, 930, 935) 2

– Face in lateral aspect more strongly convex. Body comparatively broad and gibbous. Aedeagus in lateral aspect narrow (Text-figs. 940, 945, 949) 13

2 (1) Small species, 3.3–4.5 mm. Hind tibiae in the exterior (strongest) row with 8 (rarely 7 or 9) spines. On *Salix repens* .... 119. *impura* (Boheman)

– Larger species, 4–6 mm. Hind tibia in the exterior row rarely with less than 9 spines 3

3 (2) Corium with a fuscous transverse band including the transverse veins proximally delimiting subapical cells (Plate-fig. 51) 4

– Corium with or without one or two transverse bands; if present, they do not include the transverse veins basally delimiting subapical cells (Plate-figs. 50, 52) 6

4 (3) Veins of corium not or indistinctly darker than cell surface
118. *graminea* (Fabricius), f. *populi* Edwards

– Veins of corium much darker than cell surface 5

5 (4) Veins of corium only narrowly bordered with fuscous (Plate-fig. 51). Radial and median cells of corium almost entirely hyaline. On *Populus tremula* .... 117. *fuscinervis* (Boheman)

– Veins of corium broadly bordered with fuscous. Radial and median cells largely brown. On *Populus nigra*
118. *graminea* (Fabricius), f. *populi* Edwards

6 (3) Upper side greenish 7

– Upper side not green 10

7 (6) Colour grass green or light green. Veins of fore wings green. Head usually unspotted, very rarely with a black spot on apex. Small species, 4–5 mm .... 114. *prasina* (Boheman)

– Colour yellowish green or pale green, sometimes with a reddish tinge. Veins of fore wings not green or only faintly greenish. Head often with a black spot on apex 8

8 (7) Males .... 118. *graminea* (Fabricius)

– Females 9

9 (8) Part of ovipositor projecting beyond apex of pygofer (in lateral aspect) longer than broad (Text-fig. 915). On *Salix*
115. *infuscata* (J. Sahlberg), var.

– Part of ovipositor projecting beyond apex of pygofer broader than long (Text-fig. 932). On *Populus nigra*
118. graminea (Fabricius)

10 (6) Males 11
– Females 12
11 (10) Greyish yellow. Face usually strongly black spotted, spots well delimited. Aedeagus basally not or only slightly narrower than at middle (Text-fig. 913) .... 115. *infuscata* (J. Sahlberg)
– Brownish. Face unspotted or with usually diffusely delimited dark spots. Aedeagus at middle distinctly broader than basally (Text-fig. 920) .... 116. *cerea* (Germar)
12 (10) Part of ovipositor projecting beyond apex of pygofer longer than broad (Text-fig. 915) .... 115. *infuscata* (J. Sahlberg)
– Part of ovipositor projecting beyond apex of pygofer about as long as broad (Text-fig. 921) .... 116. *cerea* (Germar)
13 (1) Scutellum on each side with a triangular black spot 14
– Scutellum without triangular black spots, at most with indistinct fuscous markings .... 122. *megerlei* (Fieber)
14 (13) Commissural border of fore wing dark .... 121. *scutellata* (Boheman)
– Commissural border of fore wing light .... 120. *fuscula* (Zetterstedt)

114. ***Macropsis prasina*** (Boheman, 1852)
Plate-fig. 80, text-figs. 901–907.

Iassus prasinus Boheman, 1852b: 123.
Pediopsis virescens J. Sahlberg, 1871: 127 (nec Fabricius, 1794).

A small, comparatively finely shaped species with narrow fore wings. Cells of fore wings narrow, nearly parallel-sided. Male vividly grass green, female light green, fore body in both sexes normally without black markings, scutellum in male sometimes with darker basal triangles. Abdominal tergum in male transversely black-striped. Hind tibiae on outside usually with a black spot. Apical part of fore wing somewhat smoky, especially in males. Appendage of male pygofer as in Text-figs. 901, 902, genital style as in Text-fig. 903, aedeagus as in Text-figs. 904, 905, apical part of female abdomen as in Text-fig. 906. Overall length of males 4–4.7 mm, of females 4.5–5.1 mm. – Last instar larva green, in lateral aspect as in Text-fig. 907.

Distribution. Somewhat local but widespread in Denmark. – Widespread and locally common in Sweden (Sk.–Öl., Ög., Boh., Nrk., Upl., Vstm., Vrm., Gstr., Jmt.). – Norway: so far only found in AK: Oslo, "Tveten" (Siebke). – East Fennoscandia: only found in Vib: Muolaa, and in Kr: Räisälä and Käkisalmi. – Owing to identification difficulties the synonymy of *M. prasina* is confused and many records possibly referring to this species must be disregarded as dubious. The following may be regarded as reliable: England, Netherlands, German D.R. and F.R., Estonia, Lithuania, Latvia, Poland, Bohemia, Moravia, Slovakia, Romania.

Biology. On *Salix caprea, cinerea, aurita,* in one case also *viminalis* (Wagner, 1950). Adults July–September.

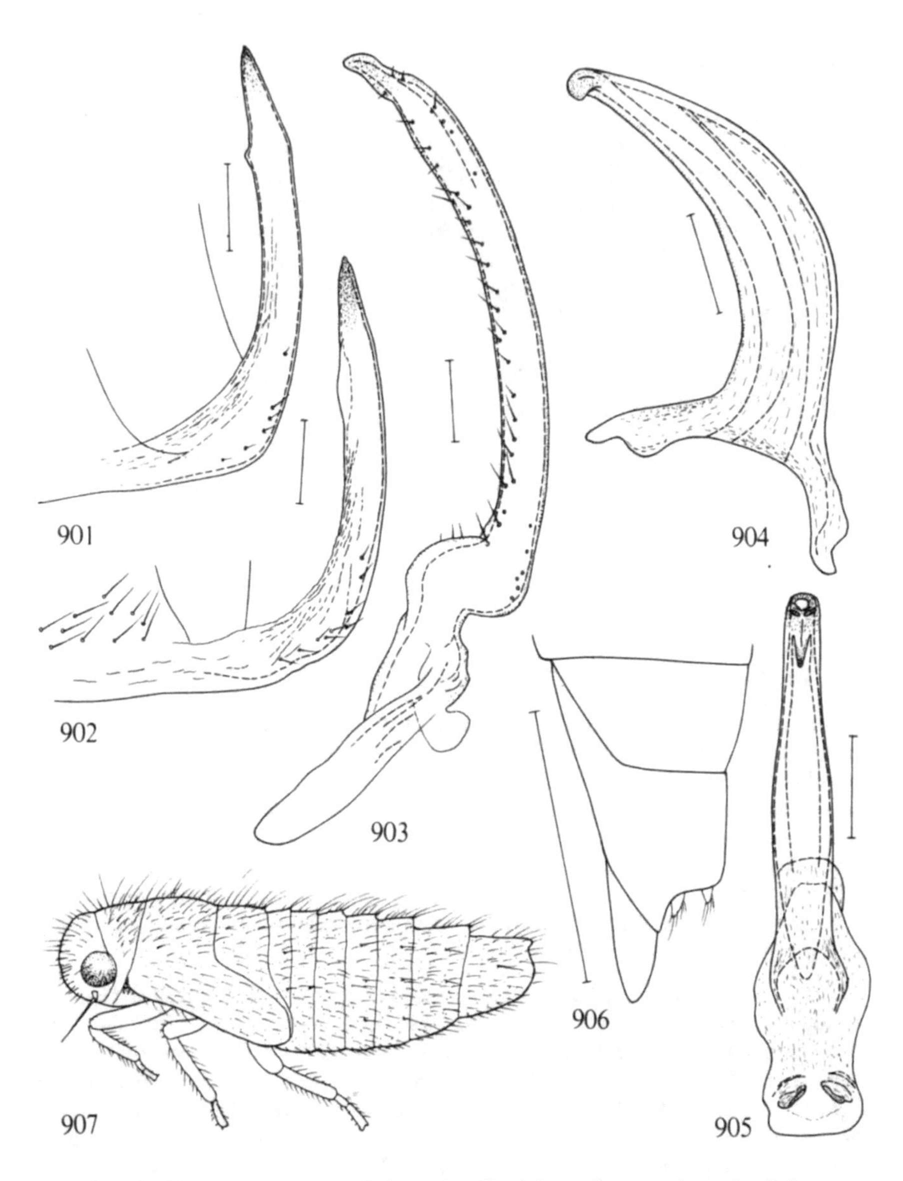

Text-figs. 901–907. *Macropsis prasina* (Boheman). – 901: left pygofer appendage of male from outside; 902: same of another specimen; 903: left genital style from outside; 904: aedeagus from the left; 905: aedeagus in ventral aspect; 906: apical part of female abdomen from the left; 907: last instar larva. Scale: 0.1 mm for 901–905, 0.5 mm for 906 and 907.

115. ***Macropsis infuscata*** (J. Sahlberg, 1871)
Plate-fig. 50, text-figs. 908–916.

Pediopsis infuscatus J. Sahlberg, 1871: 118.
Pediopsis nassatus, var. a, b, c, d, e J. Sahlberg, 1871: 132 (nec Germar, 1837).
Pediopsis distinctus Scott, 1874a: 191.
Macropsis cerea, var. *insolitus* W. Wagner, 1941: 113.
Macropsis cerea, var. *Kästneri* W. Wagner, 1941: 113.
Macropsis cerea Ossiannilsson, 1946c: 88 (nec Germar, 1834).

Body elongate, strongly shaped, usually greyish yellow, venter largely black. Face more or less strongly pigmented, often but not always with discoidal, ocellar, and thyridial spots (cf. Text-fig. 942). Central spot always absent but usually with an apical spot (Text-figs. 908-910). Pronotum with a black spot behind each eye. Scutellum with a pair of lateral spots and often a black longitudinal stripe between these. Fore wings unicolorous or with two indistinctly marked dark transverse bands not placed on the transverse veins (Plate-fig. 50). The fore wings and their cells are broader than in *prasina,* the cells less often parallel-sided. Legs, especially hind tibiae, dark-striped, the latter with the usual black spot on outside near basis. In f. *kaestneri* W. Wagner (♀) the body and fore wings are pale green or yellowish green. In f. *insolita* W. Wagner (♀) the fore wings are pale yellow with brownish spots. Appendage of male pygofer as in Text-fig. 911, genital style as in Text-fig. 912, aedeagus as in Text-figs. 913, 914. Apical part of female abdomen as in Text-fig. 915. Length of male 4.1–5.1 mm, of female 5.1–5.8 mm. - Lateral aspect of last instar larva as in Text-fig. 916, colour of larva green or brownish, the green form sometimes with a pair of dark lateral bands on fore part of body.

Distribution. Rare in Denmark, only found in NEJ: Agersted 7.VIII.1965 (Trolle) and LFM: Vindeholme Lindeskov 10.VII.1966 (N. Møller Andersen). - Locally common in Sweden, Sk. - Med. - Apparently rare in Norway, found in AK: Hovin 18.VIII.1851 by Siebke, and in Bø: Ringerike 28.VII.1906 by Warloe. - Common in East Fennoscandia up to LkW. - Austria, Bulgaria, Bohemia, Moravia, Slovakia, France, England, Netherlands, German D.R. and F.R., Poland, Estonia, Latvia, Lithuania, Switzerland, Italy, n. Russia, Ukraine, Kazakhstan.

Biology. Monophagous on *Salix caprea* (Wagner, 1950). Adults in June–September.

116. ***Macropsis cerea*** (Germar, 1837)
Plate-fig. 52, text-figs. 917–921.

Jassus cereus Germar, 1837: pl.14.
Pediopsis planicollis Thomson, 1870: 320.
Macropsis haupti Ossiannilsson, 1946c: 90 (nec W. Wagner, 1941).

Resembling *infuscata* but somewhat less elongate, predominating colour more brownish. Face usually without or with indistinct dark markings, rarely with an apical

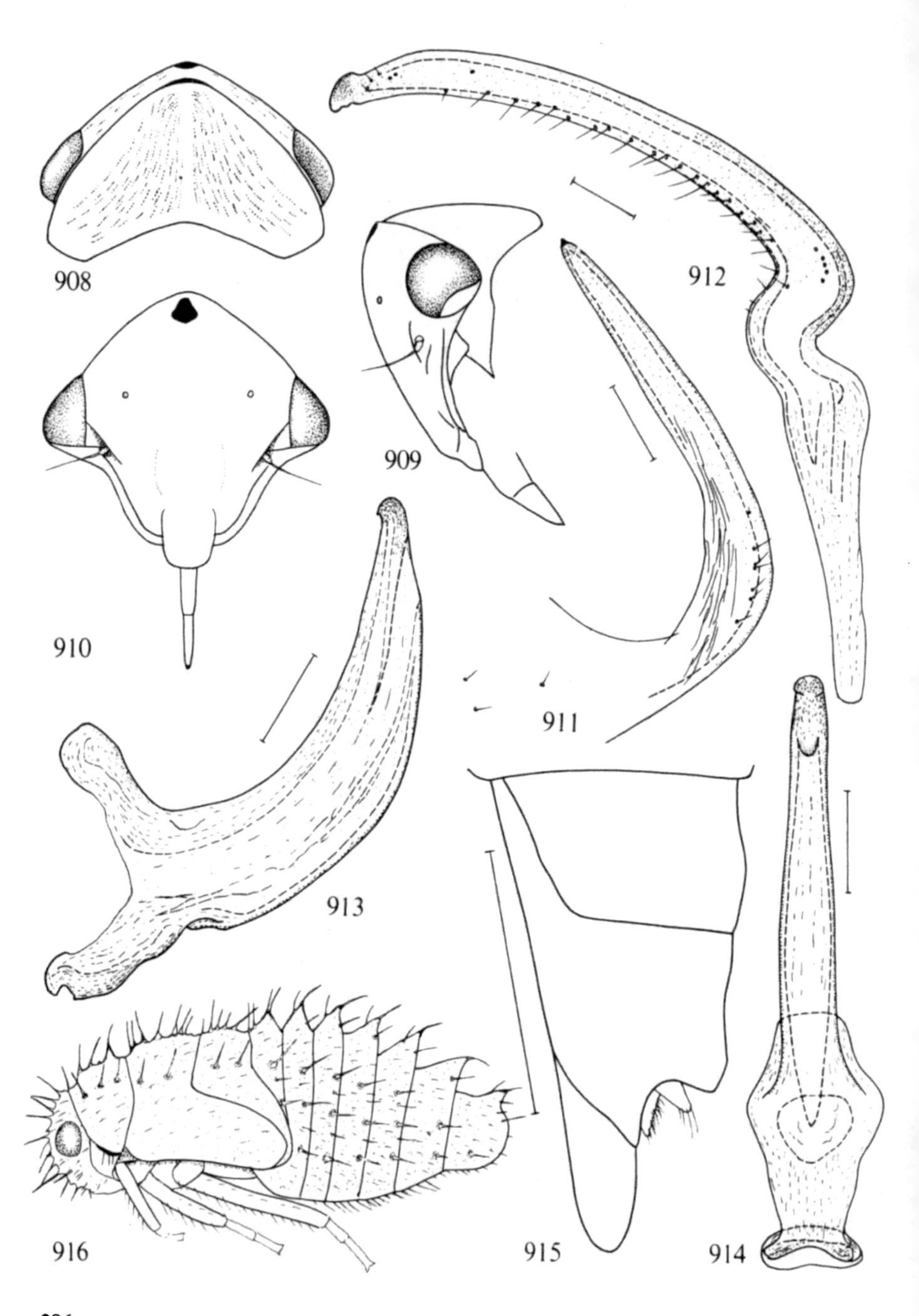

908
909
910
911
912
913
914
915
916

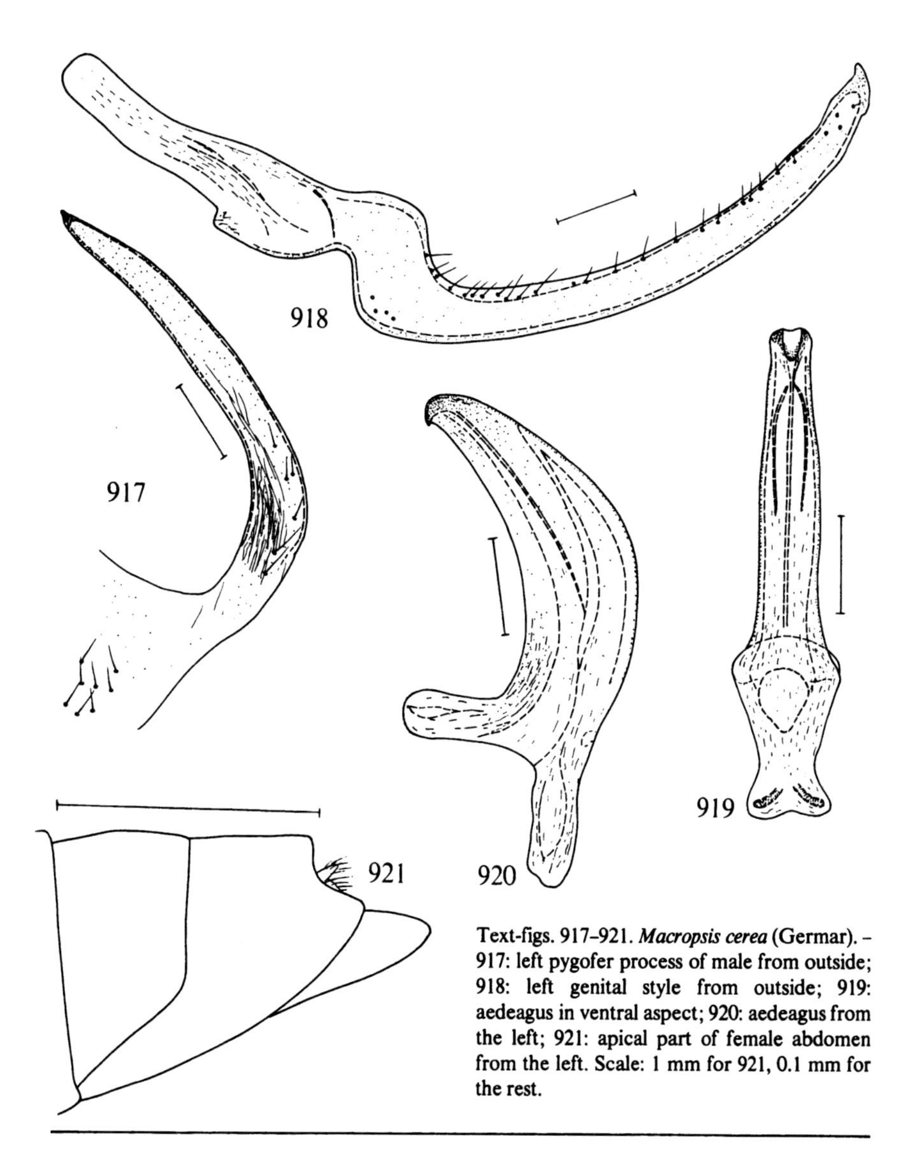

Text-figs. 917–921. *Macropsis cerea* (Germar). – 917: left pygofer process of male from outside; 918: left genital style from outside; 919: aedeagus in ventral aspect; 920: aedeagus from the left; 921: apical part of female abdomen from the left. Scale: 1 mm for 921, 0.1 mm for the rest.

Text-figs. 908–916. *Macropsis infuscata* (J. Sahlberg). – 908: head and pronotum from above; 909: head and prothorax from the left; 910: face; 911: left pygofer process in male from outside; 912: left genital style from outside; 913: aedeagus from the left; 914: aedeagus in ventral aspect; 915: apical part of female abdomen from the left; 916: last instar larva. Scale: 0.1 mm for 908–914, 1 mm for 915 and 916.

black spot. Also pronotum and scutellum often devoid of black markings. Fore wings in males usually without, in females usually with an irregular dark transverse band across middle (Plate-fig.52). Within this band the veins are often conspicuously darker than the adjacent wing surface. In females there is often an additional dark transverse band across the subapical cells. Transverse veins light. Venter concolorous, with or without black spots. Basal spot of hind tibia often indistinct. There is no green variety of this species. Appendage of male pygofer as in Text-fig. 917, genital style as in Text-fig. 918, aedeagus as in Text-figs. 919, 920, apical part of female abdomen as in Text-fig. 921. Length of male 4–4.6 mm, of female 4.6–5.5 mm. – Larva greyish to black, tergum unarmed, in lateral aspect smooth.

Distribution. Fairly common in Denmark. – Common in Sweden, Sk. – Hls., Vb. – Probably common also in Norway, at least in the south. I have seen specimens from AK, HEs, Os, Bø, Bv, VE, AAy. – Common in southern and central East Fennoscandia, found up to ObN, also in Vib and Kr. – Austria, Belgium, Bohemia, Moravia, Slovakia, France, German D.R. and F.R., England, Scotland, Wales, Ireland, Netherlands, Poland, Romania, Estonia, Latvia, Lithuania, Italy, n. and m. Russia, Ukraine, m. Siberia, Mongolia, Altai Mts.

Biology. On *Salix caprea, cinerea, aurita, triandra, purpurea* (Wagner, 1950). I found *M. cerea* also on *Salix pentandra.* Adults in June–August.

117. ***Macropsis fuscinervis*** (Boheman, 1845)
Plate-fig. 51, text-figs. 922–927.

Jassus fuscinervis Boheman, 1845b: 163.
Pediopsis nassatus Var. f. J. Sahlberg, 1871: 132 (nec Germar, 1834).

Yellowish to brownish yellow, face spotted as in *infuscata.* Pronotum brownish grey or brownish yellow, at fore border with a pair of black spots behind the eyes. Scutellum with the same black markings as in *infuscata,* these often confluent. Males with fore wing veins bordered with black or fuscous, and with a fuscous transverse band over the subapical transverse veins. Often the entire subcostal cell is fuscous (Plate-fig. 51). In females the veins are less distinctly darker than the brownish yellow fore wing membrane and the dark transverse band is also rather pale but usually discernible. Venter of males usually strongly dark-spotted, in females largely light. Legs as in *infuscata.* Appendage of male pygofer as in Text-fig. 922, genital style as in Text-fig. 923, aedeagus as in Text-figs. 924, 925, apical part of female abdomen as in Text-fig. 926. Length of males 4–4.8 mm, of females 4.8–5.8 mm. Larva (Text-fig. 927) light green, brown or mottled brownish, dorsum and tibiae oversprinkled with minute black dots.

Distribution. Fairly common in Denmark (SJ, EJ, F, NEZ, B). – Common in southern and central Sweden, found in most provinces up to Ång. – Norway: found in AK: Dal (Holgersen); Bø: Drammen (Warloe), Bingen, Modum (Holgersen); VAy; Kristiansand (Warloe). – Scarce in East Fennoscandia, found in Al, Ab, N, Ta, Sa; Kr. – Austria, Belgium, Bulgaria, Bohemia, France, England, German D.R. and F.R.,

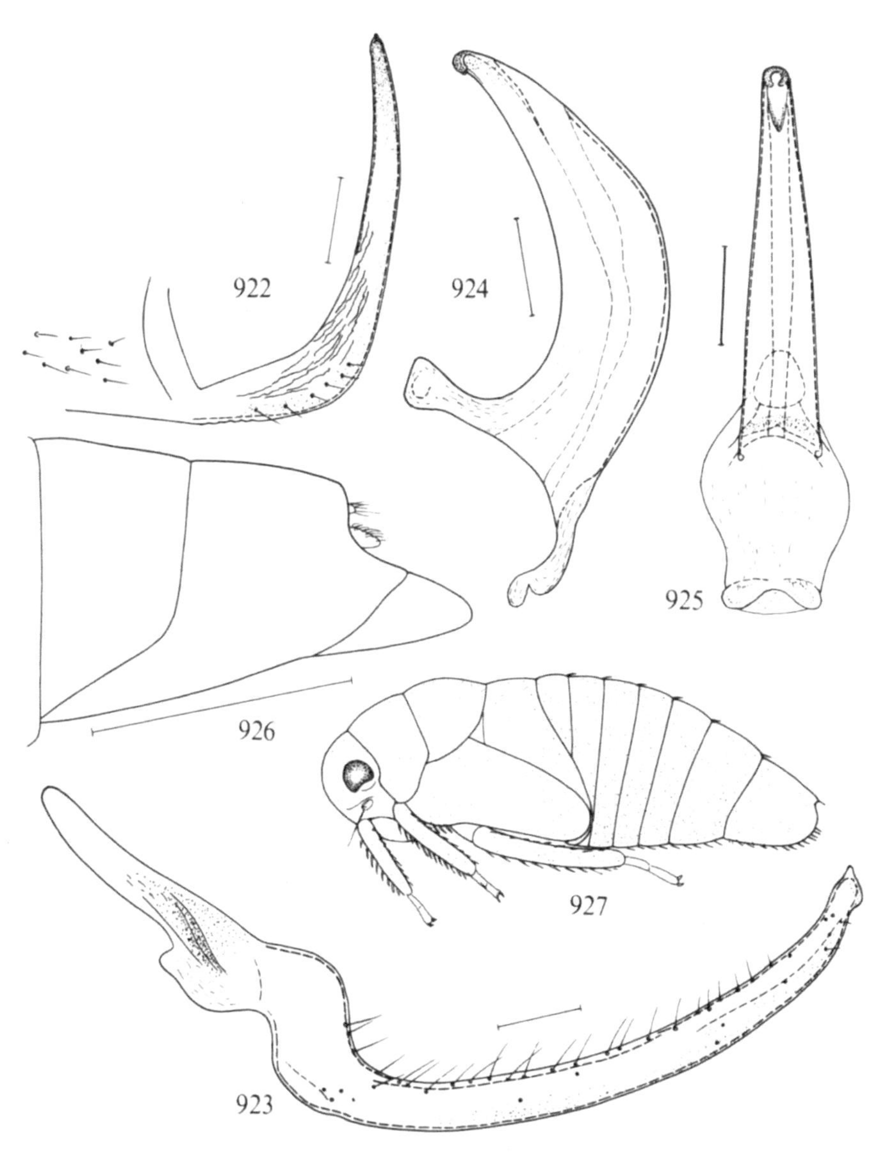

Text-figs 922–927. *Macropsis fuscinervis* (Boheman). – 922: left pygofer appendage of male from outside; 923: left genital style from outside; 924: aedeagus from the left; 925: aedeagus in ventral aspect; 926: apical part of female abdomen from the left; 927: last instar larva. Scale: 1 mm for 926, 0.1 mm for 922–925.

Hungary, Italy, Netherlands, Poland, Estonia, Latvia, Lithuania, Romania, Yugoslavia, n. and m. Russia, Ukraine, Moldavia, Kazakhstan, w. Siberia, China.

Biology. On *Populus tremula,* adults in June–August.

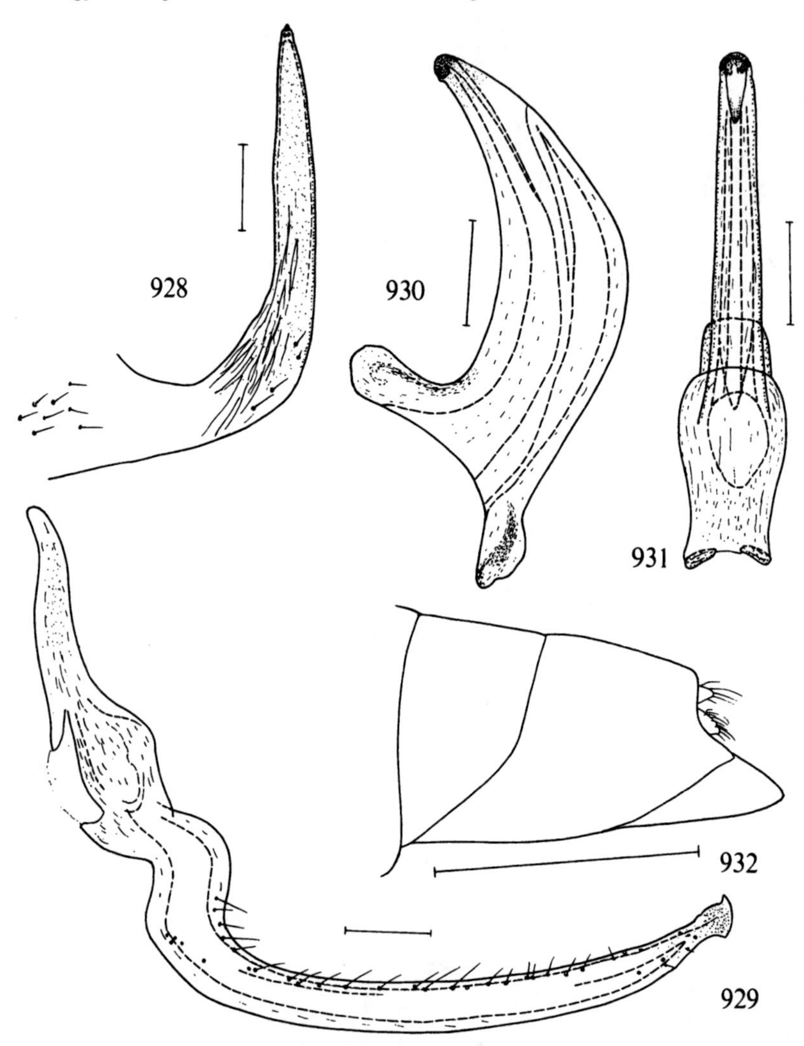

Text-figs. 928–932. *Macropsis graminea* (Fabricus). – 928: left pygofer process of male from outside; 929: left genital style from outside; 930: aedeagus from the left; 931: aedeagus in ventral aspect; 932: apical part of female abdomen from the left. Scale: 1 mm for 932, 0,1 mm for the rest.

118. ***Macropsis graminea*** (Fabricius, 1798)
Text-figs. 928–932.

Cicada graminea Fabricius, 1798: 521.
Macropsis populi Edwards, 1919: 56.

In the typical form the fore body is greenish yellow, scutellum often conspicuously lighter, sulphur yellow. Head with or without a black apical spot much varying in size. Fore wings pale yellowish or pale greenish with a reddish tinge, veins concolorous or darker, brownish or partly indistinctly greenish, costal margin largely green. In f. *populi* Edwards the pronotum is brownish in varying gradation, sometimes with a pair of dark spots along fore margin. Scutellum with two brown or black basal triangles. Fore wing veins dark, cells partly brownish, and there is a dark transverse band including subapical transverse veins as in *fuscinervis*. Process of male pygofer as in Text-fig. 928, genital style as in Text-fig. 929, aedeagus as in Text-figs. 930, 931, apical part of female abdomen as in Text-fig. 932. Length (♂♀) 4.2–4.9 mm.

Distribution. Not found in Denmark and East Fennoscandia. I have seen a few old specimens from the vicinity of Stockholm (in coll. Swedish State Museum of Natural History), and 19 specimens collected in Norway, AK: Oslo, Tøyen, by Holgersen. – Austria, Bulgaria, Bohemia, Moravia, Slovakia, France, German, D.R. and F.R., England, Hungary, Italy, Netherlands, Poland, Anatolia, Altai Mts., Armenia, Moldavia, Ukraine, Yugoslavia, Mongolia.

Biology. On *Populus nigra* and *P. nigra italica.* Adults in July–September (Le Quesne, 1965).

119. ***Macropsis impura*** (Boheman, 1847)
Text-figs. 933–937.

Jassus impurus Boheman, 1847b: 265.

Resembling the greyish yellow form of *infuscata* but much smaller. Fore wings without transverse bands. Female genital segment comparatively shorter than in *infuscata.* Process of male pygofer as in Text-fig. 933, genital style as in Text-fig. 934, aedeagus as in Text-figs. 935, 936, apical part of female abdomen as in Text-fig. 937. Length of males 3.3–4.2 mm, of females 4.2–4.5 mm. – Abdominal sterna of last instar larvae whitish, conspicuously lighter than dorsum.

Distribution. Common and widespread in Denmark and Sweden (Sk.–Lu. Lpm.). – Norway: found in AAy, VAy, and Ry. – Scarce and sporadic in southern and central East Fennoscandia (Al, Ab, N, St, Ta, Sa, Sb, Kb). – Austria, Belgium, Slovakia, France, England, Scotland, Wales, Ireland, Netherlands, German D.R. and F.R., Hungary, Yugoslavia, Poland, Estonia, Latvia, Lithuania, m. Russia, Ukraine, Altai Mts., Kazakhstan, Mongolia.

Biology. On *Salix repens,* adults in June–September.

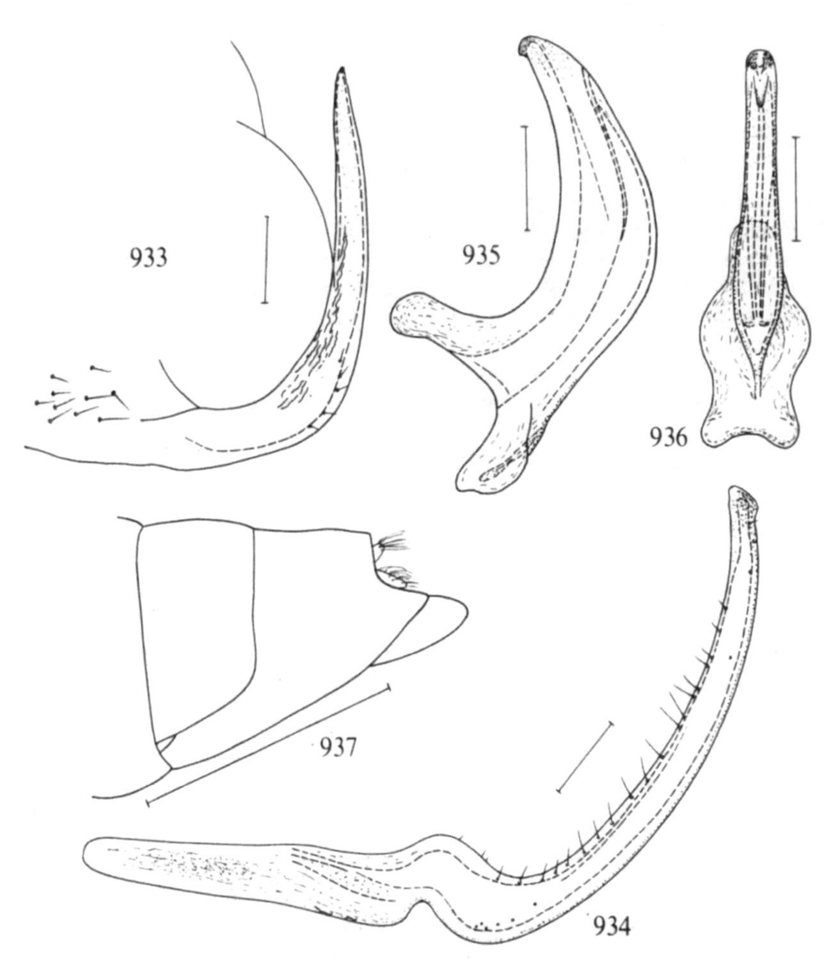

Text-figs. 933–937. *Macropsis impura* (Boheman). – 933: left pygofer process of male from outside; 934: left genital style from outside; 935: aedeagus from the left; 936: aedeagus in ventral aspect; 937: apical part of female abdomen from the left. Scale: 1 mm for 937, 0.1 mm for the rest.

120. ***Macropsis fuscula*** (Zetterstedt, 1828)
Text-figs. 938–943.

Jassus fruticola var. *fusculus* Zetterstedt, 1828: 544.
Jassus rubi Boheman, 1845b: 162,

Roundish, coarsely shaped. Greyish yellow, often with a faintly greenish tinge. Face

(Text-fig. 942) usually with well developed black markings: discoidal, ocellar, thyridial, and central spots, the last-mentioned spot often missing or indistinct. Pronotum along fore border with some black spots varying in shape and size. Scutellum with a pair of triangular black spots at its lateral corners, one unpaired black spot between these, and

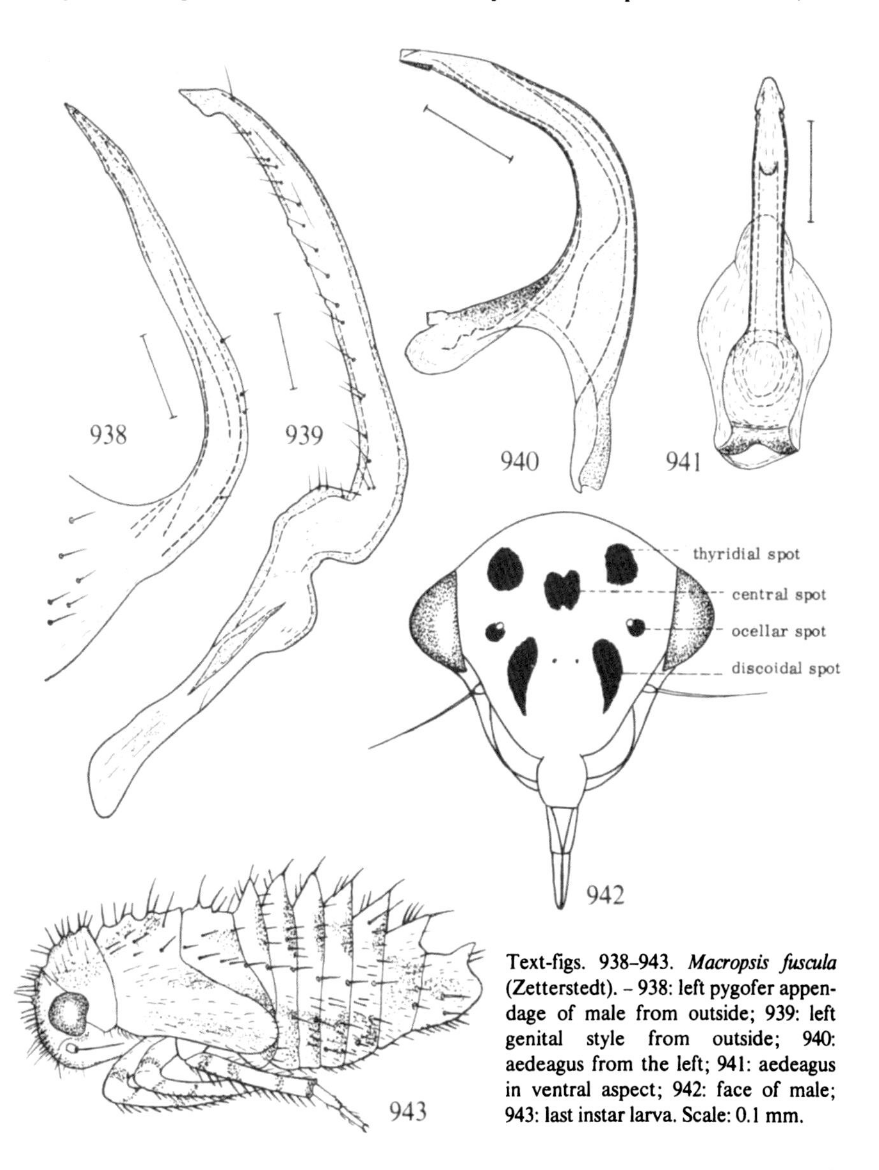

Text-figs. 938–943. *Macropsis fuscula* (Zetterstedt). – 938: left pygofer appendage of male from outside; 939: left genital style from outside; 940: aedeagus from the left; 941: aedeagus in ventral aspect; 942: face of male; 943: last instar larva. Scale: 0.1 mm.

a pair of transversely oval black spots medially and somewhat distally of the caudal apices of the lateral spots. These spots may be more or less confluent or partly reduced, the lateral spots being more constantly present than the others. Fore wings semihyaline or more or less smoke-coloured, their apical cells broad, partly polygonal. Veins concolorous or fuscous, commissural border always light. Venter light, more or less dark-spotted; propleura each with a black spot behind compound eye, abdomen of male largely dark. Legs largely light, hind tibia on outside near basis with a roundish black spot. Process of male pygofer as in Text-fig. 938, genital style as in Text-fig. 939, aedeagus as in Text-figs. 940, 941. Length of male 4–4.7 mm, of female 4.7–5.5 mm. – Larva in lateral aspect as in Text-fig. 943, greyish yellow or greenish.

Distribution. Common in Denmark (SJ, EJ, F, LFM, B). – Common and widespread in Sweden (Sk.–Med., Ång., T.Lpm.). – Norway: found in Ø, AK, HEs, Bø, VAy, HOi, and SFi. – Scarce in East Fennoscandia, found in Ab, N, St, Ta, Sa, Oa, Tb, Sb, ObN; Vib, Kr. – Widespread in Europe, also found in Armenia, Iraq, Kazakhstan, m. Siberia, Tadzhikistan, Japan, and Nearctic region.

Biology. On *Rubus idaeus* and several other *Rubus* spp. Univoltine, hibernates in the egg stage (Müller, 1956). Adults in July–September.

Economic importance. Vector of *Rubus* stunt disease virus (in laboratory tests).

121. ***Macropsis scutellata*** (Boheman, 1845)
Text-figs. 944–947.

Jassus scutellatus Boheman, 1845b: 162.
Pediopsis tibialis Scott 1874a: 195.

Resembling *fuscula* but a little larger. Lateral margins of lora divergent, directed towards antennal bases. Central spot of face, if present, usually consisting of a pair of indistinct confluent longitudinal streaks. Commissural border of fore wing dark. Process of male pygofer as in Text-fig. 944, genital style as in Text-fig. 945, aedeagus as in Text-figs. 946, 947. Length of male 4.4–4.6 mm, of female 5.2–5.5 mm.

Distribution. Rare in Denmark, so far found in LFM: Lysemose, Maribo 11.VIII. 1965 (N. Møller Andersen) and NWZ: Jyderup 17.VII.1917 (C. C. R. Larsen). – Rare in Sweden, first found in Scania by Zetterstedt, later in Lund by Thomson. I captured one specimen in Sk., Svalöf 13.VII.1936, and 21 specimens in Öl., Mörbylånga, Beteby 22.VII.1950. H. Andersson and R. Danielsson found one female in Öl., Högsrum, "Halltorps hage" 3.–6.VIII.1976. – Not found in Norway, nor in East Fennoscandia. – Widespread in Europe, also recorded from Morocco, Anatolia, Palestine, Syria, Iran, Afghanistan, Azerbaijan, Georgia, Uzbekistan, w. and m. Siberia, China, Japan, Korean Peninsula, and Oriental region.

Biology. On *Urtica dioica*. Adults in July–October (Le Quesne, 1965).

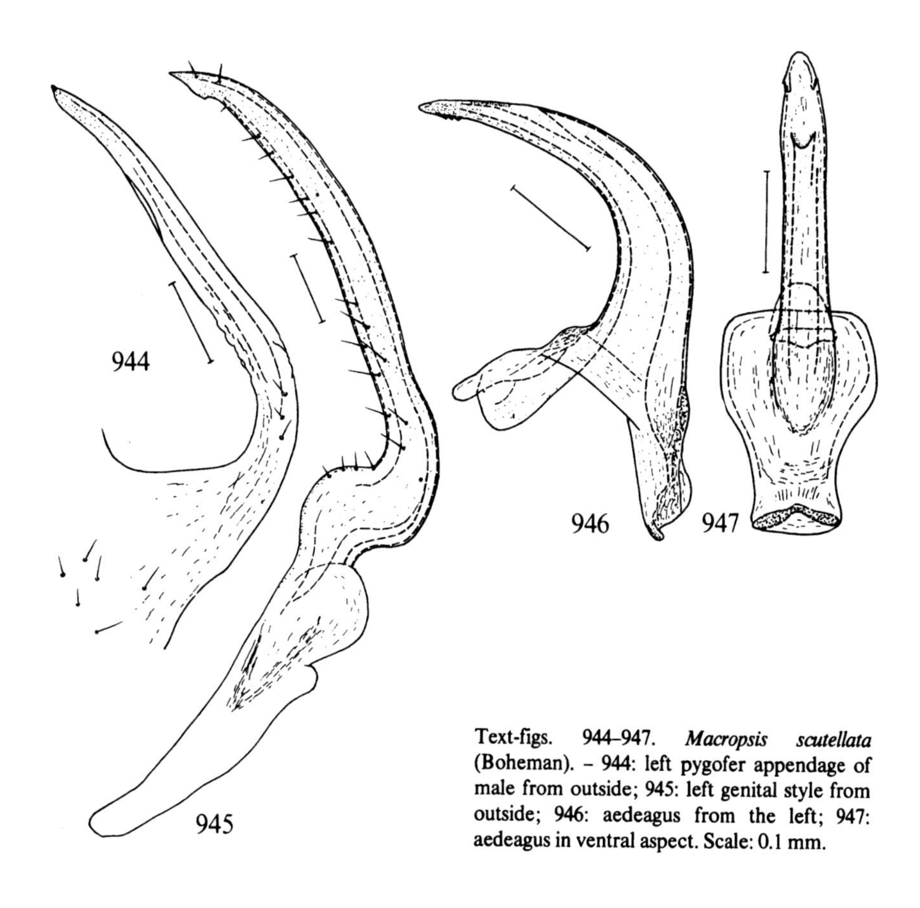

Text-figs. 944–947. *Macropsis scutellata* (Boheman). – 944: left pygofer appendage of male from outside; 945: left genital style from outside; 946: aedeagus from the left; 947: aedeagus in ventral aspect. Scale: 0.1 mm.

122. ***Macropsis megerlei*** (Fieber, 1868)
Text-figs. 948–951.

Pediopsis megerlei Fieber, 1868: 460.

Face yellowish, sparsely spotted, in male at most with thyridial, ocellar, and discoidal spots, central spot always absent, in female unspotted or with small thyridial spots, sometimes also indistinct discoidal spots. Pronotum yellowish, hind border more or less broadly brownish, the brown colour gradually lighter towards the front. Scutellum yellow, unspotted or with diffuse brownish markings. Fore wings in male semi-hyaline, greyish, apically more or less fuscous, with fuscous veins, in female brownish yellow with concolorous veins. Venter black spotted in males, unspotted or with one black spot on each propleurum in females. Legs entirely yellow (♀) or black spotted (♂), fore and middle tibiae in males each with two more or less well developed ring-shaped transverse bands. Process of male pygofer as in Text-fig. 948, genital style as in Text-fig. 949, aedeagus as in Text-figs. 950 and 951. Length of male 4–4.3 mm, of female 4.5–5 mm.

Distribution. Not in Denmark, Sweden and East Fennoscandia. - Norway: AK: V. Aker, 1 ♂ (Esmark); Bø: Ringerike 1.VIII.1899, 1 ♀ (Warloe). - Austria, Bulgaria, Bohemia, Slovakia, France, German D.R. and F.R., Greece, Hungary, Netherlands, Poland, Romania, Spain, Yugoslavia; Algeria, Tunisia, Anatolia, Georgia, Kazakhstan, w. Siberia.

Biology. On wild roses, *Rosa pimpinellifolia* (Wagner, 1964). On *Acer* and *Quercus* (Jankovic, 1966).

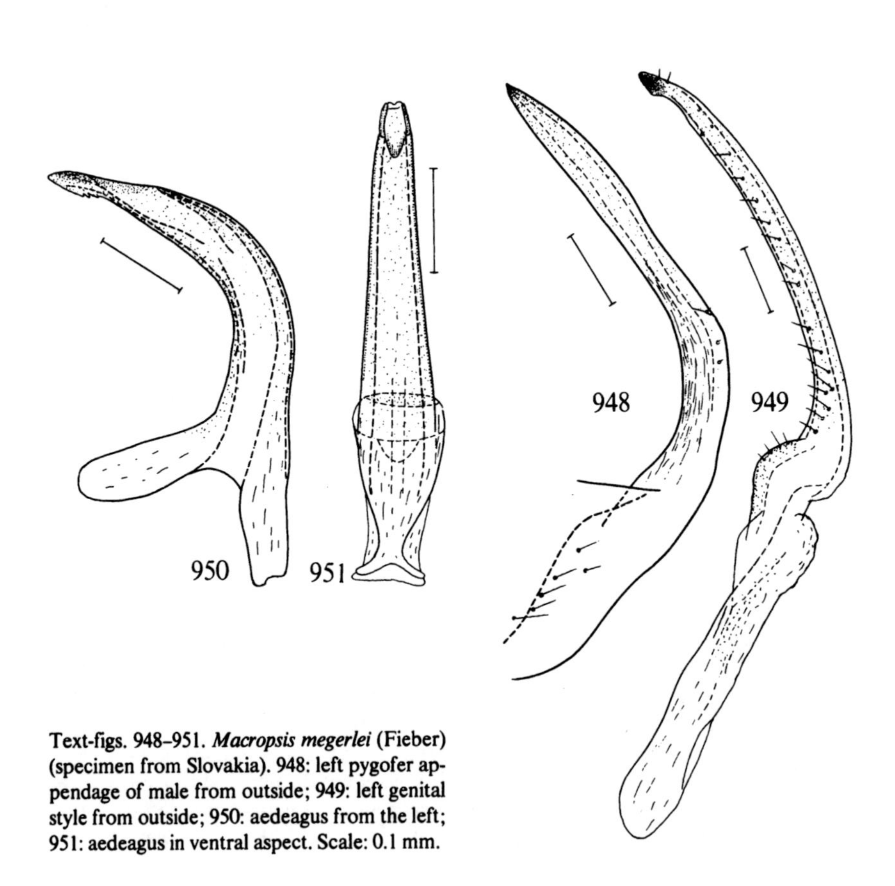

Text-figs. 948–951. *Macropsis megerlei* (Fieber) (specimen from Slovakia). 948: left pygofer appendage of male from outside; 949: left genital style from outside; 950: aedeagus from the left; 951: aedeagus in ventral aspect. Scale: 0.1 mm.

Genus *Hephathus* Ribaut, 1952

Hephathus Ribaut, 1952: 437.
Type-species: *Bythoscopus nanus* Herrich-Schäffer, 1835, by original designation.

As *Macropsis*, differing principally by the absence of a process at the ventral border of the male pygofer. Pronotum broader than the head. Anteclypeus basally not dilated. In Northern Europe one species.

123. ***Hephathus nanus*** (Herrich-Schäffer, 1835)
Text-figs. 952–955.

Bythoscopus nanus Herrich-Schäffer, 1835a: 69.

Body short and broad. Face strongly convex. Upper side greyish yellow, with black pigmentation much varying in extension. Anteclypeus black, apex and lateral margins light. Frontoclypeus usually largely black up to ocelli. In light specimens there are separate ocellar and thyridial spots but usually these coalesce with each others and with

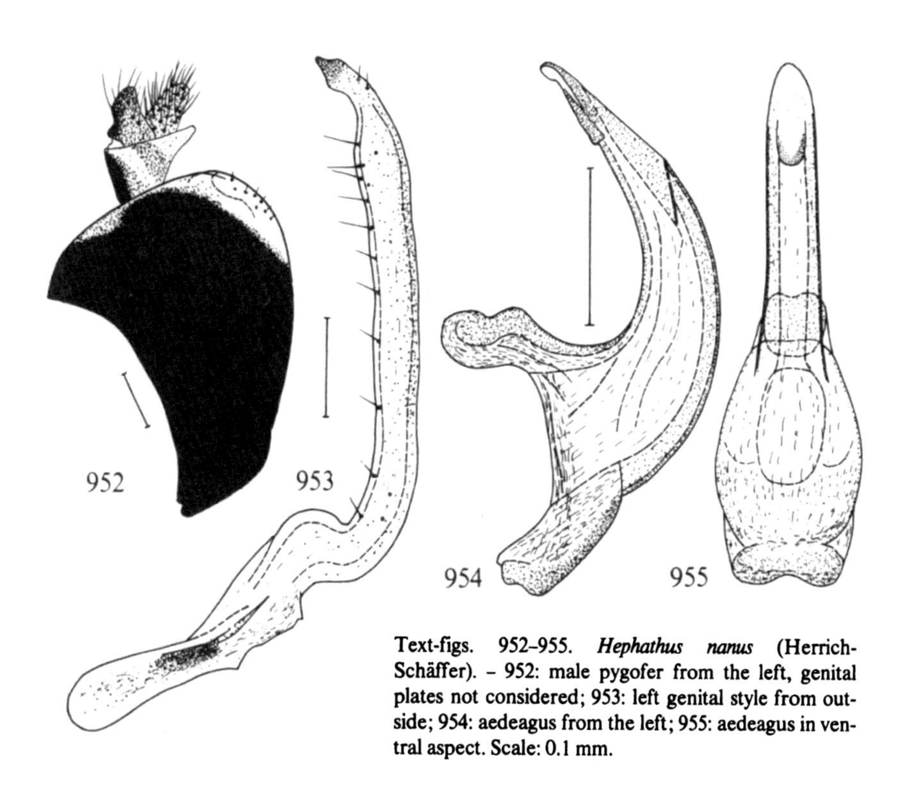

Text-figs. 952–955. *Hephathus nanus* (Herrich-Schäffer). – 952: male pygofer from the left, genital plates not considered; 953: left genital style from outside; 954: aedeagus from the left; 955: aedeagus in ventral aspect. Scale: 0.1 mm.

the large black spot on lower part of face. Sometimes the face is entirely black. Pronotum greyish yellow with a few black spots at fore border, or largely fuscous with lateral and posterior margins light, or even entirely black. Scutellum yellowish-brown with two black basal triangles, often with additional dark markings, or largely black. Fore wings hyaline, veins pale or fuscous, in apex of wing often bordered with fuscous. Venter and abdomen largely black, the latter with segmental margins light, genital plates of male yellowish. Anterior femora light with a broad black band on upper side, intermediate and posterior femora black with light apices, tibiae light, each with a black spot on outside near basis. Male pygofer and anal tube as in Text-fig. 952, genital style as in Text-fig. 953, aedeagus as in Text-figs. 954, 955. Length of male 2.8–3.2 mm, of female 3–3.5 mm.

Distribution. Not found in Denmark, Sweden and Norway. – Rare in eastern East Fennoscandia, found in Kb: Kontiolahti by Håkan Lindberg and in Kb: Hammaslahti by Kontkanen, also recorded from Vib and Kr. – Widespread in Europe, also found in Morocco, Jordan, Palestine, Syria, Armenia, Georgia, Kazakhstan, Kirghizia, Azerbaijan, w. Siberia, Uzbekistan.

Biology. "Auf trocknen Wiesen" (Lindberg, 1947). "Bewohner von Trockenrasen an sonnigen Hängen" (Wagner & Franz, 1961). "In short grass in dry places" (Le Quesne, 1965). Univoltine, hibernation in the egg stage (Schiemenz, 1969). Adults in July and August.

SUBFAMILY AGALLIINAE

Vertex smoothly rounded into face. Head broad, anterior outline in dorsal aspect arcuate. Epicranial suture distinct. Frontal sutures obsolete. Ocelli situated on face. Sides of pronotum short. Hind tibiae with numerous setae but without strong spines. In Northern Europe one genus.

Genus *Agallia* Curtis, 1833

Agallia Curtis, 1833: 193.

Type-species: *Agallia consobrina* Curtis, 1833, by monotypy.

Anaceratagallia Zachvatkin, 1946: 159, 160.

Type-species: *Cicada venosa* Fourcroy, 1785, by original designation.

Small but coarsely built species. Head as seen from above short, anteriorly rounded, broader than pronotum. Fore border of pronotum arcuate or obtuse angled, hind border almost straight. Fore wings in macropters longer than abdomen, membranous, in brachypters leathery. Apical membrane of fore wings narrow or indistinct. Hind tibiae with several rows of strong setae. In Denmark and Fennoscandia four species.

Key to species of *Agallia*

1 Pronotum shagreened .... 124. *consobrina* Curtis
– Pronotum transversely striolate 2
2 (1) Normally brachypterous, fore wings less than half as long as abdomen, apically truncate with rounded angles (Plate-fig. 53) 127. *brachyptera* (Boheman)
– Macropterous, fore wings as long as abdomen or longer 3
3 (2) Upper surface of fore wings densely covered with coarse tubercles of approximately the same size in all cells. (Under surface with more sparsely arranged minute spinules). Appendages of anal collar in male as in Text-fig. 970 .... 126. *ribauti* Ossiannilsson
– Tubercles on upper surface of fore wing reduced or more or less missing in an area mainly consisting of apical parts of radial and median cells and subapical cells 1 and 2. (Under surface spinules present as in *ribauti*). Appendages of anal collar in male as in Text-figs. 964, 965 .... 125. *venosa* (Fourcroy)

124. ***Agallia consobrina*** Curtis, 1833
Text-figs. 956–961.

Agallia consobrina Curtis, 1833: 193.
Jassus puncticeps Germar, 1837: pl.12.
Agallia versicolor Flor, 1861a: 550.

Greyish yellow with brownish and black markings. Face medially with a brownish, in lower part broader, double stripe. On each side medially of the clypeal suture there is an irregular row of small black spots or transverse streaks (muscle traces). A large rounded black spot on each side of vertex half-way between middle line and eye. Ocelli below black-edged, surrounded by a loop-shaped brownish figure. Pronotum anteriorly often with a pair of black spots near middle line, a blackish or brownish coil-shaped figure proceeding from each of them to the side. Pronotum with a light brownish, caudally broader, median longitudinal stripe, caudally on each side of the latter with a triangular light brownish spot. Scutellum with a spot in each lateral corner. Fore wing veins in clavus and basal quarter of corium whitish with partly dark-mottled interspaces, in distal three quarters of corium fuscous and more or less broadly bordered with fuscous. Legs with dark longitudinal stripes. Abdomen above blackish with light margins and segmental borders, ventrally entirely light or dark-spotted. First anal segment of male as in Text-fig. 956, genital style as in Text-fig. 957, aedeagus as in Text-figs. 958, 959, 2nd abdominal sternum in male as in Text-figs. 960, 961. Length of male 3.4–3.7 mm, of female 3.4–4.0 mm.

Distribution. Rare in Denmark (SJ, EJ). – Very rare in Sweden, found only in Öl. where Boheman first caught it. Later found in Borgholm by Kemner and Ossiannilsson, in Böda, "Neptuni åkrar", by E. Olofsson, and in Högsrum, "Halltorps hage" by

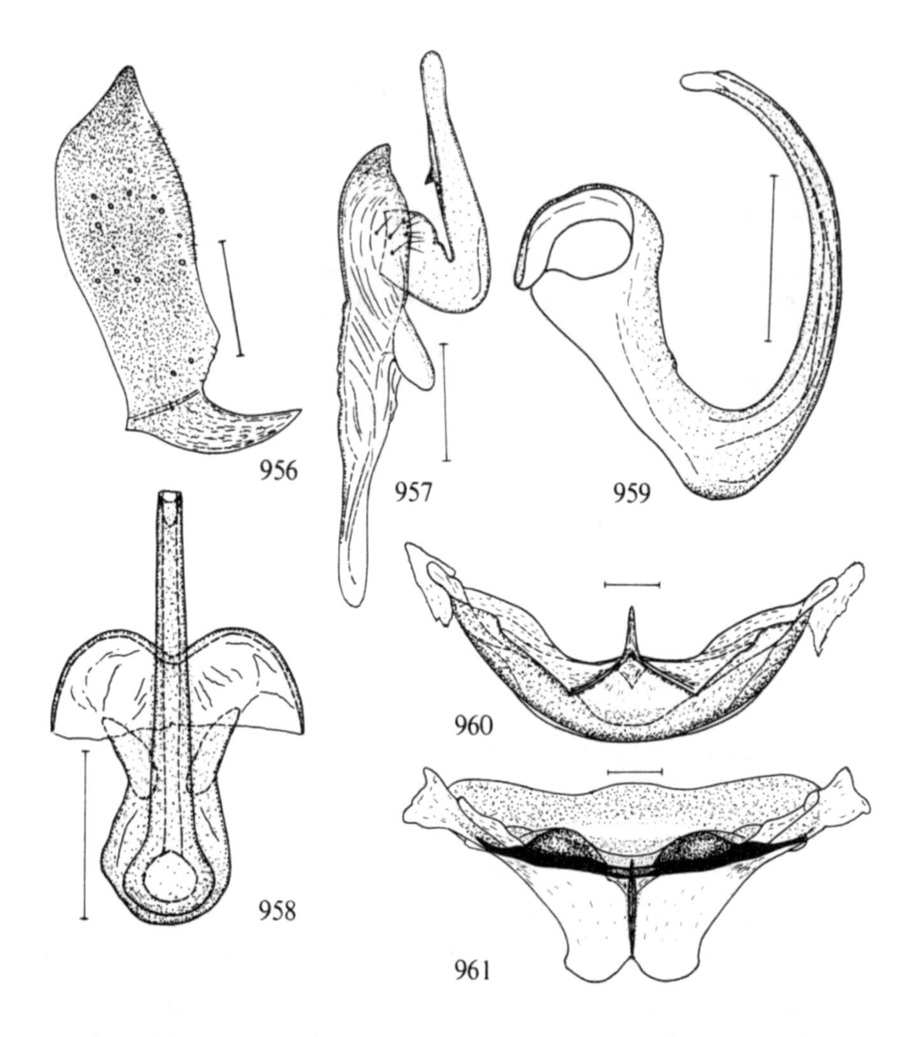

Text-figs. 956–961. *Agallia consobrina* Curtis. – 956: 1st anal segment of male from the left; 957: right genital style from above; 958: aedeagus in ventral aspect; 959: aedeagus from the left; 960: 2nd abdominal sternum in male from behind; 961: same from above. Scale: 0.1 mm.

Text-figs. 962–969. *Agallia venosa* (Fourcroy). – 962: apex of male abdomen from below; 963: male pygofer from the left; 964: male anal tube with left anal collar appendage (a) from the left; 965: left anal collar appendage of another specimen; 966: right genital style from above; 967: aedeagus in ventral aspect; 968: aedeagus from the left; 969: 2nd and 3rd abdominal sterna in male from above. Scale: 0.1 mm.

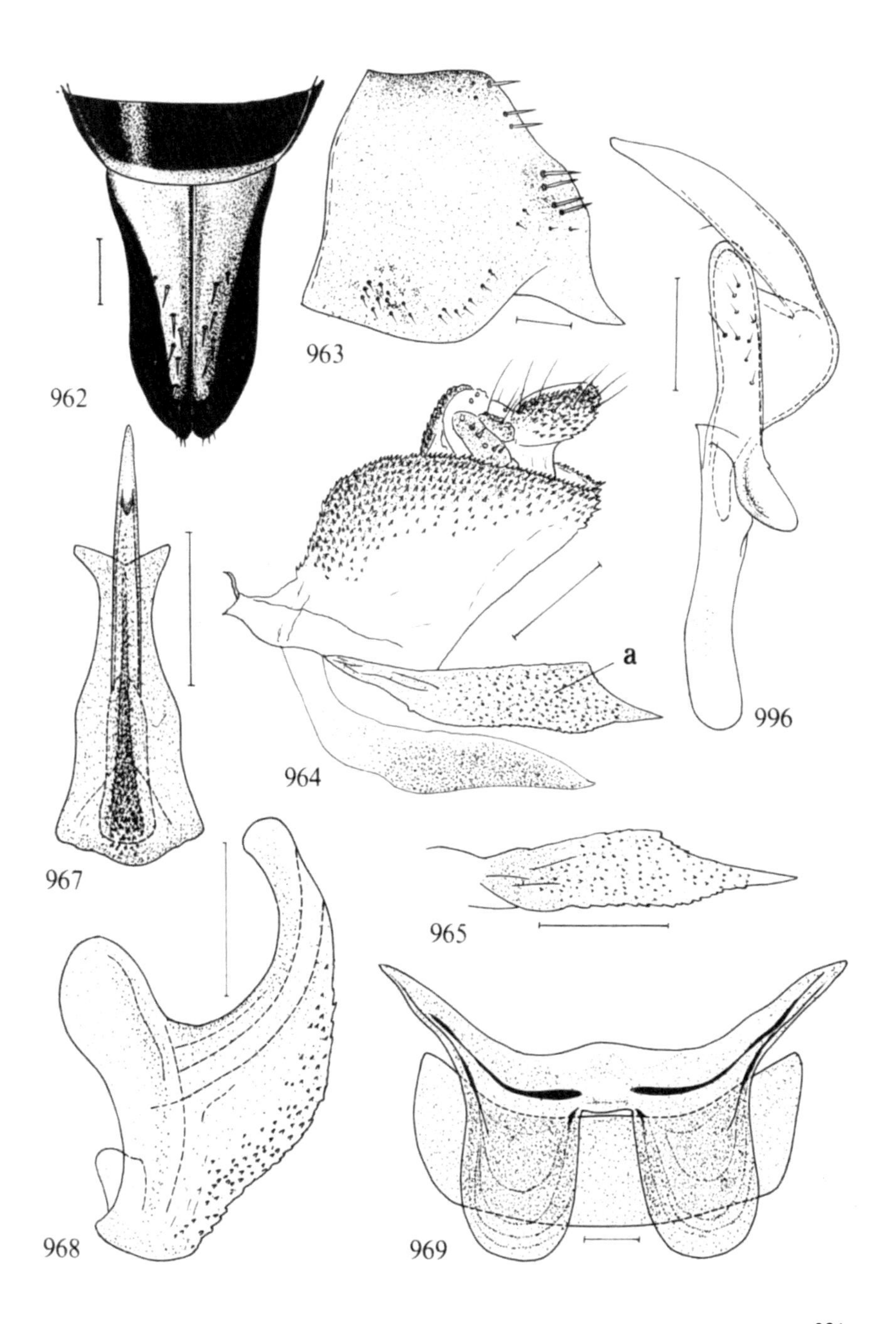
962
963
964
a
965
967
968
969
996

H. Andersson and R. Danielsson. – Not found in Norway, nor in East Fennoscandia. – Austria, Belgium, Hungaria, Bohemia, Slovakia, France, German D.R. and F.R., England, Scotland, Wales, Greece, Hungary, Italy, Netherlands, Poland, Portugal, Spain, Yugoslavia, s. Russia, Azerbaijan.

Biology, "An *Urtica dioica*" (Kuntze, 1937). "In Laubwäldern auf Lichtungen" (Wagner & Franz, 1961). "Common among low plants" (Le Quesne, 1965). Adults in February, April, May, July–November (Le Quesne, l.c.).

125. ***Agallia venosa*** (Fourcroy, 1785)
Plate-fig. 58, text-figs. 962–969.

Cicada venosa Fourcroy, 1785: 188.
Cicada venosa Fallén, 1806: 25.
Agallia aspera Ribaut, 1935a: 36.

Resembling *consobrina* but with more extended black markings. Vertex in male (Plate-fig. 58) with two roundish or oval black spots and a black, sometimes double, median streak continuing on face and dividing in two strongly diverging branches below ocelli. Frontoclypeus with more or less confluent black transverse streaks (muscle traces). An irregular black spot on each side between eye and ocellus. Genae and lora partly black. Pronotum with black median line and on each side an irregularly hour-glass-shaped black figure. Scutellum with a black spot on each side corner and a central figure consisting of small black spots. Veins of fore wings black, bordered with black. In the female the dark markings are brownish and less extended. Apex of male abdomen from below as in Text-fig. 962, pygofer as in Text-fig. 963, male anal tube with appendage as in Text-fig. 964, appendage as in Text-fig. 965, genital style as in Text-fig. 966, aedeagus as in Text-figs. 967, 968, 2nd and 3rd abdominal sterna in male as in Text-fig. 969. Overall length of male 2.5–3.2 mm, of female 3.1–3.5 mm.

Distribution. Common and widespread in Denmark as well as in Sweden up to Nb. – Appears to have a southern and eastern distribution in Norway, so far found in AK, HEs, Os, Bø, TEi, AAy, and VAy. – Common in southern and central East Fennoscandia (Al, Ab, N–Oa, Kb, Om, Ks; Vib, Kr.) – Austria, Belgium, Bohemia, Moravia, Slovakia, France, England, Scotland, Wales, Netherlands, Italy, German D.R. and F.R., Romania, Switzerland, Poland, Estonia, Latvia, Lithuania, n., m., and s. Russia, Ukraine, Azerbaijan, Uzbekistan, Kirghizia.

Biology. On or in "Sandfeldern, besonnten Hängen, Wiesen" (Kuntze, 1937). In dryish fields, moist sloping meadows, rich swampy woods, rich moist grass-herb woods, and dry *Vaccinium* pine woods (Linnavuori, 1952a). In "Waldsteppen" (Okáli, 1960). Hibernation takes place in the egg stage (Schiemenz, 1969; Müller, 1978). Monovoltine (Müller, 1978). In Sweden adults are found in July–September.

126. ***Agallia ribauti*** Ossiannilsson, 1938
Text-figs. 970–974.

Agallia venosa Ribaut, 1935: 32 (nec Fourcroy, 1785).
Agallia ribauti Ossiannilsson, 1938b: 77, 78 (n.n.).

Resembling *venosa,* differing by characters given in the key. Appendage of male anal collar as in Text-fig. 970, genital style as in Text-fig. 971, aedeagus as in Text-figs. 972, 973, 2nd and 3rd abdominal sterna in male as in Text-fig. 974. Overall length of male 2.7–3.1 mm, of female 2.8–3.4 mm.

Distribution. Denmark: so far reliable records from EJ: Jernhatten 18.IX.1977 (Trolle), and B: Rutsker Højlyng 14.VIII.1972 (Trolle). – Sweden: less common than *venosa* but widespread (Sk., Bl., Öl., Gtl., Ög., Nrk., Sdm., Upl., Dlr., Hls., Ång.). –

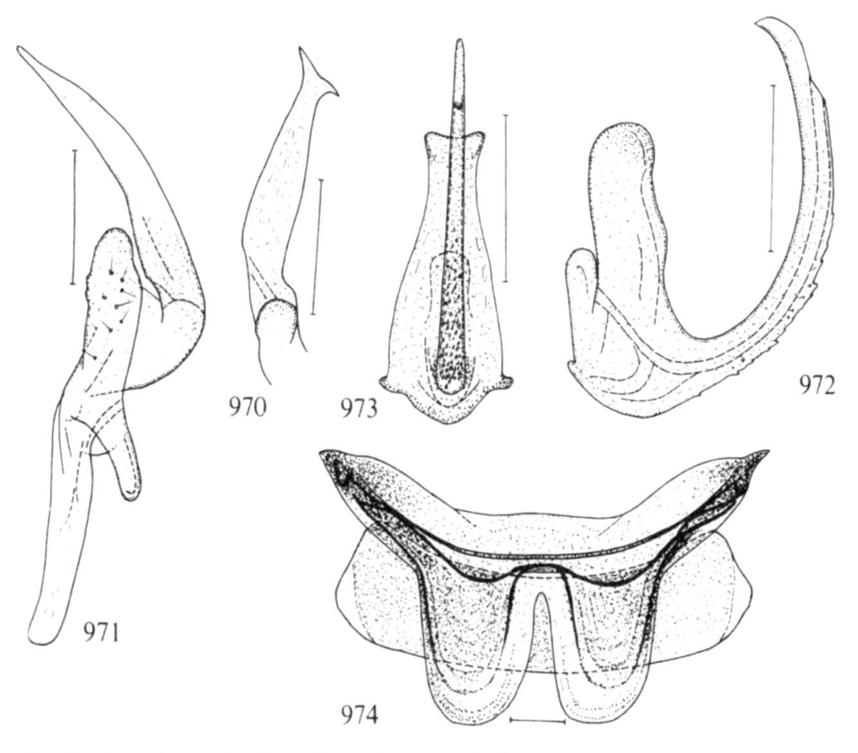

Text-figs. 970–974. *Agallia ribauti* Ossiannilsson. – 970: right anal collar appendage in male from the right; 971: right genital style from above; 972: aedeagus from the left; 973: aedeagus in ventral aspect; 974: 2nd and 3rd abdominal sterna in male from above. Scale: 0.1 mm.

Norway: so far only two records, viz. BØ: Ringerike 2.VIII.1899 (Warloe), and TEi: Smedodden, Kviteseid 1.VIII.1943 (Holgersen). – East Fennoscandia: found in Al, Ta, Sa, Sb, Kb, Om. According to Linnavuori (1969) common in the eastern part, sporadic in the rest of its area of distribution. – Widespread in Europe, also in Anatolia, Armenia, Altai Mts.

Biology. On "Stranddünen, Binnendünen, Sandfeldern, besonnten Hängen, Wiesen" (Kuntze, 1937). "Auf Kräutern an warmen trockenen Standorten" (Wagner & Franz, 1961). Hibernation by adult females (Schiemenz 1964, 1969). Monovoltine (Müller, 1978). In Sweden adults have been found in March and April and in July–October.

127. ***Agallia brachyptera*** (Boheman, 1847)
Plate-figs. 53, 64, text-figs. 975–981.

Athysanus brachypterus Boheman, 1847b: 264.

Normally brachypterous (Plate-fig. 53), fore wings covering only basis of abdomen. Finely pilose, whitish yellow to greyish yellow with the following black or brownish markings: a roundish spot on the head on each side of vertex, a longitudinal line between ocelli, a spot near each ocellus, an arched, above angular streak medially of each clypeal suture, a central spot on anteclypeus, spots on genae and lora; on pronotum a narrow median streak and on each side one anterior and one posterior spot; on scutellum one spot at each lateral corner; on each fore wing 4 or 5 broad bands along the veins; on abdomen a narrow median line, and a transverse band on or near fore border of each segment. Especially in the male these markings may coalesce into more extended black surfaces. Venter more or less dark-spotted, legs with dark streaks and dots. – One single macropterous female has been described by W. Wagner, 1937a (p.73) as follows: "Vorder- und Hinterflügel – – – überragen die Spitze des Abdomens, so dass das Tier auf den ersten Blick an eine Art aus der *A.-venosa* Fall.-Gruppe erinnert. Die Vorderflügel zeigen aber eine deutlichere pflastersteinartige Mikroskulptur und sind auch anders gezeichnet. Die Clavusnerven sind weiss und weiss gerandet. Die Zellen des Clavus sind in der Mitte durch braune Längsstreifen ausgefüllt, die bis an die Schlussnaht reichen. Letztere und ebenso der Innenrand des Clavus neben dem Schildchen sind dunkelbraun. Radius und Media sind im ersten Drittel weiss. Alle anderen Nerven im Corium sind braun. Die Costalzelle trägt in der vorderen Hälfte mitten einen braunen Längsstreifen. Die 1. Medialzelle ist nicht geteilt, die Postradialzelle fehlt. Kopf und Pronotum sind wie bei stark tingierten Exemplaren der forma *brachyptera* gezeichnet." – Male anal collar as in Text-fig. 975, genital style as in Text-fig. 976, aedeagus as in Text-figs. 977, 978, 2nd abdominal sternum of male as in Text-fig. 979, 2nd abdominal tergum of male as in Text-fig. 980, 3rd abdominal tergum of male as in Text-fig. 981. Overall length of brachypterous male 2.25–2.8 mm, of brachypterous female 2.6–3.4 mm. – Larva whitish or pale yellow with black markings, see Plate-fig. 64.

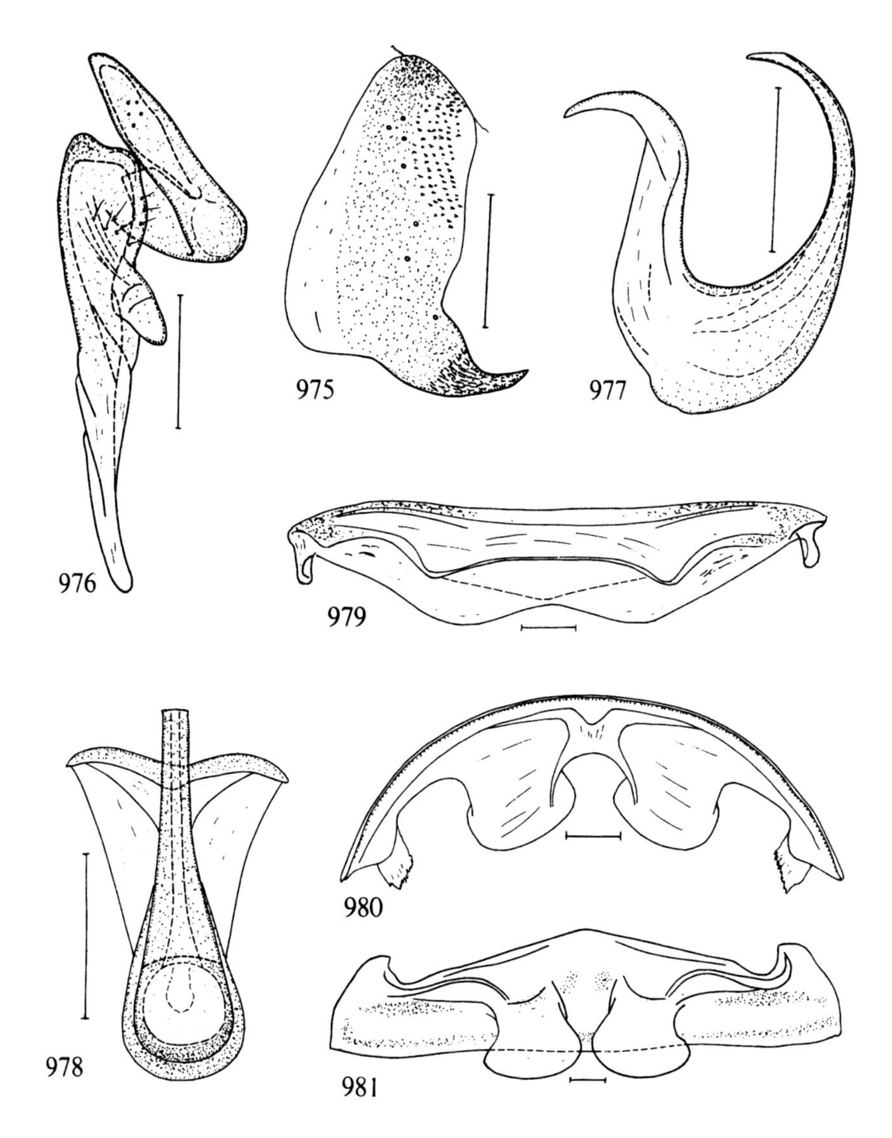

Text-figs. 975–981. *Agallia brachyptera* (Boheman). – 975: 1st anal segment of male from the left; 976: right genital style from above; 977: aedeagus from the left; 978: aedeagus in ventral aspect; 979: 2nd abdominal sternum in male from above; 980: 2nd abdominal tergum in male from behind; 981: 3rd abdominal tergum in male from below. Scale: 0.1 mm.

Distribution. Scarce in Denmark (SJ, EJ, LFM, NEZ). – Common and widespread in southern and central Sweden (Sk. – Hls.). – Norway: so far found in Ø, AK, HEs, Bø, and TEy. – Common in southern and central East Fennoscandia (Al, Ab, N, Ta, Sa, Tb, Sb, Kb, Om; Vib, Kr.). – Widespread in Europe, also found in Algeria, Morocco, Afghanistan, Kazakhstan, Altai Mts.

Biology. In "Flachmooren, Wäldern, Waldlichtungen, Wiesen" (Kuntze, 1937). In dryish fields, moist sloping meadows, cultivated fields, rich swampy woods, and moist grass-herb woods (Linnavuori, 1952a). "Meso-hygrophile Wiesenart" (Schiemenz, 1964). "Not a single adult specimen was found before 21.VII, when the species had their maxima" (Törmälä & Raatikainen, 1976). In Sweden adults have been found from 12.VII. to 3.X. I found larvae of *Agallia brachyptera* under *Rumex acetosella, Trifolium repens,* and *Taraxacum* sp. in June and July. They were observed feeding on these plants and also on *Achillea millefolium.* The species hibernates in the egg stage (Remane, 1958; Törmälä & Raatikainen, l.c.).

SUBFAMILY IDIOCERINAE

Vertex without a distinct boundary merging into the face. Head broad, its anterior outline arched. Antennae of males often with a terminal so-called palette (Text-fig. 6). Ocelli situated on the face, frontal sutures distinct. Wings always well developed, fore wings long and narrow with a broad apical membrane (Text-fig. 982). Hind wings with a complete peripheric vein. Fore and middle tibiae without dorsal spines, fore and middle femora each with two apical spines. Hind tibiae with many strong setae. 7th abdominal sternum of male short, laterally fused with the pygofer. In Denmark and Fennoscandia six genera.

Key to genera of Idiocerinae

1 Vertex and pronotum deeply and coarsely transversely wrinkled .... *Rhytidodus* Fieber (p.307)
– Vertex and pronotum not transversely wrinkled 2
2(1) Vertex finely punctured. Fore wing without a closed radial cell (Text-fig. 1133). Genital plates of male strongly reduced, almost entirely covered by the 7th abdominal sternum
Sahlbergotettix Zachvatkin (p.351)
– Vertex and pronotum shagreened. A closed radial cell present. Genital plates well developed, long 3
3(2) Face broad, angle of side margins 90° or greater 4
– Face less broad, angle of side margins less than 90° 5
4(3) Distance between ocelli only slightly greater than distance between one ocellus and the nearest compound eye (Text-fig. 1097) .... *Metidiocerus* n. gen. (p.318)

- Distance between ocelli distinctly greater than distance between an ocellus and nearest compound eye .... *Idiocerus* Lewis (p.310)

5(3) Margin of face behind each compound eye with about 4 short setae. Frontoclypeus more than twice as long as broad *Populicerus* Dlabola (p.323)

- Margin of face behind eyes devoid of short setae. Frontoclypeus less than twice as long as broad 6

6(5) Maxillar plates apically protracted, apex angular (Text-fig. 1097) or rounded. Genital style of male slender, parallel-sided, without subapical strong setae. Male antenna without palette *Tremulicerus* Dlabola (p.339)

- Maxillary plates apically not protracted, lateral margin with a shallow indentation (Text-fig. 1124). Genital style of male broad, evenly curved, with a few strong subapical setae. Male antennae with a slender palette .... *Stenidiocerus* n.gen. (p.348)

Genus *Rhytidodus* Fieber, 1872

Bythoscopus (Rhytidodus) Fieber, 1872:8.

Type-species: *Idiocerus germari* Fieber, 1868, by subsequent designation.

Vertex and pronotum coarsely transversely striated. Fore wings with two or three subapical cells, leathery, cells wrinkled. Complex eyes comparatively strongly convex. Fore border of vertex almost parallel with hind border. Face almost twice as broad as long, below with a short pilosity. Male antennae without palette. Frontoclypeus broader than long. In Northern Europe one species.

128. ***Rhytidodus decimusquartus*** (Schrank, 1776)
Text-figs. 982–989.

Cicada decimaquarta Schrank, 1776:76.
Jassus scurra Germar, 1837:pl. 11.
Idiocerus germari Fieber, 1868:451.

Strongly built, upper side shining. Lower part of face in both sexes with a white pilosity. Distance between ocelli almost twice as long as distance between an ocellus and the nearest compound eye. Setae of hind tibiae comparatively weak. Yellowish with black markings, the latter more sparse in females. Males usually with a broad black or brownish transverse band on face between eyes. This band is often more or less disintegrated in smaller transverse streaks. Frontoclypeus below the transverse band just mentioned usually with a horseshoe-shaped figure consisting of short black or brownish transverse streaks. This is absent in females and in weakly pigmented males. Pronotum and scutellum in males sordid yellow with more or less extended irregular black or brownish markings, sometimes largely black, in female usually brownish yellow, unspotted or with indistinct brownish spots. Fore wings in male semi-hyaline,

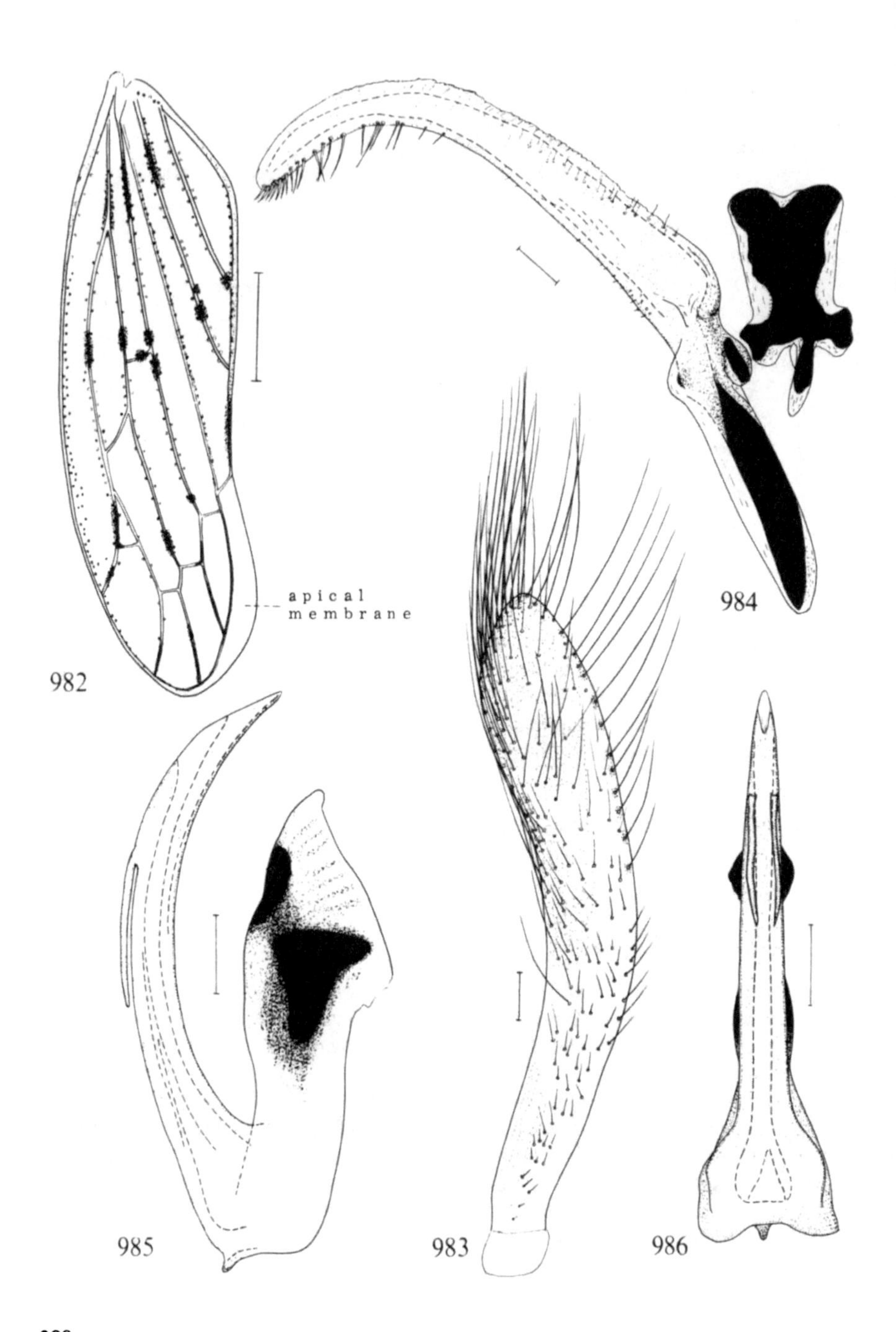
apical
membrane
982
984
985
983
986

greyish or sordid yellowish, veins whitish or yellowish, here and there with black spots (Text-fig. 982), in female usually brownish yellow with some white spots, veins concolorous, sometimes narrowly bordered with fuscous. Thoracal venter and abdominal dorsum in males usually black with light margins of thoracal pleura and abdominal terga, abdominal venter largely light. Thoracal venter and abdomen in female yellowish, legs pale in both sexes. Male genital plate as in Text-fig. 983, genital style and connective as in Text-fig. 984, aedeagus as in Text-figs. 985, 986, 1st and 2nd abdominal sterna in male as in Text-figs. 987, 988, 2nd abdominal tergum in male as in Text-fig. 989. Length of male 5.5–6.3 mm, of female 6.8–7.5 mm.

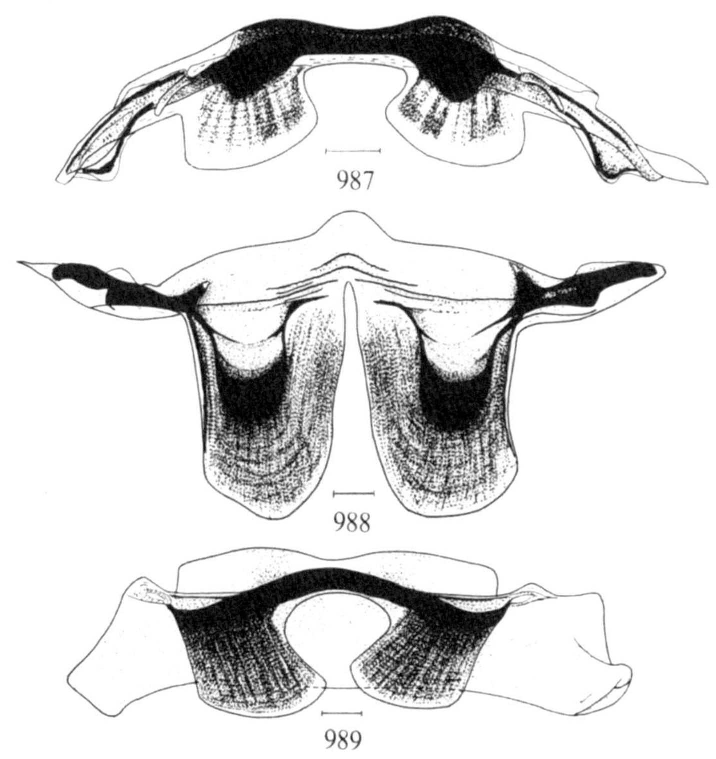

Text-figs. 987–989. *Rhytidodus decimusquartus* (Schrank). – 987: 1st abdominal sternum in male from behind; 988: 2nd abdominal sternum in male from above; 989: 2nd abdominal tergum in male from below. Scale: 0.1 mm.

Text-figs. 982–986. *Rhytidodus decimusquartus* (Schrank). – 982: left fore wing of male; 983: left genital plate from the left; 984: right genital style from above; 985: aedeagus from the right; 986: aedeagus in ventral aspect. Scale: 1 mm for 982, 0.1 mm for the rest.

Distribution. R. Remane (in litt.) found *Rh. decimusquartus* in Gotland. The species has not been found elsewhere in Sweden, nor in Denmark, Norway and East Fennoscandia. – Widespread in Europe except the northern part, also found in Tunisia, Armenia, Azerbaijan, Georgia, Kazakhstan, Uzbekistan. Introduced into North America and New Zealand.

Biology. On *Populus nigra* and *P. nigra italica* (Kuntze, 1937; Wagner & Franz, 1961; Le Quesne, 1965), also on *Populus deltoides* (Müller, 1956). "Eiablage in Gruppen von 3–6 unter die Rinde von Pappelzweigen in Knospennähe. Larven schlüpfen Ende Mai, L_1 bis L_3 saugen an Blättern, die älteren Larven an kleinen Zweigen, stets kopfabwärts. Ad. ab Ende Juni. ♀♀ der ersten Generation legen Anfang Juli, die der zweiten Ende August ab" (Müller, 1956). Hibernates in the adult stage and in the egg stage. (Müller, 1957).

Economic importance. "Durch Weissverfärbung der Blattoberseite schädlich, in Frankreich auch für krebsartige Erkrankungen der Schwarzpappeln verantwortlich gemacht" (Müller, 1956).

Genus *Idiocerus* Lewis, 1834

Idiocerus Lewis, 1834:47.
Type-species: *Idiocerus stigmaticalis* Lewis, 1834, by monotypy.

Face with eyes broader than long. Genae below with a light pilosity. Frontoclypeus almost roundish, upper half much broader than the lower part. Lateral margins of face straight, forming an angle of over 90° with each other, below eyes with a shallow incision. Vertex medially shorter than near eyes, about half as long as pronotum. Veins of fore wings strong. Costal margin of fore wing in basal 2/3 deflected, especially in males. Male antenna always with palette. In Denmark and Fennoscandia three species.

Key to species of *Idiocerus*

1 Transverse veins proximally delimiting first and second apical cells in fore wing level or nearly so (Plate-fig. 66) .... 129. *stigmaticalis* Lewis
– Transverse vein proximally delimiting first apical cell situated considerably nearer base of fore wing than transverse vein belonging to second apical cell (Plate-fig. 69).. 2

Text-figs. 990-996. *Idiocerus stigmaticalis* Lewis. – 990: male pygofer and anal segment from the left; 991: same from above; 992: right genital style and connective from above; 993: aedeagus from the right; 994: aedeagus in ventral aspect; 995: 1st and 2nd abdominal sterna in male from above (2nd abdominal sternum broken at arrow. In reality the inner margins of the median apodemes are partly contiguous, approximately parallel); 996: 2nd and 3rd abdominal terga in male in ventral aspect. Scale: 0.5 mm for 990 and 991, 0.1 mm for the rest.

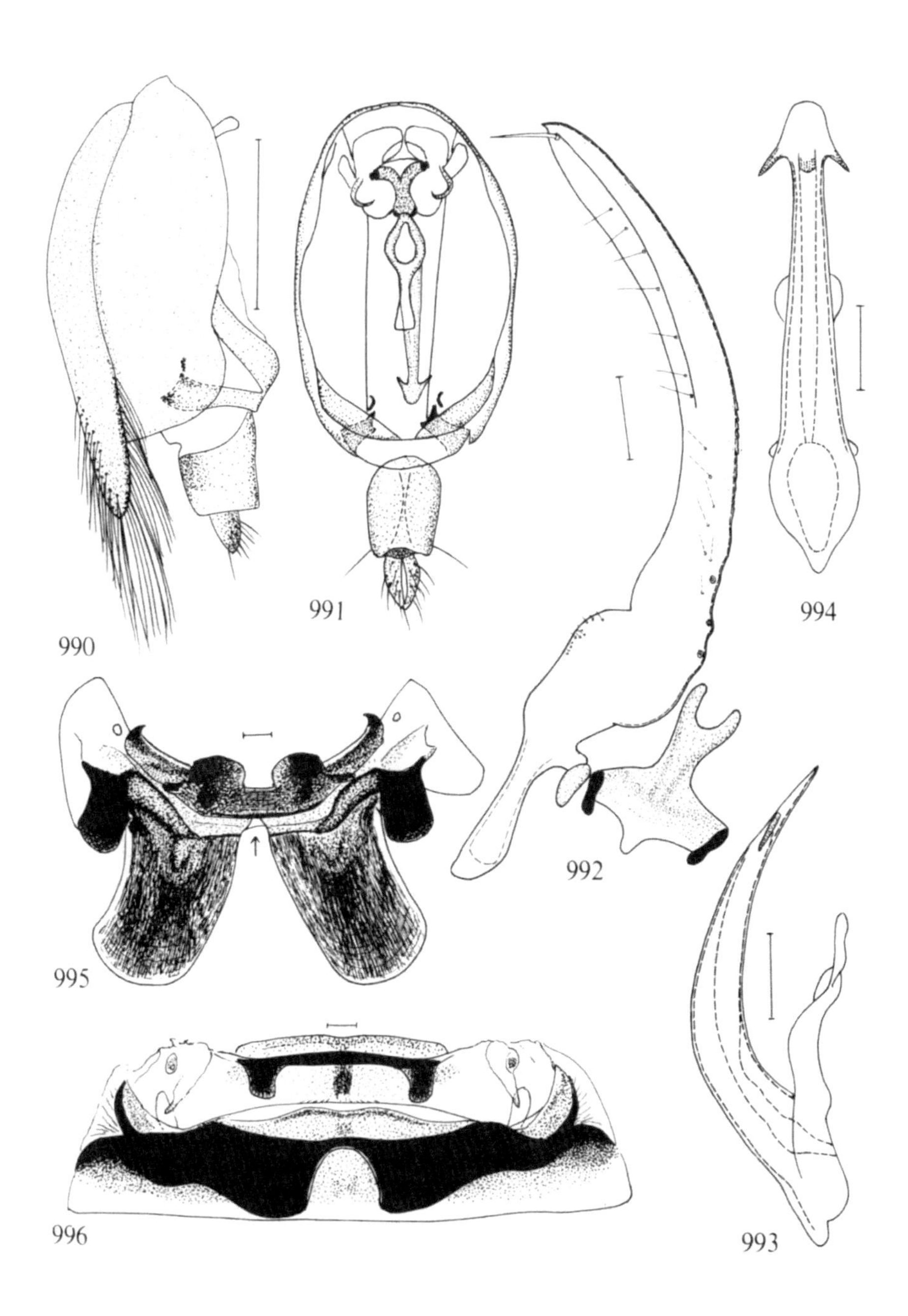
990
991
992
993
994
995
996

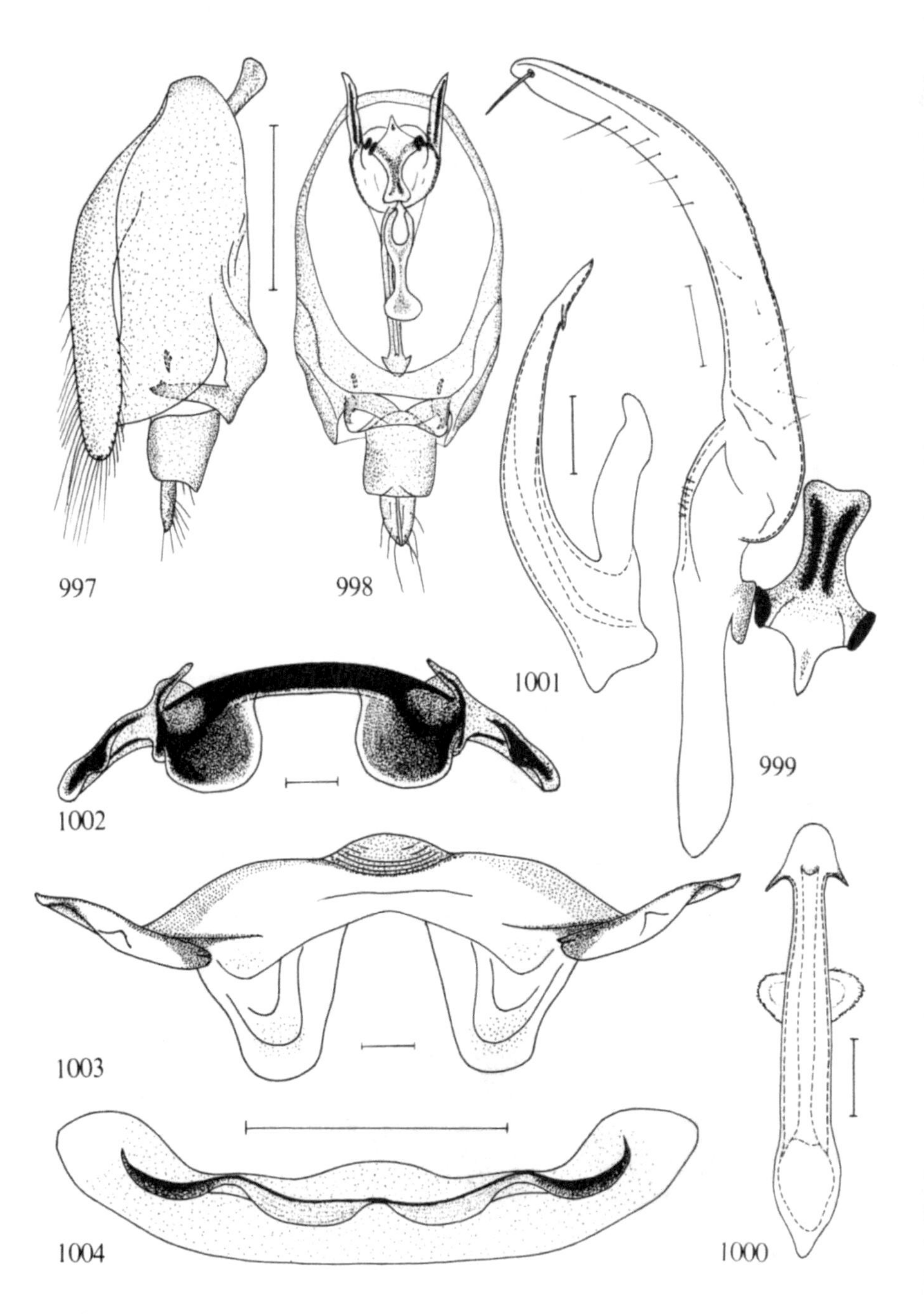
997
998
1001
999
1002
1003
1000
1004

2(1) Face in male with a strong, in female with a less strong whitish pilosity. Veins of fore wing black, the black colour being interrupted by numerous (18 - 30, fore border not considered) whitish sections (Plate-fig. 69)...................................... 131. *herrichii* Kirschbaum

– Face in both sexes only indistinctly pilose. Veins of fore wing dark with in all about 6 light sections.................................. 130. *lituratus* (Fallén)

129. ***Idiocerus stigmaticalis*** Lewis, 1834
Plate-fig. 66, text-figs. 990–996.

Idiocerus stigmaticalis Lewis, 1834:48.
Bythoscopus adustus Herrich-Schäffer, 1837:9.

Whitish yellow or greyish yellow. Antenna of male with a small palette. Genae white-pilose (pilosity shorter in females). Face whitish or orange brown, with or without dark markings. Pronotum mottled with brownish and greyish white, with 3 greyish white longitudinal streaks and at fore border with a transverse row of black spots. Scutellum with a triangular black spot on each side corner, a black longitudinal streak branching on mesoscutellum, and a rounded black spot on each side of the median line. Costal margin of fore wing in basal half extended downwards, in male with a row of tubercles (Plate-fig. 66). Fore wing in male milky semi-transparent with black-bordered veins. Their black colour is especially strong in a smoky oblique transverse band, distally of which there is a whitish transverse band where also the veins are white. Fore border of fore wing in the area of the dark transverse band broadly black. Here the deflected part of costal margin is broadest, in the male usually partly bright orange brown. Fore wings in female semi-transparent with brownish veins, the latter with some pale patches. Dorsum of abdomen and male genital segment black, venter pale yellowish. Legs usually light, femora sometimes largely black, tibiae on outside with a more or less distinct black longitudinal streak. Male pygofer as in Text-figs. 990, 991, genital style as in Text-fig. 992, aedeagus as in Text-figs. 993, 994, 1st and 2nd abdominal sterna in male as in Text-fig. 995, 2nd and 3rd abdominal terga in male as in Text-fig. 996. Overall length of male 6.0–6.5 mm, of female 6.5–7.2 mm. – Last instar larva light green or pale flesh-coloured, with black markings much varying in extension, and fairly densely covered with long, partly black, erect setae. In pale specimens part of these arise from small circular black dots.

Distribution. Fairly common in Denmark as well as in Sweden up to Gstr.–Appears

---

Text-figs. 997–1004. *Idiocerus lituratus* (Fallén). – 997: male pygofer from the left; 998: male pygofer from above, venter not considered; 999: right genital style and connective from above; 1000: aedeagus in ventral aspect; 1001: aedeagus from the right; 1002: 1st abdominal sternum in male from above; 1003: 2nd abdominal sternum in male from above; 1004: 3rd abdominal tergum in male from below. Scale: 0.5 mm for 997 and 998, 1 mm for 1004, 0.1 mm for the rest.

to be scarce in Norway, found in AK, Bø, VE, and VAy.–Scarce also in East Fennoscandia, found in Al, Ab, N, Ka, Tb. – Austria, Belgium, Bulgaria, Bohemia, Moravia, Slovakia, France, German D.R. and F.R., England, Scotland, Hungary, Italy, Netherlands, Poland, Spain, Yugoslavia, Latvia, Estonia, Lithuania, Ukraine, Algeria; also in the Nearctic region.

Biology. "Auf *Salix*-Arten, besonders *Salix alba*" (Wagner & Franz, 1961). On *Salix*, especially *S. alba* and *S. fragilis* (Le Quesne, 1965). I found *I. stigmaticalis* also on *Salix caprea* and *S. pentandra*. Adults in July – October.

130. ***Idiocerus lituratus*** (Fallén, 1806)
Plate-fig. 65, text-figs. 997–1004.

Jassus lituratus Fallén, 1806b:117.

Male antenna with palette. Whitish yellow or greyish white with yellowish brown and black markings. Face usually entirely light, sometimes tinged with orange, vertex usually brownish, thyridia black. Pronotum greyish brown, cephalad lighter with some small black spots in a transverse row, the dark part with light mottlings. Scutellum much as in *stigmaticalis*. Fore wings of male (Plate-fig. 65) without tubercles on costal border, with strong fuscous veins and a more or less distinct irregular fuscous transverse band over middle, the dark colour of the veins being interrupted in a small number of pale patches. Fore wing veins in females less marked, brownish, with whitish patches as in male. Venter largely light, legs light with black or brownish longitudinal streaks. Pygofer of male as in Text-figs. 997, 998, genital style as in Text-fig. 999, aedeagus as in Text-figs. 1000, 1001, 1st and 2nd abdominal sterna in male as in Text-figs. 1002, 1003, 3rd abdominal tergum in male as in Text-fig. 1004. Overall length of male 5.0–6.1 mm, of female 6.0–6.7 mm. – Last instar larvae olive green, light green or sordid yellow, pigmentation and pilosity as in *stigmaticalis*.

Distribution. Common and widespread in Denmark and Sweden (Sk.–T.Lpm.).–Probably common also in Norway but so far recorded only from AK, AAy, VAy, and Ry.–Common in southern and central East Fennoscandia, found up to ObN and Ks; also in Vib and Kr. – Widespread in Europe, also recorded from Morocco, Anatolia, Armenia, Azerbaijan, Kazakhstan, m.Siberia, Maritime Territory.

Biology. On *Salix*. "Vor allem auf *Salix aurita*" (Kuntze, 1937). "Lebt auf *Salix aurita* und anderen *Salix*-Arten, im Gebiete besonders auch auf *Salix purpurea*" (Wagner & Franz, 1961). On *Salix caprea, cinerea, repens*, and *fragilis* (Le Quesne, 1965). I collected the species on *Salix aurita, cinerea, repens*, and *pentandra*. Adults in July – October.

131. ***Idiocerus herrichii*** Kirschbaum, 1868
Plate-fig. 69, text-figs. 1005–1013.

Idiocerus herrichii Kirschbaum, 1868a:7.

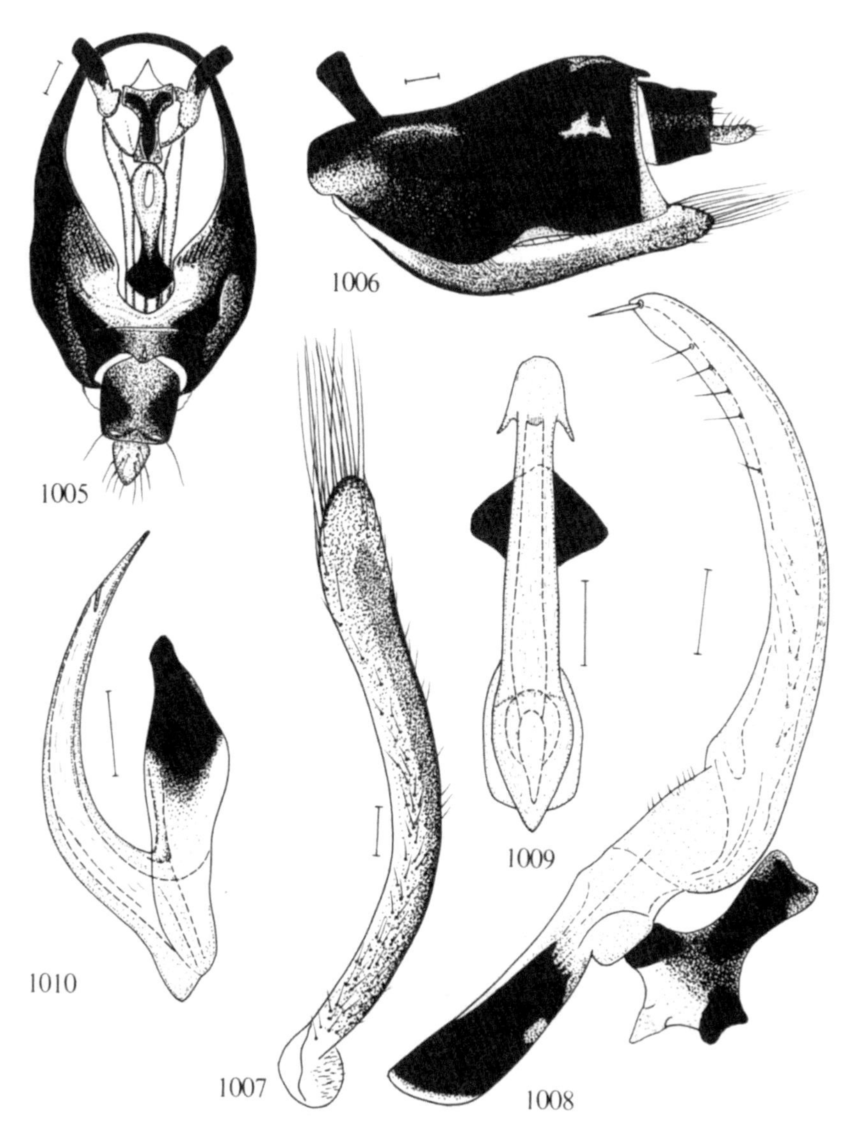

Text-figs. 1005–1010. *Idiocerus herrichii* Kirschbaum. – 1005: male pygofer from above, venter not considered; 1006: male pygofer from the left; 1007: left genital plate from outside; 1008: right genital style and connective from above; 1009: aedeagus in ventral aspect; 1010: aedeagus from the right. Scale: 0.1 mm.

Whitish, strongly variegated by more or less extended black markings. Antenna of male with palette. Genae in male with a long and dense, in female with a shorter white pubescence. Head black-spotted or largely black, above in dark individuals usually with a light median spot and a light spot at each eye. Pronotum with more or less extended black or fuscous markings, usually with a broad light median longitudinal band. Scutellum with five oblong longitudinal spots, three at fore margin and two behind them, or scutellum largely black with a few small light dots. Fore wings in both sexes but especially in male (Plate-fig. 69) with strong black and black-bordered veins, the black colour being interrupted in numerous ivory-white spots. Costal margin without tubercles. Legs with black longitudinal streaks. Male pygofer as in Text-figs. 1005, 1006, genital plate as in Text-fig. 1007, genital style and connective as in Text-fig. 1008, aedeagus as in Text-figs. 1009, 1010, 1st and 2nd abdominal sterna in male as in Text-fig. 1011, 2nd and 3rd abdominal terga in male as in Text-figs. 1012, 1013. Overall length of male 6.0–6.2, of female 6.4–7.0 mm.

Distribution. So far not found in Denmark. – Rare in Sweden; first found in Sm. by C. G. Thomson. Per Brinck got a male in Sk., Degeberga. Found in Uppsala (Upl.) by Wallengren, Johansson, Haglund and Ossiannilsson. I also collected a few specimens in

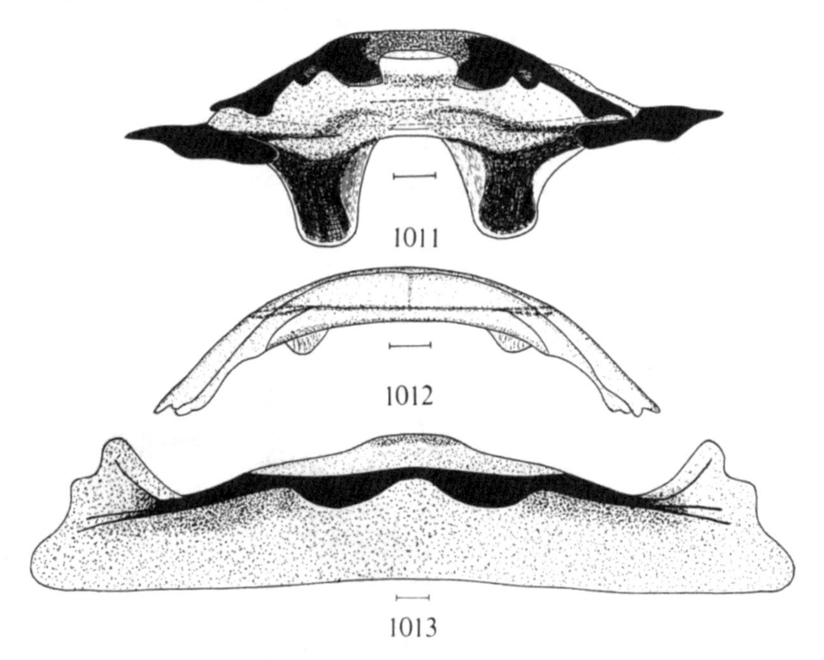

Text-figs. 1011–1013. *Idiocerus herrichii* Kirschbaum. – 1011: 1st and 2nd abdominal sterna in male from above; 1012: 2nd abdominal tergum in male from behind; 1013: 3rd abdominal tergum in male from below. Scale: 0.1 mm.

Upl.: Börstil, St. Torrön, and in Skogstibble, Degermossen, and in Nysätra, Oxdjupet. - Norway: so far found only in AK: Oslo, Tøyen, by Siebke. - Rare also in East Fennoscandia, found in Al, Ab, N, St, Ta; Kr. - Austria, Bulgaria, Bohemia, Moravia, Slovakia, France, German D. R. and F. R., England, Hungary, Italy, Poland, Romania, Switzerland, Yugoslavia, n., m., and s. Russia, Ukraine, Moldavia, Armenia, Georgia, Afghanistan, Kazakhstan, Tadzhikistan, Uzbekistan, Kirghizia.

Biology. On *Salix alba* (Le Quesne, 1965). I found *I. herrichii* on *Salix pentandra*. Adults in May, June, July, August, September, October.

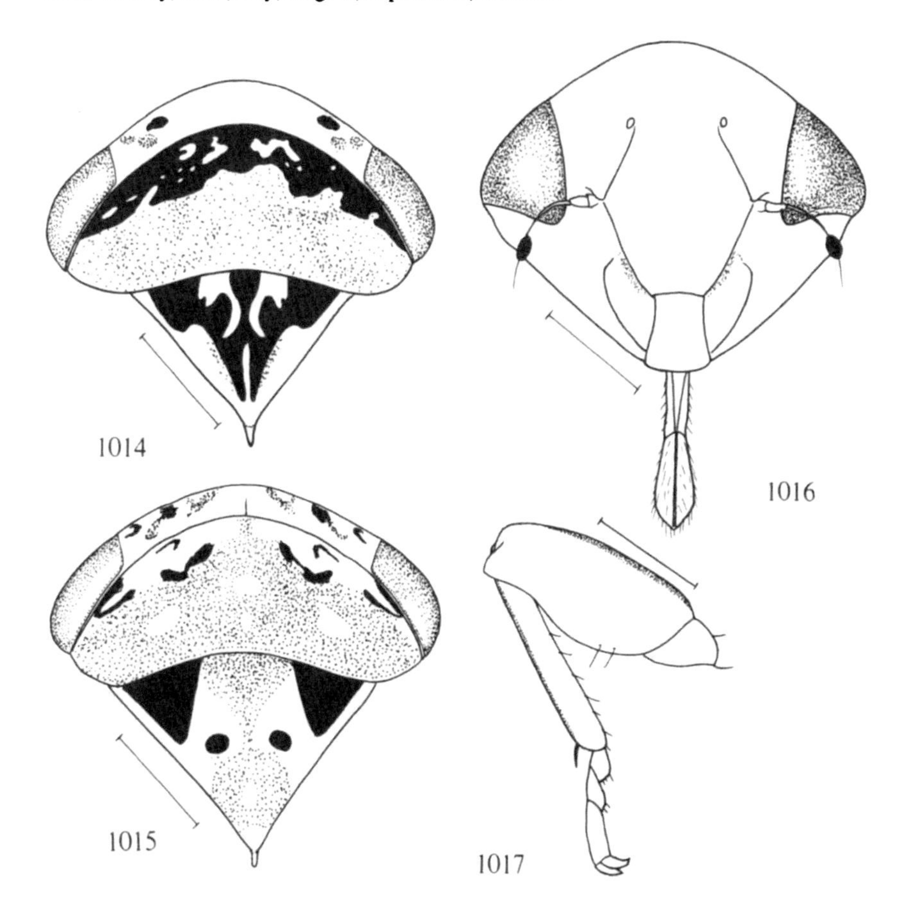

Text-figs. 1014–1017. *Metidiocerus crassipes* (J. Sahlberg). - 1014: head, pronotum and scutellum of male from above; 1015: head, pronotum and scutellum of female from above; 1016: face of male; 1017: right middle leg of male from below. Scale: 0.5 mm.

Genus *Metidiocerus* Ossiannilsson, new genus

Type-species: *Idiocerus crassipes* J. Sahlberg, 1871.

Face with eyes broader than long. Genae not pilose or indistinctly pilose. Frontoclypeus hexagonal (Text-fig. 1016). Distance between ocelli not or little greater than distance between one ocellus and the nearest compound eye. Vertex anteriorly less broadly rounded in *Idiocerus,* medially not or little shorter than near eyes (Text-figs. 1014, 1015). Costal border of fore wing not deflected or deflected part very narrow. In Fennoscandia two species; in addition to these also the non–Scandinavian species *impressifrons* Kirschbaum and *rutilans* Kirschbaum seemingly belong to this genus.

Key to species of *Metidiocerus*

1 Fore wings without distinct transverse bands. Vertex of male without a median black spot. Middle femora of male (Text-fig. 1017) strongly swollen, their maximal width approximately equals width of an eye.. 132. *crassipes* (J. Sahlberg)

– Fore wings distally of middle with a light transverse band (Plate-fig. 68). Vertex of male with a median black spot. Middle femora of male less swollen... 133. *elegans* (Flor)

132. ***Metidiocerus crassipes*** (J. Sahlberg, 1871) comb. n.
Text-figs. 1014–1025.

Idiocerus crassipes J. Sahlberg, 1871:143.

Superficially resembling a small *Idiocerus lituratus,* easily distinguished by characters given in the key to genera. Head as seen from above rounded obtuse angular (Text-figs. 1014, 1015). Antennae short, in male with a large palette (Text-fig. 1016). Middle femora swollen also in female though less strongly than in male. Pigmentation on dorsum of fore part of body more or less as in Text-figs. 1014 (♂) and 1015 (♀), but owing to the rarity of this species, its variation has not been studied. Face of male pale yellow, of female orange mottled. Dorsum of fore body pale yellow (♂) or light yellow to orange yellow (♀) with dark markings, pronotum with some small milk-coloured spots. Fore wings without distinct transverse bands; the black or brownish colour of their veins is interrupted in a few small milk-coloured spots. As usual there is a somewhat larger white spot surrounding the distal end of the longer claval vein. Thoracal venter and

---

Text-figs. 1018–1025. *Metidiocerus crassipes* (J. Sahlberg). – 1018: male pygofer from above, venter not considered; 1019: male genitalia from the left; 1020: right genital plate from outside; 1021: left genital style and connective from below; 1022: aedeagus from the right; 1023: aedeagus in ventral aspect; 1024: 1st abdominal sternum in male from above; 1025: 2nd abdominal sternum in male from above. Scale: 0.5 mm for 1018 and 1019, 0.2 mm for 1024, 0.1 mm for the rest.

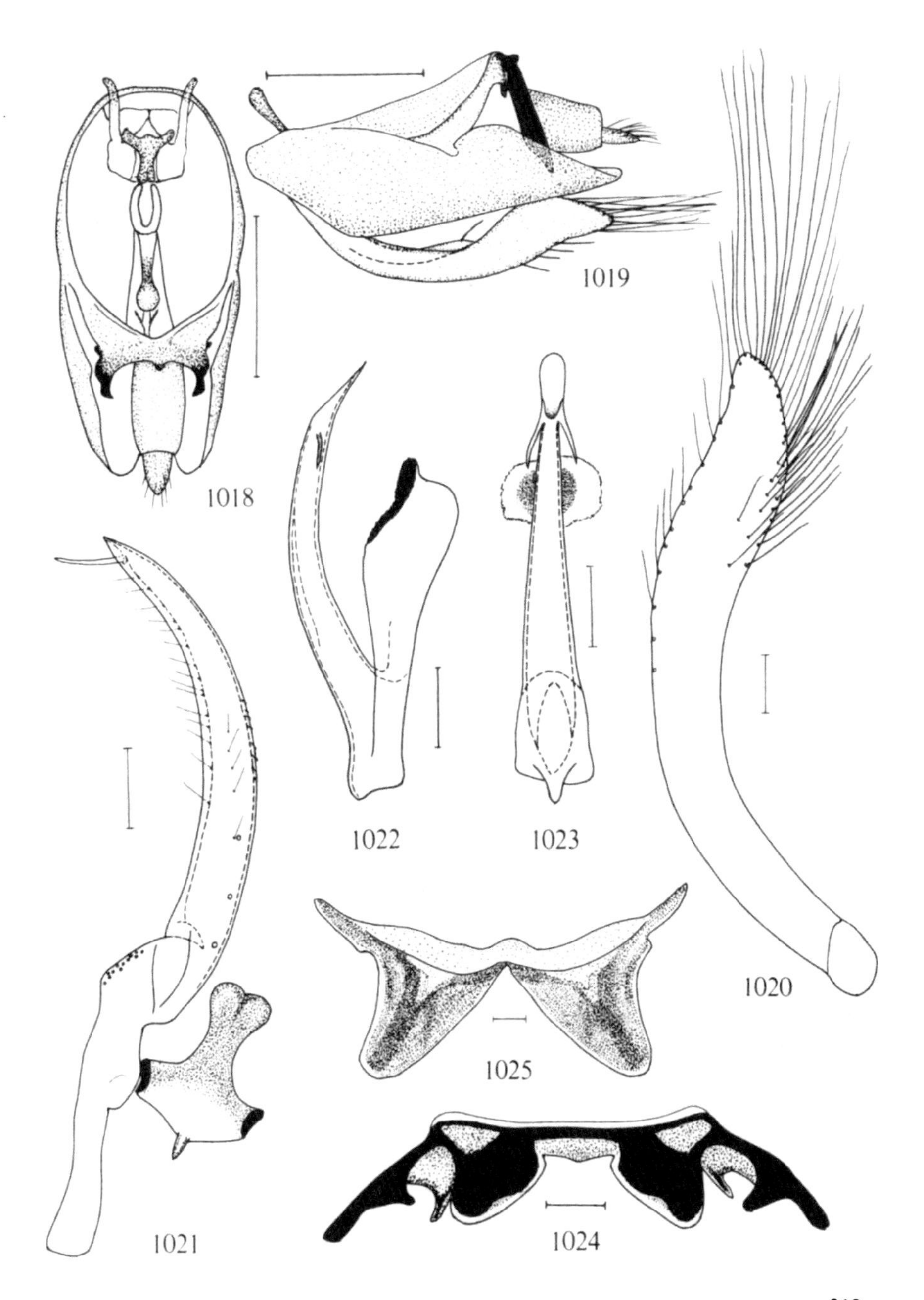
1018
1019
1020
1021
1022
1023
1024
1025

abdomen largely black or partly light, legs pale yellow or orange yellow with black longitudinal streaks. Male pygofer as in Text-figs. 1018, 1019, genital plate as in Text-fig. 1020, genital style and connective as in Text-fig. 1021, aedeagus as in Text-figs. 1022, 1023, 1st and 2nd abdominal sterna in male as in Text-figs. 1024, 1025. Overall length ♂♀ 5–5.2 mm.

Distribution. Not in Denmark, nor in Norway so far. – Rare in central and northern Sweden. Found in Dlr.: Sollerön (Tullgren), Floda, Sångån (T. Tjeder), Rättvik, Glisstjärn (T. Tjeder), Rättvik, Råberget and Ångtjärnen (T. Tjeder), Orsa, Läftomäck, Kvarnborg (T. Tjeder); Hls.: Edsbyn, Hobergstjärnsvägen, Acktjärn, and Börnasbo (Bo Henriksson); Ång.: Sollefteå, Fanby (E. Runquist); Nb.: Boden (F. Nordström). I have seen in all 6 ♂♂ and 7 ♀♀ of this very rare species. – East Fennoscandia: only found in Kr: Tiudie 14. VIII. 1869 by J. Sahlberg. – Outside Fennoscandia only recorded from Olonets in nw. Russia.

Biology. On *Salix caprea*, according to Sahlberg (1871). Our adults were collected in May, June, August, and September.

133. ***Metidiocerus elegans*** (Flor, 1861) comb. n.
Plate-fig. 68, text-figs. 1026–1034.

Idiocerus elegans Flor, 1861a:174.
Idiocerus frontalis Melichar, 1896:160.

Male antenna with a small palette. Venter largely yellowish white or yellow with orange tinge, dorsum brownish with black and milk-white markings. Head of male medially with an oblong black spot on fore border; head of female marbled with reddish or brownish yellow, yellowish white and blackish brown, above often with a 5- or 4-radiated figure and laterally of this figure two roundish black spots on thyridia. These markings may be more or less extended. Pronotum brownish with some pale whitish spots among which an incomplete median longitudinal band is noticeable; anterior part of pronotum in male often largely black, in female dotted with black. Fore wings brown or red-brown with darker veins (Plate-fig. 68), basally lighter, distally of middle with a whitish transverse band. Within this band as well as in the usual light spot at the distal end of the longer claval vein, and in a small spot on Cu level with the claval spot, the veins are whitish. Legs yellowish with fuscous longitudinal streaks. In both sexes the middle femora are thicker than fore and hind femora. Male pygofer as in Text-figs. 1026, 1027, genital plate as in Text-fig. 1028, genital style and connective as in Text-fig. 1029, aedeagus as in Text-figs. 1030, 1031, 1st abdominal sternum in male as in Text-fig.

Text-figs. 1026–1031. *Metidiocerus elegans* (Flor). – 1026: male pygofer from above, genital plates not considered; 1027: male genitalia from the left; 1028: left genital plate from outside; 1029: right genital style and connective from above; 1030: aedeagus from the right; 1031: aedeagus in ventral aspect. Scale: 0.1 mm.

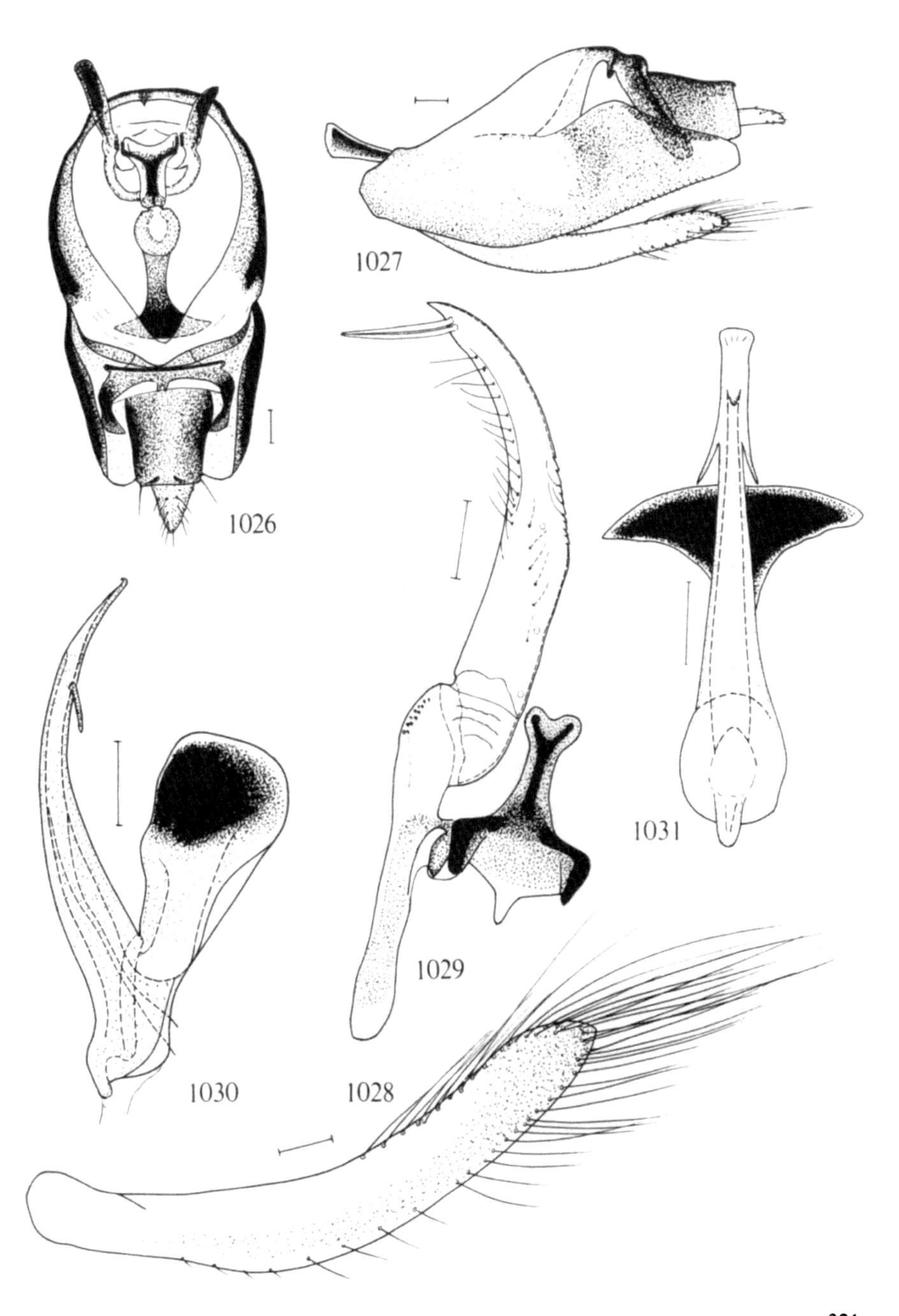
1026
1027
1028
1029
1030
1031

1032, 2nd abdominal sternum in male as in Text-fig. 1033, 3rd abdominal tergum in male as in Text-fig. 1034. Overall length: ♂ 5.2–5.5 mm, ♀ 5.5–6.0 mm. – Last instar larva brownish yellow with more or less extensive black markings, fore body without long hairs, abdominal terga each with a few semi-erect black or fuscous setae, lateral setae longer than those on median part of abdomen.

Distribution. Denmark: so far only one record from B: Bastemose 15. VII. 1978 (Trolle). – Widespread and comparatively common in Sweden, Sk.–Lu.Lpm. – Norway: so far recorded only from AK, Bø, and TEi. – Fairly common in southern and central East Fennoscandia, up to ObN and Ks, also in Vib and Kr. – Austria, Belgium, Bohemia, Slovakia, France, German D.R. and F.R., England, Scotland, Wales,

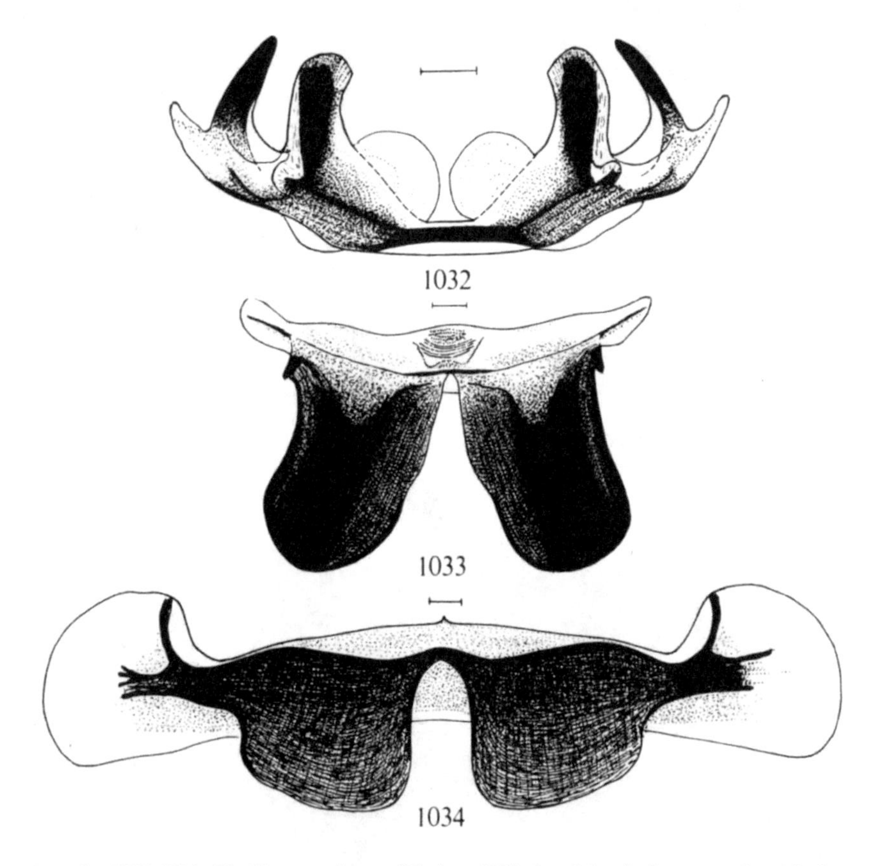

Text-figs.1032–1034. *Metidiocerus elegans* (Flor). – 1032: 1st abdominal sternum in male from behind; 1033: 2nd abdominal sternum in male from above; 1034: 3rd abdominal tergum in ventral aspect. Scale: 0.1 mm.

Netherlands, Italy, Portugal, Poland, Estonia, Latvia, Lithuania, n. Russia, Ukraine, Tunisia, Kazakhstan, Mongolia.

Biology. On *Salix*, especially *S.caprea*, *S.cinerea*, and *S.purpurea* (Le Quesne, 1965). I collected *M. elegans* on *Salix aurita*, *myrsinifolia*, and *lapponum*. Adults in April, July, August, September, and October.

Genus *Populicerus* Dlabola, 1974

Populicerus Dlabola, 1974:62.
Type-species: *Cicada populi* Linné, 1761, by original designation.

Face comparatively narrow, lateral margins together making an angle of less than 90°. Vertex medially shorter than near eyes. Eyes only slightly inflated. Face slightly convex, below without whitish pilosity. Genae laterally below eyes with a few short hairs. Frontoclypeus oblong, oval, almost twice as long as broad. Distance between ocelli only slightly longer than between ocellus and eye. Male antenna with palette. Fore wing normally with three subapical cells. In Denmark and Fennoscandia 5 species.

Key to species of *Populicerus*

1 Fore wing with a light transverse band distally of middle. Ventral margin of male genital plate distally strongly concave (Text-fig. 1049) .... 135. *laminatus* (Flor)
– Fore wing without a transverse band. Ventral margin of male genital plate distally not or weakly concave 2
2(1) Fore wing milky white or tinged greenish, not shiny, with or without smoky shades. On *Populus alba* .... 138. *albicans* (Kirschbaum)
– Fore wings strongly shiny 3
3(2) Fore wings and pronotum partly or largely fuscous. Female ovipositor projecting beyond pygofer by about 1 ½ times its own width. On *Salix* .... 137. *confusus* (Flor), f. *nigricans* (Ossiannilsson)
– Fore wings not fuscous 4
4(3) Dorsum pale yellow or pale greenish yellow. Scutellum usually – not always – entirely light. Female ovipositor projecting beyond pygofer by about 1 ½ times its own width (Text-fig. 1080). On *Salix* .... 137. *confusus* (Flor)
– Fore wings yellowish brown or smoke-brown. Female ovipositor projecting beyond pygofer by about its own width (Text-fig. 1044). On *Populus* 5
5(4) Veins in apical part of fore wing fuscous. On *Populus nigra* 136. *nitidissimus* (Herrich-Schäffer)
– Veins in apical part of fore wing not conspicuously darker than in rest of wing. On *Populus tremula* .... 134. *populi* (Linné)

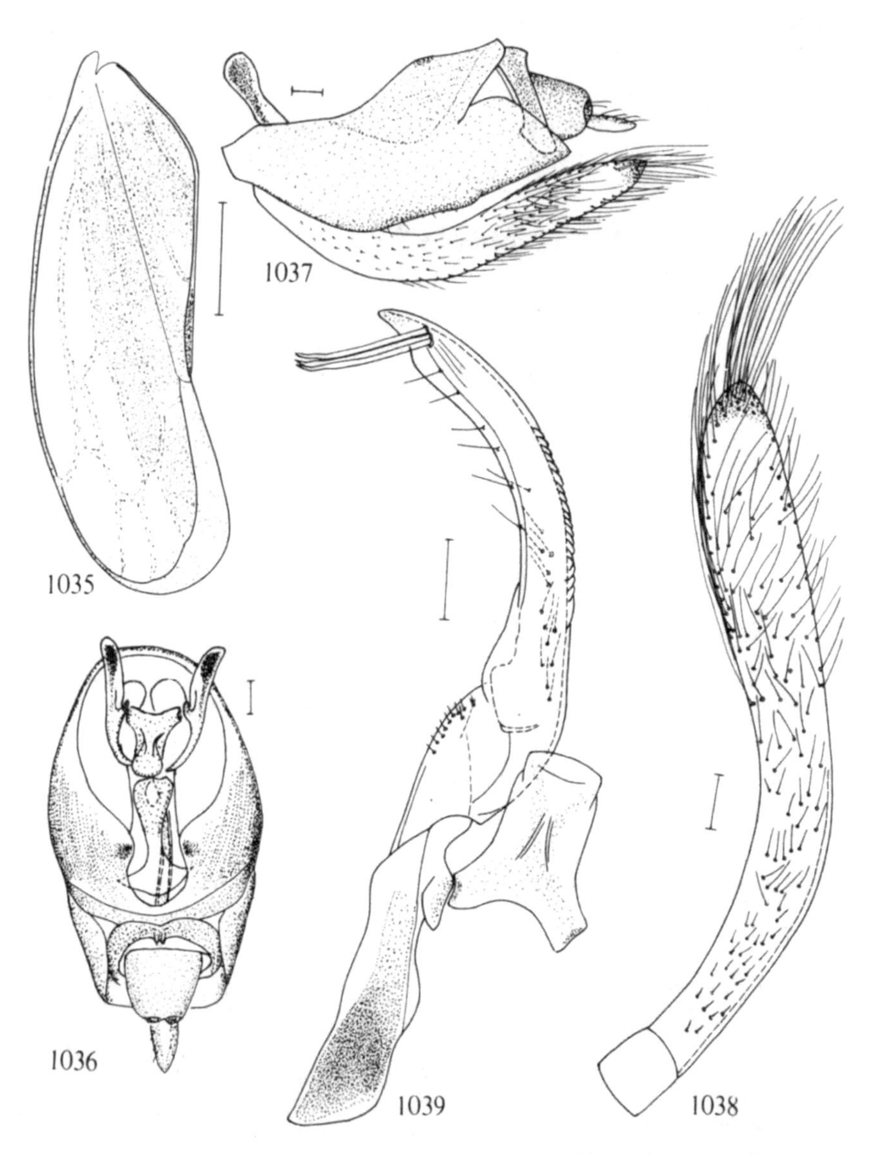

Text-figs. 1035–1039. *Populicerus populi* (Linné). – 1035: left fore wing; 1036: male pygofer from above, sternal parts not considered; 1037: male genital segment from the left; 1038: left genital plate from the left; 1039: right genital style with connective from above. Scale: 1 mm for 1035, 0.1 mm for the rest.

134. ***Populicerus populi*** (Linné, 1761)
Plate-fig. 79, text-figs. 6, 14, 1035–1046.

Cicada populi Linné, 1761:242.

Male antenna with palette (Text-fig. 6). Bright yellow, yellowish white, or greenish. Face often with a broad orange median band. Sometimes with a pair of small dark spots near eyes. Pronotum yellowish brown or leather brown, often with a few small dark spots along fore border. Scutellum with a pair of triangular black spots on lateral corners. Fore wings hyaline, yellowish brown with a strong golden lustre, veins not or only slightly darker. Veins little protruding (Text-fig. 1035), transverse veins often more or less obsolete, seemingly reducing the number of subapical cells (cf. Text-fig. 14). A whitish spot in clavus on the distal end of the first vein. Legs in male often reddish yellow. Venter light, male genital segment partly black, abdomen above largely dark. Genital plates of male (Text-fig. 1038) not broader than hind tibiae and not or only slightly longer than hind femora, apically indistinctly darker. Male pygofer as in Text-figs. 1036, 1037, genital style as in Text-fig. 1039, aedeagus as in Text-fig. 1042, 3rd abdominal tergum in male as in Text-fig. 1043. Female ovipositor projecting beyond pygofer by about its own width (Text-fig. 1044), 2nd valvulae of ovipositor as in Text-figs. 1045 and 1046. Overall length of male 5.0–5.55 mm, of female 4.9–6.5 mm. - 5th instar larva pale yellow to light green to light flesh-coloured with black markings on head, thorax and abdomen, sometimes entirely light green, dorsum hairless.

Distribution. Common and widespread in Denmark as well as in Sweden (Sk.–Nb.), and in Norway (up to NTy). - Very common in southern and central East Fennoscandia, up to ObN, also in Vib and Kr. - Widespread in Europe, found also in Algeria, Palestine, Altai Mts., Azerbaijan, Kazakhstan, Sakhalin, m. and w. Siberia, Manchuria, Maritime Territory, Japan.

Biology. On *Populus tremula*. Hibernation takes place in the egg stage (Müller, 1957). Adults July - October.

135. ***Populicerus laminatus*** (Flor, 1861)
Text-figs. 1047–1057.

Idiocerus laminatus Flor, 1861a:171.

Male antenna with palette. Light yellow or yellowish white. Head entirely light with brownish thyridia, or face with indistinct pale brownish patches. Pronotum greyish yellow to greyish brown. Scutellum with two triangular black spots and usually a diffusely marked brownish median line varying in width on scutum. Fore wings shining smoke-brown with dark veins and a whitish spot round the distal end of the first claval vein and a diffusely limited whitish transverse band distally of the claval apex. Within these light surfaces the veins are whitish. Legs with or without dark longitudinal streaks. Venter entirely or largely light, male pygofer dark. Male pygofer as in Text-fig.

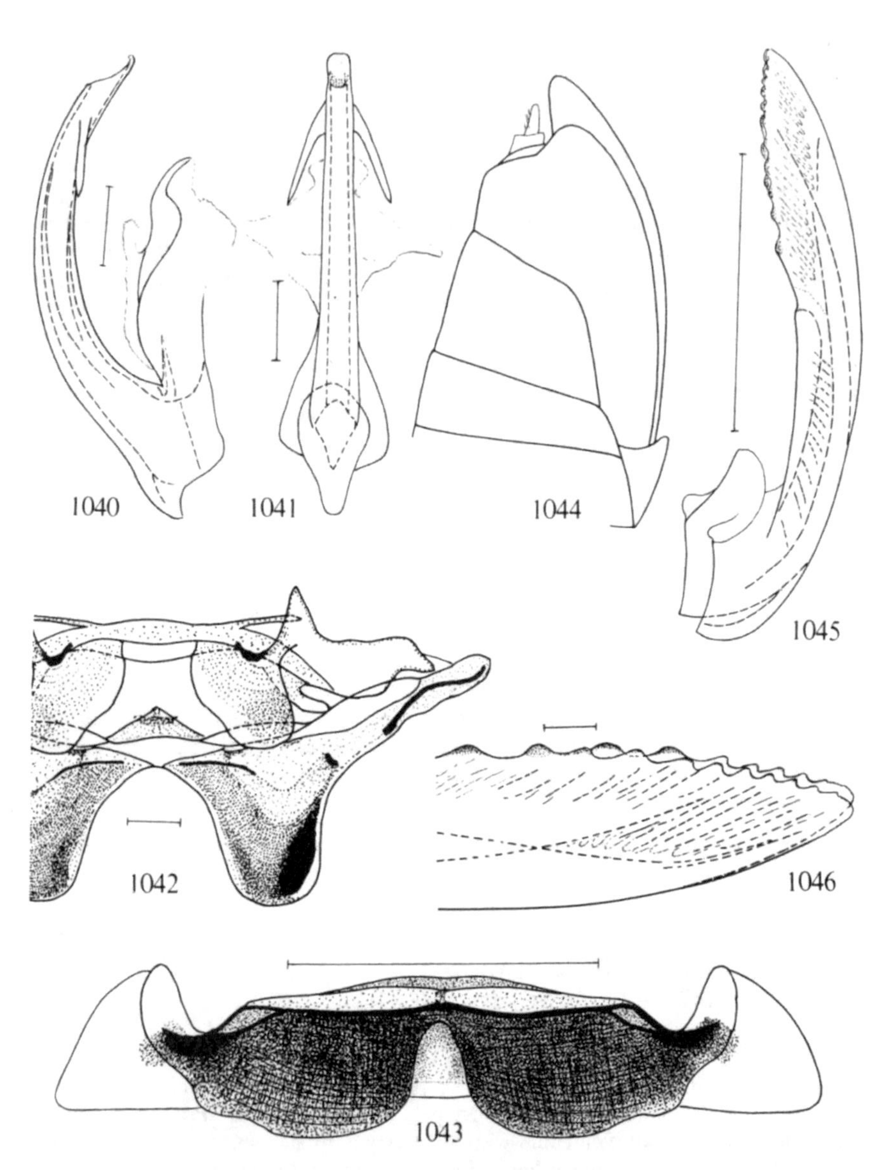

Text-figs. 1040–1046. *Populicerus populi* (Linné). – 1040: aedeagus from the right; 1041: aedeagus in ventral aspect; 1042: 1st and 2nd abdominal sterna in male from above; 1043: 3rd abdominal tergum in male from below; 1044: apex of female abdomen from the left; 1045: 2nd valvulae of ovipositor from the left; 1046: apex of 2nd valvulae of ovipositor from the left. Scale: 1 mm for 1043–1045, 0.1 mm for the rest.

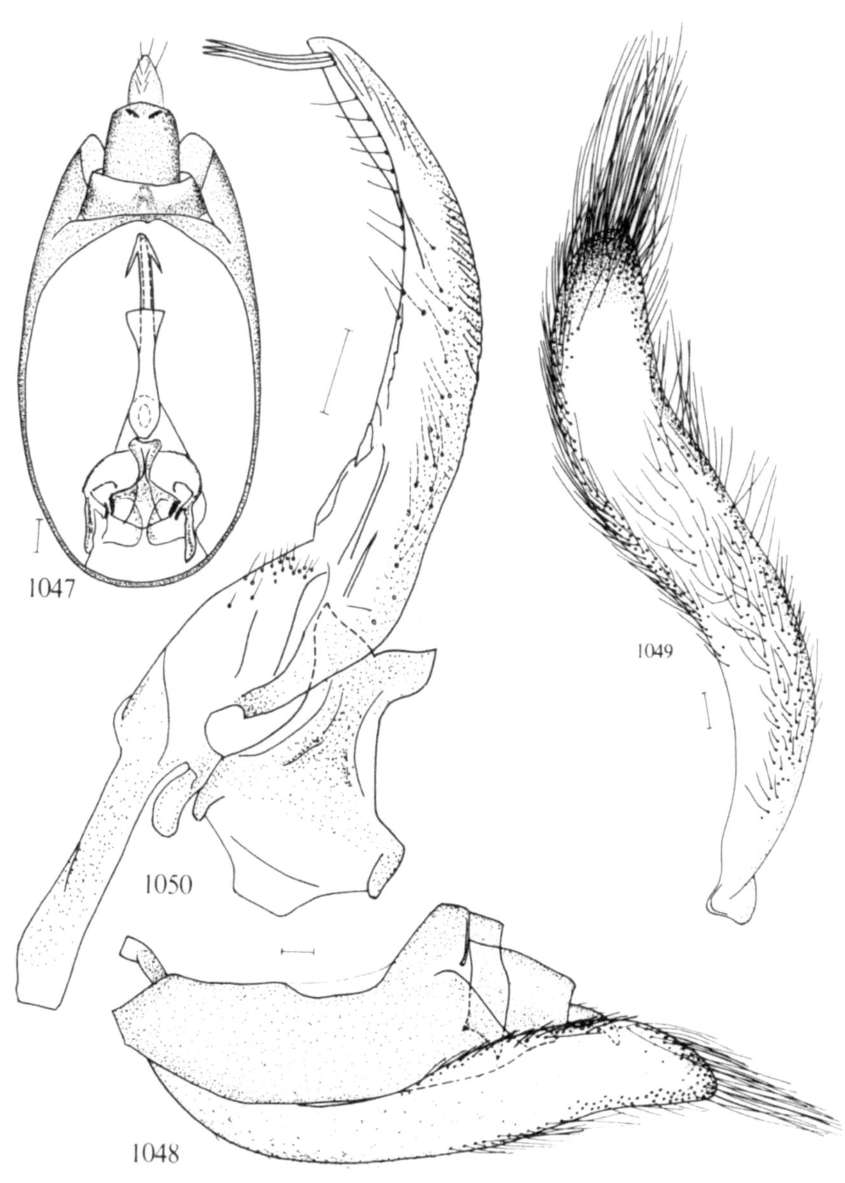

Text-figs. 1047–1050. *Populicerus laminatus* (Flor). – 1047: male pygofer from above, sternal parts not considered; 1048: male genital segment from the left; 1049: left genital plate from the left; 1050: right genital style with connective from above. Scale: 0.1 mm.

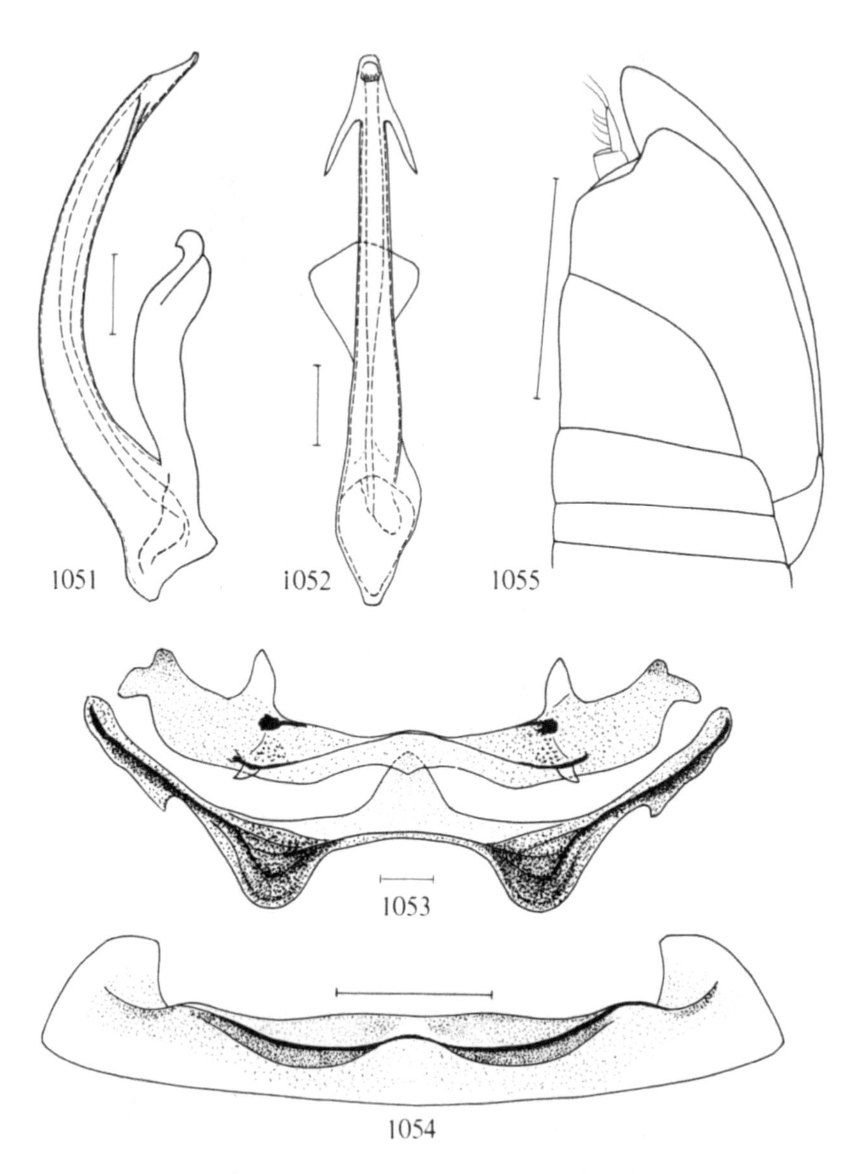

Text-figs. 1051–1055. *Populicerus laminatus* (Flor). – 1051: aedeagus from the right; 1052: aedeagus in ventral aspect; 1053: 1st and 2nd abdominal sterna in male from above; 1054: 3rd abdominal tergum in male from below; 1055: apex of female abdomen from the left. Scale: 1 mm for 1055, 0.5 mm for 1054, 0.1 mm for the rest.

1047, 1048, genital plate as in Text-fig. 1049, genital style and connective as in Text-fig. 1050, aedeagus as in Text-figs. 1051, 1052, 1st and 2nd abdominal sterna in male as in Text-fig. 1053, 3rd abdominal tergum in male as in Text-fig. 1054. Caudal part of female abdomen as in Text-fig. 1055, second valvulae of ovipositor as in Text-figs. 1056, 1057. Overall length of male 5.8–6 mm, of female 6.1–7 mm.

Distribution. Not uncommon in Denmark (SJ, EJ,NEJ, F, LFM) and in Sweden (Sk.–Lu.Lpm.). – Probably not uncommon in southern Norway, so far found in AK, Os, On, Bø, TEi, and Ry. -Scarce in southern and central East Fennoscandia, recorded from Al, Ab, N, St, Ta, Sb; Vib, Kr. – Austria, Belgium, Slovakia, France, German D.R. and F.R., England, Scotland, Italy, Spain, Switzerland, Latvia, n. and m. Russia, Ukraine, Mongolia.

Biology. On *Populus tremula* and *P. canescens*. Adults in June – October.

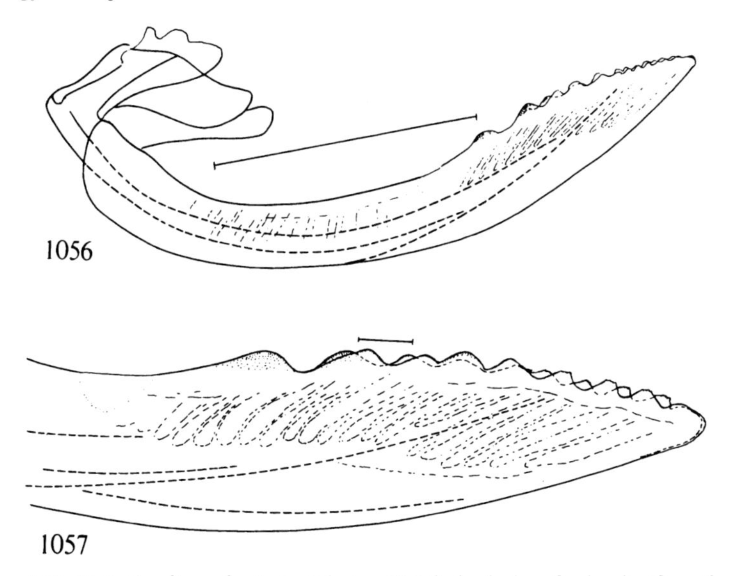

Text-figs. 1056, 1057. *Populicerus laminatus* (Flor). – 1056: 2nd valvulae of ovipositor from the left; 1057: apex of same. Scale: 1 mm for 1056, 0.1 mm for 1057.

136. ***Populicerus nitidissimus*** (Herrich-Schäffer, 1835)
Text-figs. 1058–1067.

Bythoscopus nitidissimus Herrich-Schäffer, 1835:68.
Idiocerus fulgidus auct., including Ossiannilsson, 1946c:101, nec Fabricius, 1775.

Male antenna with palette. Resembling *P. populi*, less strongly shining(!), in life often

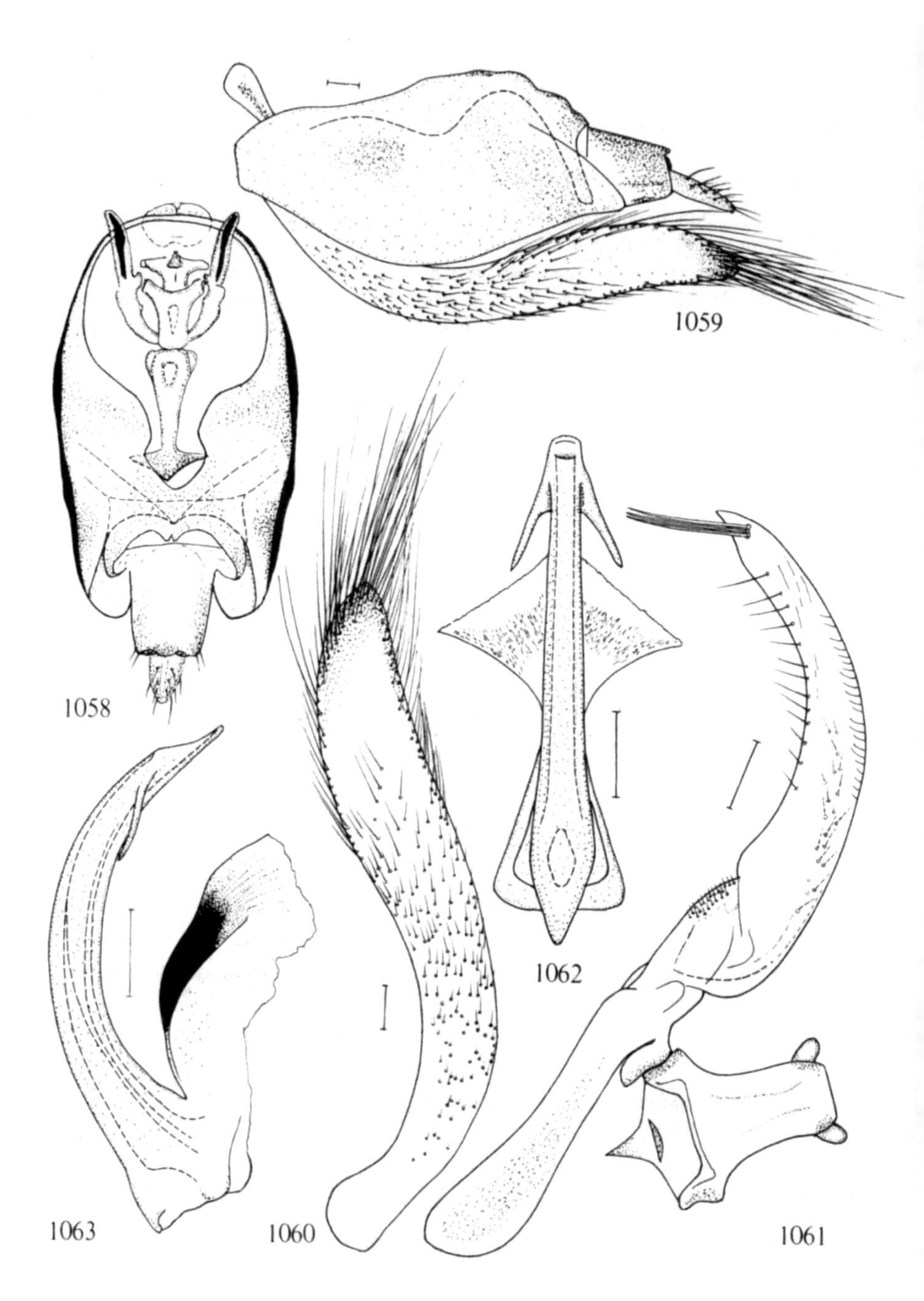
1059
1058
1062
1063
1060
1061

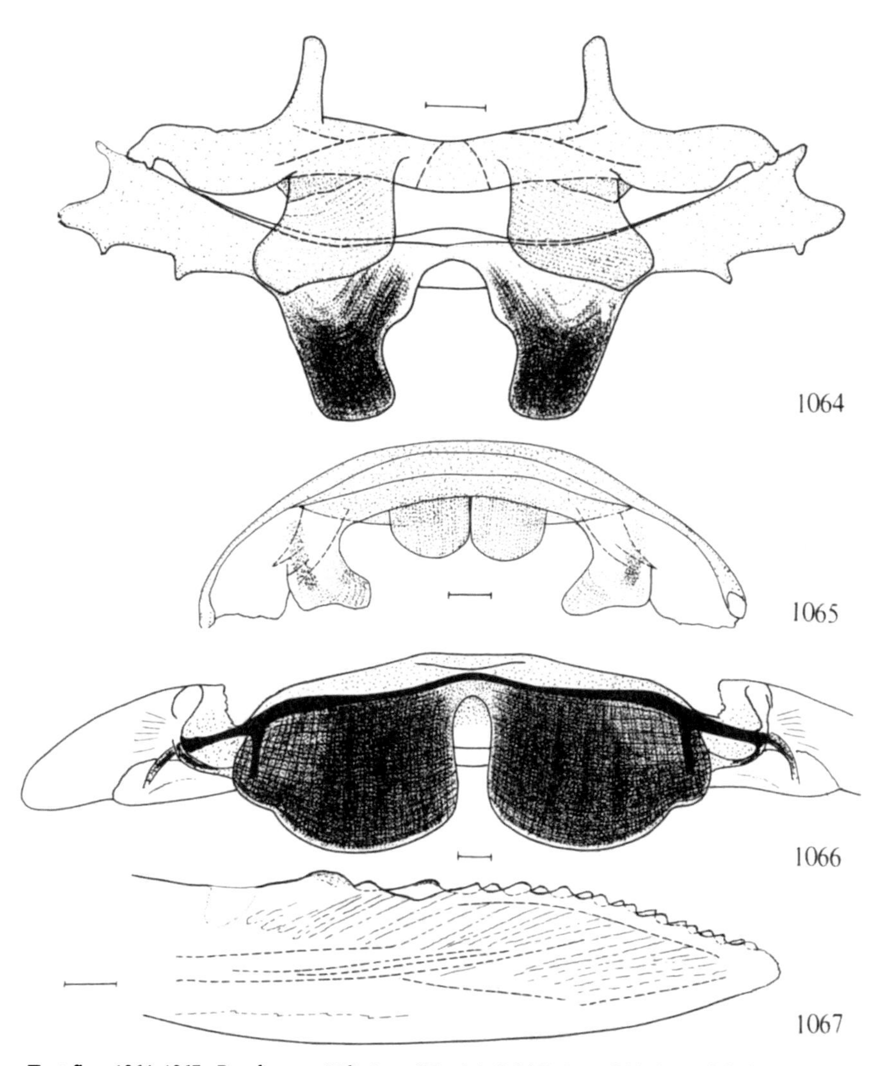

Text-figs. 1064–1067. *Populicerus nitidissimus* (Herrich-Schäffer). - 1064: 1st and 2nd abdominal sterna in male from above; 1065: 2nd abdominal tergum in male from behind; 1066: 3rd abdominal tergum in male from below; 1067: apex of 2nd valvulae of ovipositor from the left.

Text-figs. 1058–1063. *Populicerus nitidissimus* (Herrich-Schäffer). - 1058: male pygofer from above, genital plates not considered; 1059: male genital segment from the left; 1060: male genital plate from the left; 1061: right genital style and connective from above; 1062: aedeagus in ventral aspect; 1063: aedeagus from the right. Scale: 0.1 mm.

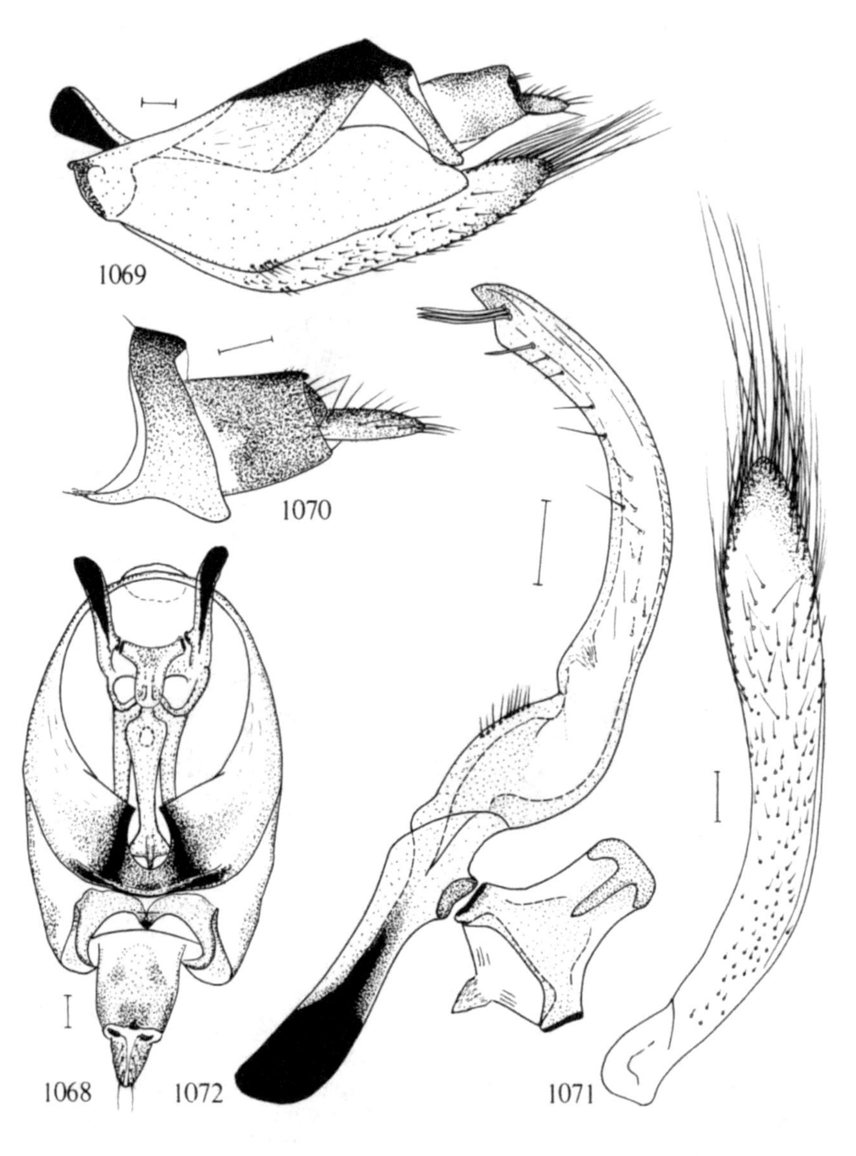

Text-figs. 1068–1072. *Populicerus confusus* (Flor). – 1068: male pygofer in dorsal aspect, genital plates not considered; 1069: male genital segment from the left; 1070: male anal apparatus from the left, left pygofer lobe removed; 1071: left genital plate from outside; 1072: right genital style with connective from above. Scale: 0.1 mm.

greenish, after death yellowish. Side corners of scutellum with or without black spots. Male pygofer as in Text-figs. 1058, 1059, genital plates (Text-fig. 1060) narrow, apically darker, genital styles and connective as in Text-fig. 1061, aedeagus as in Text-figs. 1062, 1063. 1st and 2nd abdominal sterna in male as in Text-fig. 1064, 2nd abdominal tergum in male as in Text-fig. 1065, 3rd abdominal tergum of male as in Text-fig. 1066. Apex of 2nd valvulae of female ovipositor as in Text-fig. 1067. Overall length of male 5.5–6 mm, of female 6.3–6.8 mm. - 5th instar larva more or less as in *populi*.

Note. Ribaut (1952) used the number of strong setae in the subapical group on the male genital style as a character for separating the species later placed in *Populicerus*. This character is untenable in Scandinavian material.

Distribution. So far not recorded from Denmark. - Its distribution in Sweden has been little studied. E. Sylvén caught several specimens in light traps in Sk., Åkarp, in August, 1952. Gyllensvärd collected 5 specimens in Bl., Karlskrona, 12.IX.1971. I found 11 males and 2 females on *Populus nigra italica* in the Bergian Garden in Stockholm (Upl.), 8.IX.1942. The species is usually abundant on the same hostplant in Upl., Uppsala, Kungsängen. - Norway: AK: Tøien, Oslo, 29.VII.1943 (Holgersen). - East Fennoscandia: found in Ab: Åbo by Linnavuori, and in N: Helsingfors 18–20.VIII.1946 by Håkan Lindberg. - Belgium, Bulgaria, Bohemia, Moravia, Slovakia, France, German D.R. and F.R., England, Scotland, Italy, Netherlands, Poland, Romania, Spain, Latvia, n. and m. Russia, Ukraine, Altai Mts., Kazakhstan, m. Siberia, Mongolia.

Biology. On *Populus nigra* and *P. nigra italica*. Hibernation in the egg stage (Müller, 1957). Adults in July - October.

137. ***Populicerus confusus*** (Flor, 1861)
Text-figs. 1068–1086.

Idiocerus confusus Flor, 1861a:179.
Idiocerus confusus nigricans Ossiannilsson, 1942b:114.

Male antenna with palette. Pale green or pale yellow, shining. Scutellum of f., *typica* usually without black spots. Fore wings sometimes with a dark tinge in clavus, especially in males. Veins in apical part of male fore wings somewhat darker. Tergum of male abdomen black with light segment borders. The male of f. *nigricans* (Ossiannilsson) is dorsally largely fuscous, the female considerably lighter but with lateral spots on scutellum. Male pygofer as in Text-figs. 1068, 1069, 1083, male anal apparatus as in Text-figs. 1070, 1084, genital plate as in Text-figs. 1071, 1085, genital style and connective as in Text-figs. 1072, 1086, aedeagus as in Text-figs. 1073, 1074, 1st abdominal sternum in male as in Text-figs. 1075, 1076, 2nd abdominal sternum in male as in Text-fig. 1077, 2nd abdominal tergum in male as in Text-fig. 1078, 3rd abdominal tergum in male as in Text-fig. 1079. Apex of female abdomen as in Text-fig. 1080, 2nd valvulae of female ovipositor as in Text-figs. 1081, 1082. Overall length of male 5.0–5.9

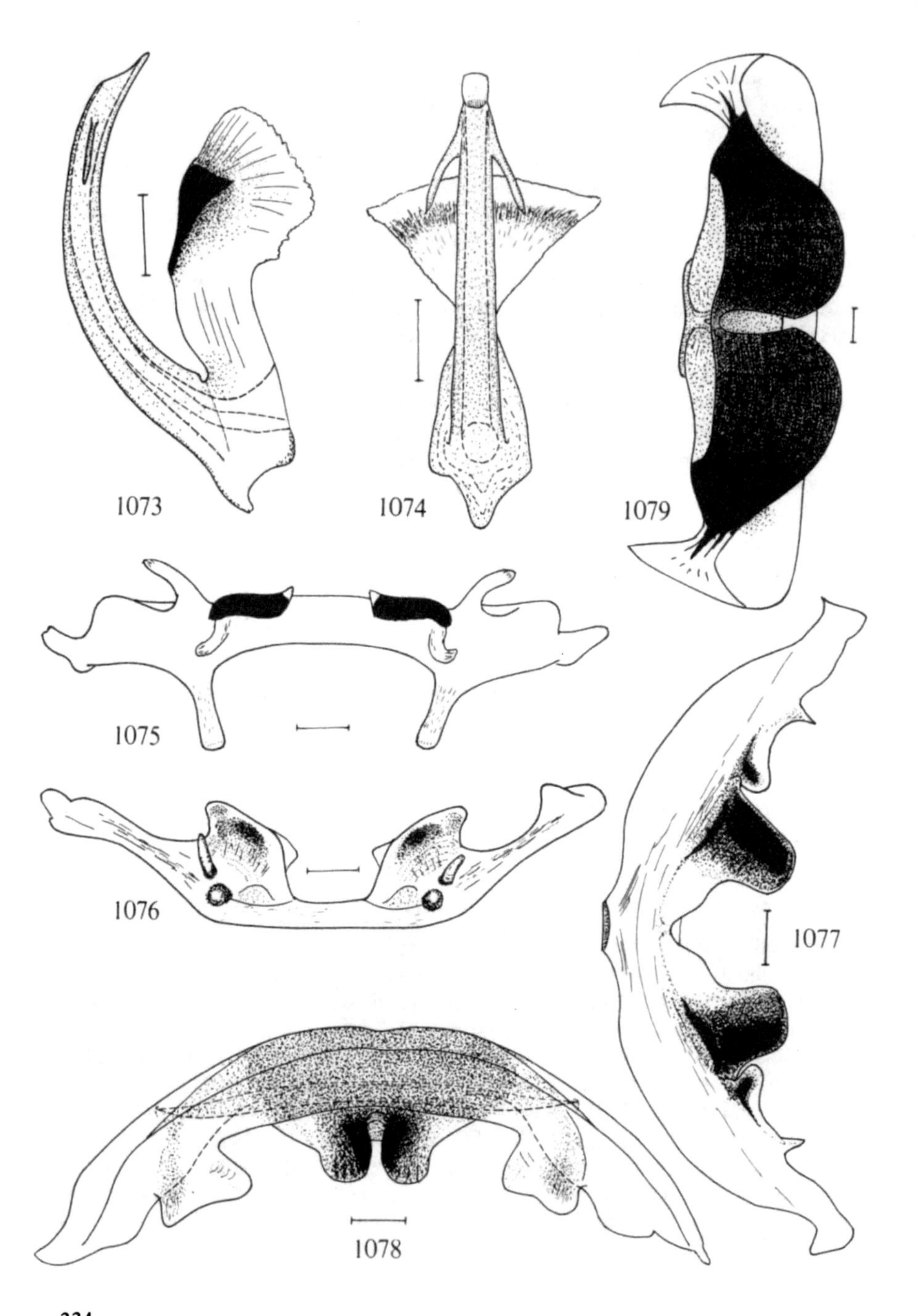
1073
1074
1079
1075
1076
1077
1078

mm, of female 6.1–7.0 mm. – Last instar larva usually light green, dorsum with an indistinct fuscous pattern consisting of narrow transverse streaks near abdominal segment borders, and a longitudinal streak on each wing bud. Pilosity sparse, light, inconspicuous.

Distribution. Common in Denmark and in Sweden up to Nb.–Norway: found in AK, Os, Bø, TEi, AAy, VAy, Ry, HOy, HOi, and SFi. – Common in southern and central

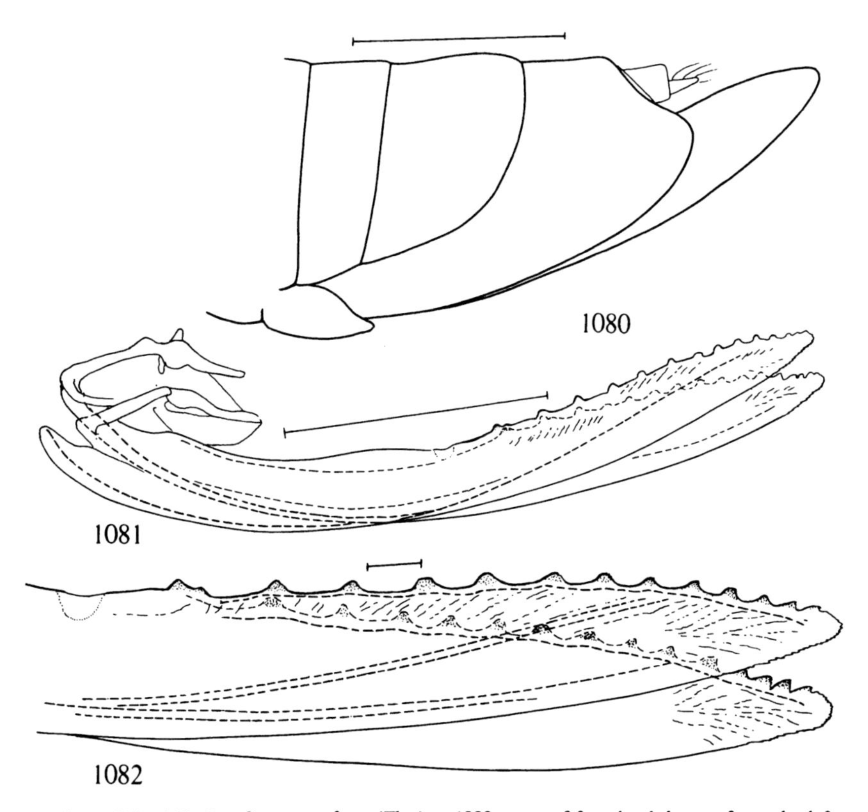

Text-figs. 1080–1082. *Populicerus confusus* (Flor). – 1080: apex of female abdomen from the left; 1081: 2nd valvulae of ovipositor from the left; 1082: apices of same from the left. Scale: 1 mm for 1080 and 1081, 0.1 mm for 1082.

Text-figs. 1073–1079. *Populicerus confusus* (Flor). – 1073: aedeagus from the right; 1074: aedeagus in ventral aspect; 1075: 1st abdominal sternum in male from above; 1076: same from behind; 1077: 2nd abdominal sternum in male from above; 1078: 2nd abdominal tergum in male from behind; 1079: 3rd abdominal tergum in male from below. Scale: 0.1 mm.

East Fennoscandia, up to ObN; also in Vib and Kr. F. *nigricans* has been found in Sweden: Dlr., Jmt., Vb., and Lu.Lpm., and in Finland: Ta. – *P. confusus* is widespread in Europe, also found in Altai Mts., Kazakhstan, Tadzhikistan, m. Siberia, Kirghizia, Mongolia, and Maritime Territory. F. *nigricans* has been recorded from Poland and Austria.

Biology. On *Salix* (e.g. *caprea, cinerea, aurita, myrsinifolia*). Adults in June – September. Hibernation in the egg stage (Müller, 1957).

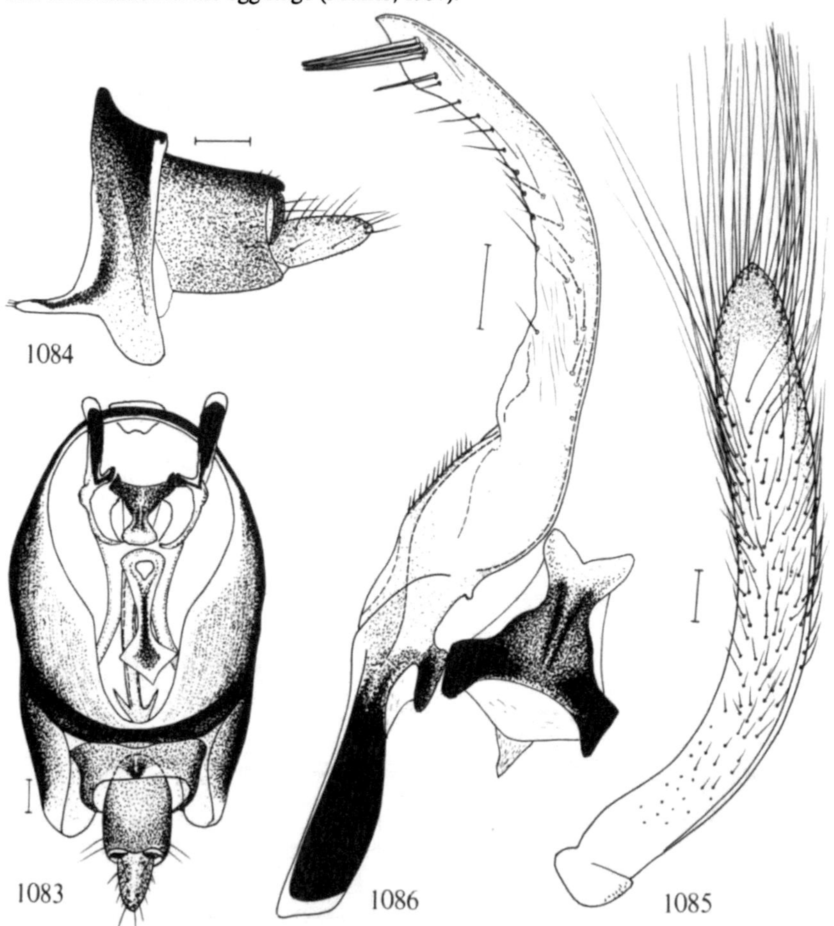

Text-figs. 1083–1086. *Populicerus confusus, f. nigricans* (Ossiannilsson). – 1083: male pygofer in dorsal aspect, genital plates not considered; 1084: male anal apparatus from the left, left pygofer lobe removed; 1085: left genital plate from the left; 1086: right genital style with connective from above. Scale: 0.1 mm.

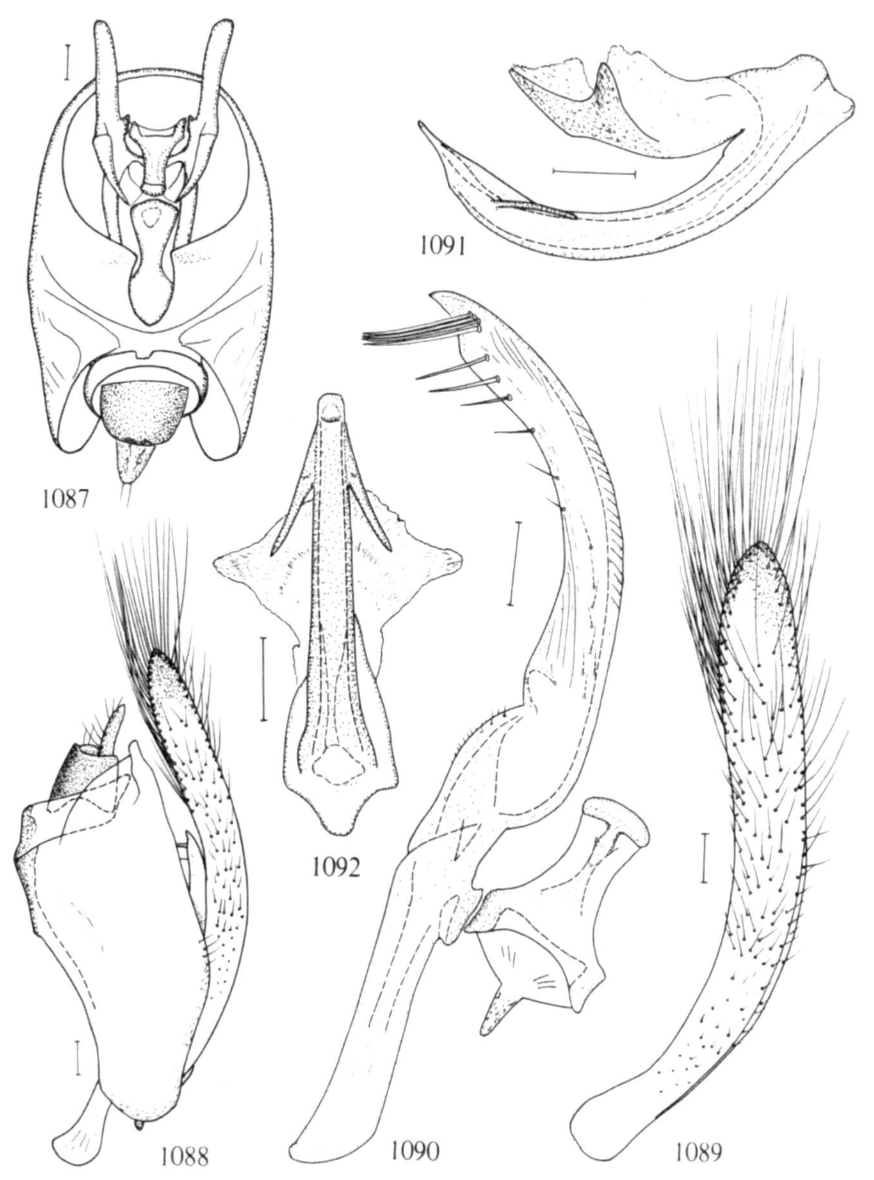

Text-figs. 1087–1092. *Populicerus albicans* (Kirschbaum). – 1087: male pygofer from above, genital plates not considered; 1088: male genital segment from the left; 1089: male genital plate from outside; 1090: genital style and connective from above; 1091: aedeagus from the right; 1092: aedeagus in ventral aspect. Scale: 0.1 mm.

138. ***Populicerus albicans*** (Kirschbaum, 1868)
Text-figs. 1087–1096.

Idiocerus albicans Kirschbaum, 1868a:17.

Male antenna with palette. Faintly shining, pale yellow to whitish, often with a greenish or bluish tinge. Upper side of male often with diffuse dark areas on pronotum, in clavus

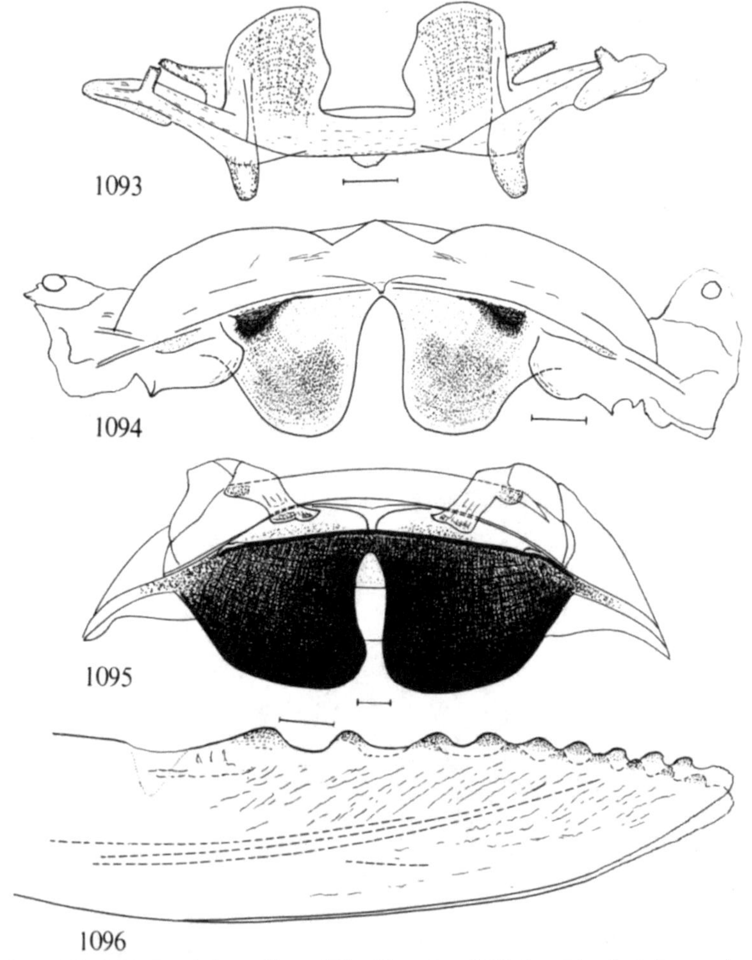

Text-figs. 1093–1096. *Populicerus albicans* (Kirschbaum). – 1093: 1st abdominal sternum in male from behind; 1094: 2nd abdominal sternum in male from above; 1095: 2nd and 3rd abdominal terga in male from below; 1096: apices of 2nd valvulae of ovipositor from the left. Scale: 0.1 mm.

and in apical part of fore wing where the veins usually are fuscous, especially in the male. Scutellum on lateral corners with or without black spots. Abdomen of male dorsally black with light segment margins. Male pygofer as in Text-figs. 1087, 1088, genital plates as in Text-fig. 1089, genital style and connective as in Text-fig. 1090, aedeagus as in Text-figs. 1091, 1092, 1st and 2nd abdominal sterna in male as in Text-figs. 1093, 1094, 2nd and 3rd abdominal terga in male as in Text-fig. 1095. Apex of 2nd valvulae of female ovipositor as in Text-fig. 1096. Overall length of male 5.4–5.9 mm, of female 6.4–7.0 mm. – 5th instar larva pale yellow, hairs on wing buds and abdominal dorsum short, concolorous, inconspicuous, abdominal segments only laterally with some a little longer light hairs.

Distribution. Locally common in Denmark, found in SJ, EJ, F, LFM, NEZ. – Locally common in the south of Sweden, found in Sk., Bl., Hall., and Öl. – Not found in Norway and East Fennoscandia. – Austria, Bohemia, Moravia, Slovakia, France, German D.R. and F.R., England, Wales, Ireland, Netherlands, Italy, Spain, n. Yugoslavia, Poland, Latvia, m. and s. Russia, Ukraine, Hungary, Moldavia, Kazakhstan, Tadzhikistan; Nearctic region.

Biology. On *Populus alba,* adults July – September.

Genus *Tremulicerus* Dlabola, 1974

Tremulicerus Dlabola, 1974:63.
Type-species: *Idiocerus fasciatus* Fieber, 1868, by original designation.

Comparatively small, narrow species. Face convex, not flattened, about as broad as long (Text-fig. 1097). Frontoclypeus oblong, moderately broad, in some species shorter, in others longer than broad. Vertex medially as long as laterally, sometimes shorter. Male antenna without palette (in our species). Side margins of face almost straight, together making an angle of less than 90°. Maxillar plates apically protracted, apex angular (see Text-fig. 1097, at arrow) or rounded, more distinctly angular in males. Fore wings with three subapical cells. Male genital style without strong subapical setae. Three species have been recorded from Denmark and Fennoscandia.

Key to species of *Tremulicerus*

1 Basal part of antennal flagellum fuscous. Scutellum in male with black triangular lateral spots. Thyridia pale in male, dark in female .... 140. *vitreus* (Fabricius)
– Antennal flagellum entirely light 2
2(1) Thyridia pale. Fore wings across middle with a broad well-defined dark transverse band reaching both costal margin and commissural border (Plate-fig. 70) .... 139. *tremulae* (Estlund)
– Thyridia dark in both sexes. Transverse band on fore wing less well-defined, not reaching costal margin 141. *distinguendus* (Kirschbaum)

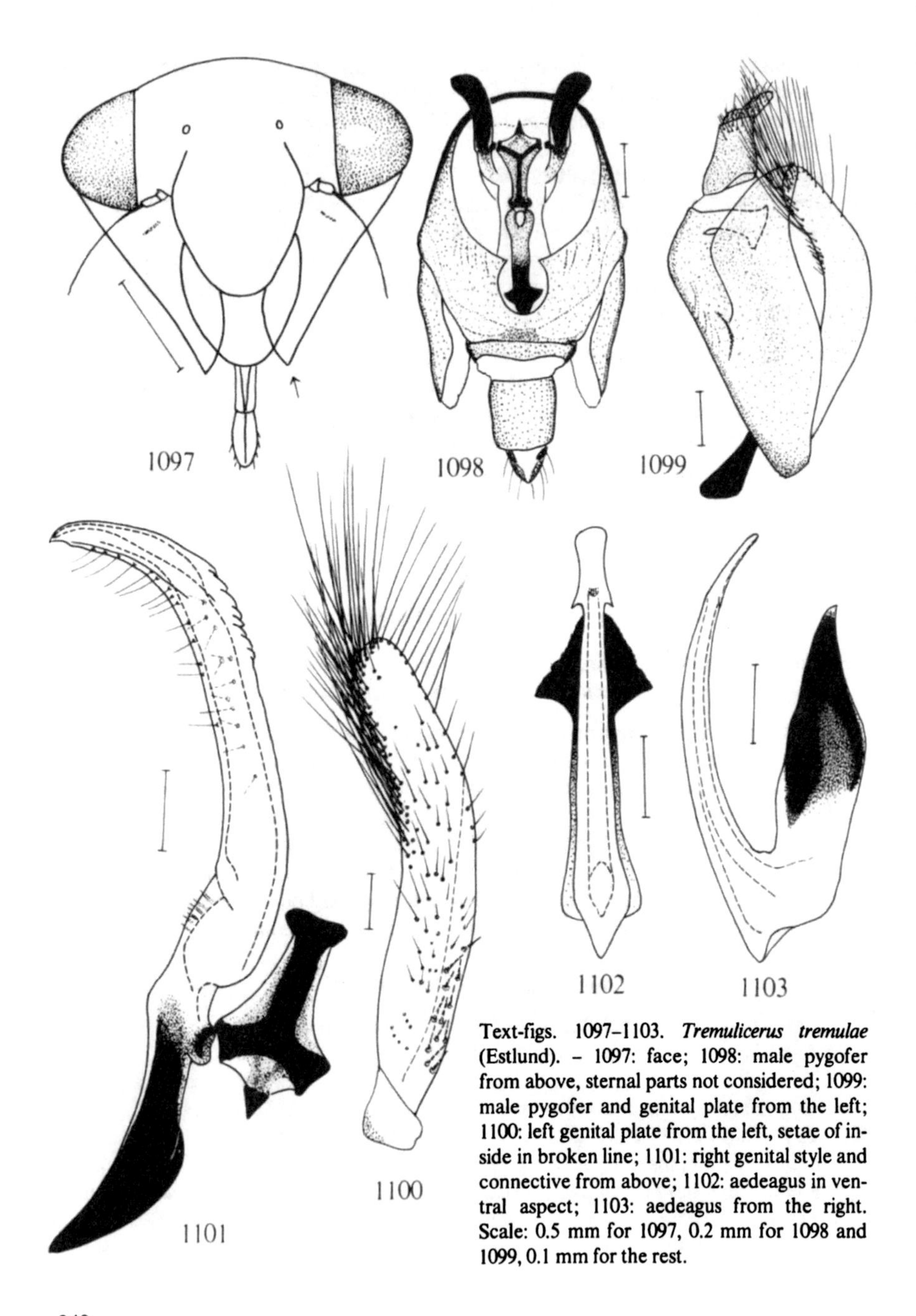

Text-figs. 1097–1103. *Tremulicerus tremulae* (Estlund). – 1097: face; 1098: male pygofer from above, sternal parts not considered; 1099: male pygofer and genital plate from the left; 1100: left genital plate from the left, setae of inside in broken line; 1101: right genital style and connective from above; 1102: aedeagus in ventral aspect; 1103: aedeagus from the right. Scale: 0.5 mm for 1097, 0.2 mm for 1098 and 1099, 0.1 mm for the rest.

139. ***Tremulicerus tremulae*** (Estlund, 1796)
Plate-fig. 70, text-figs. 1097–1106.

Cicada tremulae Estlund, 1796:129.

Body whitish to lively yellow. Head of male usually entirely light, face in females above with a marking consisting of a pair of light spots more or less broadly framed in dark brown patches laterally of ocelli. Pronotum with four diffusely defined brown spots, scutellum with two triangular brown spots at its lateral corners and usually a longitudinal oblong spot between these. Fore wings brownish with two whitish transverse bands within which also veins are white. The dark area between these bands reaches the costal margin. Legs pale. Male pygofer as in Text-figs. 1098, 1099, genital plate as in Text-fig. 1100, genital style and connective as in Text-fig. 1101, aedeagus as in Text-figs. 1102, 1103. 1st and 2nd abdominal sterna in male as in Text-figs. 1104, 1105. 2nd abdominal tergum with a pair of small semilunar unpigmented phragma lobes. 3rd abdominal tergum in male as in Text-fig. 1106. Overall length 4.7–5.0 mm, of female 5.1–6.0 mm. – 5th instar larva usually entirely light green, dorsal pilosity sparse, light, inconspicuous.

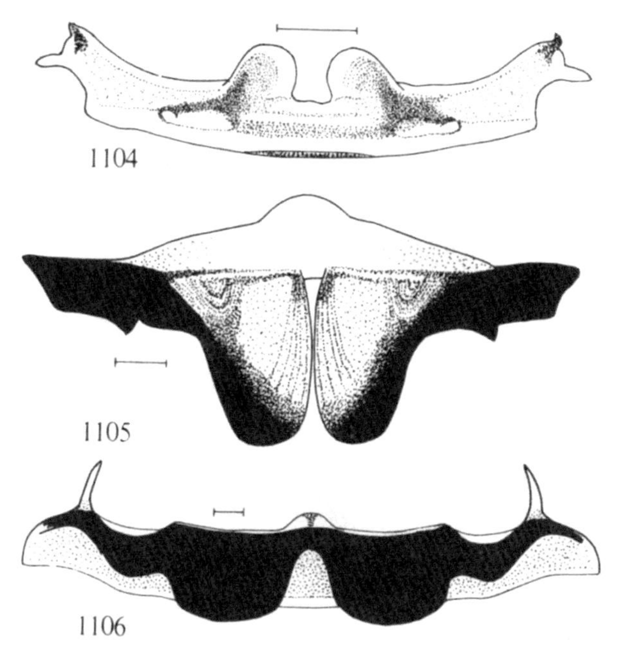

Text-figs. 1104–1106. *Tremulicerus tremulae* (Estlund). – 1104: 1st abdominal sternum in male from behind; 1105: 2nd abdominal sternum in male from above; 1106: 3rd abdominal tergum in male from below. Scale: 0.1 mm.

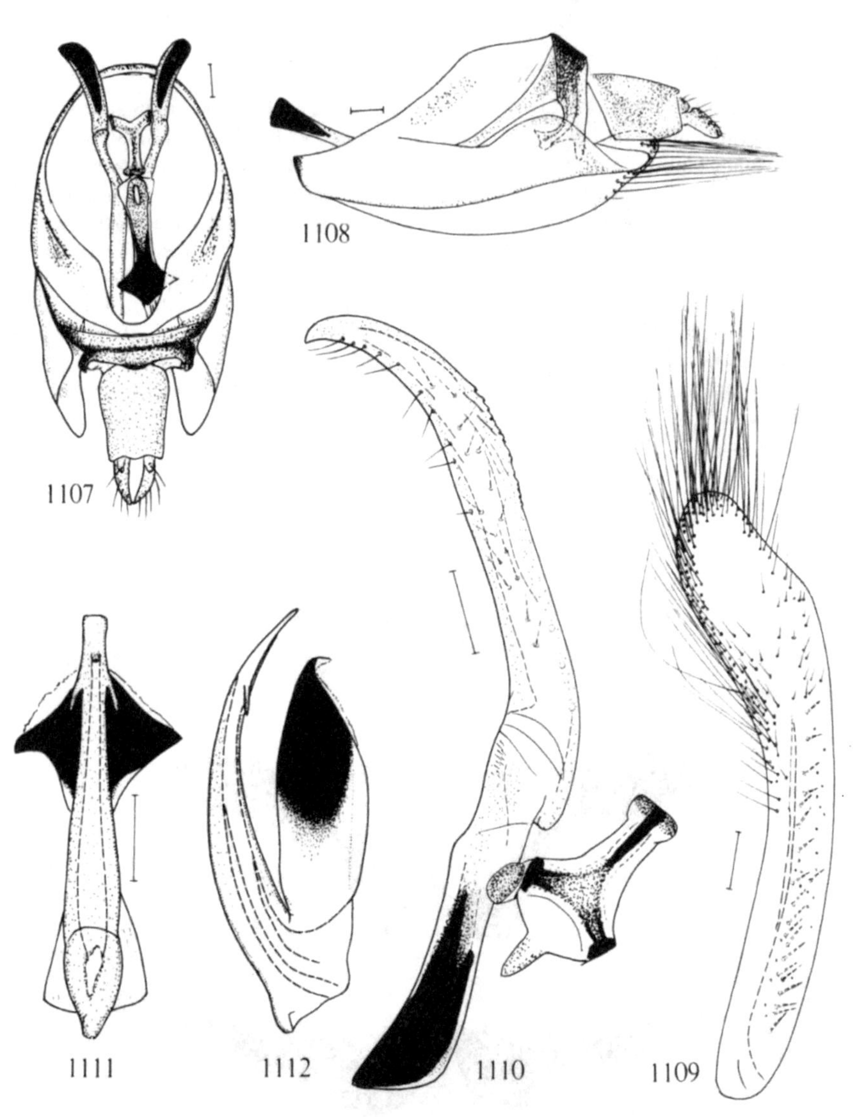

Text-figs. 1107–1112. *Tremulicerus vitreus* (Fabricius). – 1107: male pygofer from above, genital plates not considered; 1108: male pygofer with genital plate from the left; 1109: left genital plate from outside, setae of inside in broken line; 1110: right genital style and connective from above; 1111: aedeagus in ventral aspect; 1112: aedeagus from the right. Scale: 0.1 mm.

Distribution. Rare in Denmark, only recorded once: LFM: Rødbyhavn 6.X. 1978 (Trolle). – Fairly common in southern and central Sweden, Sk.-Hls. -Norway: found in AK, Os, Bø, TEi, AAy, AAi, VAy.-Fairly common in southern and central East Fennoscandia, found in Al, Ab, N, Ka, Ta; Vib, Kr. – Austria, Belgium, Bohemia, Moravia, Slovakia, France, German D.R. and F.R., England, Wales, Italy, Poland, Latvia, Lithuania, n.Russia, Ukraine, Kazakhstan.

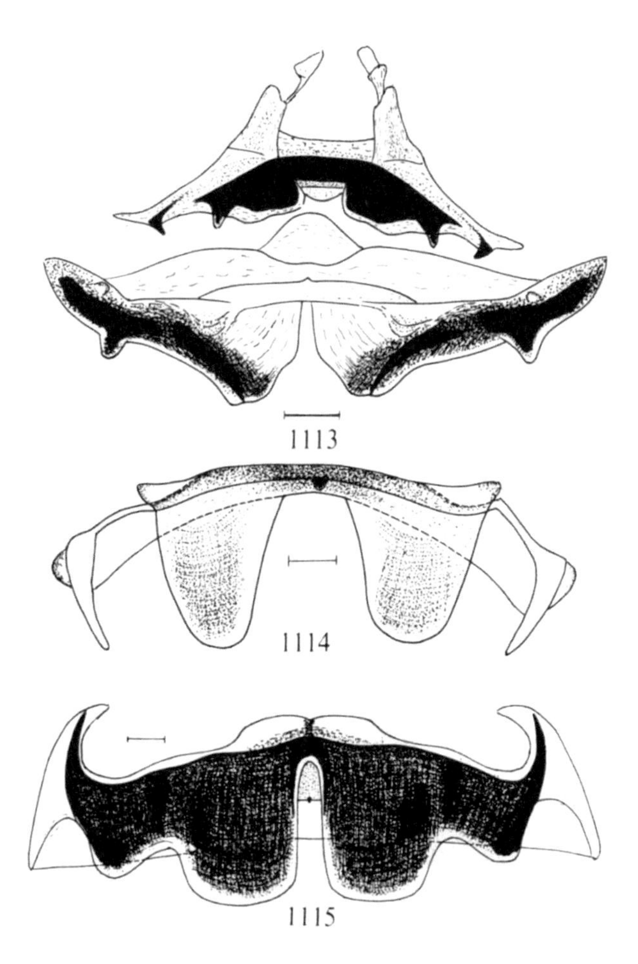

Text-figs. 1113–1115. *Tremulicerus vitreus* (Fabricius). – 1113: 1st and 2nd abdominal sterna in male from above; 1114: 2nd abdominal tergum in male from below; 1115: 3rd abdominal tergum in male from below. Scale: 0.1 mm.

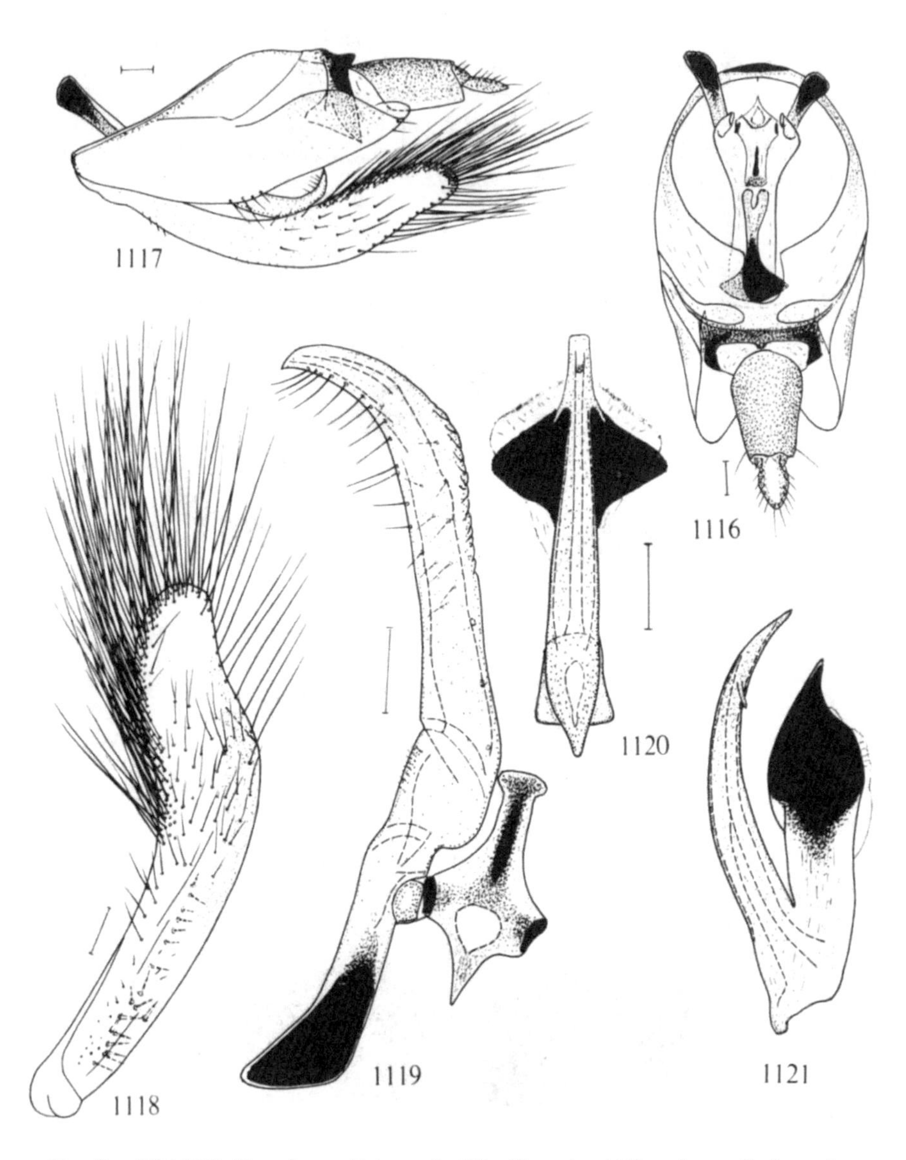

Text-figs. 1116–1121. *Tremulicerus distinguendus* (Kirschbaum). - 1116: male pygofer from above, genital plates not considered; 1117: male pygofer with genital plate from the left; 1118: genital plate from the left, setae of inside in broken line; 1119: right genital style and connective from above; 1120: aedeagus in ventral aspect; 1121: aedeagus from the right. Scale: 0.1 mm.

140. ***Tremulicerus vitreus*** (Fabricius, 1803)
Text-figs. 1107–1115.

Cicada vitrea Fabricius, 1803:79.
Idiocerus H-album Fieber, 1868:455.
Idiocerus auronitens Kirschbaum, 1868a:5.

Light greyish yellow, sometimes with a greenish tinge. Face in male entirely whitish yellow, in female transversely banded and spotted with orange or brownish, vertex in female with a median whitish spot. Pronotum whitish with two broad orange or brownish yellow longitudinal bands and a large spot of the same colour more laterally on each side. Lateral triangular spots on scutellum black in male, fuscous in female; scutellum between these spots often with a brownish longitudinal band and/or two smaller dark spots. Fore wings dirty brownish yellow with an indistinct fuscous transverse band across middle within which the veins are fuscous. Distally of this band there is an indistinct light transverse band enclosing whitish vein sections. There is a whitish or hyaline spot on distal end of first claval vein and another surrounding the proximal transverse vein between M and Cu. Veins in apical part of fore wing fuscous. Venter, abdomen and legs light, abdominal dorsum with segmental transverse black bands. Male pygofer as in Text-figs. 1107, 1108, genital plate as in Text-fig. 1109, genital style and connective as in Text-fig. 1110, aedeagus as in Text-figs. 1111, 1112, 1st and 2nd abdominal sterna in male as in Text-fig. 1113, 2nd and 3rd abdominal terga in male as in Text-figs. 1114,1115. Length of males 4.7–5.1 mm, of females 5–5.6 mm.

Distribution. Jensen-Haarup (1920) recorded this species from Denmark, NEZ: København 5.IX.1870. - Not found in Sweden, Norway and East Fennoscandia. - Austria, Belgium, Bohemia, France, German D.R. and F.R., England, Italy, Netherlands, Ukraine.

Biology. »On poplars, especially *P. nigra* L., and *italica* Moench, sometimes on sallows« (Le Quesne, 1965). Hibernation takes place in the egg stage (Müller, 1957).

141. ***Tremulicerus distinguendus*** (Kirschbaum, 1868)
Text-figs. 1116–1123.

Idiocerus distinguendus Kirschbaum, 1868a:5.
Idiocerus cognatus Fieber, 1868:455.

Face whitish, vertex, pronotum and scutellum yellowish white, thyridia black. Pronotum with a more or less distinct pale median band and a few small dark spots near fore border. Scutellum with the usual lateral triangular spots black, fuscous or yellowish, between them with or without a pair of small dark spots and a more or less distinct median marking. Fore wings dirty yellow or pale fuscous with two pale transverse bands, one proximally and one distally of middle, veins within these bands white. The area between these light bands is fuscous with dark veins but the dark colour does not reach costal margin. Veins fuscous also in apical part of fore wing. Venter and legs whitish yellow or sordid yellow, dorsum of abdomen as in *vitreus*. Male pygofer as in Text-figs. 1116, 1117, genital plate as in Text-fig. 1118, genital style and connective as in Text-fig. 1119, aedeagus as in Text-figs. 1120, 1121, 1st and 2nd abdominal sterna in male as in Text-fig. 1122, 2nd and 3rd abdominal terga in male as in Text-fig. 1123. Length of male 4.4–4.9 mm, of female 5.1–5.5 mm.

Distribution. Rare in Denmark, found in LFM, NEZ, and B. – Not found in Sweden, Norway and East Fennoscandia. – Austria, Belgium, ? Bohemia,? Moravia, France, German D.R. and F.R., England, Hungary, Italy, Netherlands, Poland, Ukraine; Nearctic region.

Biology. On *Populus alba*. "Eiablage einzeln in Juni unter die Rinde des neuen Zuwachses und der Endzweige. L_1 schlüpfen Ende Mai des folgenden Jahres und leben zunächst auf der Oberseite der nicht entfalteten Blätter, die späteren Stadien auf den Blattunterseiten. Larvalzeit ca. 1 Monat dauernd. Ad. ab Ende Juni bis Ende Oktober. Eine Generation" (Müller, 1956).

Economic importance. "Nach Amerika (New York, New Jersey) mit Silberpappeln eingeschleppt und auf diesen nur bei starkem Befall leichte Missbildungen und Welke der Blätter infolge des Saftentzuges hervorrufend" (Müller, 1956).

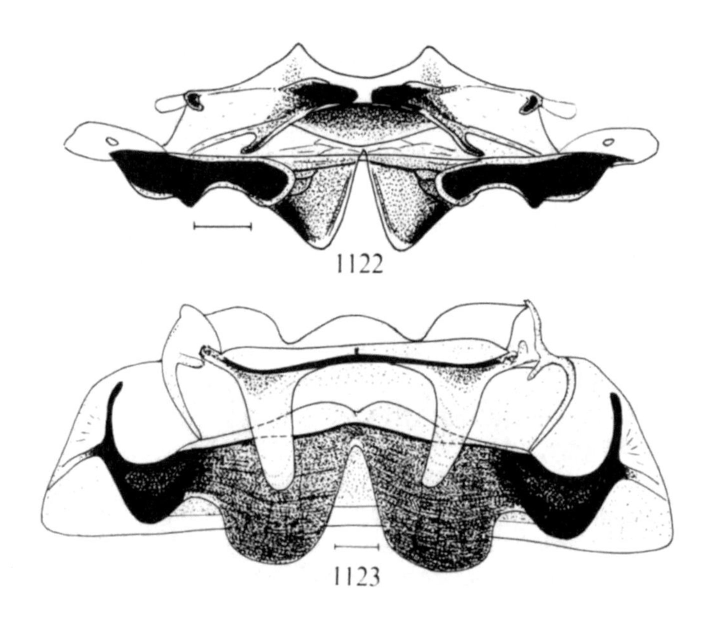

Text-figs. 1122, 1123. *Tremulicerus distinguendus* (Kirschbaum). – 1122: 1st and 2nd abdominal sterna in male from above; 1123: 2nd and 3rd abdominal terga in male from below. Scale: 0.1 mm.

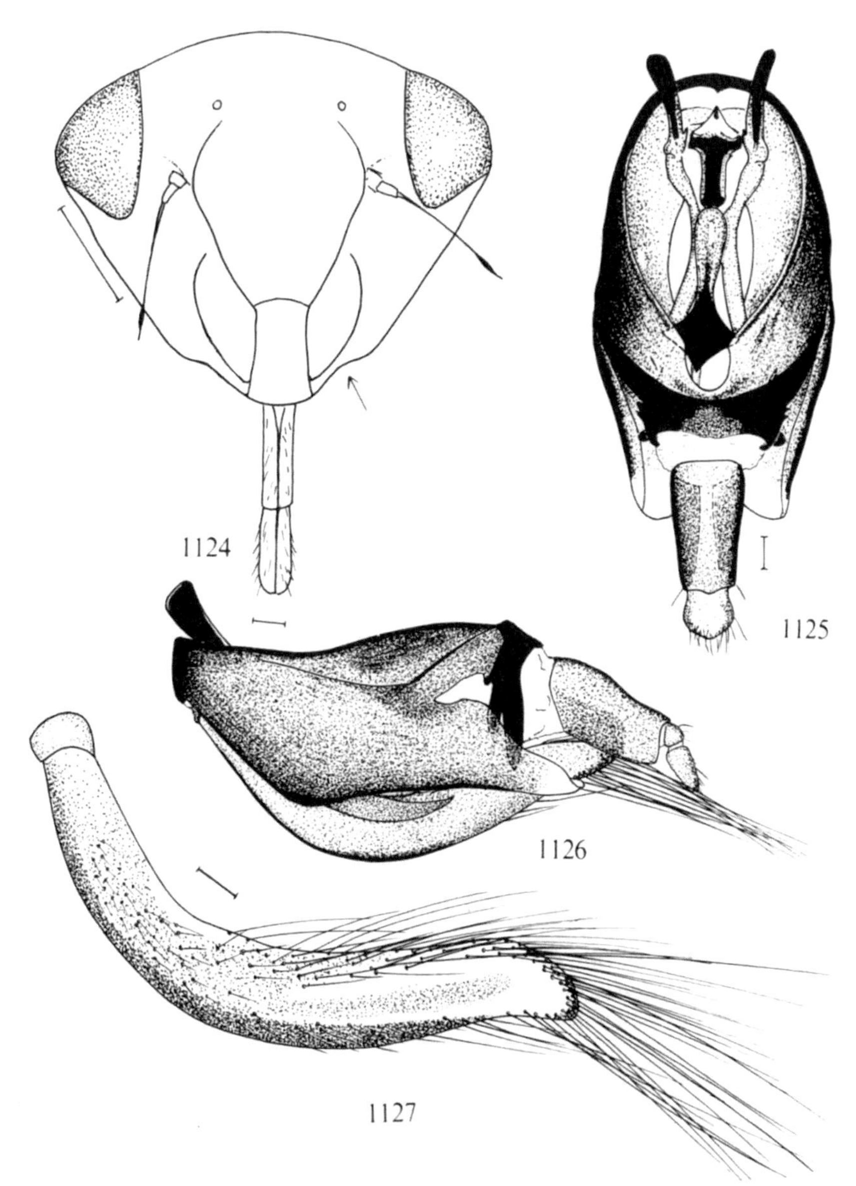

Text-figs. 1124–1127. *Stenidiocerus poecilus* (Herrich-Schäffer). – 1124: face of male; 1125: male pygofer from above, genital plates not considered; 1126: male pygofer with genital plates from outside; 1127: left genital plate from outside. Scale: 0.5 mm for 1124, 0.1 mm for the rest.

Genus *Stenidiocerus* Ossiannilsson, new genus

Type-species: *Bythoscopus poecilus* Herrich-Schäffer, 1835.

Moderately large. Face with eyes broader than long, angle of side margins less than 90°, maxillary plates not protracted, lateral margin with a shallow indentation (Text-fig. 1124, at arrow). Male antenna with a narrow palette. Vertex medially almost as long as laterally. Distance between ocelli almost twice distance between ocellus and nearest compound eye. Fore wings with strong veins, three subapical cells present. Male genital style broad, evenly curved, gradually tapering towards apex, with some moderately long and thick subapical setae. One species.

142. ***Stenidiocerus poecilus*** (Herrich-Schäffer, 1835) comb. n.
Plate-fig. 67, text-figs. 1124–1132.

Bythoscopus poecilus Herrich-Schäffer, 1835:69.
Bythoscopus (Jassus) falcifer Boheman, 1845b:161.
Idiocerus discolor Flor, 1861a:164.

Resembling *Idiocerus herrichii*. Genae glabrous. Frontoclypeus of male, often also female, with two black longitudinal bands. Upper part of face mottled black or fuscous. Thyridia black. Vertex and pronotum marbled with brown, black and milky white and with a common broad yellowish white median stripe. Scutellum largely black, the light ground-colour being superseded into smaller spots. Fore wing veins dark, here and there broadly dark-bordered, with several light spots, i.a. both claval veins being distally light (Plate-fig. 67). Venter variegated or largely black. Femora above black, below light, tibiae with black longitudinal stripes. Male pygofer as in Text-figs. 1125, 1126, genital plate as in Text-fig. 1127, genital style and connective as in Text-fig. 1128, aedeagus as in Text-figs. 1129, 1130, 1st and 2nd abdominal sterna in male as in Text-fig. 1131, 2nd and 3rd abdominal terga in male as in Text-fig. 1132. Overall length ♂♀ 4.8–6 mm.

Distribution. So far not found in Denmark. – Scarce and sporadic in Sweden, found in Sm., Ög., Upl., Nrk., Dlr., Hls., Vb. – Rare in Norway, recorded from AK and TEy. – Rare also in East Fennoscandia, found in Ab and N. – Austria, Belgium, Bulgaria, Bohemia, Moravia, Slovakia, France, German D.R. and F.R., England, Hungary, Italy, Poland, Switzerland, Greece, Yugoslavia, Latvia, Tunisia, Iraq, Ukraine, Kazakhstan, Tadzhikistan, Morocco, m. Siberia.

Biology. On *Populus nigra* (including *italica*) (Wagner & Franz, 1961; Le Quesne, 1965), and on *Populus tremula* (Sahlberg, 1871). In Sweden adults have been found in March, May, August, and November. Sahlberg (1871) found two specimens under the bark of *Populus tremula* in the winter, so hibernation apparently takes place in the adult stage.

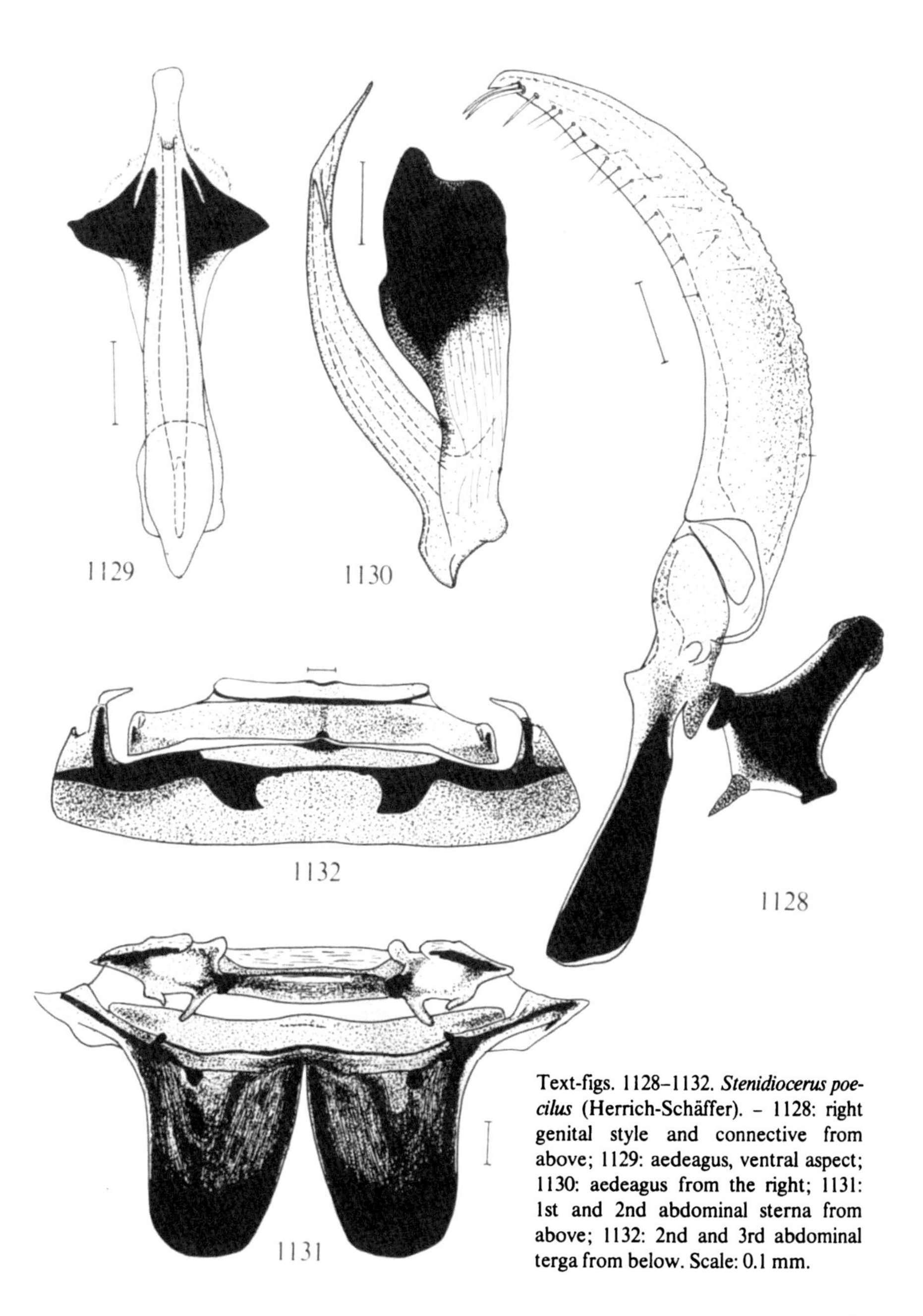

Text-figs. 1128–1132. *Stenidiocerus poecilus* (Herrich-Schäffer). – 1128: right genital style and connective from above; 1129: aedeagus, ventral aspect; 1130: aedeagus from the right; 1131: 1st and 2nd abdominal sterna from above; 1132: 2nd and 3rd abdominal terga from below. Scale: 0.1 mm.

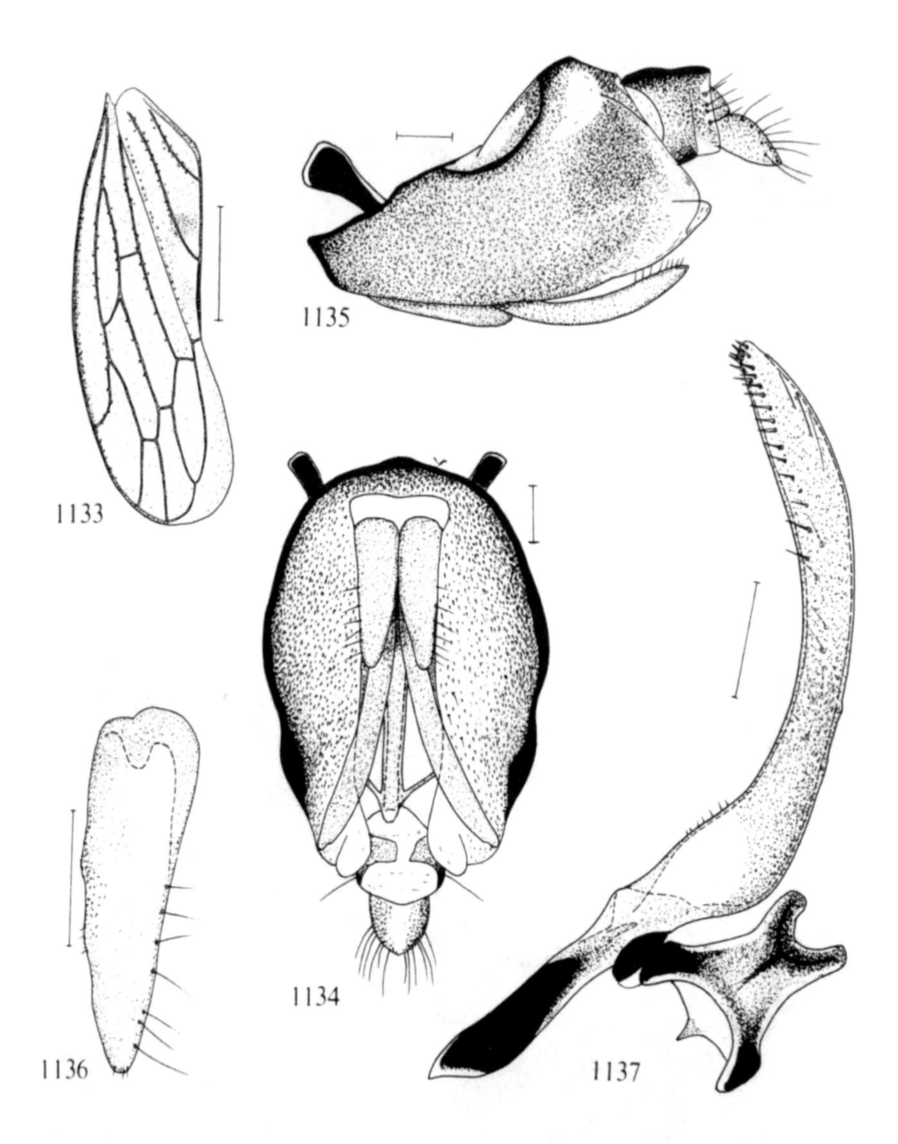

Text-figs. 1133–1137. *Sahlbergotettix salicicola* (Flor). – 1133: left fore wing of male; 1134: male pygofer in ventral aspect, 8th sternum removed; 1135: male pygofer from the left; 1136: left genital plate from below; 1137: right genital style and connective from above. Scale: 1 mm for 1133, 0,1 mm for the rest.

Genus *Sahlbergotettix* Zachvatkin, 1953

Sahlbergotettix Zachvatkin, 1953b:213.
Type-species: *Idiocerus salicicola*, Flor, 1861, by original designation.

Male genital plates strongly reduced. Fore wings with two subapical cells (Text-fig. 1133). Vertex finely punctured. Frontoclypeus convex, side margins rounded. Male antenna with a small palette. One European species.

143. ***Sahlbergotettix salicicola*** (Flor, 1861)
Text-figs. 1133–1141.

Idiocerus salicicola Flor, 1861:163.

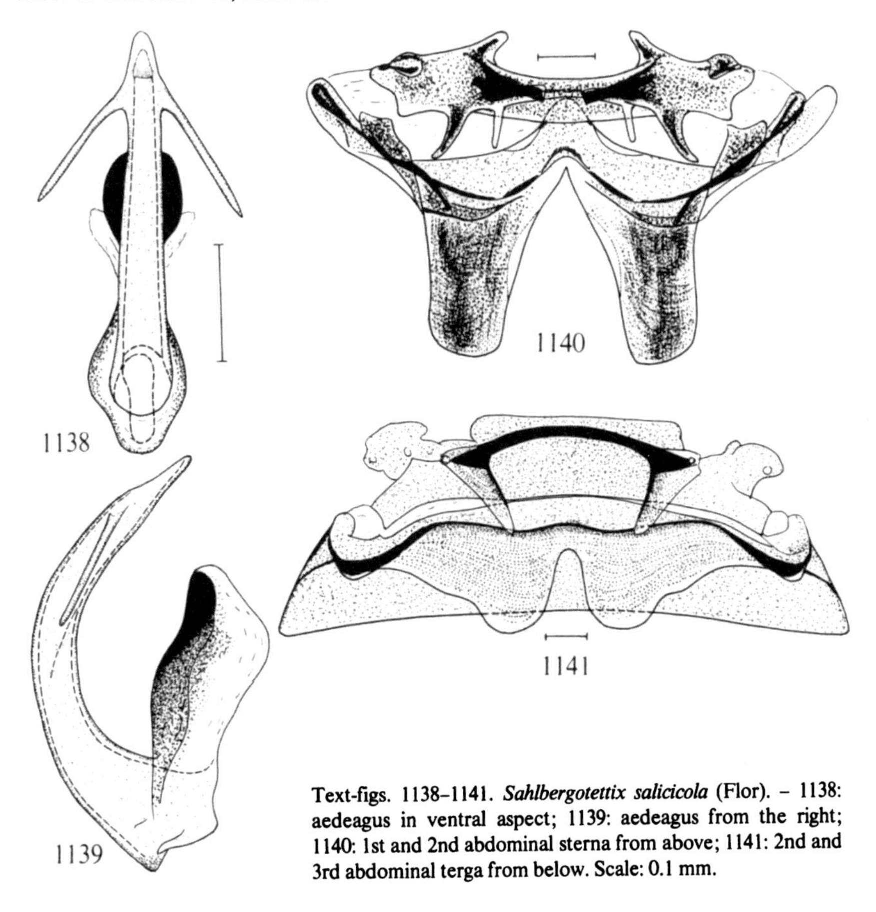

Text-figs. 1138–1141. *Sahlbergotettix salicicola* (Flor). – 1138: aedeagus in ventral aspect; 1139: aedeagus from the right; 1140: 1st and 2nd abdominal sterna from above; 1141: 2nd and 3rd abdominal terga from below. Scale: 0.1 mm.

Dirty yellow or brownish yellow. Head and pronotum more or less irregularly spotted with brownish or blackish. Frontoclypeus in lower part sometimes entirely black or brown. Thyridia black. Pronotum on anterior part with a few black spots varying in extension. In strongly pigmented specimens they may coalesce, leaving a pale median band. Scutellum with the usual lateral triangular spots brownish or black. Between these often a pair of black spots and a median longitudinal streak, these markings more or less extending in dark specimens. Abdominal tergum black, venter also black, or yellowish with black transverse bands. Fore wings whitish, hyaline, veins comparatively strong, sometimes dark brown, sometimes pale. Claval cells sometimes fuscous. A whitish spot surrounding distal end of first claval vein. Legs yellowish, tarsi sometimes fuscous. Male pygofer as in Text-figs. 1134, 1135, genital plate as in Text-fig. 1136, genital style and connective as in Text-fig. 1137, aedeagus as in Text-figs 1138, 1139, 1st and 2nd abdominal sterna in male as in Text-fig. 1140, 2nd and 3rd abdominal terga in male as in Text-fig. 1141. Length ♂♀ 4.25–4.6 mm.

Distribution. Not in Denmark, Norway, and Sweden. – East Fennoscandia: only found in St: Yläne 20.IX. 1869 by J. Sahlberg. – Slovakia, France, Hungary, Poland, Romania, Estonia, m.Russia, Kazakhstan, Tadzhikistan, Mongolia, Maritime Territory.

Biology. On *Salix rosmarinifolia* (Sahlberg, 1871); on *Salix incana* (Ribaut, 1952).

SUBFAMILY IASSINAE

Head evenly convex. Labium short or moderately long. Anteclypeus distinct from postclypeus. Maxillary plates wide. A distinct epistomal suture absent. Frontoclypeus approximately circular. Antennal ledges transverse or slightly oblique. Pronotum broad. Hind tibiae flattened and curved, with two or three rows of strong marginal setae. Fore wings in repose tectiform. Apical membrane of fore wing of moderate width. In Northern Europe two genera.

Key to genera of Iassinae

1 Hind femora apically with five spines. Male without an anal collar, genital styles well developed (Text-fig. 1154). *Batracomorphus* Lewis (p.355)

– Hind femora apically with three spines. Male with a large anal collar (Text-fig. 1154), genital styles reduced (Text-fig. 1147) *Iassus* Fabricius (p.352)

Genus *Iassus* Fabricius, 1803

Iassus Fabricius, 1803:85.
Type-species: *Cicada lanio* Linné, 1761, by subsequent designation.
Bythoscopus Germar, 1833:180.
Type-species: *Cicada lanio* Linné, 1761, by subsequent designation.

Body robust. Head slightly narrower than pronotum. Vertex short, its anterior outline broadly rounded. Pronotum with a fine transverse striation, its lateral margins a little shorter than caudal margin of an eye. Apical membrane of fore wing narrow. Fore wings slightly longer than abdomen, with several supernumerary transverse veinlets between costal border and R. Legs strong, transverse section of tibiae square, hind tibiae with many strong setae. Dorsum of fore and middle tibiae broadly flattened, on each side with a low longitudinal border. In Denmark and Fennoscandia one species.

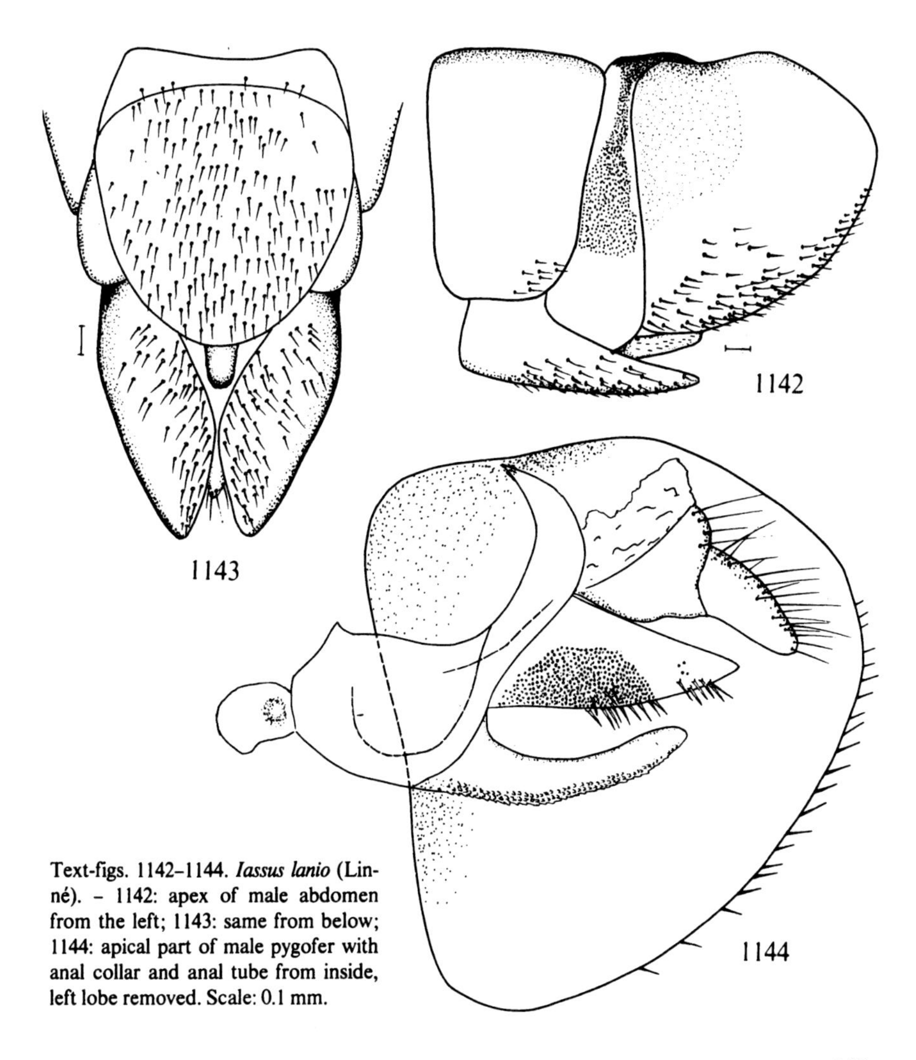

Text-figs. 1142–1144. *Iassus lanio* (Linné). – 1142: apex of male abdomen from the left; 1143: same from below; 1144: apical part of male pygofer with anal collar and anal tube from inside, left lobe removed. Scale: 0.1 mm.

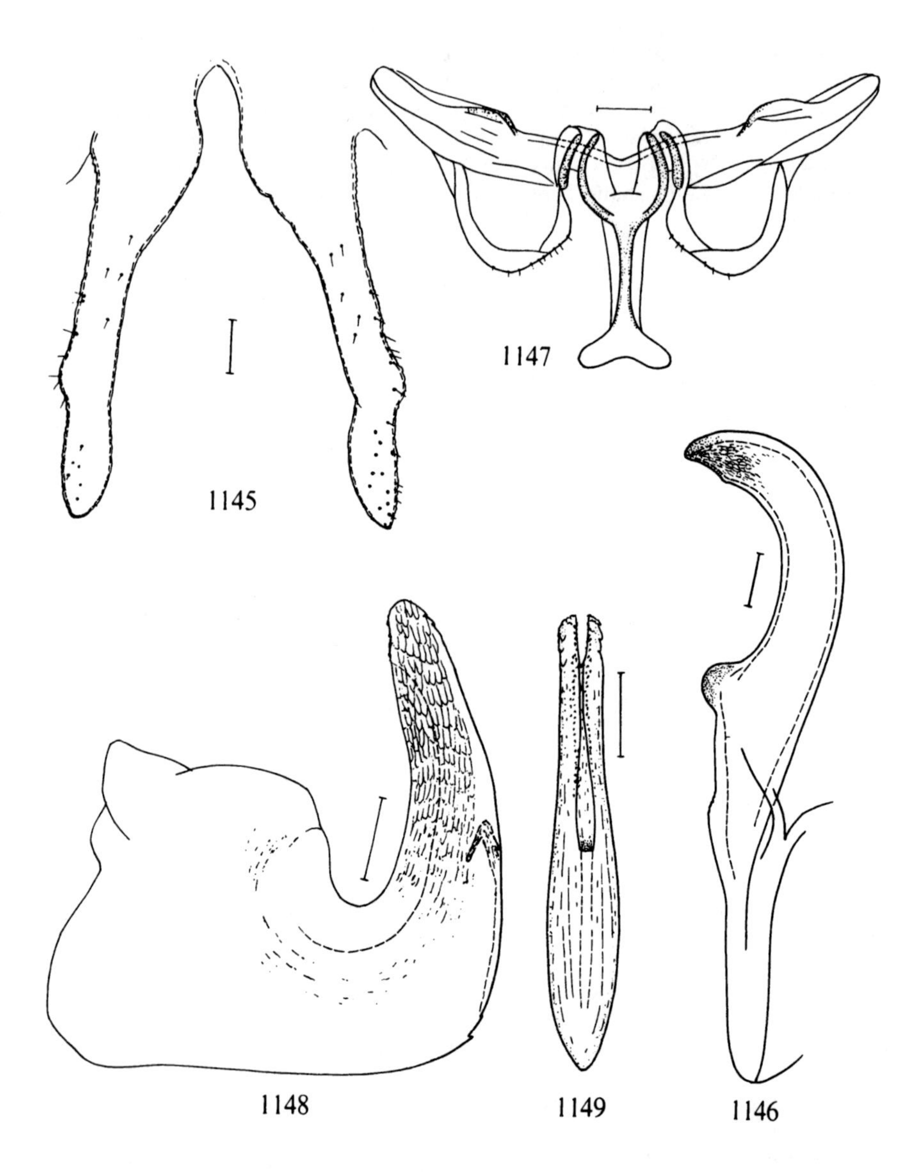

Text-figs. 1145–1149. *Iassus lanio* (Linné). – 1145: genital plates in ventral aspect; 1146: right ventral process of male pygofer; 1147: styles and connective from above; 1148: aedeagus from the left; 1149: aedeagus in ventral aspect. Scale: 0.1 mm.

144. ***Iassus lanio*** (Linné, 1761)
Plate-fig. 81, text-figs. 1142–1149.

Cicada lanio Linné, 1761:242.
Cicada brunnea Fabricius, 1794:43.

Light green, yellowish or yellowish brown. Face, vertex, pronotum, and scutellum with fine, more or less dense reddish or brownish mottlings. Pronotum and scutellum brown or (in immature individuals) green, finely transversely wrinkled. Fore wings green (f. *typica*) or brown (f. *brunnea* Fabricius), in the former case often reddish or brownish along scutellum and/or the apical membrane, with a wrinkled punctuation. Venter and legs green, yellowish, or brownish. Apex of male abdomen as in Text-figs. 1142, 1143, apical part of male pygofer with anal apparatus as in Text-fig. 1144, genital plates as in Text-fig. 1145, ventral process of pygofer as in Text-fig. 1146, styles and connective as in Text-fig. 1147, aedeagus as in Text-figs. 1148 and 1149. Overall length of male 7.0–7.3 mm, of female 7.7–8.5 mm. – Larva green or brownish, with a depressed pilosity consisting of short scales, abdominal segments laterally with some longer setae.

Distribution. Common and widespread in Denmark, as well as in Sweden up to Vrm. – Scarce in Norway, found in Ø, AK, AAy, VAy, and Ry. – East Fennoscandia: only found in Ab: Lojo and the Åbo district, and in N: Ekenäs, Gullö. Comparatively common in the Åbo district. – Widespread in Europe, also recorded from Kurile Islands.

Biology. On *Quercus*, adults in July – October. Hibernation in the egg stage (Müller, 1957).

Genus *Batracomorphus* Lewis, 1834

Batracomorphus Lewis, 1834:51.
Type-species: *Batracomorphus irroratus* Lewis, 1834, by monotypy.

As *Iassus*, differing by characters given in the key. Dorsum of fore and middle tibiae with a longitudinal carina anteriorly flanked by a narrow and shallow furrow. Sides of pronotum as long as or longer than hind margin of a compound eye. In Denmark and Fennoscandia one species.

145. ***Batracomorphus allionii*** (Turton, 1802)
Text-figs. 1150–1158.

Cicada prasina Fabricius, 1794:38 (nec Pallas, 1773).
Cicada allionii Turton, 1802:594.
Batrachomorphus [sic] *fabricii* Metcalf, 1955:266.

Green, dorsum of abdomen light reddish yellow. Head and scutellum finely transversely striated, striations of pronotum somewhat coarser. Fore wings hyaline or semi-

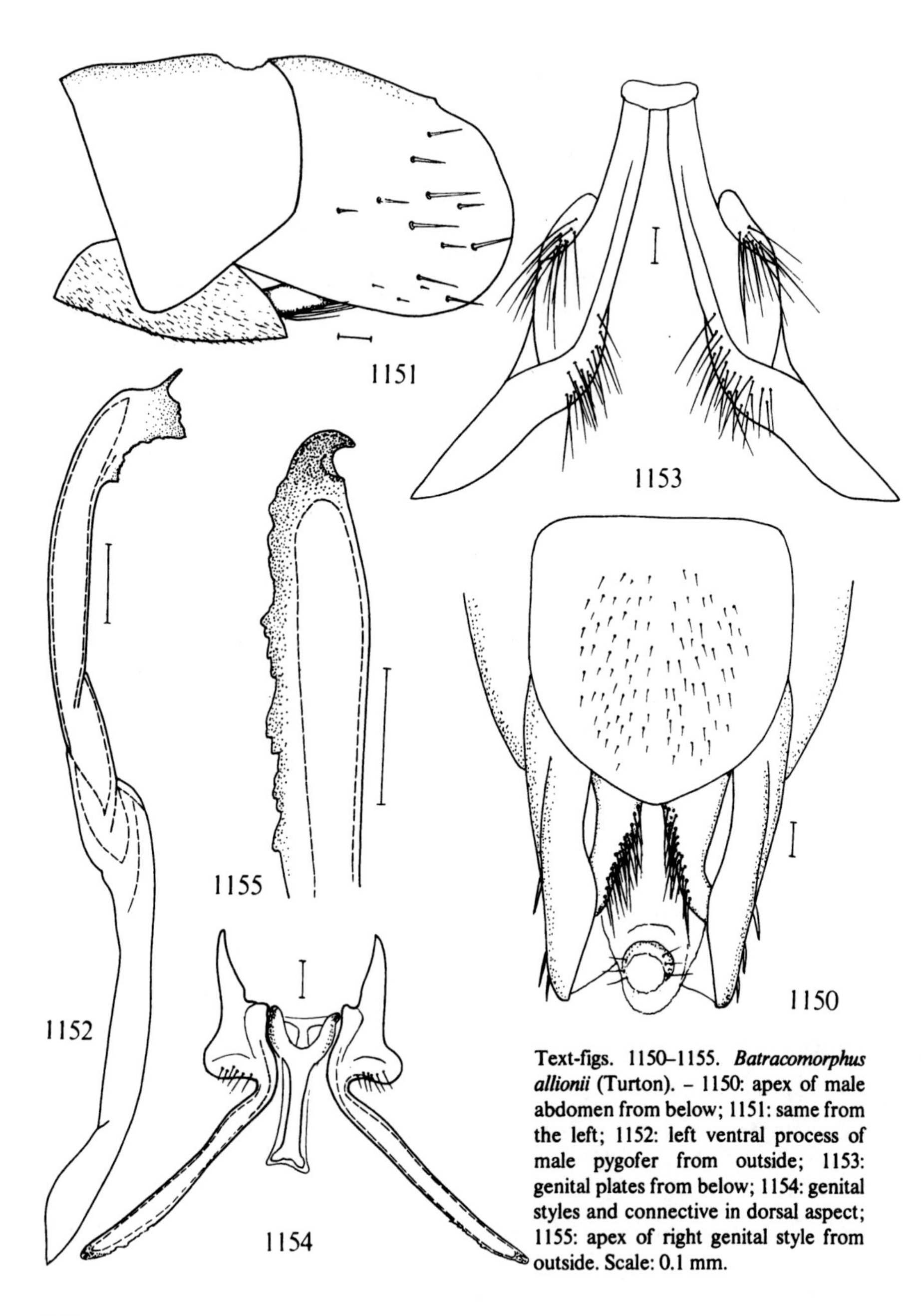

Text-figs. 1150–1155. ***Batracomorphus allionii*** (Turton). – 1150: apex of male abdomen from below; 1151: same from the left; 1152: left ventral process of male pygofer from outside; 1153: genital plates from below; 1154: genital styles and connective in dorsal aspect; 1155: apex of right genital style from outside. Scale: 0.1 mm.

hyaline, with a small dark spot in proximal end of apical membrane, veins concolorous or a little darker towards wing apex, little prominent, punctuation of fore wing sparse, shallow. Venter and legs pale greenish. Apex of male abdomen as in Text-figs. 1150, 1151, pygofer on inside with a pair of long ventral processes (Text-fig. 1152). Genital plates as in Text-fig. 1153, styles and connective as in Text-fig. 1154, apex of style as in Text-fig. 1155, aedeagus as in Text-figs. 1156, 1157. Ventral aspect of apical part of female abdomen as in Text-fig. 1158. Length 7–8 mm.

Distribution. Rare in Denmark, found only a few times in east and central Jutland. - Not in Norway, nor in Sweden. - Rare and sporadic in southern East Fennoscandia, found in Al: Kökar (Håkan Lindberg), Ab: Pargas (Reuter), Tenala 16.VII. 1920 (Håkan Lindberg), Karislojo (J. Sahlberg), Lojo 17.VII. - 21. VIII (Håkan Lindberg), Karis 20.VIII. 1944 (P. H. Lindberg), N: Lappvik (Håkan Lindberg), Ekenäs, Notholm (Håkan Lindberg), Ta: Sysmä (Hellén); Kr: Parikkala (J. Sahlberg). - Austria, Belgium, France, German D.R. and F.R., Italy, Netherlands, Poland, Romania, Spain, Bohemia, Moravia, Lithuania, n. and m. Russia, Azerbaijan, Altai Mts., m. Siberia, Mongolia, Maritime Territory.

Biology. In dry meadows and edges of woods (Linnavuori, 1969). "In Norddeutschland lebt diese Art auf *Sarothamnus scoparius*" (Wagner, 1939).

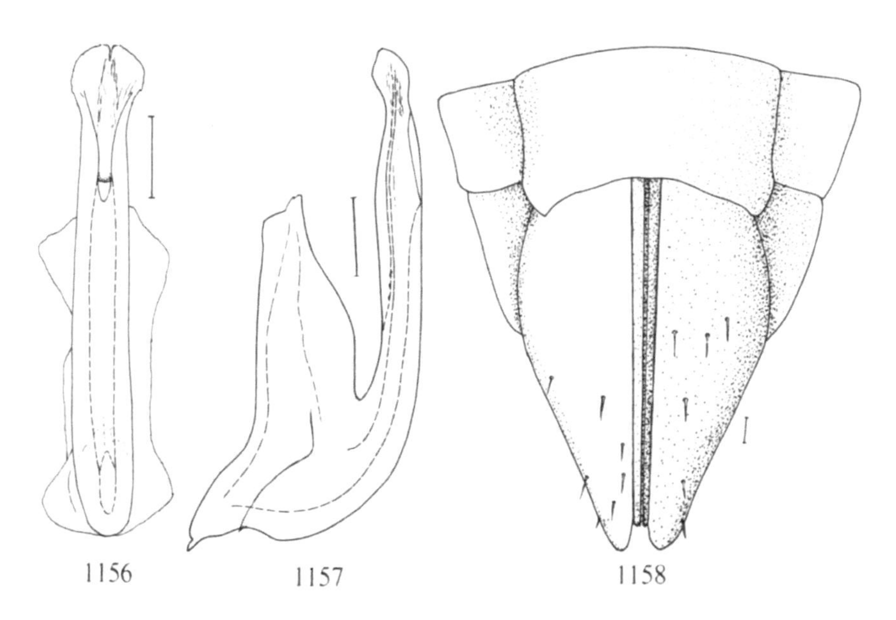

Text-figs. 1156–1158. *Batracomorphus allionii* (Turton). - 1156: aedeagus in ventral aspect; 1157: aedeagus from the left; 1158: ventral aspect of abdominal apex in female. Scale: 0.1 mm.

SUBFAMILY DORYCEPHALINAE

Face broadly keel-shaped, vertex with a longitudinal median carina. Ocelli situated on dorsal surface of head but quite near fore margin, widely separated from eyes. Antennal ridges absent. Pronotum on each side with two longitudinal carinae. Fore wings in repose held roof-like over abdomen, long and narrow. Setae of hind tibiae fairly weak and sparsely set. In Denmark and Fennoscandia one genus.

Genus *Eupelix* Germar, 1821

Eupelix Germar, 1821:94.
Type-species: *Cicada cuspidata* Fabricius, 1775, by monotypy.

Body oblong. Head anteriorly angular, depressed, lamelliform, dorsally and ventrally with a strong median carina. The sharp lateral margin of the head divides the anterior part of the compound eye in an upper and a lower part. Anteclypeus and frontoclypeus narrow. Pronotum with a median carina and two longitudinal carinae caudally of each eye, its fore border convex, obtuse-angled, hind border concave, obtuse-angled. Scutellum large. Fore wings about as long as abdomen or longer, with strong elevated veins, costal border deflected. Hind tibiae with several rows of fairly small spines. In Denmark and Fennoscandia one species.

146. ***Eupelix cuspidata*** (Fabricius, 1775)
Plate-figs. 77, 78, text-figs. 1159–1166.

Cicada cuspidata Fabricius, 1775:687.
Cicada depressa Fabricius, 1803:66.
Eupelix spathulata Germar, 1838:pl.25, figs. a–b.
Eupelix producta Germar, 1838:pl. 24, figs. a–c.

Male greyish brown to greyish yellow. Head above and below with black mottlings. Pronotum brown, caudally finely transversely wrinkled, with scattered black dots. Scutellum with two small dark marks. Fore wings greyish white with small dark dots and streaks especially along veins. Abdomen dorsally and ventrally with black markings arranged in rows. Female straw-coloured with indistinct mottlings on fore body, fore wings usually entirely light, head longer than in male. F. *depressa* (Fabricius) (= *producta* Germar) differs from f. *typica* by the longer head in males (Plate-fig. 78). Male pygofer as in Text-fig. 1159, genital plates, valve and styles as in Text-figs. 1160, 1161, aedeagus as in Text-figs. 1162, 1163, connective as in Text-fig. 1164, 1st and 2nd abdominal sterna as in Text-figs. 1165, 1166. Overall length in males 5.3–5.8 mm, in females 5.6–7.2 mm. – Body shape of larvae resembling that of adults, male larvae dark mottled, female larvae light with small dots, the latter arranged in longitudinal rows on abdomen.

Distribution. Denmark: common and widespread, though not abundant. - Common and widespread in the south of Sweden, found Sk.-Dlr. - Apparently rare in Norway, found in Ø, AK, On, Bø, AAy, Ry, HOy. - Scarce in southern and central East Fennoscandia, found in Al, Ab, N, Ka, St, Ta, Sa, Kb, Om; Vib, Kr. - Widespread in Europe, also found in Algeria, Cyprus, Egypt, Morocco, Tunisia, Iraq, Israel, Azerbaijan, Georgia, Kazakhstan, Tadzhikistan, Uzbekistan, Kirghizia, n. Siberia, Mongolia.

Biology. In "Salzstellen, Stranddünen, Binnendünen, Sandfeldern, besonnten Hängen, Flachmooren, Wiesen" (Kuntze, 1937). Belongs to the "Corynephoretum agrostidetosum aridae" (Marchand, 1953). In "Trockenrasen und Heiden" (Wagner & Franz, 1961). Hibernation takes place in the adult stage (Remane, 1958; Schiemenz, 1969b).

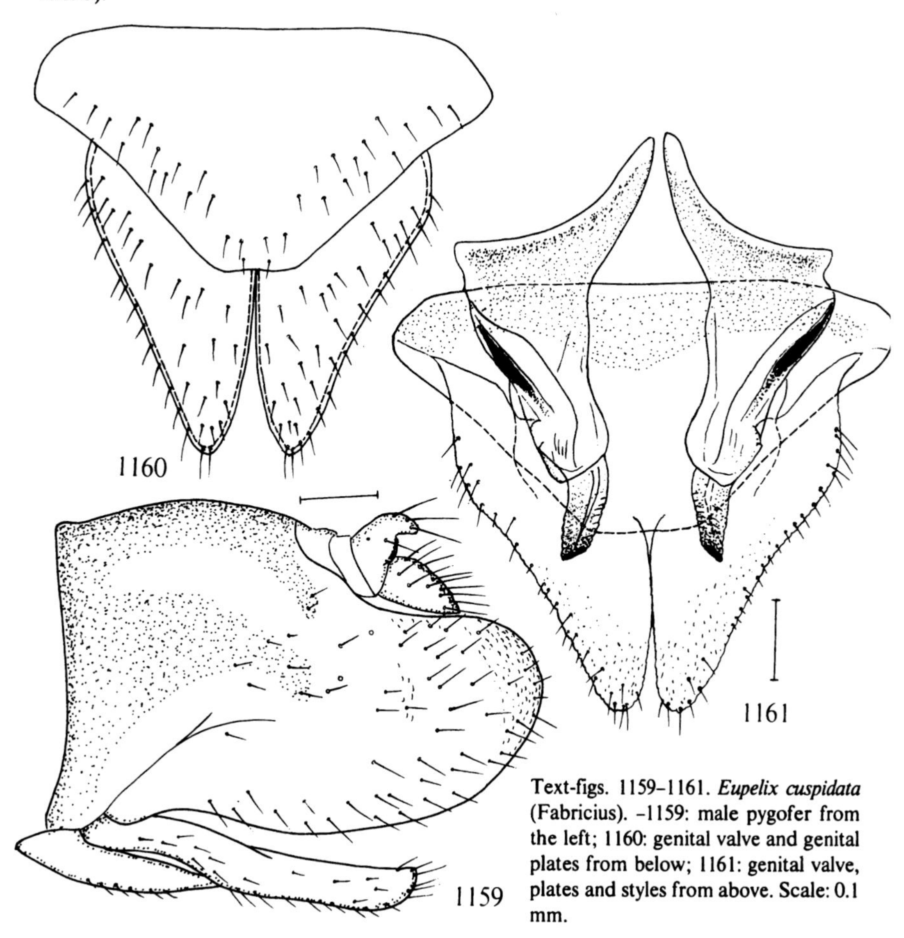

Text-figs. 1159-1161. *Eupelix cuspidata* (Fabricius). -1159: male pygofer from the left; 1160: genital valve and genital plates from below; 1161: genital valve, plates and styles from above. Scale: 0.1 mm.

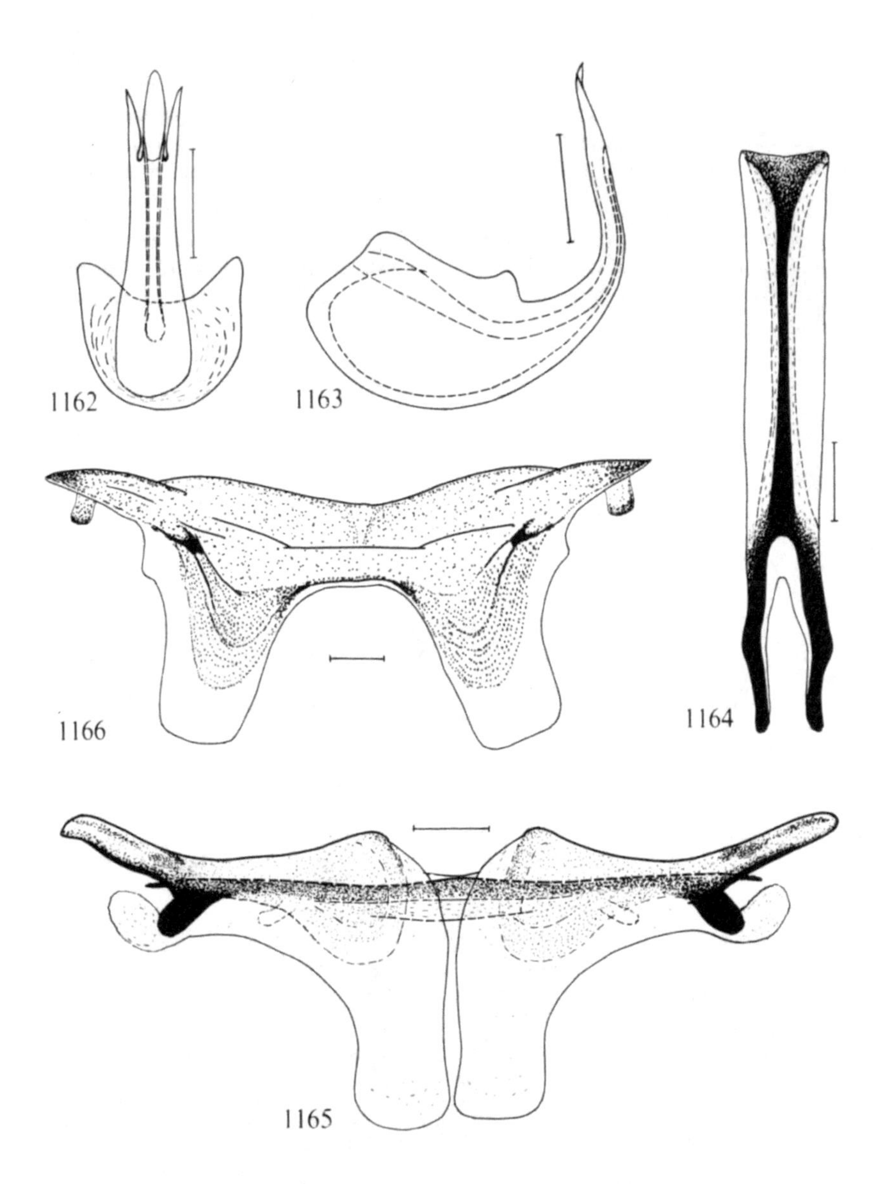

Text-figs. 1162–1166. ***Eupelix cuspidata*** (Fabricius). – 1162: aedeagus in ventral aspect; 1163: aedeagus from the left; 1164: connective from above; 1165: 1st abdominal sternum in male from above; 1166: 2nd abdominal sternum in male from above. Scale: 0.1 mm.

SUBFAMILY APHRODINAE

Small or medium-sized leafhoppers. Head usually emarginate beneath eyes. Antennal ledges short but usually distinct. Frontal and postfrontal sutures sometimes retained, coronal suture present or replaced by a carina. Ocelli either on the disc of vertex or on or near its anterior margin. Anterior and posterior margins of pronotum approximately parallel. Fore and middle tibiae rounded, hind tibiae a little flattened, armed with strong spines. On herbaceous plants; many species live and feed on roots beneath the surface litter. In Denmark and Fennoscandia four genera.

Key to genera of Aphrodinae
(After Hamilton, 1975)

1 Median length of vertex half that of pronotum .... *Stroggylocephalus* Flor (p.379)
– Median length of vertex ¾ – 1½ times that of pronotum 2
2(1) Vertex inflated between ocelli; vertex and face not separated by a carinate edge (Text-figs. 1171, 1172)............... *Anoscopus* Kirschbaum (p.370)
– Vertex flat between ocelli; distinct carinate edge between vertex and face (Text-figs. 1167–1170) 3
3(2) Vertex declivous; female frons not inflated (Text-fig. 1168); male aedeagal shaft slender, approximately cylindrical. Male fore wings without distinct markings .... *Aphrodes* Curtis (p.361)
– Vertex horizontal; female frons inflated; male aedeagal shaft broad and flat. Male fore wings with well-defined markings
Planaphrodes Hamilton (p.366)

Genus *Aphrodes* Curtis, 1829

Aphrodes Curtis, 1829:193.
Type-species: *Cicada striata* Fabricius, 1787, by subsequent designation.
Acucephalus Germar, 1833:181.
Type-species: *Cicada bicincta* Schrank, 1776, by subsequent designation.
Pholetaera Zetterstedt, 1838:288.
Type-species: *Cercopis rustica* Fabricius, 1775, by original designation.

Vertex between ocelli flat, margin sharply carinate, laterally depressed. Eyes not notched laterally. Face flat. Hind femora near apex armed with four marginal macrosetae and one smaller dorsal seta. Hind tibiae with short macrosetae. Shaft of aedeagus slender, cylindrical, with two pairs of spiniform appendages arising near middle (in our species). Phallotreme elongate. In Denmark and Fennoscandia two species.

Key to species of *Aphrodes*

1 Stem of aedeagus in lateral aspect distinctly curved (Text-fig. 1176) Large species, males 4.5–7.3 mm, females 6–8 mm. 147. *makarovi* Zachvatkin

– Stem of aedeagus in lateral aspect straight (Text-fig. 1182). Smaller species, males 4.1–5.3 mm, females 4.7–5.5 mm .... 148. *bicincta* (Schrank)

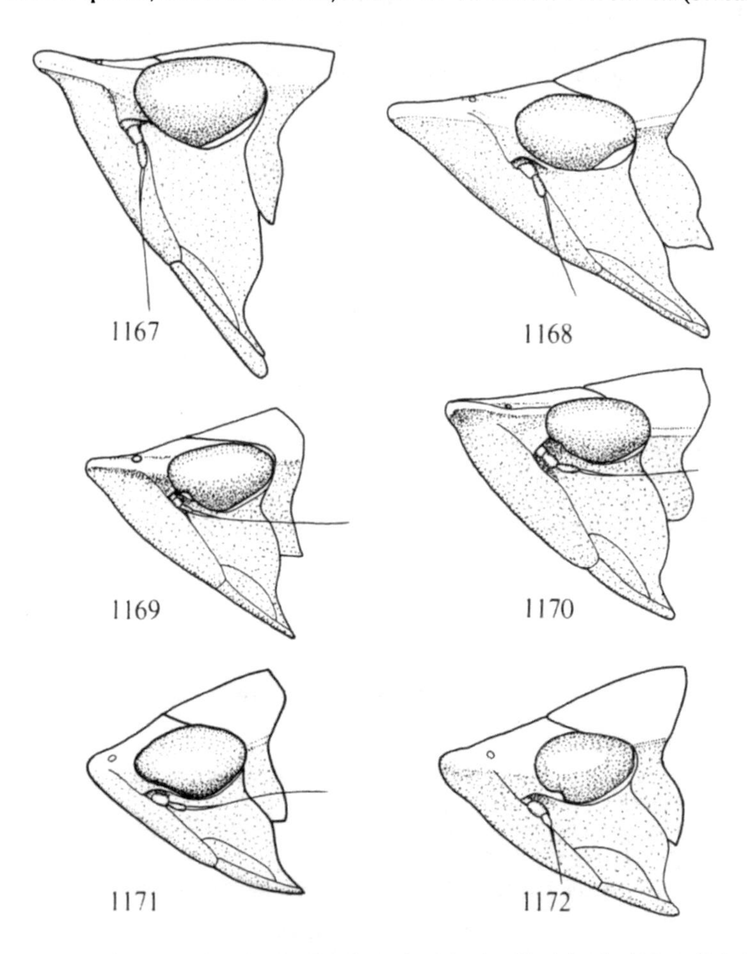

Text-figs. 1167–1172. – Head and pronotum from the left of 1167: *Aphrodes bicincta* (Schrank), male; 1168: *Aphrodes bicincta* (Schrank), female; 1169: *Planaphrodes trifasciata* (Fourcroy), male; 1170: *Planaphrodes trifasciata* (Fourcroy), female; 1171: *Anoscopus albifrons* (Linné), male; 1172: *Anoscopus albifrons* (Linné), female.

147. ***Aphrodes makarovi*** Zachvatkin, 1948
Plate-fig. 68, text-figs. 1173–1180.

? *Cicada striata* Linné, 1761:241 (Primary homonym).
? *Cercopis variegata* Fabricius, 1794:55.
? *Cercopis striatella* Fabricius, 1794:56.
? *Cercopis transversa* Fabricius, 1758:523.
? *Acucephalus pallidus* Curtis, 1836:620.
? *Acucephalus cardui* Curtis, 1836:620.
? *Acucephalus fasciatus* Curtis, 1836:620.
Aphrodes makarovi Zachvatkin, 1948b:186.
Aphrodes bicinctus Ribaut, 1952:333, nec Schrank.

Face and vertex longitudinally wrinkled, the latter with a median longitudinal carina. Male (Plate-fig. 68) with head medially a little shorter than pronotum, fore margin sharply defined. Greyish yellow or brownish yellow. Often with a black spot below each antenna, face with or without an indistinct dark band between these spots. Above this band a lighter transverse band may be present, face above the latter mottled. Vertex near hind border with a broad light transverse band. Pronotum transversely wrinkled, with a light transverse band. Scutellum mottled. Fore wing veins usually light and light-bordered, contrasting to the yellowish brown cell surfaces. Abdomen usually largely black, genital segment pale. Upper side of female entirely light yellowish brown or more or less densely dark mottled, sometimes almost entirely black. Vertex medially about as long as pronotum, sometimes slightly longer or shorter. Colour of abdomen and venter varying from entirely light to almost entirely blackish brown. Male pygofer as in Text-fig. 1173, process of pygofer as in Text-fig. 1174, genital style as in Text-fig. 1175, aedeagus as in Text-figs. 1176, 1177. Ratio length of aedeagus: width of head with eyes about 0.5. 1st abdominal sternum in male as in Text-fig. 1178, 2nd abdominal sternum in male as in Text-fig. 1179, 2nd abdominal tergum as in Text-fig. 1180. 7th abdominal sternum in female medially with a shallow angular incision. Overall length of males 4.75–6.5 mm, of females (5.5–) 6.5–8 mm.

Distribution. Common in Denmark, Sweden (Sk.-Nb.), Norway (up to SFi) and in southern and central East Fennoscandia (Al, Ab, N–Om, also in Vib and Kr.). – Widespread in the Palaearctic and Nearctic regions but many records probably refer to the following species.

Biology. In meadows and cultivated fields. Hibernation in the egg stage (Müller, 1957). Univoltine. Adults May – October.

Economic importance. Sometimes acting as a minor pest of clover and alfalfa. Vector of the stolbur virus of tomatoes etc.

Note. The specific name *makarovi* Zachvatkin for this taxon will probably be short-lived, considering the ample supply of possible older synonyms.

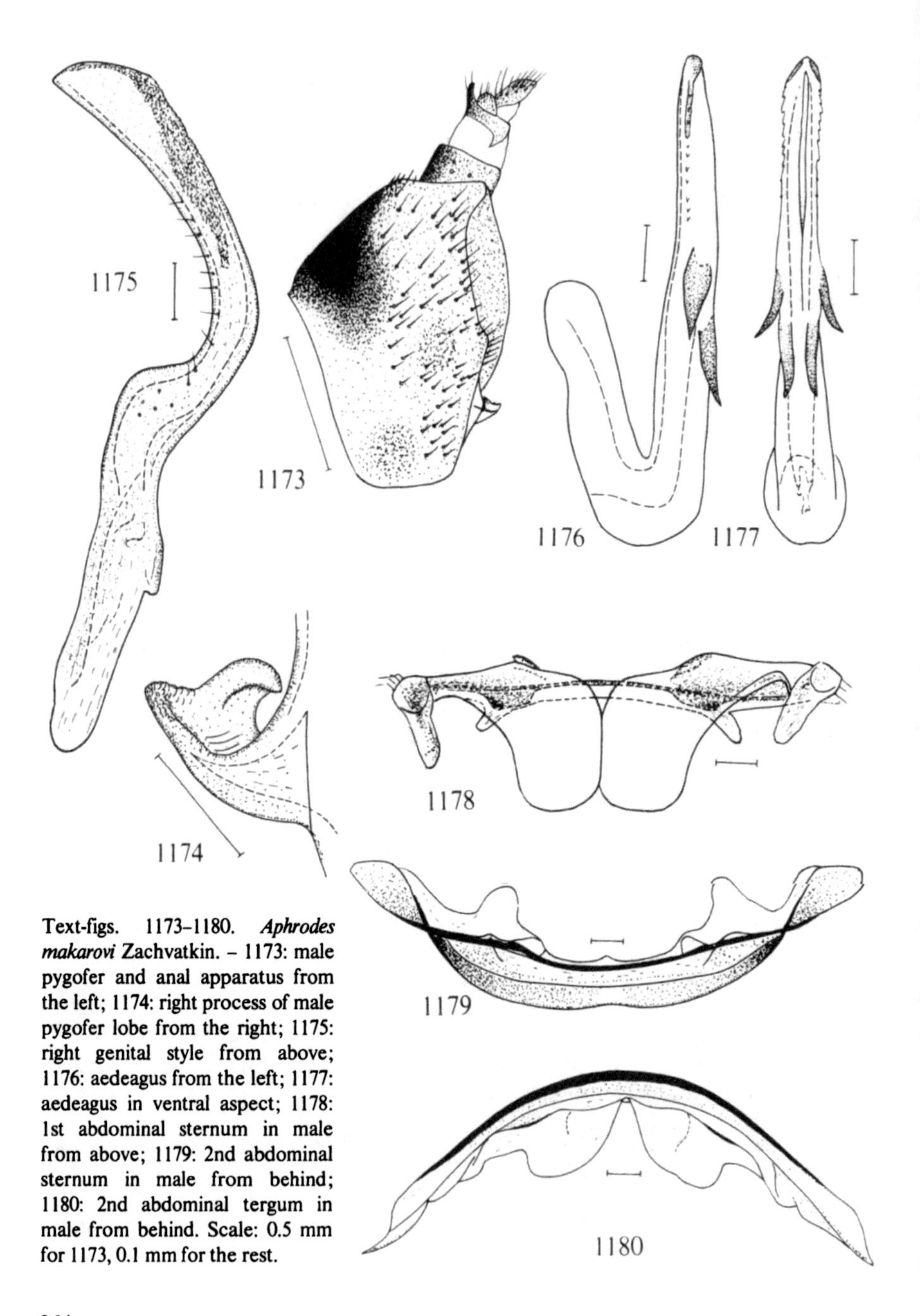

Text-figs. 1173–1180. *Aphrodes makarovi* Zachvatkin. – 1173: male pygofer and anal apparatus from the left; 1174: right process of male pygofer lobe from the right; 1175: right genital style from above; 1176: aedeagus from the left; 1177: aedeagus in ventral aspect; 1178: 1st abdominal sternum in male from above; 1179: 2nd abdominal sternum in male from behind; 1180: 2nd abdominal tergum in male from behind. Scale: 0.5 mm for 1173, 0.1 mm for the rest.

148. ***Aphrodes bicincta*** (Schrank, 1776)
Text-figs. 1167, 1168, 1181–1183.

Cicada bicincta Schrank, 1776:75.
Aphrodes bicinctus, ssp. *diminutus* Ribaut, 1952:334.

Resembling *makarovi*, body smaller, aedeagus (Text-figs. 1182, 1183) longer. Ratio length of aedeagus: width of head about 0.6. Male genital style as in Text-fig. 1181. Overall length of males 4.1–5.3 mm, of females 4.7–5.5 mm.

Note. Ribaut (1952) used the distance between the spines of aedeagus as a separating character between *"bicinctus bicinctus"* (= *makarovi*) and *"bicinctus diminutus"*. Owing to considerable individual variation this character is unreliable (cf. Nast, 1976). – The situation in *Aphrodes*, the smaller species having the longer aedeagus, recalls the conditions in the species-pair *Kelisia ribauti* – *Kelisia sabulicola* (see p. 50).

Distribution. Rare in Denmark, so far reliable records from F and NEZ. – Scarce and sporadic in Sweden, found in Sk., Hall., Sm., Öl., Gtl., Ög., Vg., Dlsl., Med.-Scarce

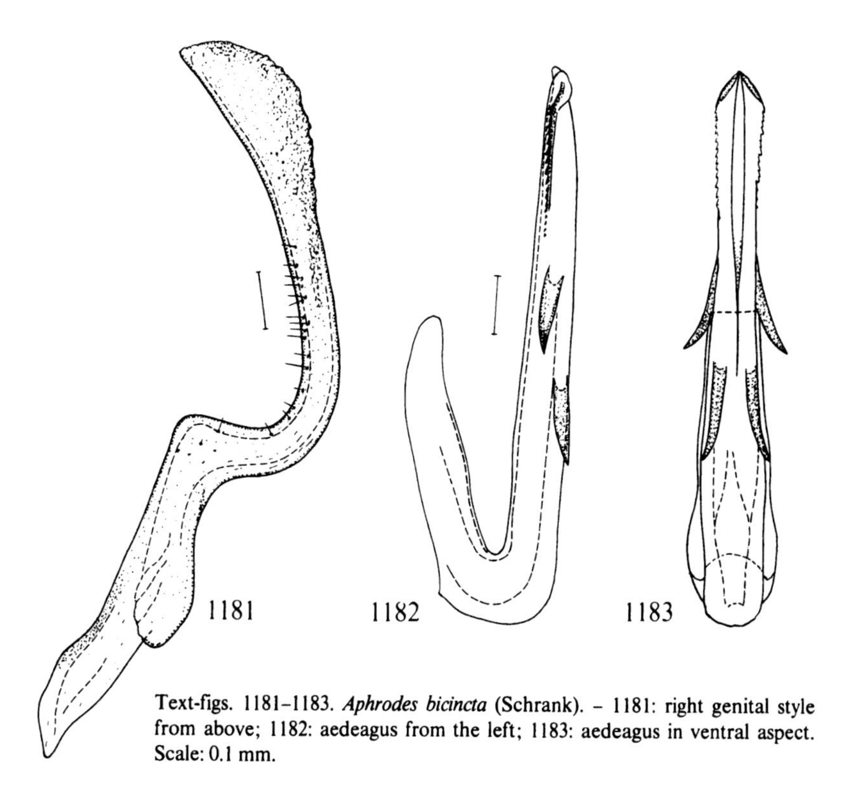

Text-figs. 1181–1183. *Aphrodes bicincta* (Schrank). – 1181: right genital style from above; 1182: aedeagus from the left; 1183: aedeagus in ventral aspect. Scale: 0.1 mm.

in Norway, found in AK, HEs, Ry. – So far not recorded from East Fennoscandia. – France, Poland, central Russia, northern Germany; North America.

Biology. In Sweden *Aphrodes bicinctus* appears to prefer dry habitats, such as the "alvar" of Öland. But in all 20 specimens were trapped in 7 different mire habitats in Norway, HEs: Eidskog (Ossiannilsson, 1977). Adults have been found in July – September.

Genus *Planaphrodes* Hamilton, 1975

Planaphrodes Hamilton, 1975:1012.

Type-species: *Cicada* (sic) *tricincta* Curtis, 1836, by original designation.

Vertex between ocelli flat, medially slightly to much longer than median length of pronotum. Margin of vertex carinately to foliaceously produced (Text-figs. 1169, 1170), sharper in males than in females, slightly to distinctly depressed laterally. Eyes broadly but shallowly notched near antennal pits. Face flat to slightly convex in female. Hind femora near apex with three subapical macrosetae and two smaller setae. Hind tibiae with short macrosetae. Shaft of aedeagus wide, strongly compressed laterally, with 2 or 3 pairs of spines arising near middle and sometimes with 2 small curved teeth near apex. Phallotreme slit-shaped. In Denmark and Fennoscandia 3 species.

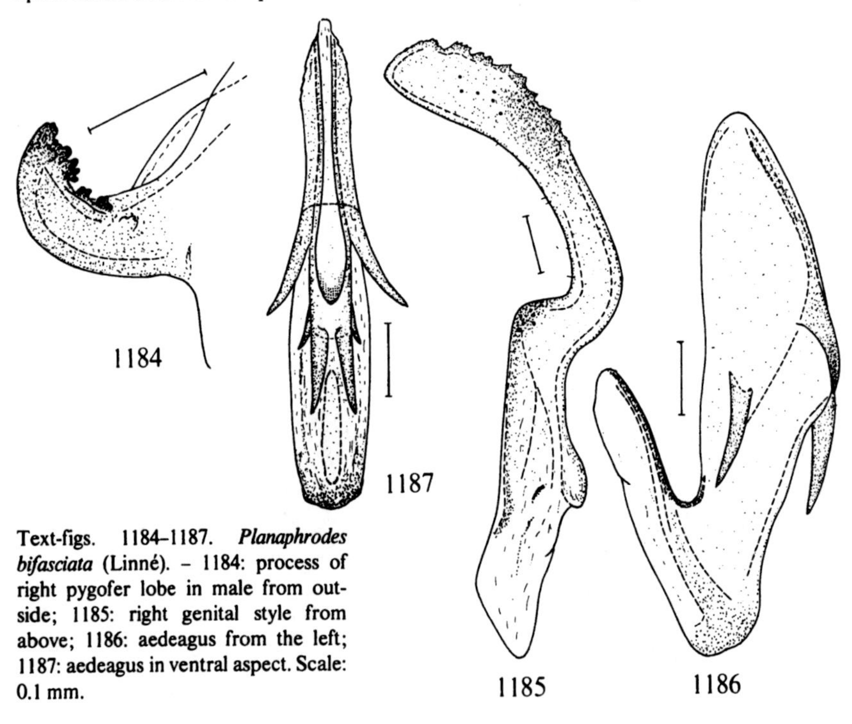

Text-figs. 1184–1187. *Planaphrodes bifasciata* (Linné). – 1184: process of right pygofer lobe in male from outside; 1185: right genital style from above; 1186: aedeagus from the left; 1187: aedeagus in ventral aspect. Scale: 0.1 mm.

Key to species of *Planaphrodes*

1	Males	2
-	Females	4
2(1)	Fore wing with two light transverse bands, apex also light. Aedeagus as in Text-figs. 1194, 1195	151. *trifasciata* (Fourcroy)
-	Fore wing with two light transverse bands, apex only rarely light (Plate-fig. 72). Aedeagus different	3
3(2)	Aedeagus as in Text-figs. 1186, 1187	149. *bifasciata* (Linné)
-	Aedeagus as in Text-figs. 1190, 1191	150. *nigrita* (Kirschbaum)
4(1)	Fore wings usually with two distinct light transverse bands or transverse rows of light spots (Plate-fig. 74). Small species, about 4 mm	151. *trifasciata* (Fourcroy)
-	Fore wings with indistinct transverse rows of light spots, or unspotted	5
5(4)	Larger, length > 5 mm. Hind wings extending beyond apex of clavus (according to Ribaut, 1952)	150. *nigrita* (Kirschbaum)
-	Smaller, length at most 5 mm. Hind wings not reaching apex of clavus	149. *bifasciata* (Linné)

149. ***Planaphrodes bifasciata*** (Linné, 1758)
Plate-figs. 72, 73, text-figs. 1184–1187.

Cicada bifasciata Linné, 1758:436.
Acucephalus tricinctus Curtis, 1836:pl.620.
Aphrodes bifasciatus, var. *simulans* Ribaut, 1952:344.

Head with fore border sharp, above with a distinct median carina and at least traces of a pair of lateral carinae. Male (Plate-fig. 72) brownish black, above with three broad ivory white transverse bands, viz. one on posterior part of pronotum (sometimes indistinct), and two across fore wings. In f. *tricincta* (Curtis) the proximal band of fore wing is divided into two spots. In. f. *simulans* (Ribaut) the apex of fore wing is white. Head medially distinctly longer than pronotum, above black with some small light spots along fore border, face along upper margin broadly brownish, for the rest usually largely yellowish white with a brownish spot below each antenna. The light colour of the face is an immediate continuation of the white transverse band of pronotum. In a similar way the proximal white band on fore wings corresponds to a broad light zone on the thoracal venter, a narrower and less distinct transverse band on the venter of the pregenital segments corresponding to the distal light transverse band of fore wings. Scutellum black, caudally sometimes paler. Head also in female (Plate-fig. 73) medially distinctly longer than pronotum. Female yellowish brown, above and below more or less densely dark mottled. Fore wings usually with only indistinct traces of the transverse bands present in males. Hind wings not extending beyond apex of clavus. Process on apical margin of male pygofer as in Text-fig. 1184, genital style as in Text-fig. 1185, aedeagus as in Text-figs. 1186, 1187. 7th abdominal sternum in female 2 – 2½ times as long as 6th sternum, medially with a distinct angular incision. Overall length of males 3.5–5 mm, of females 4.6–5 mm.

Distribution. Common and widespread in Denmark, as well as in Sweden up to Lu. Lpm. – Fairly common in Norway, found also in the north (Nnø, Nnv). – Common in southern and central East Fennoscandia, found in most provinces, including ObN and Ks; also in Vib and Kr. – Widespread in Europe, also found in Israel, Armenia, Azerbaijan, Kazakhstan, Kirghizia, Kurile Is., m. and w. Siberia, and in Newfoundland.

Biology. In dry meadows (Kontkanen, 1938). "Auf Trockenrasen" (Wagner & Franz, 1961.). "Ei-Überwinterer. Besonders am Boden" (Schiemenz, 1969). Adults (May –) June – October.

150. ***Planaphrodes nigrita*** (Kirschbaum, 1868)
Text-figs. 1188–1191.

Acocephalus nigritus Kirschbaum, 1868b:76.
Aphrodes tricinctus auct., nec Curtis, 1836.

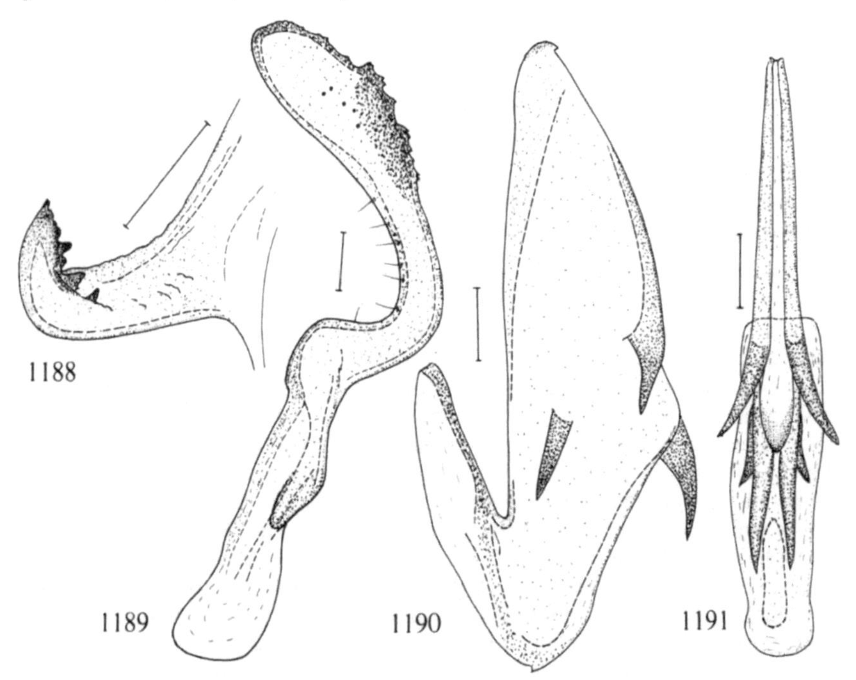

Text-figs. 1188–1191. *Planaphrodes nigrita* (Kirschbaum). – 1188: process of right pygofer lobe in male from outside; 1189: right genital style from above; 1190: aedeagus from the left; 1191: aedeagus in ventral aspect. Scale: 0.1 mm.

Resembling *bifasciata*. Proximal transverse band in male fore wing complete or broken. Female on upper part of face with 3 shallow depressions (2 in *bifasciata*), and hind wings almost as long as fore wings (according to Ribaut, 1952). Process of male pygofer lobe as in Text-fig. 1188, genital style as in Text-fig. 1189, aedeagus as in Text-figs. 1190, 1191. Overall length of males 4–4.6 mm, of females 5–5.9 mm.

Distribution. So far not found in Denmark, nor in Norway. – Very rare in Sweden, first found in "Umeå Lappmarken" by Zetterstedt. Waldén captured one male in Sk., N.Sandby, Laxbromölla 17.VII.1971. Lohmander took a male in Bl., Sölvesborg, Valje 10.VIII.1944, and Gyllensvärd found another male in Bl., Förkärla, Tromtö 5.VIII.-1956. Sundholm got two males in Lu. Lpm., Vuollerim in 1958. I found one male in Ång., Ådalsliden, Krånge, 17.VII.1966, and one male and one female in Vb., Jörn 20.VII.1939. – Very rare in East Fennoscandia, found in N: Helsingfors by Paulomo and 21.VII.1964 by Nordman; Ta: Hattula 12.VII.1945, 13.VII.1945, 18.VII.1945, and 22.VII.1945 by Nuorteva; Sa: Taipalsaari, July 1947 by Hellén, and in Tb: Jyväskylä 4–11.VII.1947 by Hackman. – Austria, Belgium, Bulgaria, Bohemia, Moravia, Slovakia, France, German D.R. and F.R., Italy, Poland, Portugal, Romania, Switzerland, Latvia, Lithuania, Yugoslavia, Ukraine; Algeria, Anatolia, Azerbaijan, Kazakhstan, Kirghizia.

Biology. Belongs to the fauna of low vegetation in woods (Schiemenz, 1965). "Ei-Überwinterer. 1 Generation. Besonders am Boden" (Schiemenz, 1969b).

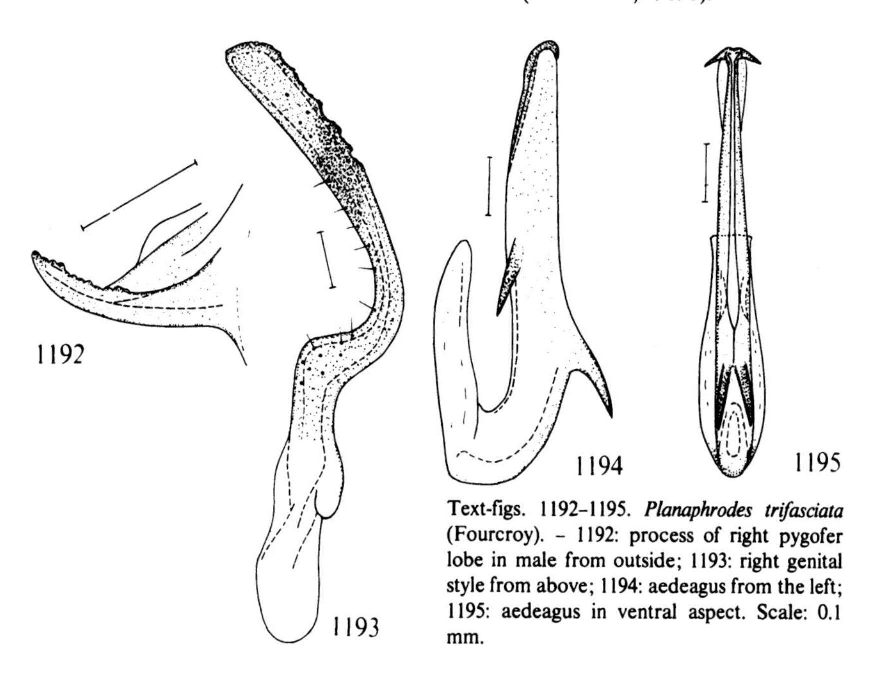

Text-figs. 1192–1195. *Planaphrodes trifasciata* (Fourcroy). – 1192: process of right pygofer lobe in male from outside; 1193: right genital style from above; 1194: aedeagus from the left; 1195: aedeagus in ventral aspect. Scale: 0.1 mm.

151. ***Planaphrodes trifasciata*** (Fourcroy, 1785)
Plate-fig. 74, text-figs. 1169, 1170, 1192–1195.

Cicada trifasciata Fourcroy, 1785:190.

Resembling *bifasciata* but smaller. Male: vertex medially as long as pronotum or slightly longer (Text-fig. 1169). Colour pattern as in male of *bifasciata* but dark surfaces usually more purely black, apex of fore wing always light. Female (Plate-fig. 74): head medially distinctly longer than pronotum (Text-fig. 1170). Colour much as in female of *bifasciata* but fore wings usually with transverse rows of distinct light spots corresponding to the transverse bands in male. Process of male pygofer lobe as in Text-fig. 1192, genital style as in Text-fig. 1193, aedeagus as in Text-figs. 1194, 1195. 7th abdominal sternum in female 2–3 times as long as 6th sternum, medially with a distinct angular incision. Overall length of males 3.5–4 mm, of females 3.6–5 mm.

Distribution. Fairly common in Denmark, found in most provinces, and in Sweden (Sk.–Nb.). – Scarce in Norway, found in AK, Ry, Ri, HOy, and SFy. – Scarce in southern and central East Fennoscandia; found in Al, Ab, N–Sa, Sb, Kb, ObN, Ks; Kr. – Austria, Belgium, Bohemia, Moravia, Slovakia, France, England, German D.R. and F.R., Netherlands, Greece, Hungary, Italy, Romania, Poland, Switzerland, Latvia, Estonia, Lithuania, n.Russia, Ukraine, Moldavia, Altai Mts., m.Siberia, Mongolia, China.

Biology. "Unter Heidekraut" (Kuntze, 1937). "Auf Heiden und Hochmooren, aber auch in Trockenrasen und Felsenheiden. Offenbar eine heliophile, aber nicht xerophile Art" (Wagner & Franz, 1961). Univoltine, hibernation in the egg stage (Remane, 1958; Schiemenz, 1969b, 1975). Adults in July – September.

Genus *Anoscopus* Kirschbaum, 1858

Acocephalus (Anoscopus) Kirschbaum, 1858b:357.
Type-species: *Cicada serratulae* Fabricius, 1775, by original designation.

Vertex curved between ocelli, more distinctly so in males. Median length of vertex approximately equalling that of pronotum. Vertex rounded to face in male without distinct edge, bluntly angled to face in female but without distinct carinate edge (Text-figs. 1171, 1172). Compound eyes with an incision near antennal pits. Hind femora with 3 large macrosetae near apex. Marginal macrosetae of hind tibiae very long. Shaft of aedeagus cylindrical, phallotreme elongate, on ventral surface. Small species. In Denmark and Fennoscandia six species.

Key to species of *Anoscopus*

1	Males	2
–	Females	7

2(1) Pronotum with distinct, well-defined transverse bands 6
– Markings of pronotum diffuse 3
3(2) Both pairs of appendages of aedeagus situated near aedeagal apex (Text-figs. 1208, 1209) .... 154. *serratulae* (Fabricius)
– Only one pair of appendages situated near apex of aedeagus 4
4(3) Proximal pair of appendages of aedeagus much longer than the distal pair (Text-figs. 1213, 1214) .... 155. *albiger* (Germar)
– Appendages of aedeagus approximately equal in length 5
5(4) Smaller, length 3–4 mm, length of aedeagus 0.63–0.7 mm. Apex of aedeagus in ventral aspect rounded (Text-fig. 1199). Ventral outline of aedeagus in lateral aspect almost straight (Text-fig. 1198) .... 152. *albifrons* (Linné)
– Larger, overall length 4.0–4.7 mm, length of aedeagus 0.74–0.82 mm. Apex of aedeagus in ventral aspect angular (Text-fig. 1204). Ventral outline of aedeagus in lateral aspect with a bulge near middle (Text-fig. 1203) .... 153. *limicola* (Edwards)
6(2) Body short and broad. Fore wings with black longitudinal bands between veins (Plate-fig. 84). Genitalia as in Text-figs. 1220–1223. .... 157. *flavostriatus* (Donovan)
– Body more elongate. Fore wings with black veins, cells light (Plate-fig. 85). Genitalia as in Text-figs. 1216–1219.... 156. *histrionicus* (Fabricius)
7(1) Veins of fore wing lighter than cell membrane, fore wings more or less distinctly longitudinally striped. Apex of head with two yellowish white spots, or with one transversely oval light spot with a small black dot on middle .... 157. *flavostriatus* (Donovan)
– Fore wings not longitudinally striped. Markings on apex of head different 8
8(7) Vertex medially distinctly shorter than pronotum. Commissural border of fore wing dark with two whitish spots, these being as long as the dark parties .... 156. *histrionicus* (Fabricius)
– Vertex not or indistinctly shorter than pronotum. Light spots on commissural border, if present, shorter than dark parts 9
9(8) 7th abdominal sternum with a distinct semicircular or arcuate median incision (Text-fig. 1215) .... 155. *albiger* (Germar)
– Median incision of 7th abdominal sternum different 10
10(9) Median incision of 7th abdominal sternum small, angular (Text-fig. 1200). Small species, length 4.4–5 mm, width of pronotum 1.5–1.6 mm .... 152. *albifrons* (Linné)
– Median incision of 7th abdominal sternum more distinctly developed (Text-figs. 1205, 1210) 11
11(10) Fore wings unicolorous, brownish yellow with concolorous veins, or finely dark mottled. Halobiont .... 153. *limicola* (Edwards)
– Fore wings usually with some well-marked blackish spots. In dry biotopes .... 154. *serratulae* (Fabricius).

152. ***Anoscopus albifrons*** (Linné, 1758)
Plate-fig. 75, text-figs. 1171, 1172, 1196–1200.

Cicada albifrons Linné, 1758:437.
Cicada subrustica Fallén, 1806:10.
Cicada affinis Fallén, 1806:13.
Cicada dispar Zetterstedt, 1828:520.

Male (Plate-fig. 75): head somewhat shorter than pronotum, vertex with a fine median carina, in lateral aspect faintly concave (Text-fig. 1171), fore border of head obtuse. Brownish yellow to reddish brown, or rosy-fuscous, colour pattern much varying. Veins of fore wings inconspicuous. In one common form the fore wings are longitudinally striped as in *histrionicus*, veins being brown and brown-bordered, interspaces pale, wing apex pale distally of a dark transverse band. In another colour form the fore wings are transversely banded more or less as in *Planaphrodes trifasciata* with two transverse bands and apex whitish; in some specimens these bands are broken in irregularly arranged spots. Transitory forms are common. Female: head as long as pronotum or slightly shorter, above plain or indistinctly concave (Text-fig. 1172). Colour varying

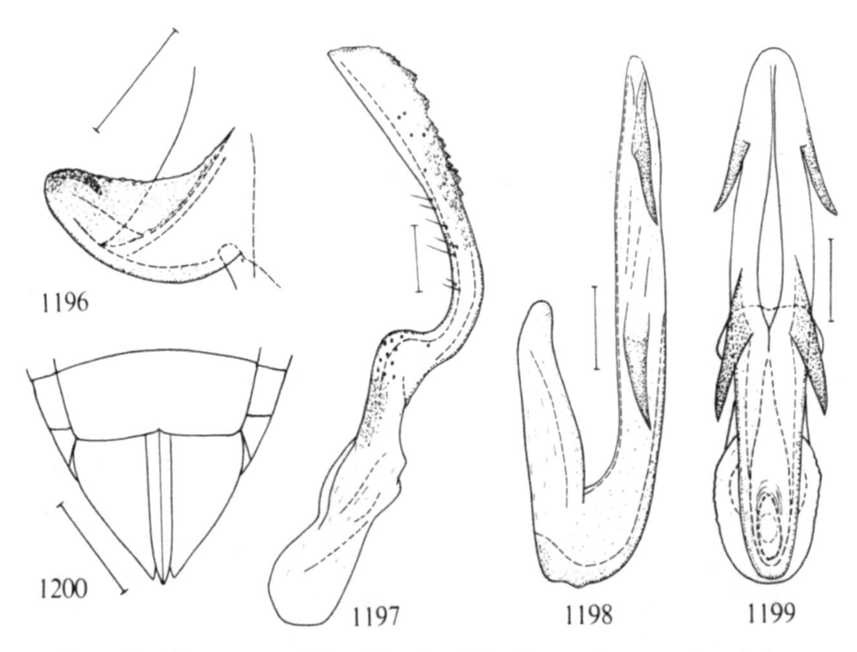

Text-figs. 1196–1200. *Anoscopus albifrons* (Linné). – 1196: right pygofer process in male from outside; 1197: right genital style from above; 1198: aedeagus from the left; 1199: aedeagus in ventral aspect; 1200: apical part of female abdomen from below. Scale: 0.5 mm for 1200, 0.1 mm for the rest.

from straw-yellow to black-brown, usually finely mottled. Process on caudal margin of male pygofer as in Text-fig. 1196, genital style as in Text-fig. 1197, aedeagus as in Text-figs. 1198, 1199. Caudal border of 7th abdominal sternum in female with a small angular median incision (Text-fig. 1200). Overall length of males 3–4 mm, of females 4.4–5 mm.

Distribution. Common in Denmark, found in all districts. - Common in the south of Sweden, Sk.-Dlr. - Sparsely spread in the south of Norway, found in AK, Bø, Bv(?), AAy, VAy, Ry, Ri, HOy, HOi, SFy, SFi, STi. - Scarce in southern and central East Fennoscandia, recorded from Al, Ab, N, Ka, St, Ta, Sa, Tb, Kb; Vib, Kr. - Widespread in Europe, also found in Morocco, Tunisia, Israel, Armenia; North America.

Biology. "In Hochmooren, Wäldern und Waldlichtungen" (Kuntze, 1937). "An Waldgräsern" (Wagner & Franz, 1961). "Mostly in rather dry localities with *Agrostis tenuis, Anthoxanthum, Luzula, Holcus mollis,* etc." (Gravestein, 1965). In "Sumpfreitgras-Beständen (*Calamagrostis canescens, Sphagnum,* sporadisch *Molinia*" (Schiemenz, 1976). Hibernation takes place in the egg stage (Müller, 1957; Remane, 1958; Schiemenz, 1969b). Univoltine. "Lebt wie alle *Aphrodes*-Arten besonders am Boden" (Schiemenz, 1969b).

153. ***Anoscopus limicola*** (Edwards, 1908)
Text-figs. 1201–1205.

Acocephalus limicola Edwards, 1908a:57.
Acocephalus limicola maculata Edwards, 1908a:58.
Aphrodes limicola W.Wagner, 1937b:66, Figs. 7a, 7b.

Resembling *Anoscopus albifrons* but larger. Male: varying in colour, main colour greyish brown. The following is taken from Gravestein (1965): "Nervures of the elytra clearly visible; mostly accentuated with black. Sharp contrast between light and dark parties. The light spots become ivory-white. The dark apical parties always touch the end of the membrane. The darkening of the elytra longitudinally. The dark coloured specimens - can be considered being the variety *maculata* as named by Edwards". In that variety the fore wings are mainly dark with light spots. Female above almost entirely brownish yellow, or finely black mottled, fore wings with veins largely pale also in dark specimens, abdomen concolorous or largely black. Process of male pygofer as in Text-figs. 1201, genital style as in Text-fig. 1202, aedeagus as in Text-figs. 1203, 1204, apical part of female abdomen as in Text-fig. 1205. Length of males 4–4.7 mm, of females 4.5–5.55 mm.

Distribution. So far not found in Denmark, Norway and East Fennoscandia. - Rare in the south of Sweden, only found in Sk., Brunnby, seashore in Mölle, 28–30.VII.1961, in all 67 specimens (Ossiannilsson), and in Hall., Värö, St. Åvan, 200 m from the river-mouth 28.VII.1970, 2 ♂♂, 1 ♀ (I. Wäreborn). - England, Ireland, Netherlands, N. Germany, Crete.

Biology. "*Aphrodes limicola* is exclusively halobiont and is only found in the Puccinellietum« (Gravestein, 1965).

154. ***Anoscopus serratulae*** (Fabricius, "775)
Plate-fig. 76, text-figs. 1206–1210.

Cicada serratulae Fabricius, 1775:686.
Cicada fuscofasciata Goeze, 1778:161.

Male (Plate-fig. 76) resembling the transversely banded form of *albifrons*. Head above not concave, about as long as pronotum, often brown or with a brown transverse band. Pronotum usually largely white or whitish. Scutellum usually brownish. Fore wings brownish, with two broad, irregular transverse bands and wing apex whitish. The band pattern of the upper side is continued on the venter. Fore and middle tibiae light with dark apices, hind tibiae largely dark. Female: head about as long as pronotum, above less distinctly concave than in female of *albifrons*. Fore wings often with well-marked black spots. Process of male pygofer as in Text-fig. 1206, genital style as in Text-fig.

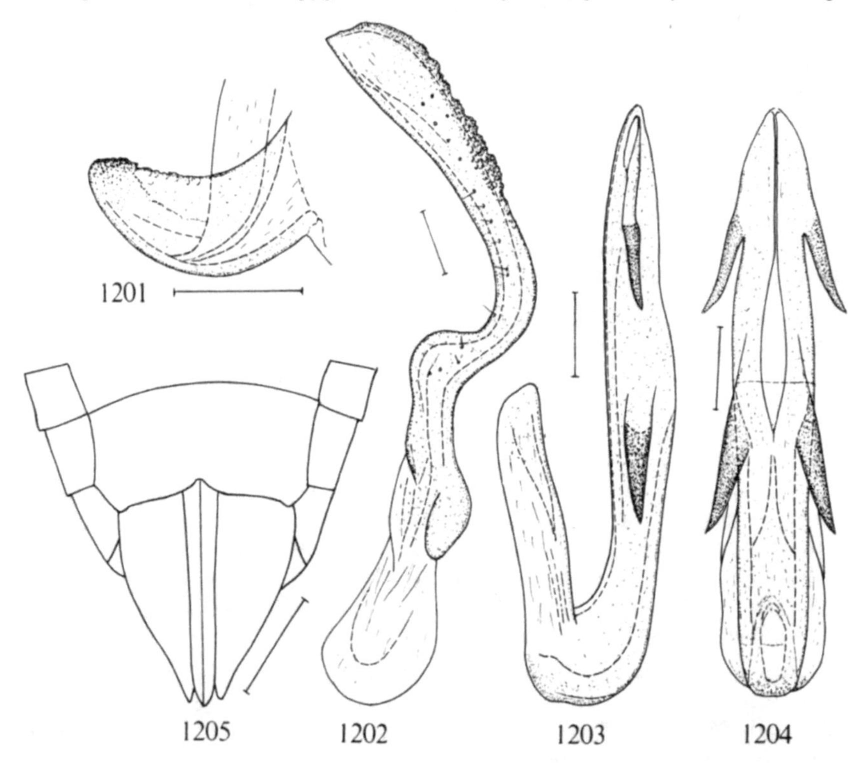

Text-figs. 1201–1205. *Anoscopus limicola* (Edwards). – 1201: right pygofer process in male from outside; 1202: right genital style from above; 1203: aedeagus from the left; 1204: aedeagus in ventral aspect; 1205: apical part of female abdomen from below. Scale: 0.5 mm for 1205, 0.1 mm for the rest.

1207, aedeagus as in Text-figs. 1208, 1209, apical part of female abdomen as in Text-fig. 1210. Overall length of male 3–4 mm, of female 4–4.7 mm.

Distribution. Uncommon in Denmark, found in EJ, NEJ, and B. - Not uncommon in the south of Sweden, found in Sk., Bl., Hall., Sm., Öl., Ög., Vg., Dlsl., Upl. - So far not found in Norway, nor in East Fennoscandia. - Widespread in Europe, also recorded from Tunisia, Georgia, and North America.

Biology. In "besonnten Hängen und Wäldern. Auch unter Steinen" (Kuntze, 1937). Belongs to the "Arrhenatheretum elatioris" (Marchand, 1953). "Auf feuchten Wiesen" (Wagner & Franz, 1961). Belongs to the "Waldsteppe" (Okáli, 1960). "In mesophilen und hygrophilen Biotopen" (Schiemenz, 1969b). Hibernation takes place in the egg stage (Remane, 1958; Schiemenz, l.c.). Univoltine (Schiemenz, l.c.). Adults in July - October.

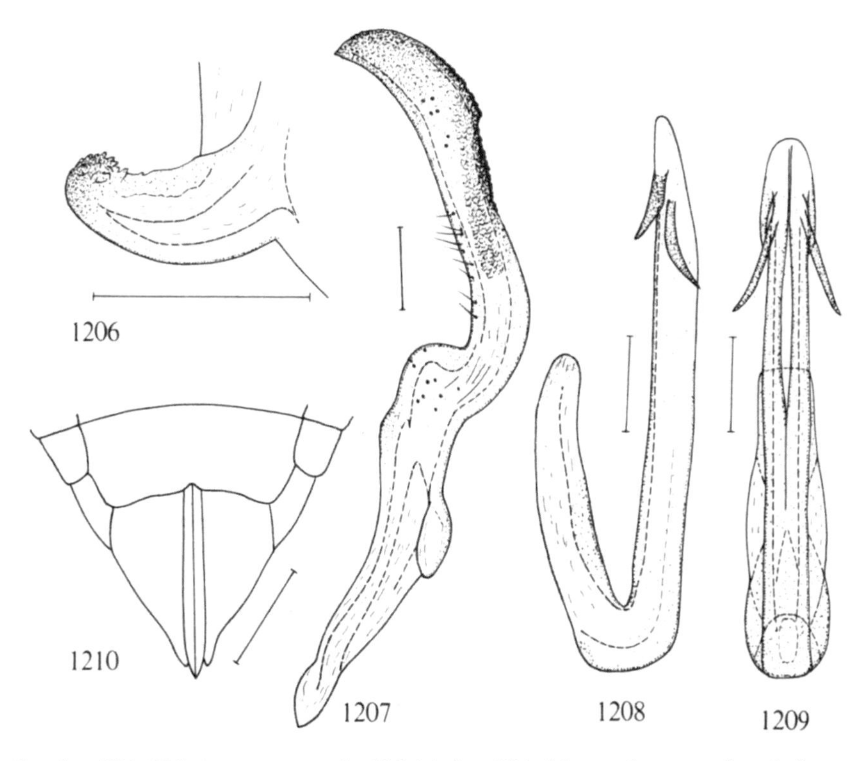

Text-figs. 1206–1210. *Anoscopus serratulae* (Fabricius). - 1206: right pygofer process in male from outside; 1207: right genital style from above; 1208: aedeagus from the left; 1209: aedeagus in ventral aspect; 1210: apical part of female abdomen from below. Scale: 0.5 mm for 1210, 0.1 mm for the rest.

155. ***Anoscopus albiger*** (Germar, 1821)
Text-figs. 1211–1215.

Jassus albiger Germar, 1821:88.

Resembling *A. albifrons*, much varying in colour. Vertex of male flat or slightly convex, not concave. Pronotum uniformly brownish mottled or with posterior half lighter. Fore wings lighter or darker brown with or without light spots, apex usually but not always light beyond a dark transverse band. Female light yellow with some dark spots along costal margin of fore wing, or black mottled. Process of male pygofer as in Text-fig. 1211, genital style as in Text-fig. 1212, aedeagus as in Text-figs. 1213, 1214, apical part of female abdomen as in Text-fig. 1215. Overall length of male 3.6–3.8 mm, of female 4.4–4.5 mm.

Distribution. Rare in Denmark, found only once – B: Ypnasted 21.VIII.1976 (Trolle). – Not found in Sweden, Norway and East Fennoscandia. – Austria, Bohemia, Moravia, Slovakia, France, England, German D.R. and F.R., Netherlands, Poland, Estonia, Latvia, Kirghizia; North America.

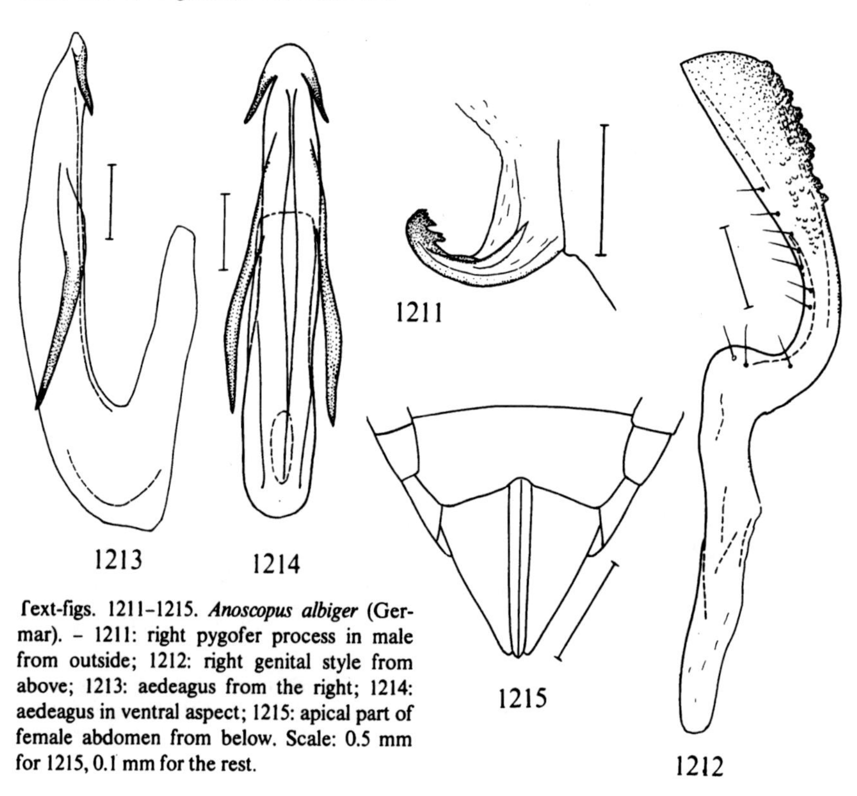

Text-figs. 1211–1215. *Anoscopus albiger* (Germar). – 1211: right pygofer process in male from outside; 1212: right genital style from above; 1213: aedeagus from the right; 1214: aedeagus in ventral aspect; 1215: apical part of female abdomen from below. Scale: 0.5 mm for 1215, 0.1 mm for the rest.

Biology. "In Norddeutschland halophil" (Wagner, 1939). "In marshy places" (Le Quesne, 1965). Hibernates in the egg stage (Remane, 1958).

156. ***Anoscopus histrionicus*** (Fabricius, 1794)
Plate-fig. 85, text-figs. 1216–1219.

Cercopis histrionica Fabricius, 1794:56.

Male (Plate-fig. 85) yellowish to greyish to white, elongate. Head distinctly shorter than pronotum, dorsal outline in lateral aspect faintly concave. Fore border of head with a large black spot enclosing a light spot on apex of head and often a small light spot below each ocellus. A black transverse band along hind border of vertex is usually connected with the black spot just mentioned by a broad median longitudinal band. A triangular black spot laterally of the longitudinal band is sometimes attached to the transverse band. Face brownish, frontoclypeus above spotted black and yellow, lora yellowish with brownish patches. Pronotum with a broad black transverse band anterior to middle and a narrower band along hind border. Scutellum largely black. Fore wings with dark and dark-bordered veins and a broad fuscous transverse band near the

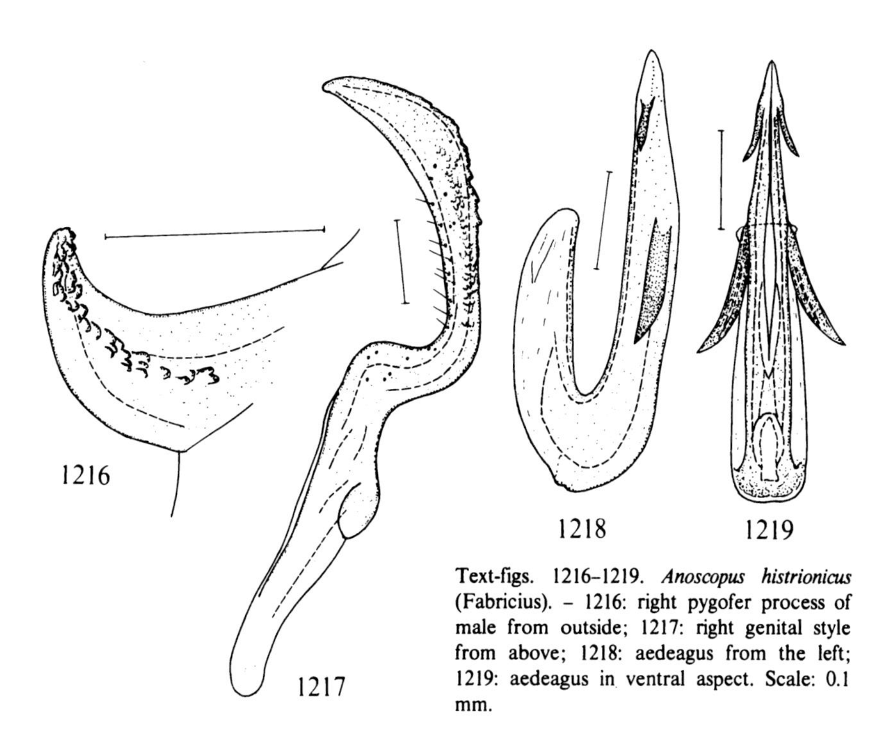

Text-figs. 1216–1219. *Anoscopus histrionicus* (Fabricius). – 1216: right pygofer process of male from outside; 1217: right genital style from above; 1218: aedeagus from the left; 1219: aedeagus in ventral aspect. Scale: 0.1 mm.

light wing apex. Venter largely dark, fore and middle tibiae with dark apices, hind tibiae largely dark. Female greyish yellow, finely dark mottled. Head shorter than pronotum, dorsal outline in lateral aspect slightly concave. Fore wings greyish white, commissural border dark with two whitish spots. spots (around distal ends of claval veins), these spots being as long as the dark parts. Process of male pygofer lobe as in Text-fig. 1216, genital style as in Text-fig. 1217, aedeagus as in Text-figs. 1218, 1219. 7th abdominal sternum in female twice as long as 6th sternum, hind border medially with a shallow angular incision. Overall length of male 3–4 mm, of female 4–5 mm.

Distribution. Scarce in Denmark, found in SJ, EJ, WJ, F, and SZ. – Rare in southern Sweden, found in Sk., Sm., Öl., Gtl., G.Sand. – Norway: only recorded from Ø: Sarpsborg 16.VIII.1867 and 14.VIII.1874 (Grimsgaard leg.). – Scarce and sporadic in southern and central East Fennoscandia, recorded from Al, Ab, N, St, Ta, Kb; Vib. – Widespread in Europe, also found in Armenia, Azerbaijan, Georgia, Kazakhstan.

Biology. In dry meadows (Sahlberg, 1871). In "Stranddünen und Waldlichtungen" (Kuntze, 1937). On grasses (Le Quesne, 1965). Adults in June – September.

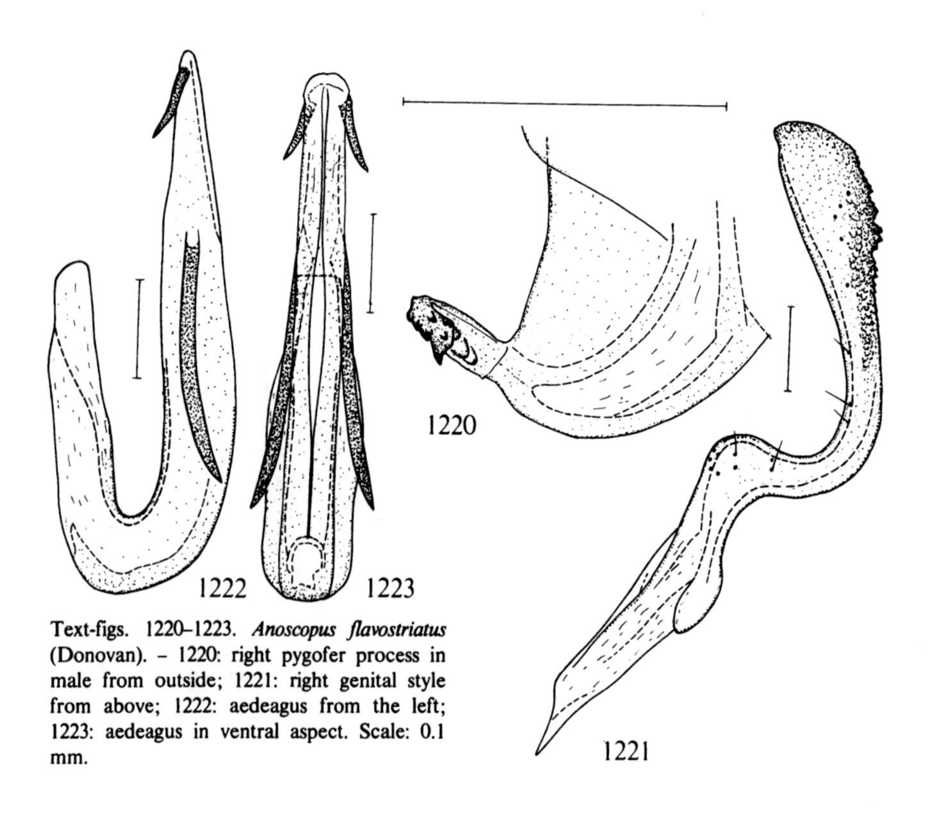

Text-figs. 1220–1223. *Anoscopus flavostriatus* (Donovan). – 1220: right pygofer process in male from outside; 1221: right genital style from above; 1222: aedeagus from the left; 1223: aedeagus in ventral aspect. Scale: 0.1 mm.

157. ***Anoscopus flavostriatus*** (Donovan, 1799)
Plate-fig. 84, text-figs. 1220–1223.

Cicada flavostriata Donovan, 1799:88.
Jassus rivularis Germar, 1821:89.

Male (Plate-fig. 84) yellowish white, body short and broad, head above convex, about as long as pronotum, vertex with a fine median carina. On the obtuse fore border of the head there is a fuscous transverse band often enclosing two or four light spots. This band does not reach eyes. Face dark-spotted. Hind border of vertex with three fuscous spots caudally usually continuous with each others, the median spot sometimes being connected with the transverse band on fore border by a dark longitudinal line. Pronotum somewhat anterior to middle with a fuscous transverse band sometimes divided into spots. Scutellum dark spotted or entirely dark. Fore wings with six dark longitudinal bands between veins. The four longitudinal bands of corium join near wing apex in an arched transverse band; costal border and wing apex light. Female yellowish brown to greyish brown, vertex as long as pronotum, convex or at least not distinctly concave, with a fine median carina. The markings on head apex mentioned in the key are characteristic. For the rest the female resembles the same sex of *albifrons*, but the colour markings on fore part of body are coarser, less finely mottled, often forming a fuzzy image of the colour pattern of the male. Fore wings more or less distinctly longitudinally striped. Process of male pygofer lobe as in Text-fig. 1220, genital style as in Text-fig. 1221, aedeagus as in Text-figs. 1222, 1223. 7th abdominal sternum in female twice as long as 6th sternum, hind border medially with a shallow angular incision. Overall length of male 2.5–3.6 mm, of female 3–4.6 mm.

Distribution. Fairly common in Denmark (SJ, EJ, WJ, LFM, SZ, NEZ, B), as well as in southern Sweden (Sk.–Gstr.). - Norway: found in AK, HEs, Ry, and NTy. - Fairly common in southern and central East Fennoscandia, recorded from Al, Ab, N, St, Ta, Tb, Kb, Om, ObS; Vib, Kr. - Widespread in Europe, also in Altai Mts., Armenia, Georgia, Uzbekistan, m. and w. Siberia, Maritime Territory, and North America.

Biology. In "Hochmooren, Wäldern, Waldlichtungen, Wiesen" (Kuntze, 1937). Hygrophilous, belonging to the "Molinietalia" (Marchand, 1953). Hibernation in the egg stage (Remane, 1958; Törmälä & Raatikainen, 1976).

Genus *Stroggylocephalus* Flor, 1861

Acocephalus (Stroggylocephalus) Flor, 1861a:210.
Type-species: *Cicada agrestis* Fallén, 1806, by monotypy.

Head anteriorly rounded, about half as long as pronotum, above anteriorly transversely wrinkled, caudally less distinctly longitudinally wrinkled. Eyes with a notch near antennal pits. Hind femora with four long macrosetae and one smaller dorsal macroseta near apex; macrosetae of hind tibiae very long. Genital valve of male distinct, aedeagal shaft with only two appendages, phallotreme short. In the Palaearctic region two species.

Key to species of *Stroggylocephalus*

1 Fore border of head very narrowly black or brownish. Pygofer of male with a forceps-shaped process (Text-fig. 1224). 7th abdominal sternum in female with a weak median incision (Text-fig. 1230) .... 158. *agrestis* (Fallén)

– Fore border of head light, head above near fore border with a dark arched line parallel with fore border. Process of male pygofer simple (Text-fig. 1231). 7th abdominal sternum in female medially with a deep angular incision (Text-fig. 1237) 159. *livens* (Zetterstedt)

158. ***Stroggylocephalus agrestis*** (Fallén, 1806)
Plate-fig. 88, text-figs. 1224–1230.

Cicada agrestis Fallén, 1806:23.

Strongly built, fairly broad, above greyish yellow. Fore border of head with a fine dark-bordered furrow running from eye to eye. Male (Plate-fig. 88) above more or less densely dark mottled. Head medially not quite twice as long as near eyes. Pronotum transversely wrinkled. Face densely dark mottled, dark pigmentation of venter more or less extended. But sometimes one finds males as light-coloured as females. Wings extending somewhat beyond apex of abdomen. Female entirely greyish yellow or indistinctly mottled. Wings approximately as long as abdomen, sometimes slightly longer or shorter. Genital plates of male together arranged as the sides of a boat. Process of male pygofer as in Text-fig. 1224, genital style as in Text-fig. 1225, aedeagus as in Text-figs. 1226, 1227, 1st abdominal sternum in male as in Text-fig. 1228, 2nd. abdominal sternum in male as in Text-fig. 1229. Apical part of female abdomen as in Text-fig. 1230. Overall length of males 5-5.9 mm, of females 5.8-6.5 mm. – Last instar larva mottled yellowish-brown. Fore border of head partly black. Notum with a pale median stripe, abdomen with three pale longitudinal stripes. Metanotum with a pair of small black spots.

Distribution. Fairly common in Denmark (SJ, EJ, F, NEZ), and in southern and central Sweden (Sk.-Gstr.). – Norway: recorded from Ø and AAy (Risør,Warloe leg.) but the record from Ø has not been verified. – Fairly common in southern and central East Fennoscandia, found in Al, Ab, N, St, Ta, Sa, Oa; Vib, Kr. – Widespread in Europe, also in palaearctic parts of Asia, and in the Oriental region.

Biology. In "Salzstellen, Hochmooren, Flachmooren. Überwintert am Boden zwischen Schilf" (Kuntze, 1937). In wet peaty meadows (Linnavuori, 1952a). In the "*Bromus racemosus – Senecio aquaticus*-Assoziation" and in the "*Cariceto canescentis – Agrostidetum caninae*-Assoziation" (Marchand, 1953). "An *Carex* auf Mooren" (Wagner & Franz, 1961). Hibernation takes place in the egg stage (Müller, 1957), in the

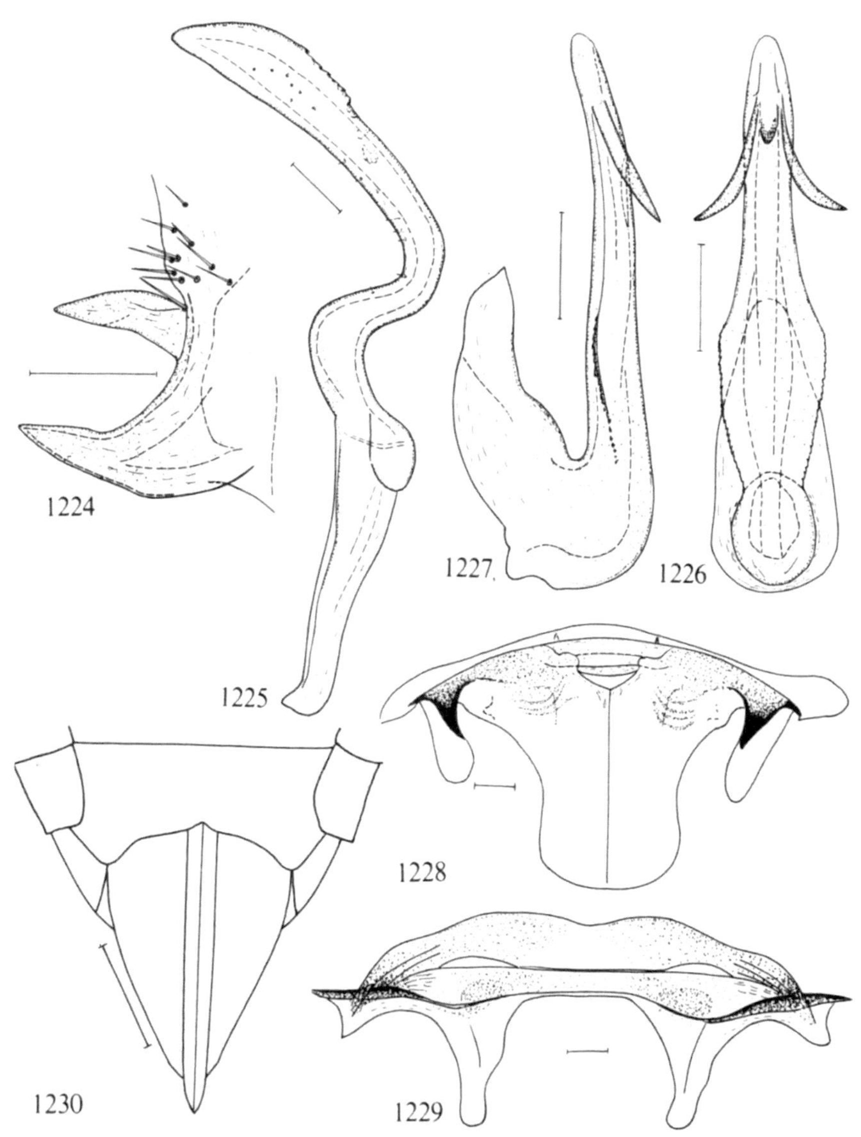

Text-figs. 1224–1230. *Stroggylocephalus agrestis* (Fallén). – 1224: right pygofer process of male from outside; 1225: right genital style from above; 1226: aedeagus in ventral aspect; 1227: aedeagus from the left; 1228: 1st abdominal sternum in male from above; 1229: 2nd abdominal sternum in male from above; 1230: apical part of female abdomen from below. Scale: 0.5 mm for 1230, 0.1 mm for the rest.

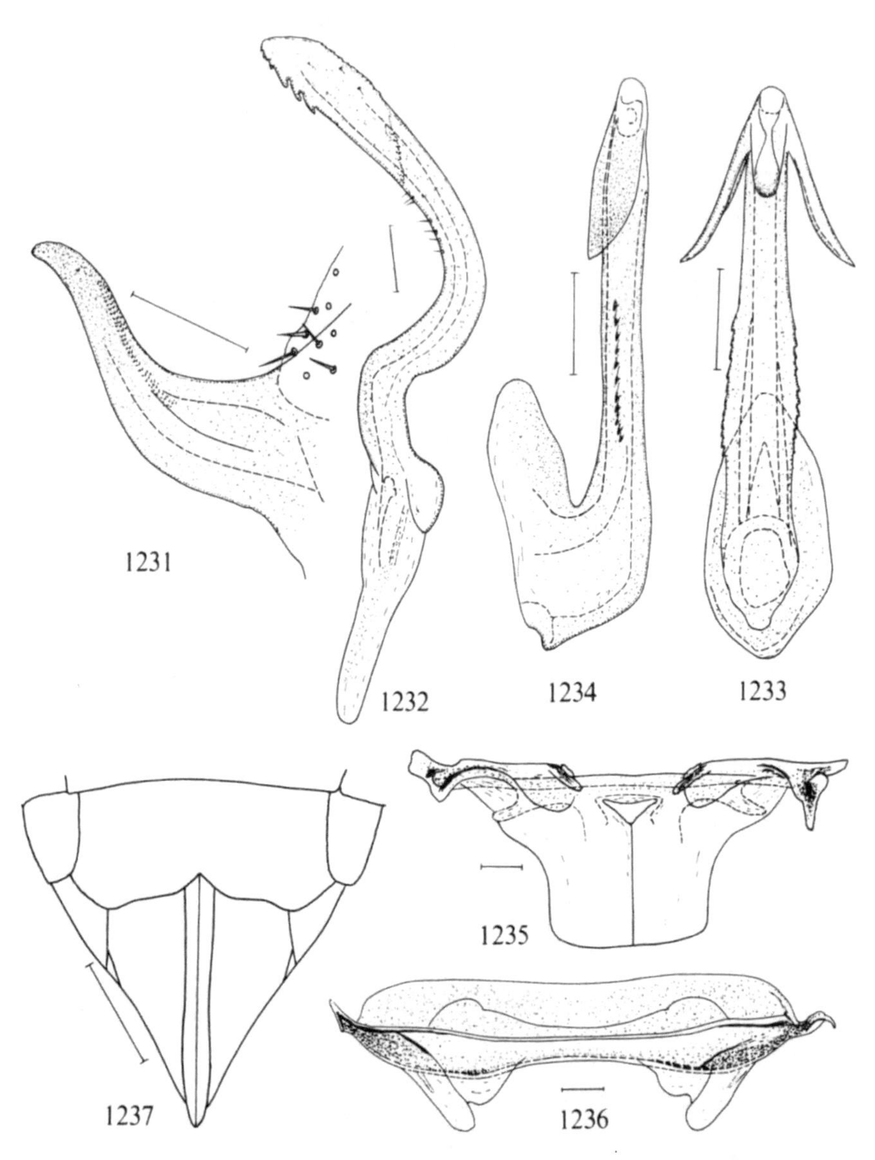

Text-figs. 1231–1237. ***Stroggylocephalus livens*** (Zetterstedt). – 1231: right pygofer process of male from outside; 1232: right genital style from above; 1233: aedeagus in ventral aspect; 1234: aedeagus from the left; 1235: 1st abdominal sternum in male from above; 1236: 2nd abdominal sternum from above; 1237: apical part of female abdomen from below. Scale: 0.5 mm for 1237, 0.1 mm for the rest.

adult stage (Schiemenz, 1976). Adults in June – October. Müller (1942) observed oviposition on *Carex riparia.*

159. ***Stroggylocephalus livens*** (Zetterstedt, 1838)
Text-figs. 1231 – 1237.

Pholetaera livens Zetterstedt, 1838:288.
Pholetaera nigropunctata Zetterstedt, 1838:288.
Strongylocephalus megerlei Scott, 1874b:122.

Resembling *agrestis,* fore border of head less sharp, without a distinct furrow. Male: dark mottlings of dorsum and fore wings coarser than in *agrestis,* venter largely black. Female above entirely brownish yellow, venter indistinctly mottled. Face above with a dark arched stripe just below and parallel with the light fore border of head. Wings comparatively slightly shorter than in *agrestis* and therefore more often shorter than abdomen. Process of male pygofer as in Text-fig. 1231, genital style as in Text-fig. 1232, aedeagus as in Text-figs. 1233, 1234, 1st abdominal sternum in male as in Text-fig. 1235, 2nd abdominal sternum in male as in Text-fig. 1236. Apical part of female abdomen as in Text-fig. 1237. Overall length of male 5–5.5 mm, of female 5.5–6.5 mm.

Distribution. Quite rare in Denmark, found in EJ and NEZ. – Widespread but comparatively scarce in Sweden, Sk.–Lu. Lpm. – Rare in Norway, found in AK: Oslo, AK: Snarøya, and AK: Drøbak, and in HEs: Eidskog. – Rare and sporadic in East Fennoscandia, found in Ab: Rusko, ObN: Aavasaksa, LkW: Kittilä; Kr: Jalguba. – Belgium, France, England, German D.R. and F.R., Italy, Netherlands, Poland, Estonia, Latvia, Lithuania, Ukraine, Mongolia, Maritime Territory.

Biology. In "Hochmooren. Überwintert im Torfmoss" (sic) (Kuntze, 1937). "In der Zwischenmoor-Bultgesellschaft. Univoltin" (Schiemenz, 1975). Hibernates in the adult stage (Remane, 1958; Schiemenz, 1975). Adults in June – November.

SUBFAMILY CICADELLINAE

Anteclypeus large, basally broad, distally narrowing and rounded. Frontoclypeus and anteclypeus swollen. Lora and genae narrow. Genae joining anteclypeus as very narrow bands. Ocelli situated on upper side of head. Frontal sutures extending over anterior margin of head to or near ocelli. In Denmark and Fennoscandia three genera.

Key to genera of Cicadellinae

1 Ocelli nearer to hind margin of head than to fore margin (Tribe **Cicadellini**) .... *Cicadella* Latreille (p.389)
– Ocelli not nearer to hind margin of head than to fore margin 2
2(1) Frontoclypeus with a median carina (Tribe **Evacanthini**) *Evacanthus* Le Peletier & Serville (p.384)
– Frontoclypeus without a median carina (Tribe **Errhomenini**) *Bathysmatophorus* J. Sahlberg (p.388)

Genus *Evacanthus* Le Peletier & Serville, 1825

Evacanthus Le Peletier & Serville, 1825:612.
Type-species: *Cicada interrupta* Linné, 1758, by monotypy.

Strongly built leafhoppers with wings a little longer (♂) or a little shorter (♀) than abdomen. Fore wings leathery. Head above with a longitudinal carina crossing a transverse carina; in front of the latter there is an arched transverse carina near fore border of head. Fore borderof pronotum moderately convex, hind border medially with a more or less distinct obtuse-angled incision. Scutellum large. Hind tibiae with four rows of strong setae. In Denmark and Fennoscandia two species.

Key to species of *Evacanthus*

1 Fore wings entirely yellow .... 160. *interruptus* (Linné), f. *xantha* Melichar
– Fore wings not entirely yellow.. 2
2(1) Main colour pale yellow to orange yellow. Pronotum with two lateral spots varying in size, sometimes coalescing. Fore wing

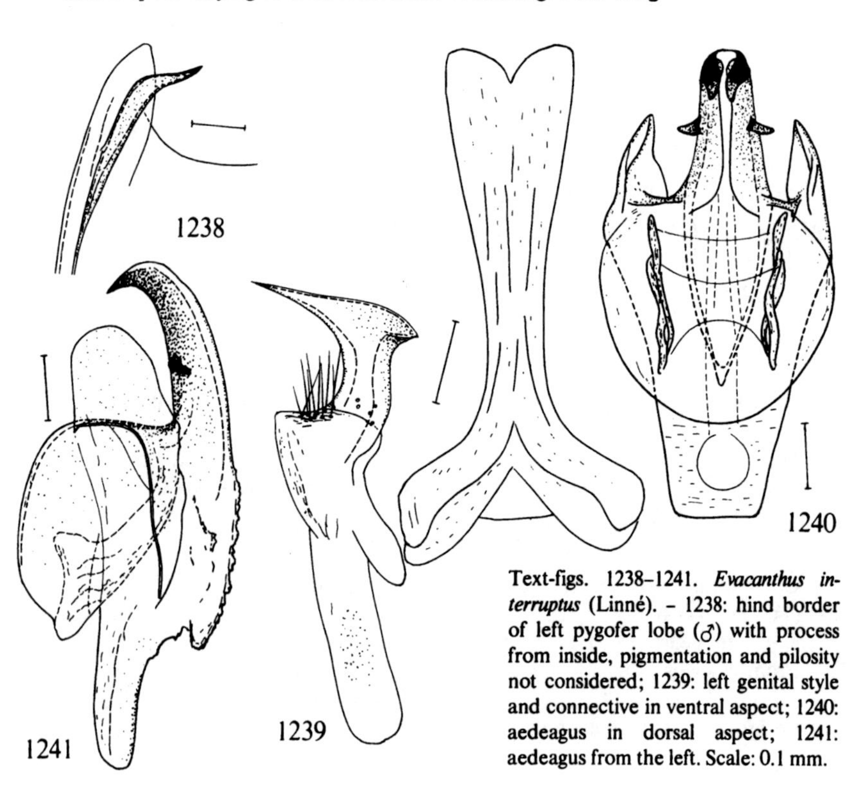

Text-figs. 1238–1241. *Evacanthus interruptus* (Linné). – 1238: hind border of left pygofer lobe (♂) with process from inside, pigmentation and pilosity not considered; 1239: left genital style and connective in ventral aspect; 1240: aedeagus in dorsal aspect; 1241: aedeagus from the left. Scale: 0.1 mm.

with a black longitudinal band interrupted by a light band along Cu (Plate-fig. 83) .... 160. *interruptus* (Linné)

– Main colour greyish yellow or brownish yellow. Dark markings of upper side diffusely limited, i.a. consisting of 3 sometimes coalescing longitudinal bands or spots on pronotum (pronotum often largely black), and some black spots on fore wing (Plate-fig. 82) .... 161. *acuminatus* (Fabricius)

160. ***Evacanthus interruptus*** (Linné, 1758)
Plate-fig. 83, text-figs. 1238-1241.

Cicada interrupta Linné, 1758:438.
Cicada moesta Zetterstedt, 1828:521.
Euacanthus interruptus xanthus Melichar, 1896:179.

Dorsum of fore body sparsely short-haired, shining. Median carina of face usually obtuse or indistinct. Head above with a black pattern consisting of some more or less confluent spots. Scutellum usually with black lateral corners. Venter almost entirely yellowish, in male with 2nd and 3rd abdominal sterna largely black, in female often with diffuse fuscous surfaces. Abdominal tergum in male largely black, in female yellow with two broad black longitudinal bands. Males are usually macropterous, females sub-brachypterous with hind wings shortened. Legs yellow, genital plates of male narrow, fuscous with a long pilosity. In f. *xantha* Melichar the fore wings are entirely yellow; on the other hand the black colour is predominating in f. *moesta* (Zetterstedt). Pygofer lobes of male on inside with a curved process (Text-fig. 1238) not visible without dissection. Genital style and connective as in Text-fig. 1239, aedeagus as in Text-figs. 1240, 1241. 7th abdominal sternum medially convex, sometimes with a small incision. Overall length of males 5–5.8 mm, of females 6.3–7 mm. – Last instar larva pale yellowish, head above with two round black spots varying in size, dorsum with two fuscous longitudinal bands extending from pronotum caudally of eyes to abdominal apex; abdomen with a sparse erect pilosity.

Distribution. Common in Denmark, found in nearly all districts. – Common in Sweden, Sk.–Nb. – Common and widespread also in Norway, found in most districts up to TRy. – Common in southern and central East Fennoscandia, found as far to the north as in ObN; also in Vib and Kr. – Widespread in Europe, also recorded from Georgia, Kazakhstan, Kirghizia, Kamchatka, Sakhalin, m. and w. Siberia, Altai Mts., China, Japan, Korean Peninsula, Maritime Territory, and Nearctic Region.

Biology. In "Flachmooren, Waldlichtungen, Wiesen" (Kuntze, 1937). Belongs to the "Arrhenatheretum" (Schwoerbel, 1957). "In Wäldern auf Kräutern und Stauden" (Wagner & Franz, 1961). "Frequent throughout the oat-growing region in eutrophic biotopes, either young mixed forests dominated by deciduous trees or meadows and fields. - Also on *Rubus idaeus*" (Törmälä & Raatikainen, 1976). Hibernation in egg stage (Müller, 1957; Remane, 1958; Törmälä & Raatikainen, l.c.). Adults in June – September.

161. ***Evacanthus acuminatus*** (Fabricius, 1794)
Plate-fig. 82, text-figs. 1242–1244.

Cicada acuminata Fabricius, 1794:36.
Cicada interstincta Fallén, 1806:29.

Median carina of face usually sharp. Pilosity of dorsum and fore wings slightly longer and denser than in *interruptus*. Scutellum dark with a light spot at each lateral corner, or entirely blackish brown. Markings of dorsum and fore wings diffusely limited, often coalescing. Face and venter more or less largely brownish. Process of male pygofer lobe almost straight (Text-fig. 1242), longer than in *interruptus*, usually visible without dissection. Aedeagus as in Text-figs. 1243, 1244. Hind border of 7th abdominal sternum in female convex, medially with an incision varying in shape. Overall length of males 5–5.6 mm, of females 5.5–6.8 mm.

Distribution. Uncommon in Denmark, especially found in the southern parts of the country. – Not uncommon in Sweden, found in most provinces as far to the north as Vb. – Scarce in Norway, found in Ø, AK, Os, Bø, VE, TEi, AAy, AAi, Ri, HOy, SFy, MRi. – Comparatively common in East Fennoscandia, found in Al, Ab, N, St, Ta, Sa, Tb, Sb, Kb; Vib, Kr. – Widespread in Europe and palaearctic parts of Asia, also present in the Nearctic Region.

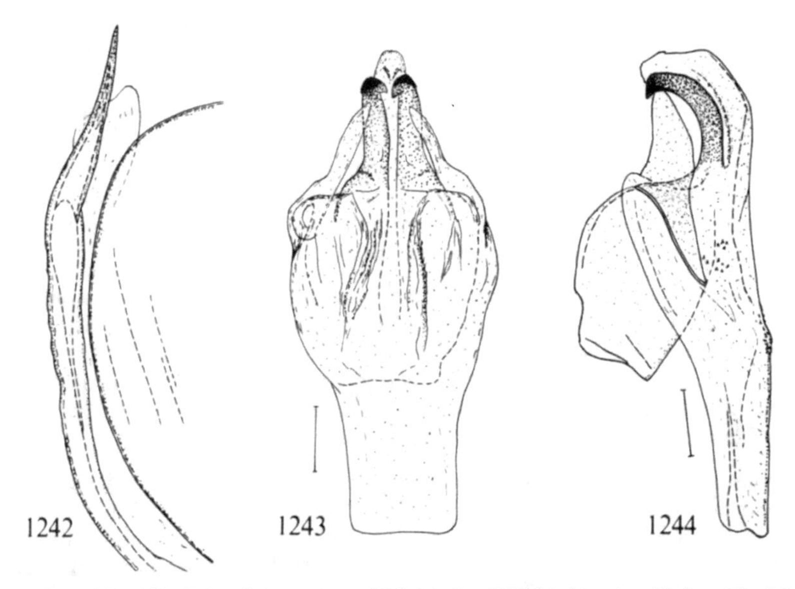

Text-figs. 1242–1244. *Evacanthus acuminatus* (Fabricius). – 1242: hind border of left pygofer lobe in male with process from inside, pigmentation and pilosity not considered; 1243: aedeagus in dorsal aspect; 1244: aedeagus from the left. Scale: 0.1 mm.

Biology. In "Waldlichtungen" (Kuntze, 1937). In sphagnous spruce woods and rich swampy woods (Linnavuori, 1952a). In the *"Querceto-Betuletum"* on the ground (Schwoerbel, 1957). "Im Auenwald und in der Weingärtenraine" (Okáli, 1960). "In Wäldern an Stauden und Kräutern, gern in Hochstaudenfluren" (Wagner & Franz, 1961). "On grasses, also on bushes" (Le Quesne, 1965). Adults in June – September.

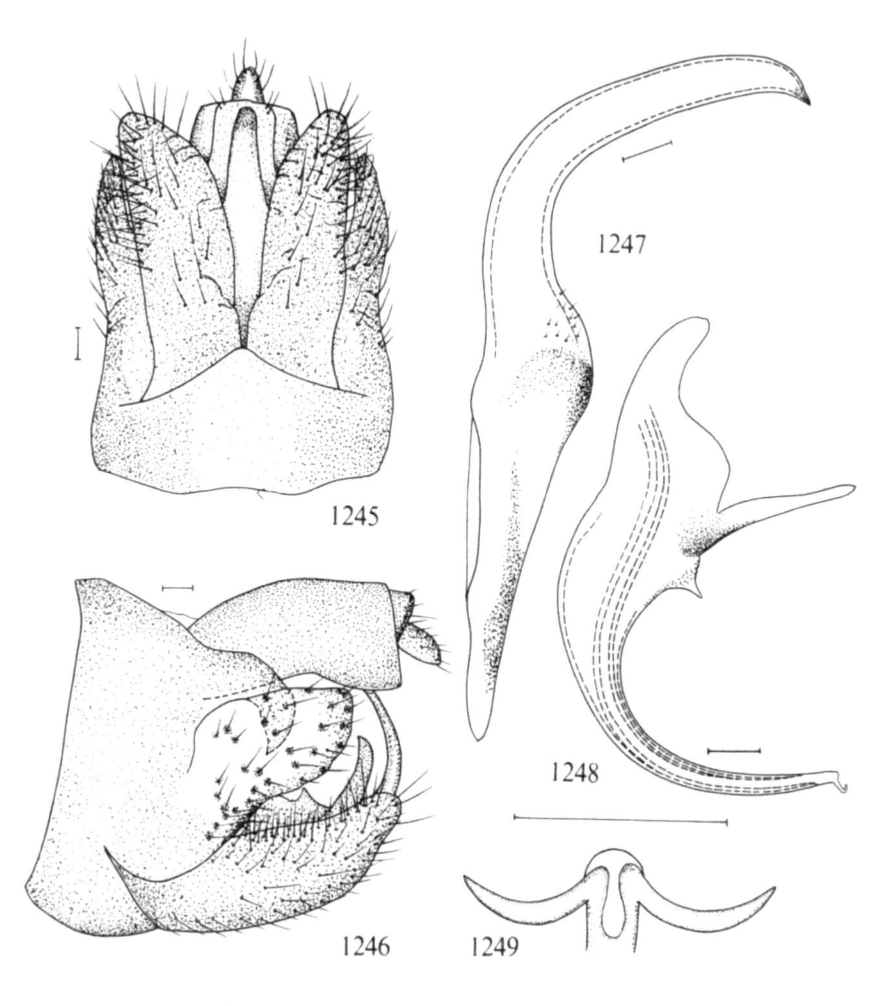

Text-figs. 1245–1249. *Bathysmatophorus reuteri* J. Sahlberg. – 1245: male genital segment from below; 1246: male genital segment from the left; 1247: right genital style from outside; 1248: aedeagus from the left; 1249: apex of aedeagus in terminal aspect. Scale: 0.1 mm.

Genus *Bathysmatophorus* J. Sahlberg, 1871

Bathysmatophorus J. Sahlberg, 1871:109.
Type-species: *Bathysmatophorus reuteri* J. Sahlberg, 1871, by monotypy.

Body strongly built, anteriorly depressed, caudally vaulted. Head frontally obtuse rounded, frontoclypeus inflated. Part of frontal suture visible from above transverse. Ocelli dorsal, situated just caudally of frontal sutures, somewhat nearer to fore than to hind border of head. Fore border of pronotum medially almost straight, hind border concave, obtuse angular, side margins little shorter than median length. Wing-dimorphous: male macropterous with wings longer than abdomen, female usually brachypterous with fore wings apically rounded, extending to caudal margin of 4th abdominal tergum. Fore and middle tibiae with three, hind tibiae with four rows of strong macrosetae. In Europe one species.

162. ***Bathysmatophorus reuteri*** J. Sahlberg, 1871
Plate-figs. 86, 87, text-figs. 1245–1249.

Bathysmatophorus reuteri J. Sahlberg, 1871:111.

Shining brownish yellow with more or less extended fuscous markings and a sparse short erect or semi-erect pilosity. Venter of male largely blackish brown, fore border of head yellowish white, dorsum black mottled. In the female the dark colour is less extended, main colour often with a greenish tinge. Fore wings in both sexes irregularly wrinkled, veins in the macropterous form (Plate-fig. 87) with alternate blackish and ivory white patches. Claval suture obliterated in the brachypterous form (Plate-fig. 86). Setae of femora and tibiae arising from partly tuberculiform blackish spots. Male genital segment as in Text-figs. 1245, 1246, genital style as in Text-fig. 1247, aedeagus as in Text-figs. 1248, 1249. 7th abdominal sternum in female medially convex. – Larvae resembling those of *Idiocerus* spp. but without erect setae (according to Sahlberg, 1871). Overall length (♂♀) 7–8 mm.

Distribution. Not found in Denmark, nor in Norway. – Very rare in Sweden, in all only four specimens found: one in Nb.: Övertorneå 9.VI.1930 (C.H. Lindroth), one male in Nb.: Boden 23.VI.1932 (K. Fahlander), one female in Nb.: Nedertorneå 22.VII.-1966 (A. Sundholm), and one in Lu. Lpm.: Pålkem 17.VII.1952 (A. Jansson). – Rare and sporadic in East Fennoscandia, found in Ks, LkE; Kr, Lr. – N. Russia, Kazakhstan, n. and w. Siberia, Mongolia.

Biology. On *Salix* (Sahlberg, 1871). "Unter Gebüsch und stattlicheren Kräutern" (Lindberg, 1947). Adults in June – August.

Genus *Cicadella* Latreille, 1817

Cicadella Latreille, 1817:406.
  Type-species; *Cicada viridis* Linné, 1758, by subsequent designation.
Cicadella Latreille, International Commission of Zoological Nomenclature, Opinion 647, 1963 (name placed on Official List of Generic Names in Zoology).

Body elongate. Frontoclypeus large, inflated. Ocelli situated nearer to compound eyes and to hind border of head than to each other. Fore border of pronotum broadly rounded, hind border almost straight. Fore wings with some transverse veins proximally of apical transverse veins. Hind tibiae with four rows of strong macrosetae. A pair of accessory structures, paraphyses, present between apex of connective and base of aedeagus. In Northern Europe two species.

Key to species of *Cicadella*

1 Frontoclypeus in both sexes with a yellowish median band between two parallel brownish longitudinal bands, laterally of which are many parallel brownish transverse lines (muscle traces). Fore wings comparatively broader, ratio length/width about 3.1–3.3, in mature males black or fuscous, rarely green, with a bluish tinge, in immature males greyish or sordid green with black veins, in females largely green. Head above in both sexes with two large quadrangular, pentagonal or hexagonal black spots on disk (Text-fig. 1250). Apodemes of 2nd abdominal sternum in male usually long (Text-fig. 1255, 1256). 163. *viridis* (Linné)

– Frontoclypeus of a more uniform brownish colour, markings indistinct. Fore wings comparatively narrower, ratio length/width about 3.8, in both sexes light green, costal margin yellow. Black spots on disk of vertex in female small, transversely oval (Text-fig. 1257) or polygonal. Apodemes of 2nd abdominal sternum in male smaller (Text-figs. 1261–1263) 164. *lasiocarpae* n.sp.

163. ***Cicadella viridis*** (Linné, 1758)
  Plate-figs. 62, 63, text-figs. 1250–1256.

Cicada viridis Linné, 1758:438.
Tettigonia arundinis Germar, 1821:71.
Tettigonia viridis var. *concolor* Haupt, 1912:184.
Amblycephalus viridis Ossiannilsson, 1946c:83.
Tettigella viridis Lindberg, 1947:27.

Body yellow, head above between ocelli with two quadrangular, pentagonal or hexagonal black spots, and a smaller black spot above each antenna (Text-fig. 1250). Frontoclypeus in female more strongly swollen than in male, in both sexes marked as

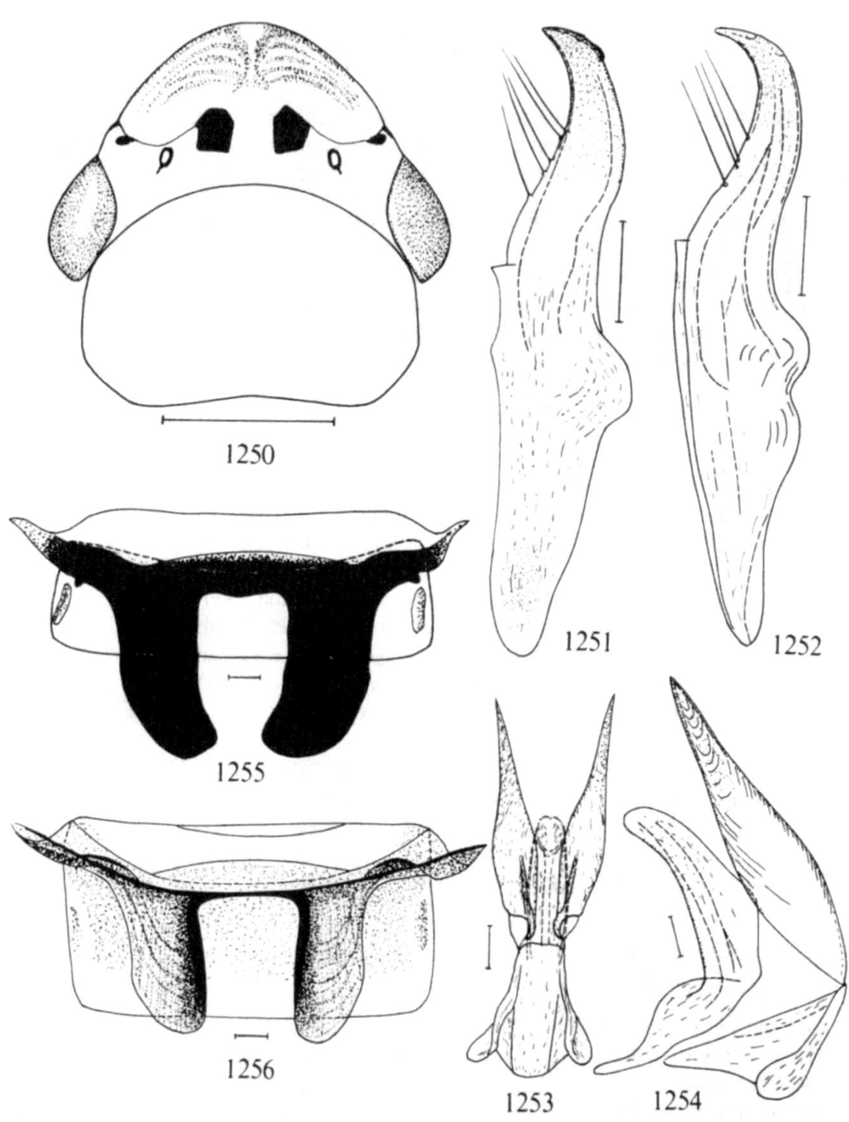

Text-figs. 1250–1256. *Cicadella viridis* (Linné). – 1250: head and pronotum of female from above; 1251: right genital style from above; 1252: same of immature specimen ("f. *arundinis*"); 1253: aedeagus, connective and paraphyses in ventral aspect; 1254: same from the left; 1255: 2nd and 3rd abdominal sterna in male from above; 1256: same of immature specimen ("*arundinis*"). Scale: 1 mm for 1250, 0.1 mm for the rest.

described in the key. Pronotum frontally with or without a transverse row of four black spots varying in size, caudally transversely wrinkled, green (except the very hind border). Fore wings in male (Plate-fig. 62) black with a bluish tinge (in living specimens coated with a whitish wax powder), sometimes green with a bluish tinge (f. *concolor* (Haupt)), apex colourless or fumose. Immature males have been described under the name of *arundinis* (Germar). In such males the fore wings are sordid greenish or greyish, veins black. Fore wings of females (Plate-fig. 63) grass-green with yellow costal border and yellow or brownish veins, apex as in male. Abdomen above bluish black or black, side borders yellow, abdominal venter in male largely black, in female usually yellow. Legs yellow or (especially in male) orange yellow. Genital style of male as in Text-figs. 1251, 1252; aedeagus, connective and paraphyses as in Text-figs. 1253, 1254, apodemes of 2nd abdominal sternum in mature males usually large, black (Text-fig. 1255), in immature males ("*arundinis*") shorter and lighter (Text-fig. 1256). Hind border of 7th abdominal sternum in female with an angular median incision. Overall length of males 5.7–7 mm, of females 7.5–9 mm. – Last instar larva yellow, frontoclypeus largely fuscous, head above with two large black spots, notum with two, abdomen with four black longitudinal bands, fore wing pads black with a light longitudinal band.

Distribution. Denmark: common, found in all districts. – Sweden: common, abundant in its typical biotope, Sk.-Lu.Lpm. – Norway: common in southern and eastern districts, found in AK, HEn, Os, Bø, VE, TEy, TEi, AAy, AAi, VAy, Ry, and MRi. – Common in southern and central East Fennoscandia, Al, Ab, N-Ok and ObN, also in Vib and Kr. – Widespread in Europe and Palaearctic part of Asia, also present in the Oriental Region but records from Canada are probably wrong (Young, 1977).

Biology. In "Flachmooren, Wäldern, Waldlichtungen, Wiesen" (Kuntze, 1937). "Insbesondere auf verschiedenen Weissmooren" (Kontkanen, 1938). In "tall-sedge bogs, wet "rimpi" bogs, quagmire marshes" (Linnavuori, 1952a). "Die Eiablage von *Cicadella viridis* erfolgt vorwiegend in die basalen Stengelteile von *Juncus* (bes. *filiformis*) und *Scirpus*, vielfach sogar durch die basale Blattscheide hindurch" (Müller, 1942). Feeds on *Holcus mollis;* hibernation in the egg stage in stems of *Juncus effusus* (Morcos, 1953). "The female makes long slits with its ovipositor in the stems of *Juncus effusus* where the eggs are laid. The egg-slits are 3.5 to 7.5 mms. long and contain 3 to 16 eggs in one row. - All of the eggs are found beneath the scale from about 2.5 to 17.5 cms. above soil level, -. The egg - is about 1.7 mm. in length. One generation per annum. Some eggs - were found parasitised by *Anagrus incarnatus* Hal." (Morcos, l.c.). In the "Molinietalia", "auf Bentgraswiesen, *Caltha*-Wiesen und extremnassen *Carex*-Wiesen" (Marchand, 1953). "Polyphag (an 166 Pflanzenarten aus 39 Familien) vorwiegend an der niederen Vegetation, besonders *Juncus, Scirpus* (und *Phragmites*) feuchter Biotope (Ufer)" (Müller, 1956). "Ausgesprochen hygrophil" (Schiemenz, 1965). Adults 10.VII.-21.X. (in Swedish conditions).

Economic importance. In spite of the normal breeding plants and food plants being wild rushes, sedges and marsh grasses, *Cicadella viridis* is known as a pest on various fruit trees and on grape-vines, also on *Sorghum*, rice, wheat, sugar-cane, sugar-beet, corn, *Phaseolus*, cabbage, *Alnus, Betula, Salix* in many countries, as France, Bulgaria, Germany, China, Japan, once also in Finland (on cabbage). The damage is caused by the oviposition of females resulting in weakening and withering of plants. See further Schmutterer (1953), Müller (1956).

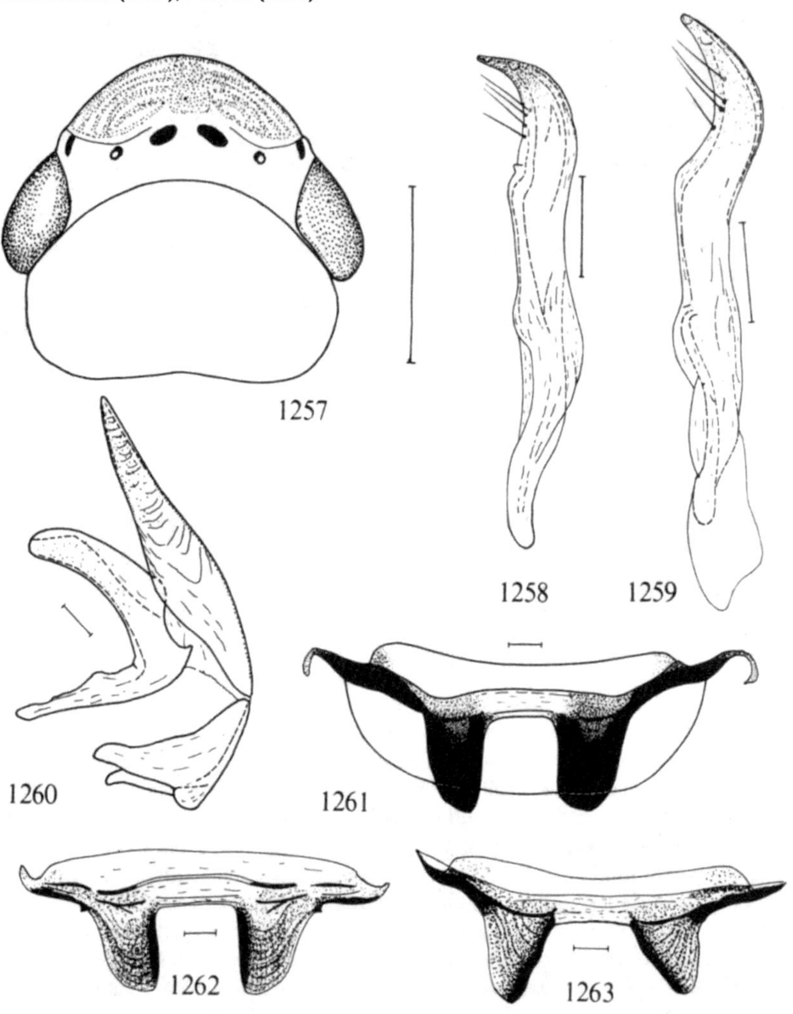

Text-figs. 1257–1263. *Cicadella lasiocarpae* n.sp. - 1257: head and pronotum of female from above; 1258, 1259: right genital styles of two specimens from above; 1260: aedeagus, connective and paraphyse from the left; 1261: 2nd and 3rd abdominal sterna in male from above; 1262, 1263: 2nd abdominal sterna of two immature males from above. Scale: 1 mm for 1257, 0.1 mm for the rest.

164. ***Cicadella lasiocarpae*** n.sp.
Text-figs. 1257–1263.

Description. Body yellow, head above with two black spots, in males usually shaped as in *viridis*, but smaller; in female distinctly smaller, polygonal or transversely oval. Head in female (Text-fig. 1257) distinctly shorter than in the same sex of *viridis*. Fore wings in both sexes light green, costal margin and veins yellow, or veins concolorous or partly fuscous. Apex of fore wing colourless or smoke-coloured. Colour of venter, abdomen and legs as in *viridis*, hind tibiae often with a fuscous tinge. Genital style of male as in Text-figs. 1258, 1259, aedeagus, connective and paraphyses as in Text-fig. 1260, apodemes of 2nd abdominal sternum in male smaller than in *viridis*, black in mature specimens (Text-fig. 1261), largely light in immature specimens (Text-figs. 1262, 1263). Median incision in caudal border of 7th abdominal sternum in female as in *viridis*. Overall length of males 5.7–8.0 mm, of females 7.3–10.0 mm.

Type material. Holotype: a micro-pinned male labelled "Suecia: Upl. Nysätra, Oxdjupet-tr.18.8.1980 Ossiannilsson leg.", and on a red label: "HOLOTYPUS Cicadella lasiocarpae Ossiannilsson". Paratypes: 18 males and 107 females collected in the type locality in August and September, 1980, labelled accordingly and with orange labels with the text: "Paratypoid Cicadella lasiocarpae Ossiannilsson". The type locality is situated 300 m SSW of the farmhouse of Oxdjupet. All types in the collection of the author.

Distribution. Denmark: so far reliable records only from B. – Sweden: comparatively scarce and sporadic, found in Sm.,Gtl.,Ög., Vg., Sdm., Upl., Gstr., Vb. – Not seen from Norway. – Rare in East Fennoscandia, found in Ta: Janakkala 28.VII.1916 (P.H. Lindberg).

Biology. In very wet bogs and quagmires, with *Carex lasiocarpa*. Females kept in captivity in late September were seen feeding on the leaves of this *Carex*, and one of them was observed ovipositing in it. The eggs were laid in small groups of 2 and 3 in a shoot 15 cm in length, about 4 cm above shoot basis, where the tissues were soft and pale yellow. The eggs are whitish yellow in colour, 1.5 mm in length, sausage-shaped, only slightly curved, their upper half somewhat thinner than the lower one. – Adults in July – September.

Note. This species was at first mistaken by the present author for *C.viridis* var. *concolor* (Haupt). By the courtesy of Prof. Dr. H. J. Müller in Jena I have been able to examine the types of *concolor*, two males and one female. They belong to a colour form of our No. 163, in accordance with the interpretation generally accepted.

SUBFAMILY TYPHLOCYBINAE

Small usually slender, fragile leafhoppers, often brightly coloured. Head anteriorly angular and rounded. Lora and genae indistinctly separated, face long and narrow.

Ocelli present or absent, frontal sutures extending to ocelli or vestiges of ocelli. Fore wings long, veins of corium not forked, often basally indistinct, transverse veins present only in apical part. Wax area of fore wing well marked. Hind tibiae with many strong macrosetae. Mostly agile fliers. In Denmark and Fennoscandia 26 genera.

Key to genera of Typhlocybinae

1 Fore wings with a distinct apical membrane (Text-fig. 1264) (Tribe **Alebrini**) .... *Alebra* Fieber (p.399)
– Fore wings without an apical membrane 2
2 (1) Apical cells of hind wings distally closed (Text-figs. 1267, 1268) 3
– Apical cells of hind wings distally open (Text-figs. 1269–1271) 12
3 (2) Hind wing with two or three apical cells (Text-fig. 1267) (Tribe **Dikraneurini**) 4
– Hind wing with only one apical cell (Text-fig. 1268) (Tribe **Empoascini**) 10
4 (3) Aedeagus with two phallotremes (Text-fig. 1336). 7th abdominal sternum in female deeply incised, leaving base of ovipositor uncovered (Text-fig. 1339) .... *Notus* Fieber (p.417)
– Aedeagus with one phallotreme. 7th abdominal sternum in female not or less deeply incised, always covering base of ovipositor 5
5 (4) Short and broad (Plate-fig. 110). Apical cells of fore wing shortened, together shorter than their width (Text-fig. 1265) *Erythria* Fieber (p.401)
– Slender. Apical cells of fore wing normal, together longer than width 6
6 (5) Genital plates of male widely separated, apically shallowly forked (Text-figs. 1321, 1327). 7th abdominal sternum in female with 2 incisions dividing caudal part in 3 lobes, the median one longer and broader (Text-fig. 1326) *Forcipata* De Long & Caldwell (p.415)
– Genital plates contiguous or nearly so, apically not forked. 7th abdominal sternum in female different 7
7 (6) Fore wings each with three well-defined black spots (Plate-fig. 94) .... *Micantulina* Anufriev (p.407)
– Fore wings without well-defined black spots 8
8 (7) Shaft of aedeagus simple, without processes. Small species. 7th abdominal sternum of female (in our species) caudally incised (Text-figs. 1314, 1320) .... *Wagneriala* Anufriev (p.410)
– Shaft of aedeagus with various processes or appendages. 7th abdominal sternum in female caudally strongly convex, not or indistinctly incised 9
9 (8) Process of male pygofer slightly curved, directed backwards (Text-fig. 1281). Shaft of aedeagus with a strong forked process near base (Text-figs. 1283, 1284). Fore part of body

above without a light longitudinal stripe .... *Emelyanoviana* Anufriev (p.403)

– Process of male pygofer directed upwards (Text-figs. 1287, 1292). Shaft of aedeagus slender, processes arising near apex (Text-figs. 1289, 1294). Fore part of body or at least head above with a light longitudinal stripe .... *Dikraneura* Hardy (p.405)

10 (3) Apical veins of fore wing all arising from median cell (Text-fig. 1452). Processes of male anal collar short (Text-fig. 1454). *Kyboasca* Zachvatkin (p.447)

– Only one or two apical veins in fore wing arising from median cell 11

11(10) Pygofer without appendages. Body and fore wings short and broad *Chlorita* Fieber (p.449)

– Male pygofer with long appendages. Body and fore wings elongate .... *Empoasca* Walsh (p.419)

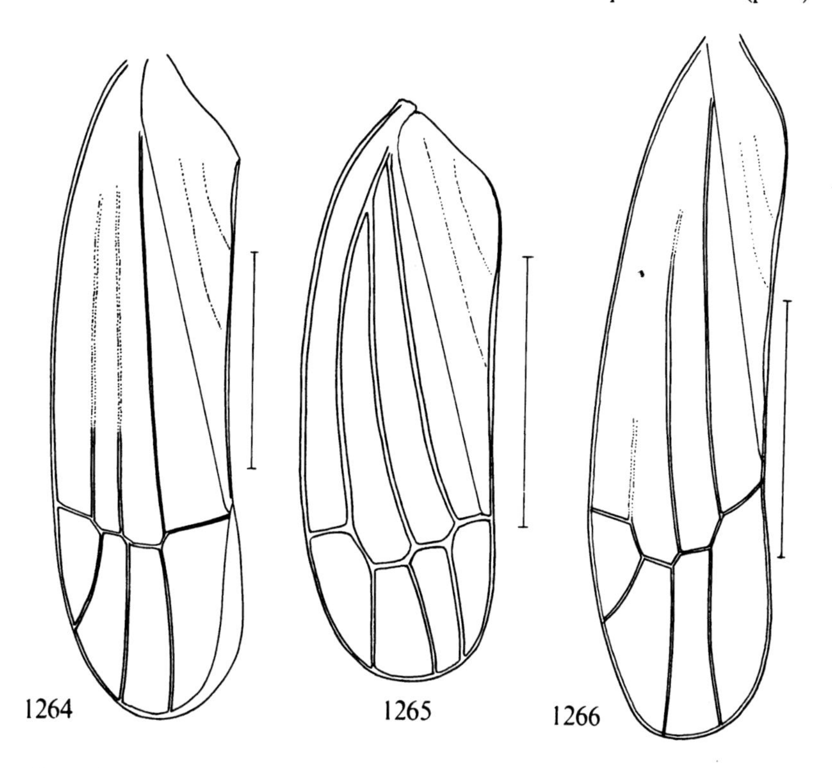

Text-figs. 1264–1266. – 1264: *Alebra albostriella* (Fallén), left fore wing of male; 1265: *Erythria aureola* (Fallén), left fore wing of female; 1266: *Alnetoidia alneti* (Dahlbom), left fore wing of male. Scale: 1 mm.

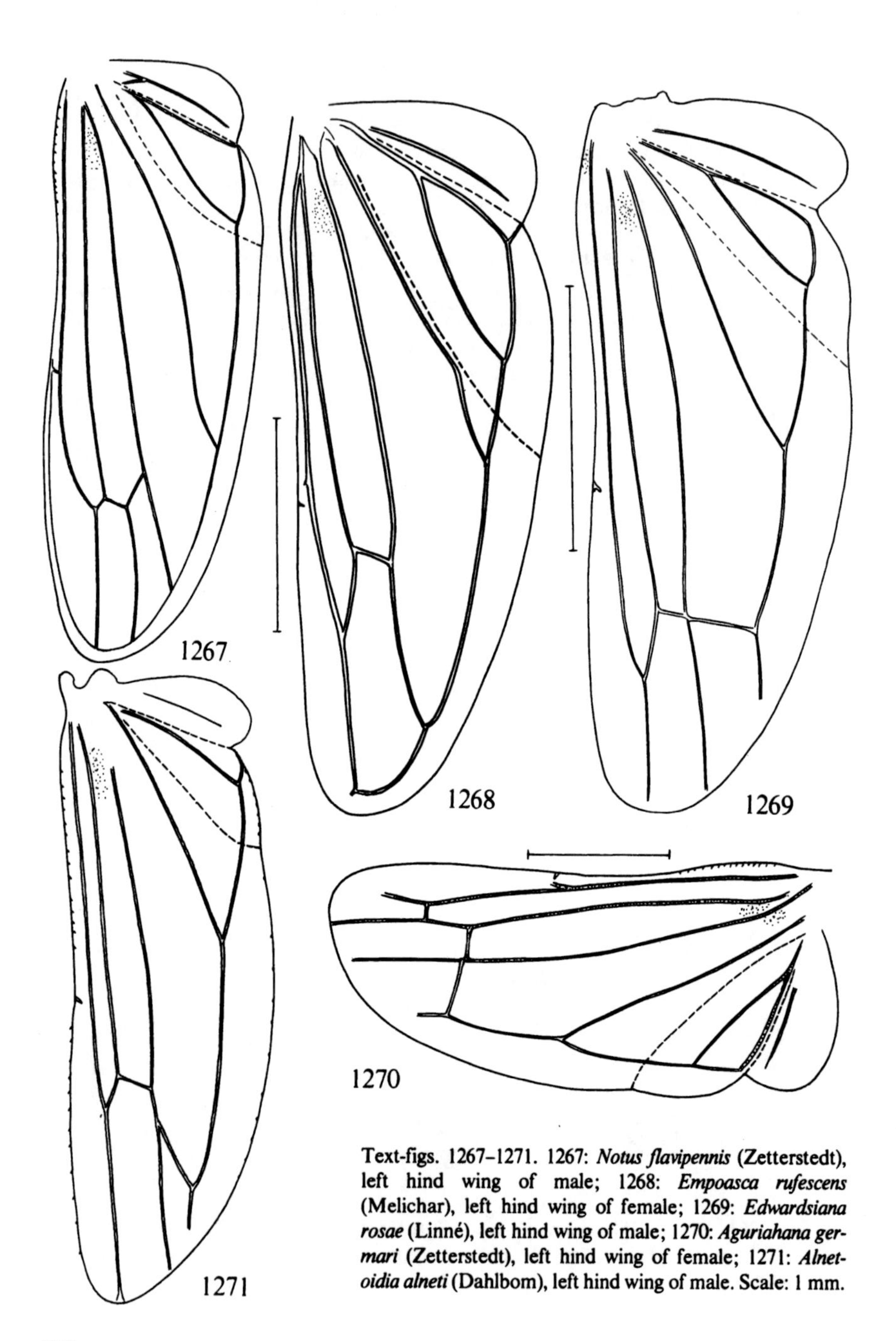

Text-figs. 1267–1271. 1267: *Notus flavipennis* (Zetterstedt), left hind wing of male; 1268: *Empoasca rufescens* (Melichar), left hind wing of female; 1269: *Edwardsiana rosae* (Linné), left hind wing of male; 1270: *Aguriahana germari* (Zetterstedt), left hind wing of female; 1271: *Alnetoidia alneti* (Dahlbom), left hind wing of male. Scale: 1 mm.

12 (2) 3rd apical vein in fore wing ending in posterior margin of wing (Text-fig. 18, Plate-figs. 95–101). 2nd apical cell in fore wing triangular, usually stalked (Tribe ***Typhlocybini***) 13

– 3rd apical vein in fore wing ending in wing apex (Text-fig. 1266). 2nd apical cell in fore wing quadrangular, never stalked (Tribe ***Erythroneurini***) 22

13(12) Sc and R in hind wing connected by a transverse vein near apex (Text-fig. 19), wing with 3 apical cells.. 14

– Sc and R in hind wing distally fused, wing with 2 apical cells (Text-fig. 1269) 16

14(13) Transverse vein connecting peripheric vein with M in hind wing reaches the latter distally of transverse vein between M and R (Text-fig. 1270). Genital plates of male each with a group of small pegs proximally of apex (reduced in *A. stellulata*) (Text-figs. 1806, 1814). "Lower part" of male genital style (i. e. the part proximally of point of conjunction with connective) short (Text-figs. 1808, 1821) .... *Aguriahana* Distant (p.542)

– Transverse vein connecting peripheric vein with M in hind wing reaches the latter proximally of or in (Text-fig. 19) the point where the transverse vein between M and R reaches the former. Genital plates without a subapical group of small pegs. "Lower" part of genital style longer 15

15(14) Apical part of fore wing narrower than fore wing at middle (Plate-figs. 96–98). Lower part of face in profile subparallel to line of dorsum .... *Eurhadina* Haupt (p.511)

– Apical part of fore wing not or only slightly narrower than fore wing at middle (Plate-figs. 99–101, 131, 132). Face evenly curved, lower part sharply divergent from line of dorsum *Eupteryx* Curtis (p.518)

16(13) Genital style of male with a long subapical spine directed laterally under an angle of 90° (Text-figs 1654, 1659). Genital plates without macrosetae. Pronotum with 4 or 6 black spots situated near fore and side margins (Plate-figs. 92, 93) (sometimes confluent)..................................... *Linnavuoriana* Dlabola (p.498)

– Genital style of male without a long subapical spine. Pronotum unspotted or markings different 17

17(16) Fore body above with strong black markings (Plate-fig. 90) *Eupterycyba* Dlabola (p.496)

– Markings of fore body different or absent.. 18

18(17) Genital plates of male without macrosetae. Dorsum of fore body and fore wings without distinct markings *Ossiannilssonola* Christian (p.458)

– Genital plates with one or more macrosetae (one near base and one or more near middle). Dorsum of fore body and fore wings with or without distinct markings 19

19(18) Genital style of male with an angular subapical extension (Text-fig. 1489) or apically forked (Text-fig. 1494). Dorsum of fore body and fore wings unspotted, at most with a pair of indistinctly limited dark patches on vertex .... *Fagocyba* Dlabola (p.453)

– Genital style of male without an angular subapical extension, not forked. Dorsum of fore body and fore wings with or without distinct markings 20

20(19) Genital plates each with more than one macroseta. Processes of aedeagus arising from base. Apical veins of fore wing, or at least some of them, each ending in a fuscous spot or streak (cf. Plate-fig. 95)... *Ribautiana* Zachvatkin (p.502)

– Genital plates each with only one macroseta arising near base. Markings of fore wing absent or different 21

21(20) Genital styles (in our species) evenly curved (Text-figs. 1684, 1691). Body and fore wings comparatively short and broad. Fore wings red-spotted, or with two broad dark brown transverse bands (Plate-figs. 140, 141) .... *Typhlocyba* Germar (p.507)

– Genital style near apex more or less abruptly bent outwards. Body and fore wings slender. Appendages of aedeagus, if present, arising near apex. Markings of fore wings, if any, different .... *Edwardsiana* Zachvatkin (p.460)

22(12) Pygofer lobes each with one dorsal and one ventral process, the dorsal one pointing backwards (Text-fig. 1825). Genital style as in Text-fig. 1826, aedeagus as in Text-figs. 1827, 1828. Dorsum of fore body and fore wings unicolorous, whitish or light yellow .... *Alnetoidia* Dlabola (p.547)

– Pygofer lobe with or without a prolongation. Genital style and aedeagus different 23

23(22) Vertex with two well-defined rounded black spots 24

– Markings of vertex different or absent 25

24(23) Genital style (in our species) apically with three angular extensions (Text-fig. 1933)...................................... *Arboridia* Zachvatkin (p.592)

– Genital style different (Text-fig. 1832) .... *Hauptidia* Dworakowska (p.549)

25(23) Process of pygofer lobe bifurcate (Text-figs. 1837, 1844). Aedeagus with a pair of processes arising from basal part. Markings on dorsum of fore body and fore wings never red nor orange... *Zyginidia* Haupt (p.551)

– Process of pygofer lobe simple or forked (Text-figs. 1853, 1925). Aedeagus without processes arising from basal part. Upper side often with red or yellow markings .... *Zygina* Fieber (p.554)

Note. A well-illustrated key to the genera of the nymphs of British woodland *Typhlocybinae* has been published by Wilson (1978). This paper has been made use of in our generic diagnoses and descriptions of species.

Genus *Alebra* Fieber, 1872

Compsus Fieber, 1866: 507 (nec Schönherr, 1826).
Type-species: *Cicada albostriella* Fallén, 1826, by subsequent designation.
Alebra Fieber, 1872: 14 (n.n.)
Type-species: *Cicada albostriella* Fallén, 1826, by subsequent designation.

Comparatively strongly built typhlocybines. Head anteriorly rounded. Ocelli distinct. Fore wing with an apical membrane. Hind wings with 3 closed apical cells. Vertex of 5th instar nymphs short, barely extending in front of eyes; spines on anterior margin of vertex shorter than 2nd antennal segment. In Denmark and Fennoscandia two species.

Key to species of *Alebra*

1 Macrosetae on outside of upper margin of hind tibia arising from dark spots. Apical cells of fore wing usually infuscate. Apodemes of 2nd abdominal sternum in male long, about twice as long as wide .... 165. *albostriella* (Fallén)
– Macrosetae on hind tibiae not arising from dark spots. Apical part of fore wing not or only slightly infuscate. Apodemes of 2nd abdominal sternum in male not longer than wide... 166. *wahlbergi* (Boheman)

165. ***Alebra albostriella*** (Fallén, 1826)
Plate-figs. 102–104, text-figs. 1264, 1272–1275.

Cicada albostriella Fallén, 1826:54.
Cicada elegantula Zetterstedt, 1828:536.
Typhlocyba discicollis Herrich-Schäffer, 1834b:8.
Alebra albostriella var. *viridis* Rey, 1894:46.
Alebra albostriella var. *diluta* Ribaut, 1936b:196
Alebra albostriella var. *Dufouri* Ribaut, 1936b:197.

Much varying in colour. The typical male (Plate-fig. 102) is above uniformly yellow, only apical part of fore wing smoke-coloured. dorsum of abdomen and fore borders of abdominal sterna black or fuscous. The corresponding female (Plate-fig. 103) is yellowish white with the following yellow, orange, or red markings: two spots on vertex, two on scutellum, two broad longitudinal bands on pronotum and three longitudinal bands on each fore wing: one along commissural border, one in corium along claval suture, and one along costal border of corium; apex of fore wing infuscate, abdomen more or less largely black. F. *diluta* Ribaut: abdomen entirely light, corium in female with only one yellow longitudinal line. F. *discicollis* (Herrich-Schäffer): fore wing with a broad dark brown transverse band on level with middle of clavus, often also with a broad brownish longitudinal band on tergum of fore body. In f. *viridis* Rey the whitish band in clavus is narrower than in f. *typica,* sometimes invading cubital cell across claval suture. F. *dufouri* Ribaut: tergum of fore body and proximal 5/7 of fore wings dark brown, thoracal venter light, abdomen largely dark. Male genital segment as in Text-fig. 1272, genital style as in Text-fig. 1273, aedeagus as in Text-figs. 1274, 1275. 7th abdominal sternum in female semicircular, hind border medially very slightly emarginate. Overall length of males 3.2–4 mm, of females 3.5–4.5 mm. – Last instar nymph pale yellow with variable brown markings, setae of head and body short, dark.

Distribution. Common in most parts of Denmark. – Common in southern and central Sweden, Sk.–Vrm. – Norway: found in Ø: Jeløy (O. Sørum), Os: Råde (Holgersen), AAy: Lillesand (Holgersen), and in VAy: Søgne (Holgersen). – Not found in East Fennoscandia. – Widespread in Europe, also in Algeria, Azores, Madeira, Anatolia, Syria, Jordan, Georgia, and in the Nearctic Region.

Biology. "An verschiedenen Laubbäumen, vor allem Eichen und Linden" (Kuntze, 1937). "Auf *Quercus*" (Schwoerbel, 1957). Monophagous on *Quercus* (Claridge & Wilson, 1976). Hibernates in the egg stage (Müller, 1957). Adults in July – September.

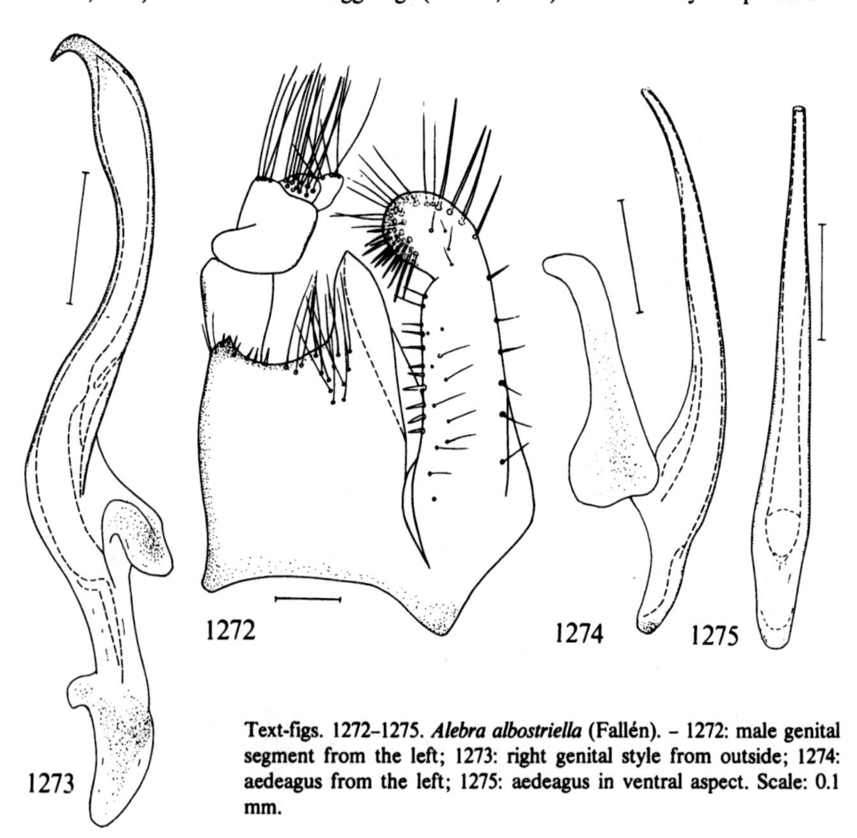

Text-figs. 1272–1275. *Alebra albostriella* (Fallén). – 1272: male genital segment from the left; 1273: right genital style from outside; 1274: aedeagus from the left; 1275: aedeagus in ventral aspect. Scale: 0.1 mm.

166. ***Alebra wahlbergi*** (Boheman, 1845)
Plate-fig. 105.

Typhlocyba wahlbergi Boheman, 1845b:160.
Alebra wahlbergi var. *pallescens* Ribaut, 1936b:198.
Alebra wahlbergi var. *brunnea* Ribaut, 1936b:198.

Venter of fore body and abdomen uniformly yellowish white. Male also above entirely pale yellow. Fore wings of female with a more or less distinct yellow longitudinal band along commissural border and another in corium. In f. *typica* (♀), the fore body has a broad dark brown longitudinal band and the fore wings have a broad transverse band of the same colour situated across apex of clavus. In f. *pallescens* Ribaut (♀) dark markings are absent. In f. *brunnea* Ribaut the fore wings are uniformly brownish in corium and clavus. Male genitalia as in *albostriella*. Overall length of males 3.6–4.2 mm, of females 4.0–4.5 mm. – Last instar larvae entirely pale yellow, setae dark.

Distribution. Denmark: less common, so far found in NEJ, F, and B. – Sweden: scarce, found in Sk., Öl., Ög., Vg., and Vstm. – Norway: found in Bø: Tofteholmen 10. VII. 1953 (Holgersen), and in HEs: Bröttum 4. IX. 1972 (G. Söderman, in litt.). – Not found in East Fennoscandia. – Austria, Belgium, Bohemia, Moravia, Slovakia, France, German D.R. and F.R., England, Netherlands, Italy, Switzerland, Hungary, Poland, Yugoslavia, Latvia, Lithuania, Estonia, Moldavia, m. Russia, Ukraine, Algeria, Anatolia.

Biology. "Hauptsächlich im Querceto – Carpinetum, wo sie auf - *Acer campestre* lebt; im Querceto – Betuletum lassen sich die Tiere dieser Art auf *Populus* feststellen" (Schwoerbel, 1957). On *Acer pseudoplatanus* and *campestre, Carpinus betulus* and *Tilia* (Wagner & Franz, 1961). Oligophagous. "*A. wahlbergi* showed an almost equally strong preference for maple -, but considerable populations were also taken from alder, hornbeam, hazel, beech and oak" (Claridge & Wilson, 1976). Boheman (1845a) found *wahlbergi* "sat copiose" on *Fraxinus*. Adults in July – September.

Genus *Erythria* Fieber, 1866

Erythria Fieber, 1866:507.
Type-species: *Cicada aureola* Fallén, 1806, by monotypy.

Body short and broad. Apical part of fore wing short. Pygofer lobes of male and aedeagus with various appendages. 7th abdominal sternum in female little specialized, caudally prolonged. In Denmark and Fennoscandia one species.

167. ***Erythria aureola*** (Fallén, 1826)
Plate-fig. 110, text-figs. 1265, 1276–1280.

Cicada aureola Fallén, 1806:25.

Short and broad. Head brownish, usually with more or less extended blood-red surfaces, above with four indistinct dark spots, muscle traces on frontoclypeus brownish. Pronotum greyish to yellowish green, often with blood-red sprinkles. Fore wings yellowish with a pine-needle greenish tinge, vax area with a bluish tinge. Hind wings with black veins. Venter of fore body and abdomen largely black. Legs dirty brownish or partly blood-red. Genital plates of male elongate, triangular, apically acuminate. Genital segment of male as in Text-fig. 1276, pygofer lobe as in Text-fig. 1277, genital style as in Text-fig. 1278, aedeagus as in Text-figs. 1279, 1280. 7th abdominal sternum in female caudally strongly convex, medially about half as long as visible part of saw-case. Overall length of males 2.35–2.55 mm, of females 2.55–2.9 mm.

Distribution. Scarce in Denmark, found in Jutland only (SJ, EJ, NWJ, NEJ). - Locally common in Sweden, found in Sk., Sm., Ög., Vg., Boh., Sdm., Upl., Vb., P. Lpm. - Norway: found in AK: Oslo, Hovin (Siebke), TEi: Øvrebø (Holgersen), HEn: Sørnesset, Sollia (Holgersen). - Common in southern and central East Fennoscandia, recorded from Ab, N, Ka, Tb, Kb; Kr. - Austria, Belgium, France, England, Scotland,

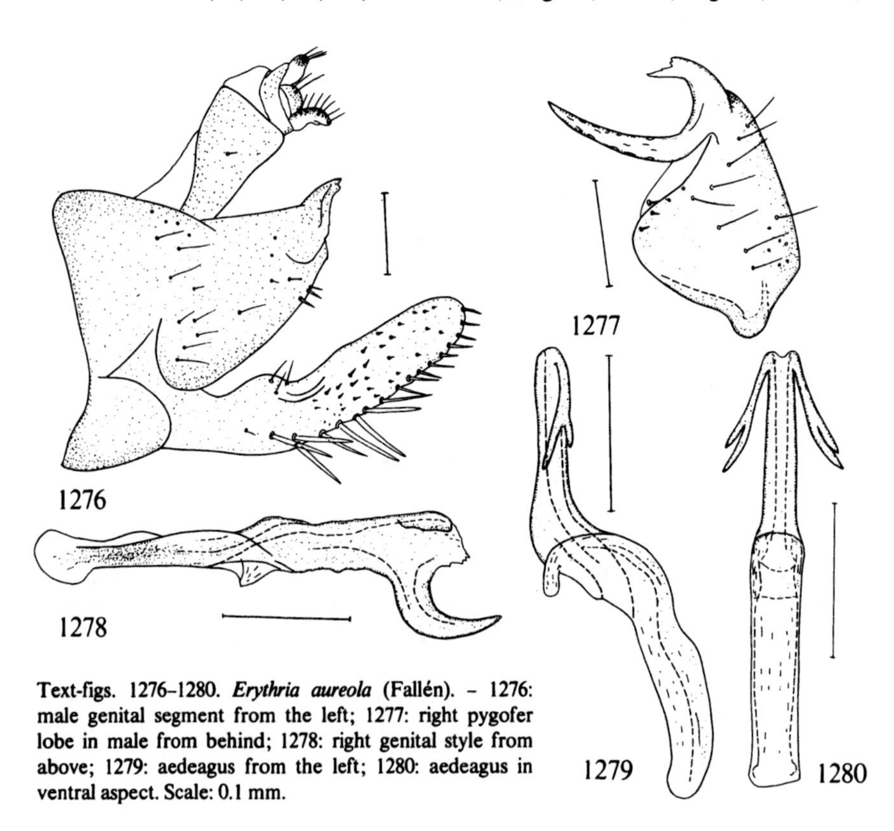

Text-figs. 1276–1280. *Erythria aureola* (Fallén). - 1276: male genital segment from the left; 1277: right pygofer lobe in male from behind; 1278: right genital style from above; 1279: aedeagus from the left; 1280: aedeagus in ventral aspect. Scale: 0.1 mm.

Netherlands, German D.R. and F.R., Switzerland, Italy, Spain, Romania, Bohemia, Moravia, Slovakia, Yugoslavia, Poland, Latvia, Lithuania, Estonia, n. and m. Russia.

Biology. "In Heiden und Wäldern. An *Calluna vulgaris*" (Kuntze, 1937). In pine bogs with undershrubs, dry *Vaccinium* pine woods, and dry *Calluna* heaths (Linnavuori, 1952a). "Eine ausgeprägt heliophile Art -. Wird von *Calluna vulgaris* und *Thymus* angegeben, Franz beobachtete sie wiederholt an *Globularia cordifolia* und *Teucrium montanum*" (Wagner & Franz, 1961). Hibernation takes place in the egg stage; two generations p.a.(in German D.R.) (Schiemenz, 1969b). Adults in July – September (in Fennoscandia).

Genus *Emelyanoviana* Anufriev, 1970

Emelyanoviana Anufriev, 1970a: 263.
Type-species: *Typhlocyba mollicula* Boheman, 1850, by original designation.

"Venation of fore- and hindwings as in *Dikraneura* Hardy, 1850, but male genitalia different. Pygophore sides with strong, well-sclerotized processes deviating from caudoventral angles; they point backwards from base and then curve upwards. Each genital plate with a row of macrosetae along outer margin in basal part and along inner margin in apical part. Connective lamella-like, of nearly equal length and width, widest at place of articulation with styles. Penis with lamella-like broadened base and short shaft provided with a strong, bifurcated process near base." (Anufriev, l.c.). In Denmark and Fennoscandia one species.

168. *Emelyanoviana mollicula* (Boheman, 1845)
Text-figs. 1281–1286.

Typhlocyba mollicula Boheman, 1845b: 160.
Typhlocyba facialis Flor, 1861a: 385.

Comparatively elongate. Head frontally obtuse angular. Yellowish white, usually with a greenish tinge. Mesosternum and dorsum of abdomen partly black, side margins of the latter broadly, hind margins of terga narrowly light. Abdominal sterna of male anteriorly more or less dark-bordered. Veins of hind wings light. Genital plates of male narrow, elongate, in ventral aspect triangular, curved upwards, side margins caudally somewhat folded upwards. Male pygofer lobe as in Text-fig. 1281, genital style as in Text-fig. 1282, aedeagus as in Text-figs. 1283, 1284, 1st–3rd abdominal sterna in male as in Text-fig. 1285. 7th abdominal sternum in female caudally strongly convex, more than half as long as visible part of saw-case, medially with a shallow incision (Text-fig. 1286). Overall length (♂♀) 3.2–3.6 mm. – Last instar larva light-coloured, antenna slightly longer than hind tibiae, tergum with some long, apically knobbed macrosetae (6 on head, 8 on pronotum, 6 on mesonotum, 4 on metanotum, 4 on each of most abdominal terga, and some on wing buds).

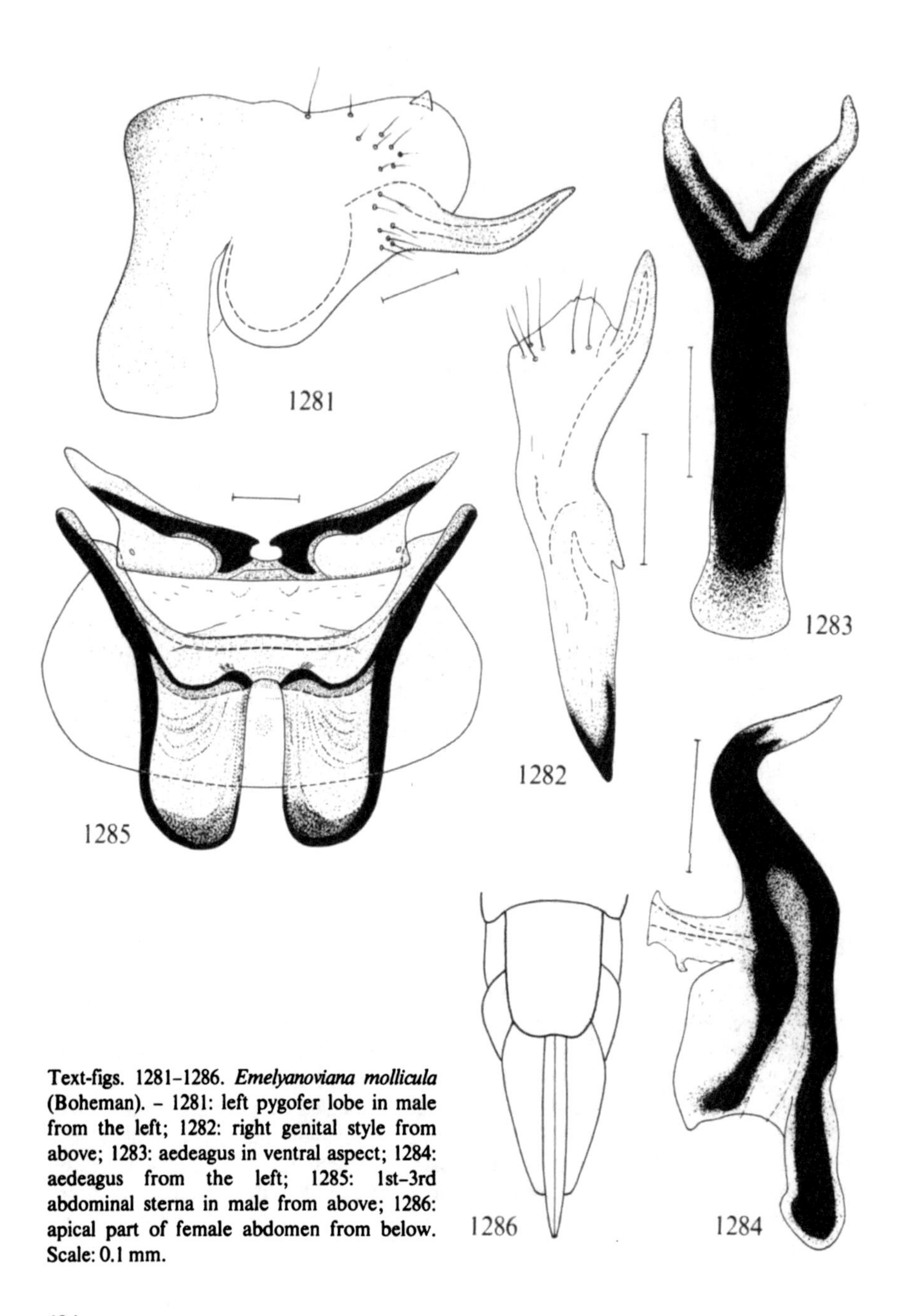

Text-figs. 1281–1286. *Emelyanoviana mollicula* (Boheman). – 1281: left pygofer lobe in male from the left; 1282: right genital style from above; 1283: aedeagus in ventral aspect; 1284: aedeagus from the left; 1285: 1st–3rd abdominal sterna in male from above; 1286: apical part of female abdomen from below. Scale: 0.1 mm.

Distribution. Rare in Denmark, only found in EJ: Silkeborg 29.V.1914 (Oluf Jacobsen) and 16.X.1915 (Jensen-Haarup) and Kongshus 1.VIII.1921 (Jensen-Haarup). - Fairly common in southern and central Sweden (Sk.–Vstm.). - Apparently not uncommon in southern Norway, found in several localities in AK, Ø, Bø, and TEi. - Rare in East Fennoscandia, only found in Kb: Pielisensuu (Håkan Lindberg), and Hammaslahti. - Widespread in Europe, also in Tunisia, Anatolia, Altai Mts., Georgia, Kazakhstan.

Biology. In "besonnten Hängen" (Kuntze, 1937). "A species of the dry biotopes. Most abundant in the mid-summer aspect and late summer aspect. Two generations. Richest in specimens in dry, warm summers" (Kontkanen, 1950). "Auf Labiaten und *Verbascum*-Arten; vorwiegend an xerothermen Standorten" (Wagner & Franz, 1961). "On *Salvia, Teucrium, Mentha, Satureja,* and other Labiatae, *Verbascum, Cannabis, Parietaria, Artemisia* and many other herbaceous plants pertaining to various families. Overwintering probably in adult-stage" (Vidano, 1965). "Ei-Überwinterer, 2 Generationen" (Schiemenz, 1969b). *Emelyanoviana mollicula* was found breeding on cultivated strawberry plants in Norway (Taksdal, 1977). Adults in June–October.

Genus *Dikraneura* Hardy, 1850

Dikraneura Hardy, 1850: 423.
Type-species: *Dikraneura variata* Hardy, 1850, by monotypy.

Body and fore wings elongate. Head frontally produced, angular. Eyes much longer than high. Apical part of fore wing well developed. Pygofer lobes in male each with a hook-like prolongation, apex directed upwards. Genital plates triangular. Aedeagus with appendages arising near apex. 7th abdominal sternum in female little specialized, caudally convex. In Denmark and East Fennoscandia two species.

Key to species of *Dikraneura*

1 Aedeagus in lateral aspect narrower, subapical appendages of aedeagus longer (Text-fig. 1290) .... 169. *aridella* (J. Sahlberg)
\- Aedeagus in lateral aspect broader, subapical appendages shorter (Text-fig. 1295) .... 170. *variata* Hardy

169. ***Dikraneura aridella*** (J. Sahlberg, 1871)
Text-figs. 1287–1291.

Typhlocyba citrinella Flor, 1861: 386 (nec Zetterstedt, 1828).
Notus aridellus J. Sahlberg, 1871: 167.

Whitish yellow with a greyish green tinge. Tergum of fore body or at least vertex with a narrow whitish longitudinal line. Fore wings with a bluish white hyaline longitudinal streak in cubital cell. Veins of hind wings only partly dark. Thoracal venter and abdomen largely black. Male genital plates narrow, triangular with lateral border concave, apically slightly bent upwards. Male pygofer lobe as in Text-fig. 1287, genital style as in Text-fig. 1288, aedeagus as in Text-figs. 1289, 1290, 2nd and 3rd abdominal sterna in male as in Text-fig. 1291. 7th abdominal sternum in female caudally strongly convex, medially indistinctly emarginate, about 1/3 as long as visible part of saw-case, the latter extending far beyond apex of pygofer. Overall length (♂♀) 2.95–3.5 mm.

Distribution. So far not found in Denmark. – Norway: AAy: Åmli 2.VIII, 1♂, 1♀ (Holgersen); one female from Bø: Drammen 5.X.1926 (Warloe) may belong to the present species (or the following). – Scarce in Sweden, found in Bl., Ög., Sdm., Upl., Vstm.,

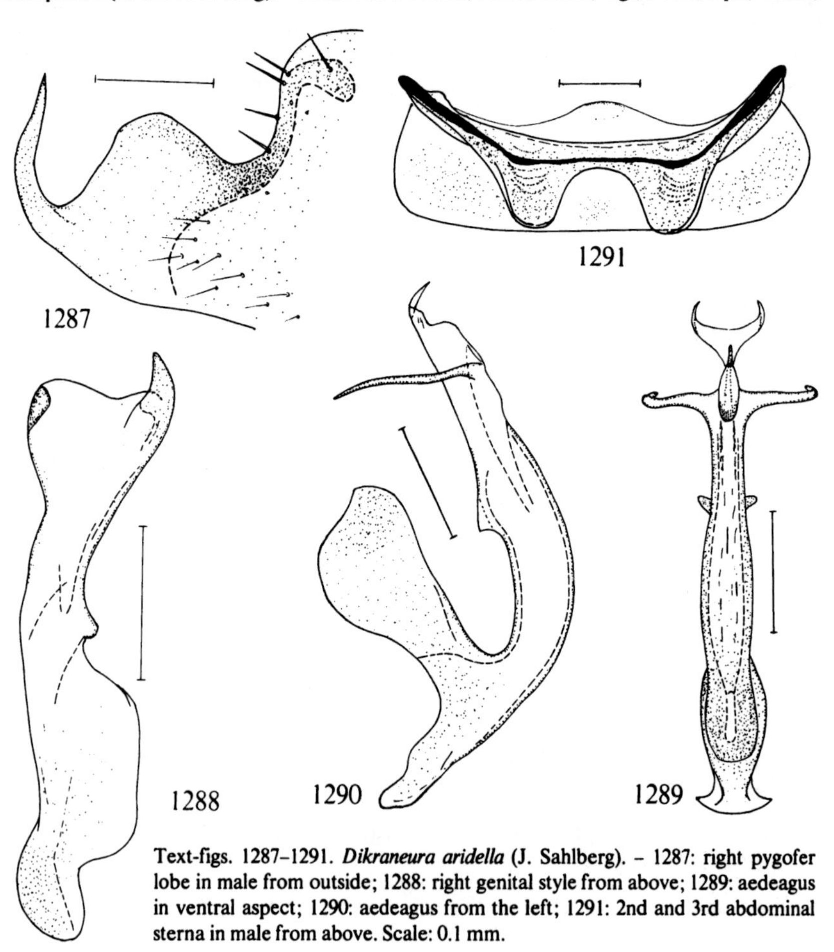

Text-figs. 1287–1291. *Dikraneura aridella* (J. Sahlberg). – 1287: right pygofer lobe in male from outside; 1288: right genital style from above; 1289: aedeagus in ventral aspect; 1290: aedeagus from the left; 1291: 2nd and 3rd abdominal sterna in male from above. Scale: 0.1 mm.

Dlr., Hls., Med., and Vb. – Scarce in southern and central East Fennoscandia, found in Al, Ab, N, Ka, Ta, Oa, Sb, Kb, Ks, Li; Kr, Lr. – Austria, German F.R., n. Italy, n. Russia, Kazakhstan, Mongolia, Maritime Territory.

Biology. In rich swampy woods and rich moist grass-herb woods (Linnavuori, 1952a). "On *Graminaceae* growing amid *Vaccinium myrtillus* in coniferous woods" (Vidano, 1965). Sahlberg (1871) found the species on dry sandy slopes among grass and other herbs from July to September. I found it in most cases on mossy ground with sparse grass vegetation, often shaded by conifers. Hibernation takes place in the adult stage (Lindberg, 1947). In Sweden adults have been found in May–September.

170. ***Dikraneura variata*** Hardy, 1850
Text-figs. 1292–1296.

Dikraneura variata Hardy, 1850: 423.
Dicraneura aridella Jensen-Haarup, 1920: 163, nec J. Sahlberg, 1871.

As *aridella*. Pale longitudinal median line on vertex often obsolete. Male pygofer lobe as in Text-fig. 1292, genital style and connective as in Text-fig. 1293, aedeagus as in Text-figs. 1294, 1295, 2nd and 3rd abdominal sterna in male as in Text-fig. 1296. Length of male 2.92–3.64 mm, of female 3.22–4.08 mm (according to Knight, 1968).

Distribution. Denmark: locally common and widespread in Jutland. – Not found in Sweden, nor in Norway. – Rare in East Fennoscandia, found in Al, Ab, Ta, Sa. – Widespread in Europe, also in Anatolia, Kazakhstan, m. Siberia, Mongolia, and Nearctic Region.

Biology. In "besonnten Hängen, Wäldern" (Kuntze, 1937). "Im Gras im Kiefern-Fichten-Forst" (Schiemenz, 1964). "On Graminaceae growing in coppices and woods particularly of *Quercus* and *Castanea*. Overwintering in adult-stage" (Vidano, 1965). Hibernation in the adult stage has also been established by Müller (1957), Remane (1958), and Schiemenz (1964). Two generations (Schiemenz, 1969).

Genus *Micantulina* Anufriev, 1970

Micantulina Anufriev, 1970a: 262.
Type-species: *Cicadula micantula* Zetterstedt, 1838, by original designation.

"Venation of fore- and hindwings as in *Dikraneura* Hardy, 1850, but the male genitalia are as follows: pygophore sides without long processes, more or less truncated behind; genital plates of moderate length, with two kinds of setae arranged in a row (basal setae long, well developed, apical one small), connective Y-shaped, base and branches of nearly equal length. Penis comparatively short, with or without basal processes. This genus consists of two subgenera – *Micantulina* s.str. and *Mulsantina* subgen.nov." (Anufriev, l.c.). In Fennoscandia two species, both belonging to subgenus *Micantulina* s.str., with the following characters: "Pygophore sides without processes. Genital plates without distinct lateral angles; only two basal setae on each plate being long. Penis with basal processes." (Anufriev, l.c.).

Key to species of *Micantulina*

1 Aedeagus as in Text-figs. 1298, 1299. 7th abdominal sternum infemale posteriorly broadly rounded, medially with or without a shallow emargination (Text-fig. 1301) . 171. *micantula* (Zetterstedt)

– Aedeagus as in Text-figs. 1304, 1305. 7th abdominal sternum in female posteriorly strongly convex, medially emarginate (Text- 1308) .... 172. *pseudomicantula* (Knight).

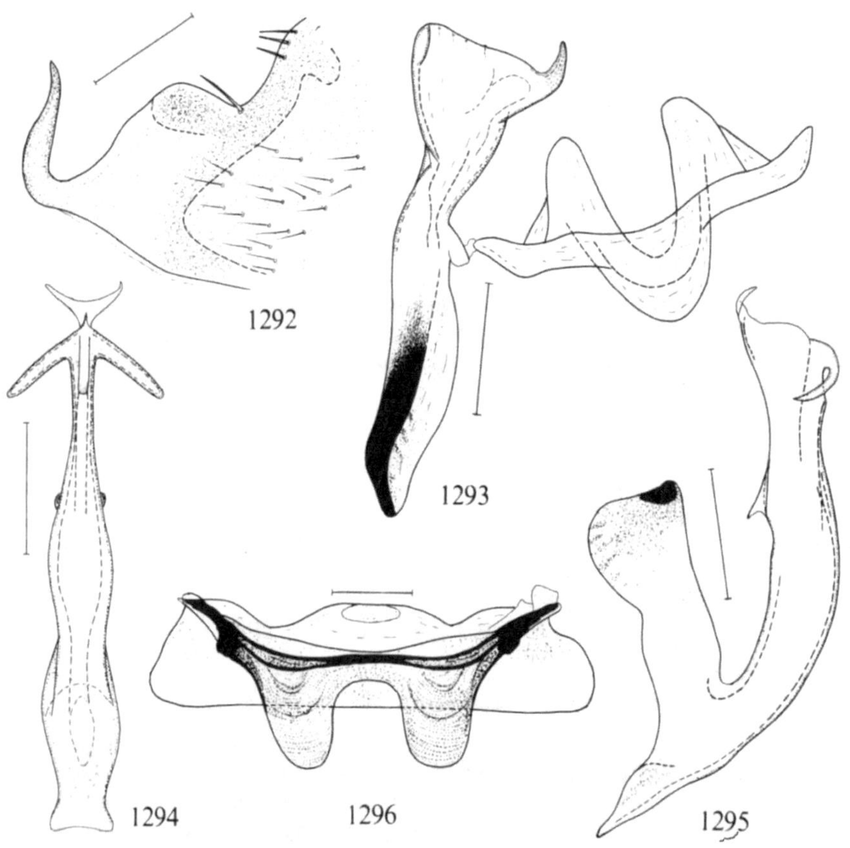

Text-figs. 1292–1296. *Dikraneura variata* Hardy. – 1292: right pygofer lobe in male from outside; 1293: right genital style and connective from above; 1294: aedeagus in ventral aspect; 1295: aedeagus from the left; 1296: 2nd and 3rd abdominal sterna in male from above. Scale: 0.1 mm.

171. ***Micantulina micantula*** (Zetterstedt, 1838)
Plate-fig. 94, text-figs. 1297–1301.

Cicadula micantula Zetterstedt, 1838: 299.

Elongate, yellowish white, shining. Head above often with two small brownish spots. Fore wings (Plate-fig. 94) each with two black spots in clavus: one near base and one at apex, and with a third black spot in corium at middle of claval suture. Thoracal venter anteriorly dark brown. Abdomen largely dark brown or black, lateral margin of tergum, lateral areas of sternum and often caudal borders of sternal segments yellowish. Tergum of female pygofer dark brown, apex of saw-case black (Text-fig.

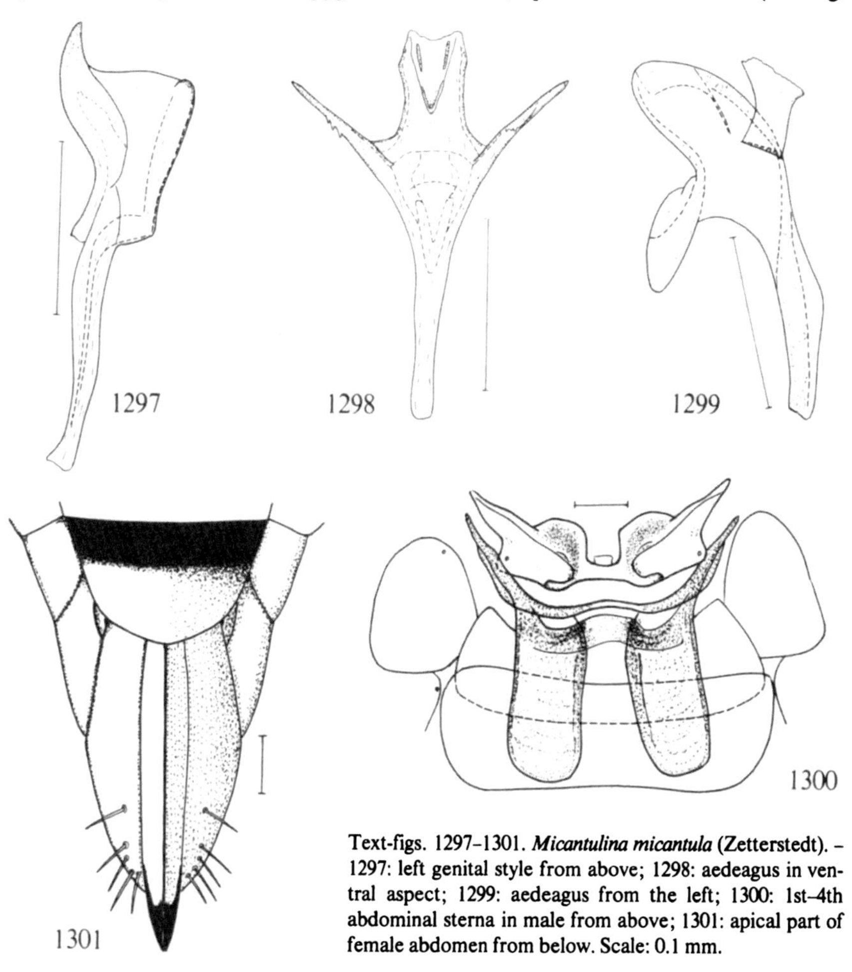

Text-figs. 1297–1301. *Micantulina micantula* (Zetterstedt). – 1297: left genital style from above; 1298: aedeagus in ventral aspect; 1299: aedeagus from the left; 1300: 1st–4th abdominal sterna in male from above; 1301: apical part of female abdomen from below. Scale: 0.1 mm.

1301). Male genital style as in Text-fig. 1297, aedeagus as in Text-figs. 1298, 1299, 1st to 4th abdominal sterna in male as in Text-fig. 1300. Overall length (♂♀) 3.3–3.75 mm.

Distribution. So far not found in Denmark. – Rare in Sweden, found near Stockholm according to Tullgren & Wahlgren (1920–22), and in Upl.: Uppsala by Haglund. I have seen in all six specimens collected in Nrk.: Örebro, Oset 4.VIII.1949, 10.IX.1949, 10.VIII.1951, 23.VIII.1955 and 28.IX.1955, by Anton Jansson, and one specimen captured by the same collector in Nrk.: örebro, Lillån 4.VIII.1955. A specimen labelled "Lpl." is present in the collection of C. G. Thomson. – Norway: Zetterstedt described the species on the basis of one male from "Lapponia Norvegica", leg. Boheman. A specimen from Dovre is preserved in the Museum of Natural History in Stockholm. – East Fennoscandia: so far only found in Al: Lemland 21.V.1978, and Finström 22.V.-1978, by Albrecht. – Austria, ? Bohemia, Yugoslavia, n. Italy, Switzerland, Poland, n. Russia, Algeria, Georgia, Altai Mts., Kazakhstan, m. Siberia, Kirghizia, Mongolia.

Biology. On *Thalictrum flavum, Th. foetidum alpestre, Th. foetidum majus,* and *Th. aquilegiifolium* (Vidano, 1965). "Overwintering probably in adult-stage" (Vidano, l.c.).

172. ***Micantulina pseudomicantula*** (Knight, 1966)
Text-figs. 1302–1308.

Dicraneura micantula Lindberg, 1947: 59 (p.p.?).
Dikraneura pseudomicantula Knight, 1966: 347.

Resembling *micantula.* Male genital segment as in Text-figs. 1302, 1303, genital style as in Text-fig. 1304, aedeagus as in Text-figs. 1305, 1306, 1st to 4th abdominal sterna in male as in Text-fig. 1307, venter of apical part of female abdomen as in Text-fig. 1308. Length (according to Knight, l.c.): male 3.28 mm, female 3.31 mm.

Distribution. So far not recorded from Denmark, Sweden and Norway. – East Fennoscandia: rare, found in Ab: Lojo 2.VI–1.IX.1920 (Håkan Lindberg, P. H. Lindberg), and in Ta: Lammi, August 1949 (Linnavuori). – Lithuania, Mongolia.

Biology. Linnavuori (1969) found *pseudomicantula* among lush vegetation, in wet groveland, abundantly overgrown with *Filipendula ulmaria,* together with *Eupteryx signatipennis.* Mesophilous (Emelyanov, 1977).

Genus *Wagneriala* Anufriev, 1970

Wagneriana Anufriev, 1970a: 262 (nec Mc Cook, 1904).
Type-species: *Notus minutus* (sic) J. Sahlberg, 1871, by original designation.
Wagneriala Anufriev, 1970b: 635 (n.n.).
Type-species: *Notus minimus* J. Sahlberg, 1871, by original designation.

"Venation of fore- and hindwings as in the genus *Dikraneura* Hardy, 1850, but male genitalia distinctly different. Genital plates turned up approximately from middle, bear-

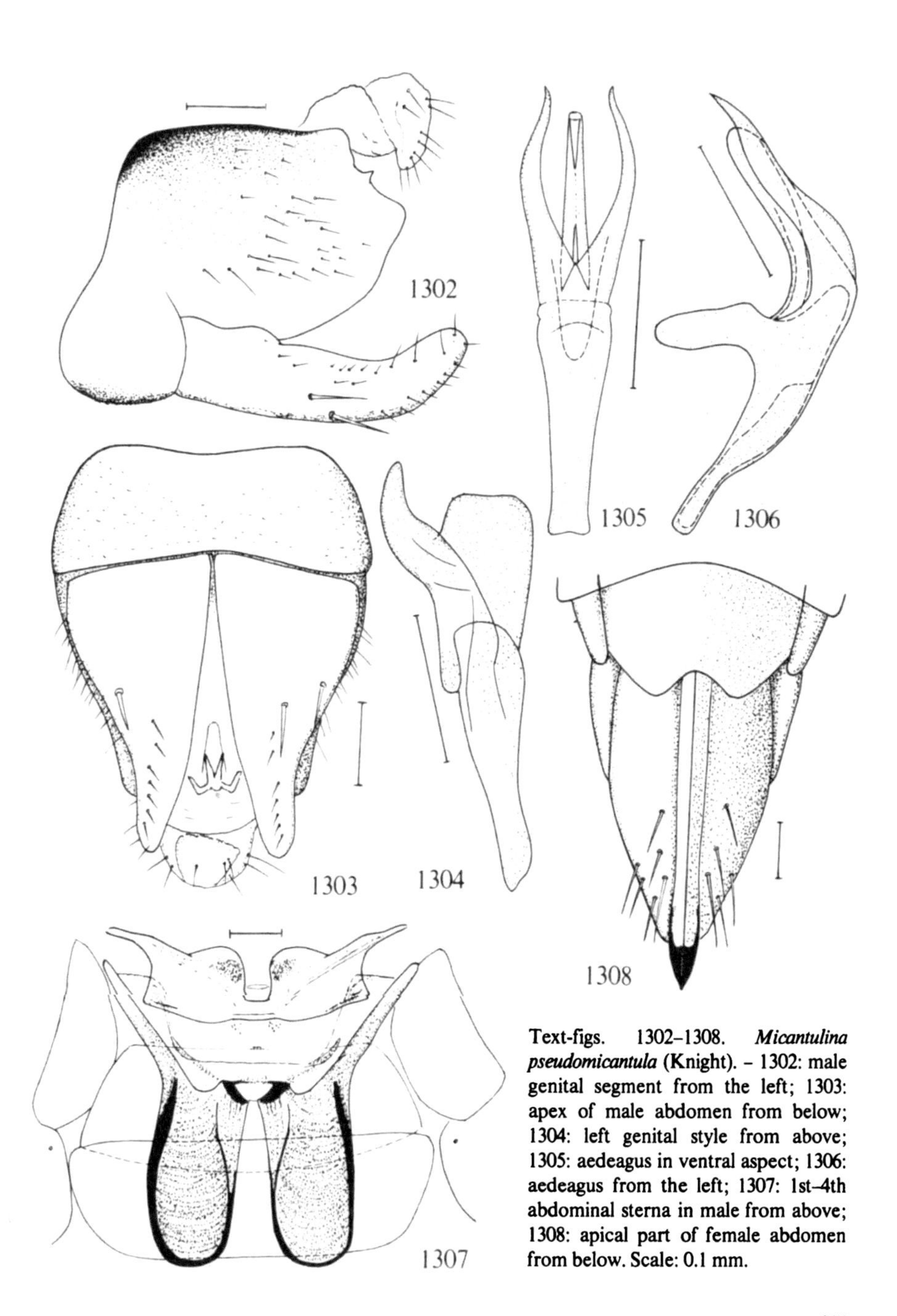

Text-figs. 1302–1308. *Micantulina pseudomicantula* (Knight). – 1302: male genital segment from the left; 1303: apex of male abdomen from below; 1304: left genital style from above; 1305: aedeagus in ventral aspect; 1306: aedeagus from the left; 1307: 1st–4th abdominal sterna in male from above; 1308: apical part of female abdomen from below. Scale: 0.1 mm.

ing some macrosetae in middle of concave lateral margins. Connective lamellate, very short and wide. Penis without processes, more or less arched, with apical gonopore" (Anufriev, 1970a). In Denmark and Fennoscandia two species.

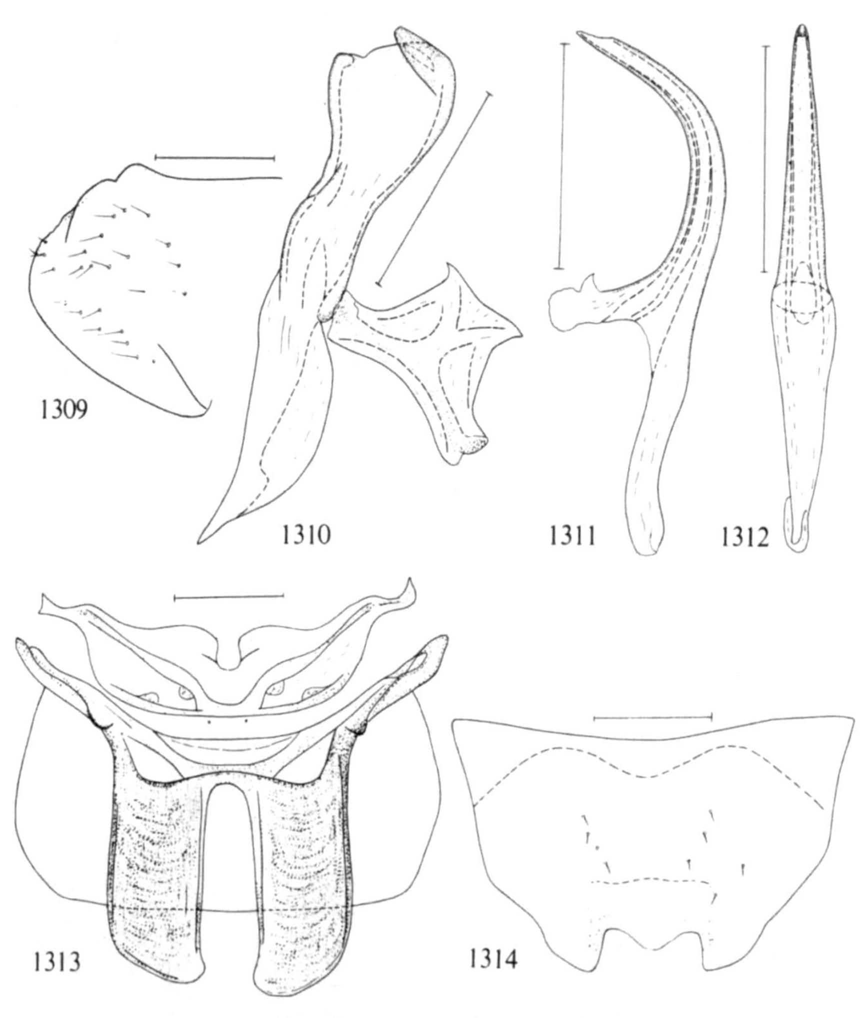

Text-figs. 1309–1314. *Wagneriala minima* (J. Sahlberg). – 1309: right pygofer lobe in male from outside; 1310: right genital style and connective from above; 1311: aedeagus from the left; 1312: aedeagus in ventral aspect; 1313: 1st–3rd abdominal sterna in male from above; 1314: 7th abdominal sternum in female from below. Scale: 0.1 mm.

Key to species of *Wagneriala*

1 Aedeagus as in Text-figs. 1311, 1312, genital style as in Text-fig. 1310. 7th abdominal sternum in female caudally with two shallow incisions (Text-fig. 1314) .... 173. *minima* (J. Sahlberg)

– Aedeagus as in Text-figs. 1317, 1318, genital style as in Text-fig. 1316. 7th abdomial sternum in female caudally with two deep incisions (Text-fig. 1320) .... 174. *incisa* (Then)

173. ***Wagneriala minima*** (J. Sahlberg, 1871)
Text-figs. 1309–1314.

Notus minimus J. Sahlberg, 1871: 168.

Fairly elongate. Head frontally strongly produced, angular. Light yellow or yellowish white. Apical part of fore wing smoke-coloured. Veins of hind wings light. Tergum of abdomen black-brown except lateral margins and posterior borders of segments. Genital plates of male long, narrow, tapering towards apices, side margins concave, apices somewhat curved upwards. Pygofer lobe of male as in Text-fig. 1309, genital style and connective as in Text-fig. 1310, aedeagus as in Text-figs. 1311, 1312, 1st–3rd abdominal sterna in male as in Text-fig. 1313, 7th abdominal sternum in female as in Text-fig. 1314. Overall length (♂♀) 2–2.5 mm.

Distribution. Very rare in Denmark, one female only found in EJ: Funder 4.VIII. 1915 (Oluf Jacobsen). – Rare in Sweden, found in Bl.: Förkärla by Gyllensvärd, in Öl.: Stenåsa, Resmo, Vickleby, and Borgholm by Ossiannilsson, and in Öl.: Resmo, Kalkstad, Dröstorp, Ekelunda, and Karlevi by H. Bornfeldt. E. Haglund collected the type specimens in Upl.: Uppsala, Polacksbacken, and A. Jansson found one specimen in Nrk.: Örebro. – In Norway according to Oshanin, 1912; I have not seen Norwegian specimens. – East Fennoscandia: Sa: Joutseno (Linnavuori). – Austria, Bohemia, German D.R. and F.R., Italy, Poland, Estonia, Latvia, Lithuania; Altai Mts., Mongolia.

Biology. "In Trockenrasen" (Wagner & Franz, 1961). On *Carex levis* in pastureland; on *Carex* sp. growing amid *Helianthemum* spp., *Teucrium montanum* (Vidano, 1965). "Bewohner von Gebüschen und Waldrändern" (Müller, 1978). The find localities on Öland are situated in the "alvar". Adults in June–August.

174. ***Wagneriala incisa*** (Then, 1897)
Text-figs. 1315–1320.

Dicraneura incisa Then, 1897: 115.

Resembling *minima*, slightly larger. Male pygofer lobes as in Text-fig. 1315, genital style as in Text-fig. 1316, aedeagus as in Text-figs. 1317, 1318, 2nd and 3rd. abdominal sterna in male as in Text-fig. 1319. 7th abdominal sternum in female as in Text-fig. 1320. Overall length (♂♀) 2.52–2.75 mm.

Distribution. So far not found in Denmark, Norway and East Fennoscandia. – Sweden: Gtl. 1974 (R. Remane, in litt.). Upl.: Uppsala Botanical Garden 21.–24.VII.-1952, 51 specimens (Ossiannilsson). – Poland, Romania, n. Yugoslavia, n. Italy.

Biology. On several species of xerophilous *Carex (verna, muricata, montana)* usually growing in *Castanea* and *Quercus* coppices. Overwintering in egg-stage (Vidano, 1965). I found *W. incisa* in cultivated tufts of *Carex pediformis*.

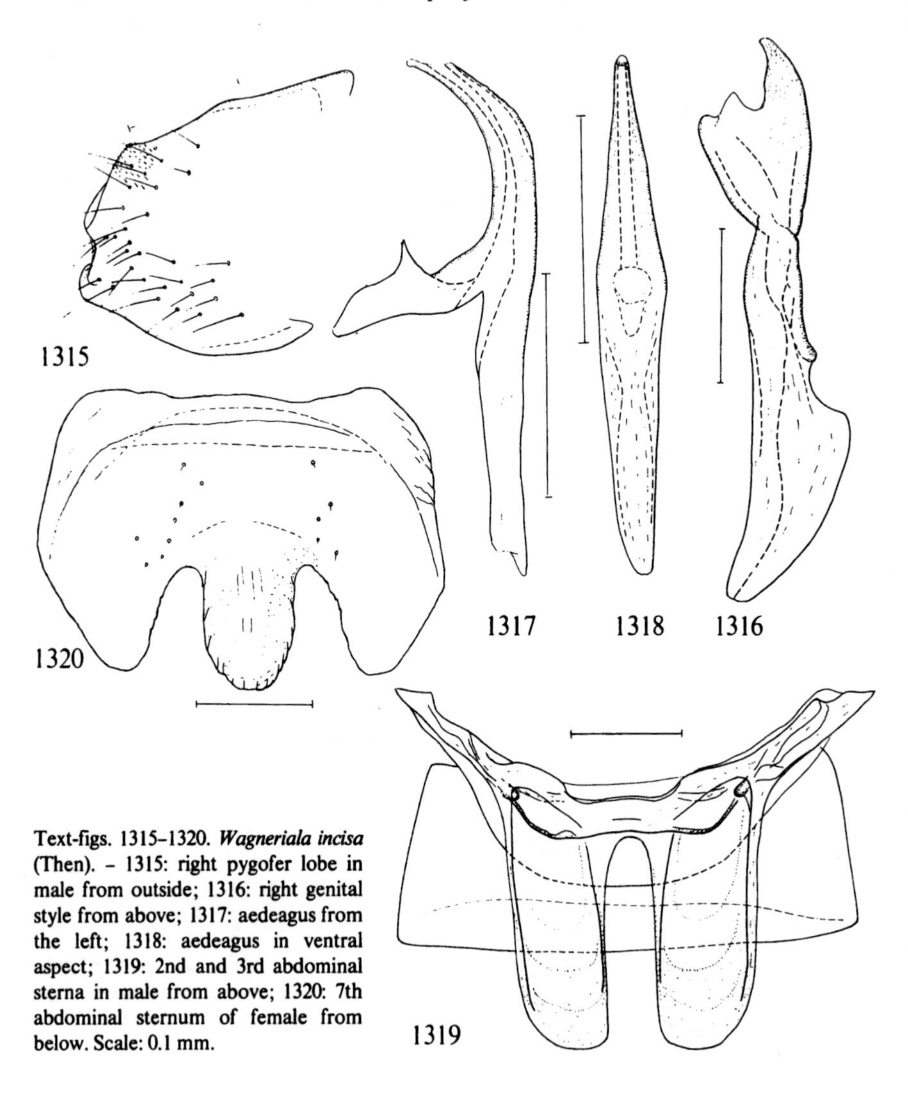

Text-figs. 1315–1320. *Wagneriala incisa* (Then). – 1315: right pygofer lobe in male from outside; 1316: right genital style from above; 1317: aedeagus from the left; 1318: aedeagus in ventral aspect; 1319: 2nd and 3rd abdominal sterna in male from above; 1320: 7th abdominal sternum of female from below. Scale: 0.1 mm.

Genus *Forcipata* De Long & Caldwell, 1936

Forcipata De Long & Caldwell, 1936: 70.
Type-species: *Forcipata loca* De Long & Caldwell, 1936, by subsequent designation.

Elongate, slender species. Head medially slightly produced, apex rounded, as wide as pronotum or slightly wider. Pronotum elongate. Fore wing long and slender, apical cells long, fourth apical cell nearly reaching apical margin of wing. Genital plates forcipate, exposing aedeagus, each with a longitudinal row of strong macrosetae. Genital valve large, on inside with a median longitudinal ridge. A distinct sclerotized connective absent. Pygofer of male without appendages. Stem of aedeagus also without appendages. 7th abdominal sternum in female with two incisions dividing the sternum in three lobes, the median one longer and broader. In Denmark and Fennoscandia two species.

Key to species of *Forcipata*

1 Genital plates slender, much longer than genital valve Text-fig. 1321) .... 175. *citrinella* (Zetterstedt)
– Genital plates short and stout, not longer than genital valve (Text-fig. 1327) .... 176. *forcipata* (Flor)

175. ***Forcipata citrinella*** (Zetterstedt, 1828)
Text-figs. 1321–1326.

Cicada citrinella Zetterstedt, 1828: 536.
Cicadula gracilis Zetterstedt, 1838: 299.
Dicranoneura similis Edwards, 1885: 229.
Notus fieberi Löw, in Then, 1886: 39 (n.n.).

Elongate, light yellow or dirty yellow, shining. Head anteriorly rounded. Veins of hind wings usually slightly dirty yellow. Abdomen above largely black, ventrally with segments anteriorly partly dark-bordered. Genital valve of male large, shield-shaped, genital plates long (Text-fig. 1321). Genital style as in Text-fig. 1322, aedeagus as in Text-figs. 1323, 1324, 1st–4th abdominal sterna in male as in Text-fig. 1325. Ventral aspect of apical part of female abdomen as in Text-fig. 1326; median lobe of 7th sternum apically with or without a shallow emargination. Overall length of males 3.2–3.5 mm, of females 3.4–3.7 mm.

Distribution. Uncommon in Denmark, found in EJ, SJ, NEZ. – Fairly common in Sweden, found in most provinces, Sk.–T.Lpm. – Probably not uncommon in Norway, found in Bv, AAy, Ry, HOi, Nsy, and TRi. – Fairly common in southern and central East Fennoscandia, found in Al, Ab, N, Kb, ObN, Ks. – Widespread in Europe, also recorded from Altai Mts., Kazakhstan, m. Siberia, Kirghizia, Mongolia, Korean Peninsula, Maritime Territory; Nearctic Region.

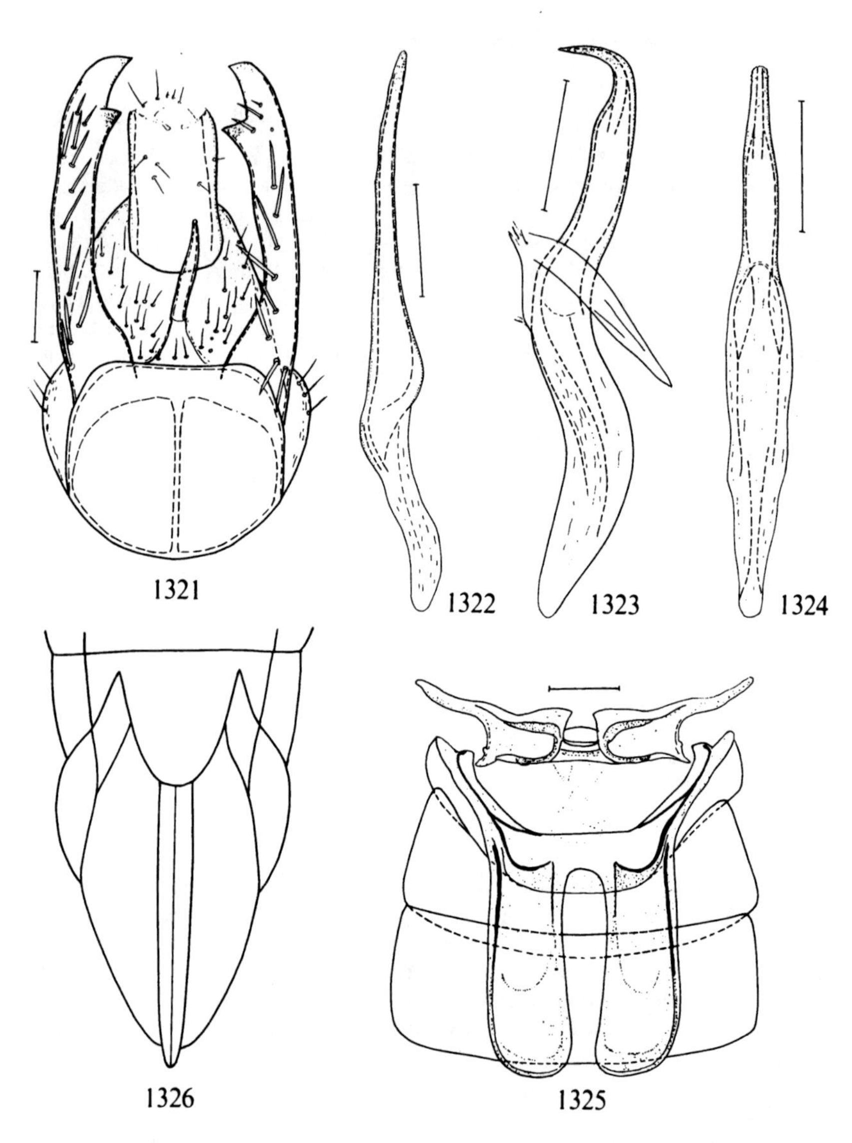

Text-figs. 1321–1226: *Forcipata citrinella* (Zetterstedt). – 1321: male genital segment in ventral aspect; 1322: right genital style from above; 1323: aedeagus from the left; 1324: aedeagus in ventral aspect; 1325: 1st–4th abdominal sterna in male from above (pigmentation of 3rd and 4th st. omitted); 1326: apical part of female abdomen from below. Scale: 0.1 mm.

Biology. "In Flachmooren, Waldlichtungen" (Kuntze, 1937). In sphagnous spruce woods, rich swampy woods, rich moist grass-herb woods, moist *Oxalis-Myrtillus* spruce woods, and moist *Myrtillus* spruce woods, also in wet peaty meadows (Linnavuori, 1952a). In the "Molinietalia", especially in "zeitweilig nassen Bentgraswiesen" (Marchand, 1953). "Nur an feuchten Orten" (Schwoerbel, 1957). "Kommt auch in extremen Trockenrasen vor" (Wagner & Franz, 1961). "On Cyperaceae and Graminaceae growing in damp meadows" (Vidano, 1965). Hibernation in the egg stage, two generations (Schiemenz, 1969b). Adults in June–September.

176. ***Forcipata forcipata*** (Flor, 1861)
Text-figs. 1327–1331.

Typhlocyba forcipata Flor, 1861: 389.
Dicraneura citrinella Jensen-Haarup, 1920: 164 (nec Zetterstedt, 1828).
Dicraneura citrinella Ribaut, 1936b: 217 (nec Zetterstedt, 1828).

Resembling *citrinella*, usually more dirty yellow. Male genital segment as in Text-fig. 1327, genital style as in Text-fig. 1328, aedeagus as in Text-figs. 1329, 1330, 1st–3rd abdominal sterna in male as in Text-fig. 1331. 7th abdominal sternum in female as in *citrinella*, median lobe slightly shorter. Overall length (♂♀) 3.2–4.1 mm.

Distribution. Scarce in Denmark, found in EJ, SZ, NEZ. - Scarce in the south of Sweden, apparently commoner in the north, found in Sk., Ög., Upl., Vstm., Dlr., Hls., Hrj., Jmt., Ång., Vb., Lu.Lpm. - Norway: found in HEn, Os, Bø, TEi, MRy. - Scarce in East Fennoscandia, found in Ab, Sb, Kb, Li. - Austria, Bohemia, Moravia, Slovakia, Yugoslavia, England, France, Belgium, Switzerland, German D.R. and F.R., Poland, Estonia, Latvia, Lithuania, m. Russia, Ukraine, Altai Mts., Kazakhstan, n. Siberia, Mongolia.

Biology. In "Hochmooren, Wäldern" (Kuntze, 1937). "Nur an den feuchtesten Plätzen" (Schwoerbel, 1957). "Bewohnt feuchte, mit *Carex*-Arten bestandene Wiesen, im Gebiet der Nördlichen Kalkalpen besonders in subalpinen bis hochsubalpinen Lagen. Lebt mit Vorliebe auch an Föhrenheidestandorten" (Wagner & Franz, 1961). Univoltine, hibernation in egg stage (Schiemenz, 1975). Adults in June–September.

Genus *Notus* Fieber, 1866

Notus Fieber, 1866: 508.
Type-species: *Cicada flavipennis* Zetterstedt, 1828, by subsequent designation.

Elongate, slender, resembling *Forcipata*. Head in dorsal aspect medially produced, apex rounded or angular. Head with eyes broader than pronotum. Pronotum elongate. Genital valve of male as long as broad or longer. Genital plates short, each with an oblique row of macrosetae and a abterminal horn-like process. Aedeagus U-shaped, with two phallotremes. 7th abdominal sternum in female divided into two lateral parts. In Denmark and Fennoscandia one species.

177. ***Notus flavipennis*** (Zetterstedt, 1828)
Plate-fig. 106, text-figs. 1267, 1332–1339.

Cicada flavipennis Zetterstedt, 1828: 525.
Cicadula orichalcea Dahlbom, 1850: 183.
Notus marginatus J. Sahlberg, 1871: 164.

Elongate, lively yellow, shining. Head medially produced, angular especially in female. Apical part of fore wing proximally hyaline, distally smoke-coloured. Hind wings of male with darker veins. Mesosternum and abdomen dorsally and ventrally partly blackish. Resembling *Forcipata citrinella* and *forcipata*, differing by the shape of the head, by the more lively yellow colour, and by the structure of genitalia in both sexes. Male genital segment as in Text-fig. 1332, genital plates as in Text-figs. 1333, 1334,

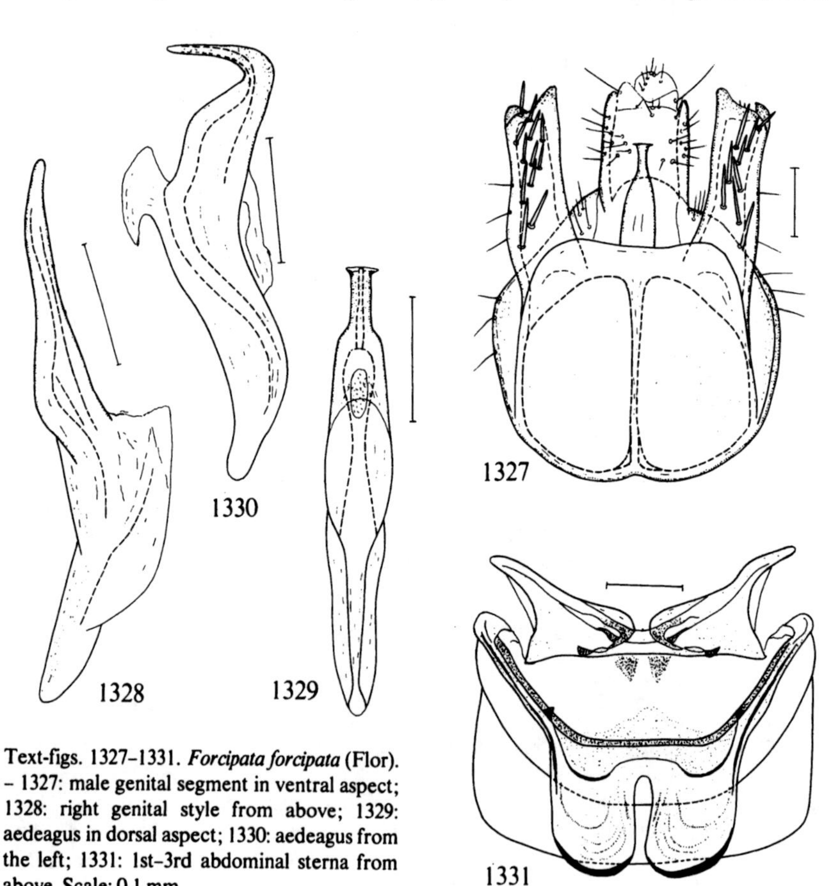

Text-figs. 1327–1331. *Forcipata forcipata* (Flor). – 1327: male genital segment in ventral aspect; 1328: right genital style from above; 1329: aedeagus in dorsal aspect; 1330: aedeagus from the left; 1331: 1st–3rd abdominal sterna from above. Scale: 0.1 mm.

genital style as in Text-fig. 1335, aedeagus as in Text-figs. 1336, 1337, 1st–4th abdominal sterna in male as in Text-fig. 1338; ventral aspect of caudal part of female abdomen as in Text-fig. 1339. Overall length (♂♀) 3.8–4.1 mm.

Distribution. Common and widespread in Denmark, Sweden and East Fennoscandia, probably also in Norway. I have seen specimens from the Norwegian districts Ø, AK, VAy, Ry, STi, Nnø, and TRi. Usually abundant in its typical biotopes. – Widespread in northern, western, central, and eastern Europe, also in Altai Mts., Kazakhstan, Uzbekistan, Kirghizia, Kurile Islands, n. Siberia, Mongolia, and Nearctic Region.

Biology. In "Salzstellen, Flachmooren, Waldlichtungen, Wiesen" (Kuntze, 1937). "Kommt auf *Sphagnum*-reichen, licht mit *Carex* bewachsenen Weissmooren in der Zeit Juli–August verhältnismässig reichlich vor" (Kontkanen, 1938). On seashores, both wet and drier peaty meadows, tall-sedge bogs, short-sedge bogs, wet "rimpi" bogs, and in sphagnous spruce woods and rich swampy woods (Linnavuori, 1952a) In the "Molinietalia" (Marchand, 1953). "In shore meadows, treeless fens and even in spruce and birch swamps; it thrives in wet meadows cleared by man" (Raatikainen & Vasarainen, 1976). Hibernation takes place in the egg stage (Törmälä & Raatikainen, 1976). Adults in June–October.

Genus *Empoasca* Walsh, 1862

Empoasca Walsh, 1862: 149.
Type-species: *Empoasca viridescens* Walsh, 1862, by subsequent designation.
Kybos Fieber, 1866: 508.
Type-species: *Cicada smaragdula* Fallén, 1806, by subsequent designation.

Body and fore wings elongate. Anterior margin of head rounded, not angular. Ocelli present or absent. Apical cell of hind wings distally closed. Hind wing with only one apical cell. Male pygofer with long appendages arising from ventral margin. Macrosetae of genital plates usually not placed in rows. Main colour in most species greenish. In Denmark and Fennoscandia 17 species. Identification of females in many cases difficult or impossible.

Key to species of *Empoasca*

1 Fore and hind margins of vertex parallel, vertex medially not or only slightly longer than near eyes. Second apical cell in fore wing usually stalked. Strongly built, large species (>3.7 mm). (Subgenus *Kybos* Fieber) 7
– Fore margin of vertex produced, vertex medially distinctly longer than near eyes. Second apical cell in fore wing normally not stalked. Slender, fragile species (Subgenus *Empoasca* Walsh) 2

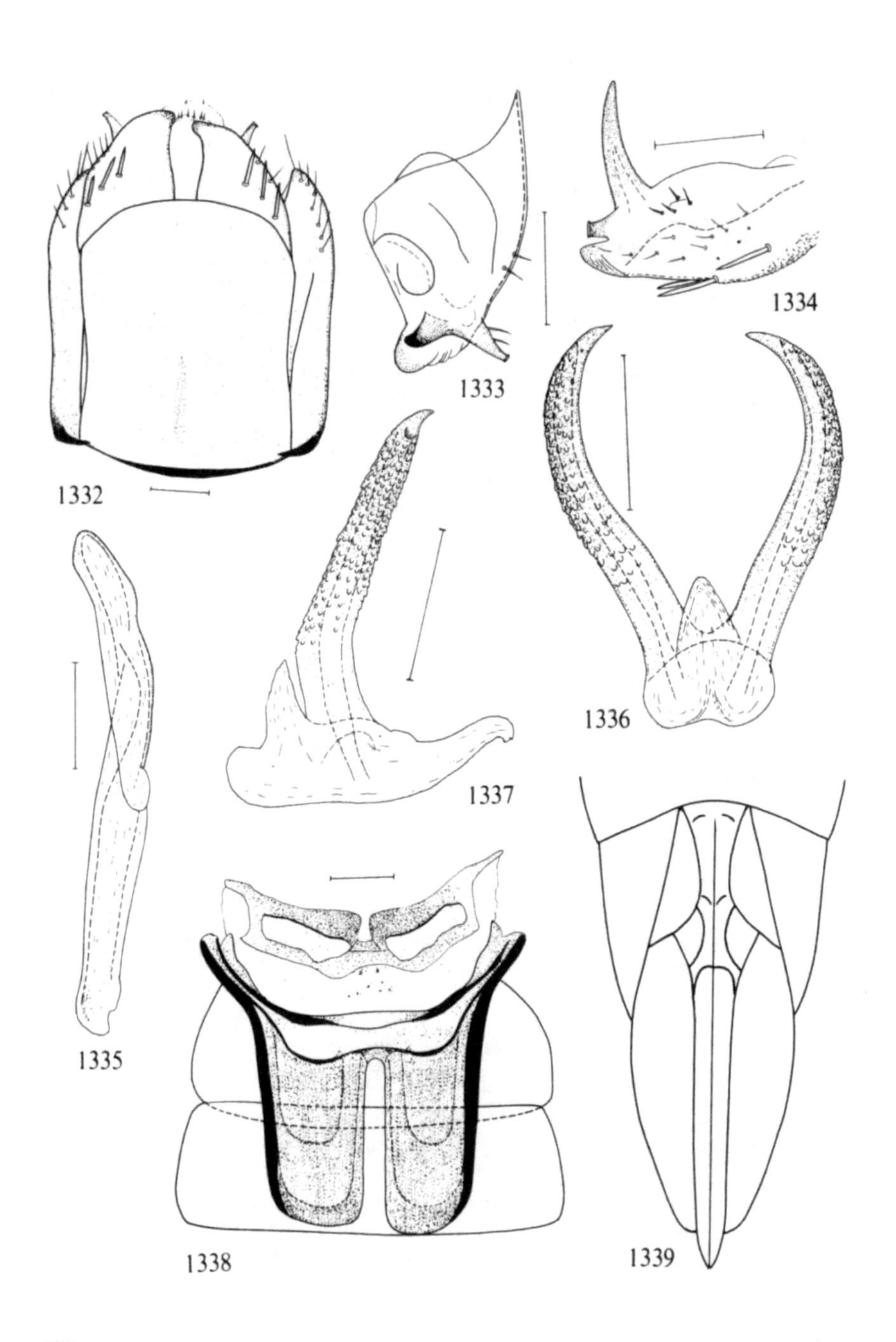

1332
1333
1334
1335
1336
1337
1338
1339

2 (1) Pale yellow. Fore wings with apical part including distal ends of median and cubital cells smoke-coloured. Appendages of anal collar acute-pointed (Text-fig. 1372). Pygofer processes as in Text-fig. 1373 .... 183. *apicalis* (Flor)

– Body and fore wings light green, apical part of fore wing colourless or only faintly fumose 3

3 (2) Fore wing with a bluish or colourless hyaline longitudinal stripe in median cell and apices of cubital and radial cells. Process of anal collar acute – pointed, curved in frontal direction (Text-fig. 1341). Appendages of pygofer as in Text-fig. 1342 .... 178. *vitis* (Göthe)

– Fore wing without a hyaline longitudinal stripe 4

4 (3) Anal tube processes acute-pointed (Text-figs. 1359, 1366) 5

– Anal tube processes blunt, towards apices granulate or denticulate (Text-figs. 1347, 1353) 6

5 (4) Pygofer appendages as in Text-figs. 1360, 1361 ... 181. *kontkaneni* Ossiannilsson

– Pygofer appendages as in Text-fig. 1367 .... 182. *ossiannilssoni* Nuorteva

6 (4) Pygofer appendages as in Text-fig. 1348 .... 179. *solani* (Curtis)

– Pygofer appendages as in Text-fig. 1354 .... 180. *decipiens* Paoli

7 (1) Fore wing along commissural border with a brown band often extending on scutellum and pronotum. Aedeagus without processes arising from base of shaft. Anal collar processes long, slender (Text-figs. 1421, 1427). On *Salix* 8

– Commissural border of fore wing dark or concolorous with wing, brown band absent 9

8 (7) Phragma lobes of 3rd abdominal tergum in male short (Text-fig. 1426). On *Salix purpurea* .... 190. *rufescens* (Melichar)

– Phragma lobes of 3rd abdominal tergum in male as long as broad (Text-fig. 1432). On *Salix* spp. (*repens, cinerea, caprea, myrsinifolia*) .... 191. *butleri* Edwards

9 (7) Corioclaval suture black or fuscous. Aedeagus with a pair of processes arising from base of shaft 10

– Corioclaval suture concolorous with fore wing, rarely fuscous. Aedeagus without basal processes 15

10 (9) Processes of anal collar short, stout (Text-figs. 1379, 1388, 1394, 1402) 11

– Processes of anal collar long, slender (Text-figs. 1408, 1414) 14

Text-figs. 1332–1339. *Notus flavipennis* (Zetterstedt). – 1332: male genital segment from below; 1333: right genital plate from above; 1334: same from outside; 1335: right genital style from above; 1336: aedeagus in dorsal aspect; 1337: aedeagus from the left; 1338: 1st–4th abdominal sterna in male from above (pigmentation of 3rd and 4th sterna omitted); 1339: apical part of female abdomen from below. Scale: 0.1 mm.

11 (10) Processes arising from base of aedeagal shaft strongly diverging. Shaft of aedeagus, often also basal processes, denticulate, armed with some teeth varying in size and arrangement (Text-figs. 1404, 1405). On *Salix* 187. *strigilifera* Ossiannilsson

– Processes of aedeagus in ventral aspect more or less parallel 12

12 (11) Apodemes of 2nd abdominal sternum in male very short, inconspicuous (Text-fig. 1384). On *Alnus* .... 184. *smaragdula* (Fallén)

– Apodemes of 2nd abdominal sternum in male distinct, about as long as broad. On *Betula* 13

13 (12) Processes of aedeagus basally widely apart from each other (Text-fig 1397) .... 186. *betulicola* W. Wagner

– Processes of aedeagus basally close to each other (Text-fig. 1390) .... 185. *lindbergi* Linnavuori

14 (10) Phragma lobes of 3rd abdominal tergum in male very short (Text-fig. 1413). Apodemes of 2nd abdominal sternum about twice as long as broad (Text-fig. 1412) .... 188. *virgator* Ribaut

– Phragma lobes of 3rd abdominal tergum longer (Text-fig. 1420). Apodemes of 2nd abdominal sternum shorter (Text-fig. 1419) .... 189. *volgensis* (Vilbaste)

15 (9) Face, sometimes also dorsum of fore body, in male largely fuscous. On *Salix* .... 193. *sordidula* Ossiannilsson

– Face green or yellowish, not fuscous. On *Populus* 16

16 (15) Pygofer process in male as in Text-fig. 1434. On *Populus tremula* .... 192. *populi* Edwards

– Pygofer process as in Text-fig. 1447. On *Populus nigra* and *P. nigra italica* .... 194. *abstrusa* Linnavuori

The following information concerning the larvae of *Empoasca* is taken from Wilson (1978). Spines on anterior margin of vertex longer than 2nd antennal segment. Thorax, wing pads and abdomen with long spines. The larvae of *Empoasca* s.str. are uniformly pale green with pale spines. (A probable exception is *apicalis*, a species not studied by Wilson). The nymphs of subgenus *Kybos* are dominantly green in colour with variable brown markings on dorsal surface. Large body spines frequently dark pigmented.

178. ***Empoasca (Empoasca) vitis*** (Göthe, 1875)

Plate-fig. 108, text-figs. 1340–1346.

Typhlocyba vitis Göthe, 1875: 397.
Typhlocyba flavescens Flor, 1861: 394 (nec Fabricius, 1794).
Empoasca vitium Paoli, 1930: 64.

Elongate, slender. Light green, often with a bluish tinge. Fore body often with whitish spots. Fore wings with a bluish white or colourless hyaline longitudinal stripe as described in the key. Apical part of fore wing colourless. Abdomen in both sexes unicolorous light. Legs partly verdigris green. Male genital segment as in Text-fig. 1340, anal apparatus as in Text-fig. 1341, pygofer appendages of male as in Text-fig. 1342, genital style and connective as in Text-fig. 1343, aedeagus in Text-figs. 1344, 1345, 1st–4th abdominal sterna in male as in Text-fig. 1346. Caudal border of 7th abdominal sternum in female moderately convex. Overall length (♂♀) 3.1–3.8 mm.

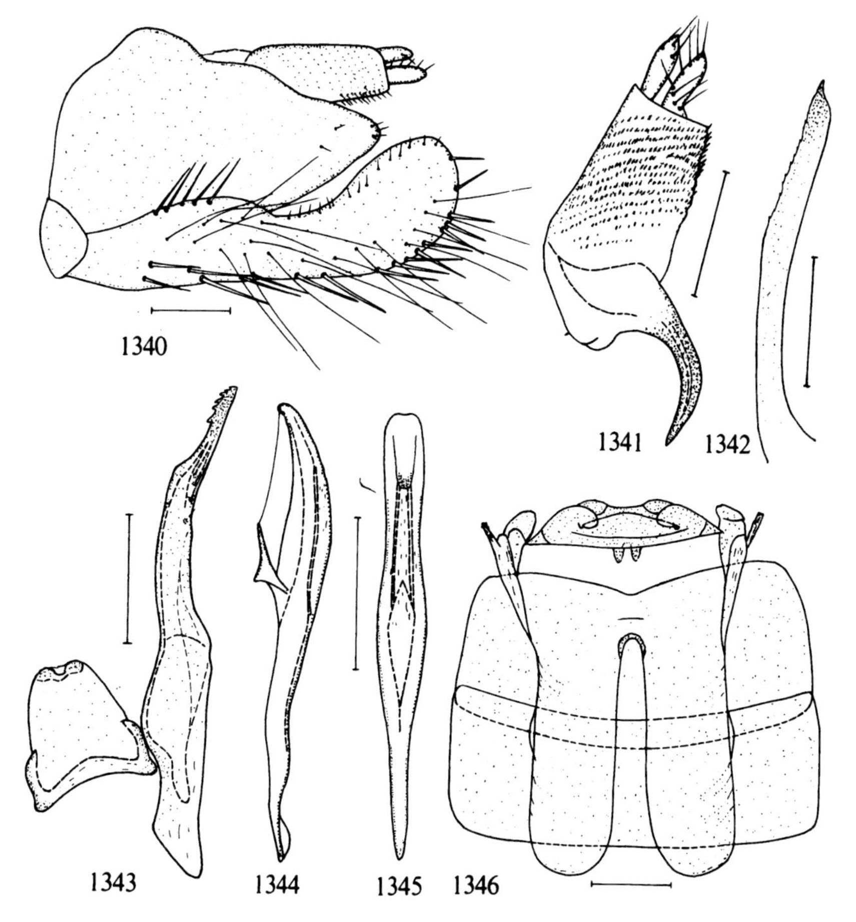

Text-figs. 1340–1346. *Empoasca vitis* (Göthe). – 1340: male genital segment from the left; 1341: male anal apparatus from the left; 1342: left pygofer appendage of male from inside; 1343: left genital style and connective from above; 1344: aedeagus from the left; 1345: aedeagus in ventral aspect; 1346: 1st–4th abdominal sterna in male from above. Scale: 0.1 mm.

Distribution. Common and widespread in Denmark, Sweden (Sk.–Vb.), southern Norway (Ø, AK–STi), and in southern and central East Fennoscandia (Al, Ab, N–Kb; Kr). – Widespread in the Palaeartic region; also in the Oriental region.

Biology. "In fast allen Lebensräumen" (Kuntze, 1937). "Polyphag an Sträuchern und Kräutern; überwintert an Koniferen" (Wagner & Franz, 1961). Ubiquitous, polyphagous; in Italy 2–3 overlapping generations. Larvae and adults usually on the underside of leaves. Phloem sucker (Vidano, 1958). "Wirtspflanzen in Zucht: Apfel, Kirsche, Reben, Baumwolle, Ackerbohne" (Günthart, 1971). Polyphagous; reared on *Acer pseudoplatanus, Alnus, Carpinus, Corylus, Fagus, Fraxinus, Quercus, Sorbus, Ulmus* (Claridge & Wilson, 1976). Hibernation in adult stage (Ossiannilsson, 1943; Müller, 1957; Remane, 1958). In the south of Sweden two overlapping generations (Ossiannilsson, 1946c). Adults can be found at any time of the year.

Economic importance. From many parts of the world reported as a dangerous pest on many cultivated plants, e.g. tea, potato, sugar-beet and sugar-cane, tobacco, cotton, hop, fruit trees. "Schädigt das Palisadenparenchym und die Leitbündel. Die Blätter kräuseln sich, bleiben im Wachstum zurück, und die Pflanzen kümmern" (Müller, 1956). Certain statements on *E. vitis* as a virus vector are regarded as doubtful (Heinze, 1959), but since the species has been shown to be a phloëm sucker (Vidano, 1958) they are not altogether improbable.

179. ***Empoasca (Empoasca) solani*** (Curtis, 1846)
Text-figs. 1347–1352.

Eupteryx solani Curtis, 1846: 388.
Typhlocyba pteridis Dahlbom, 1850: 179.
Empoasca tullgreni Ribaut, 1933: 154.
(Swedish: betstrit.)

Resembling *vitis* but more purely light green. Rarely with light spots on fore body. Fore wings comparatively shorter than in *vitis*, without a hyaline longitudinal stripe; apical part colourless. Male anal apparatus as in Text-fig. 1347, pygofer appendage as in Text-fig. 1348, genital style as in 1349, aedeagus as in Text-figs. 1350, 1351, apodemes of 2nd abdominal sternum in male as in Text-fig. 1352. Overall length (♂♀) 3–3.6 mm.

Distribution. Denmark: so far reliable records from B: Saltuna 15.VII.1974 (Trolle), and Ypnasted 21.VIII.1976 (Trolle). – Common in the south of Sweden, Sk.–Ög., Upl., Nrk., Dlr. – Norway: only found in AK: Sem, Asker 5.VI.1973 (Taksdal). – Rare in East Fennoscandia, found in Ab: Åbo according to Linnavuori (1950), and in Ab: Pojo 30.VIII.1976 (Albrecht); N: Strömfors 16.IX.1973 (Albrecht); Vib: Kuolemajärvi 22.VIII.1934 (P. H. Lindberg). – Widespread in Europe, also in Algeria, Morocco, Tunisia, Anatolia, Georgia, Kazakhstan, Mongolia.

Biology. "An Kartoffelkraut oft in Menge" (Kuntze, 1937). Not particularly common on potato in Sweden but sometimes abundant (Ossiannilsson, 1943a). In all biotopes examined ("Kultursteppe, feuchte Wiese, Auenwald, gemischter Wald, Waldschlag, Weingärtenraine, Waldsteppe") (Okáli, 1960). Hibernation by adult females (Schiemenz, 1964). Occasional phloëm sucker (Vidano, 1959). Subdominant in the herbage stratum of lucerne fields in s. Moravia in 1961 (Obrtel, 1969). "Imaginal-Überwinterer, 3 Generationen" (Schiemenz, 1969b). Reared on "Zuckerrüben, Ackerbohnen, Kartoffeln, Bohnen, Baumwolle" (Günthart, 1971). – In Sweden adults have been found in June–September, never on conifers.

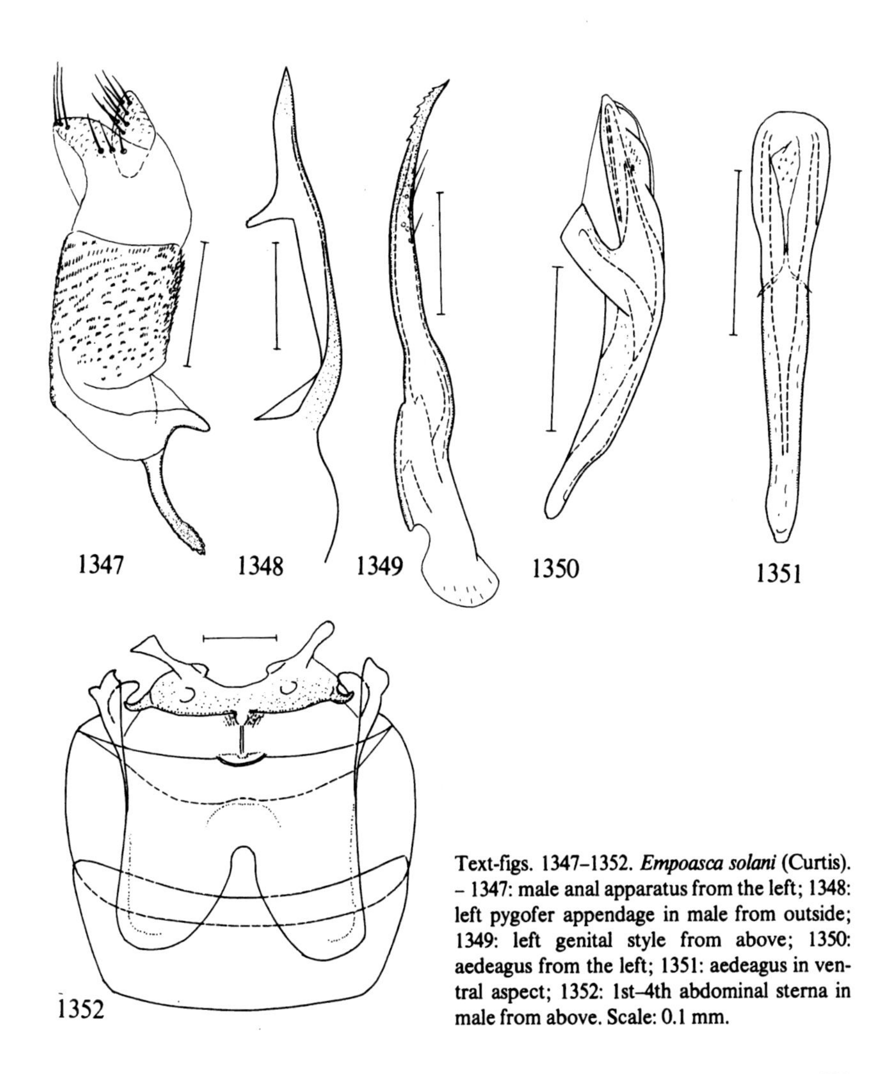

Text-figs. 1347–1352. *Empoasca solani* (Curtis). – 1347: male anal apparatus from the left; 1348: left pygofer appendage in male from outside; 1349: left genital style from above; 1350: aedeagus from the left; 1351: aedeagus in ventral aspect; 1352: 1st–4th abdominal sterna in male from above. Scale: 0.1 mm.

180. ***Empoasca (Empoasca) decipiens*** Paoli, 1930
Text-figs. 1353–1358.

Empoasca decipiens Paoli, 1930: 74.

Resembling *E. solani.* Male anal apparatus as in Text-fig. 1353, male pygofer process as in Text-fig. 1354, genital style as in Text-fig. 1355, aedeagus as in Text-figs. 1356, 1357, 1st–4th abdominal sterna in male as in Text-fig. 1358. Overall length (♂♀) 3.2–4 mm.

Distribution. Denmark: so far found only once in a garden in EJ: Århus 16.IX.1977 (L. Trolle). – Norway: TEy: Heistad, 1 ♂ (Holgersen). – Not in Sweden, nor in East Fennoscandia. – Widespread in Europe, also in Morocco, Tunisia, Egypt, Anatolia, Israel, Iran, Iraq, Jordan, Lebanon, Afghanistan, Georgia, Kazakhstan, Tadzhikistan, Turkmenia, Uzbekistan, Kirghizia; Ethiopian region.

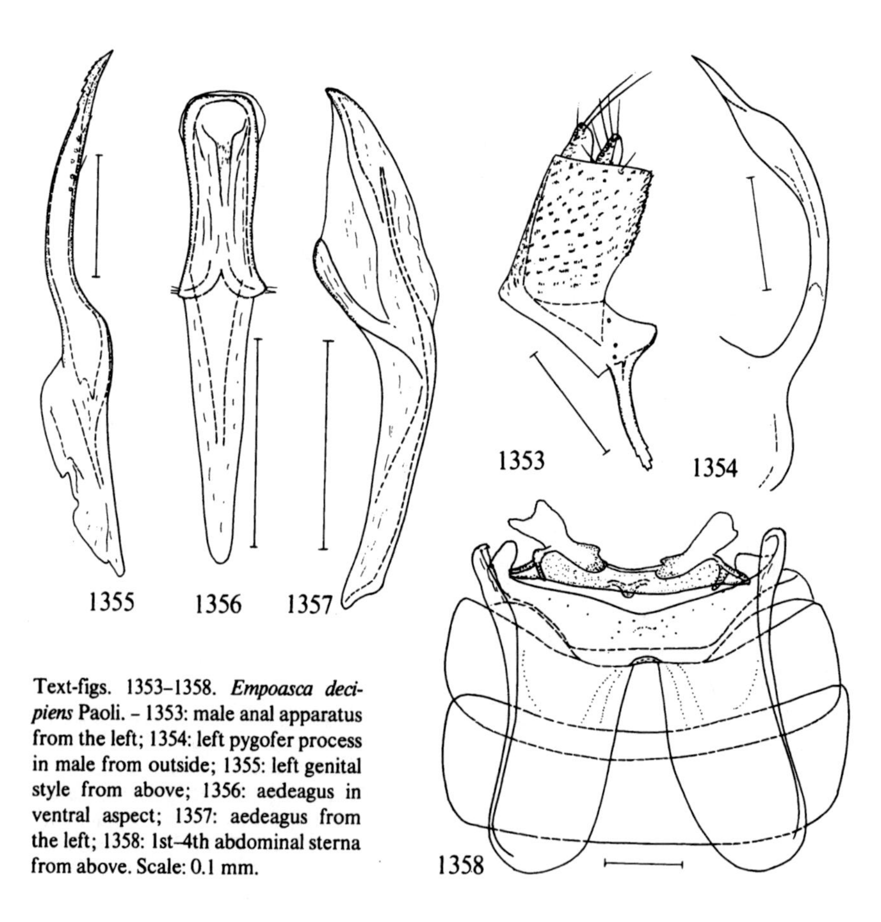

Text-figs. 1353–1358. *Empoasca decipiens* Paoli. – 1353: male anal apparatus from the left; 1354: left pygofer process in male from outside; 1355: left genital style from above; 1356: aedeagus in ventral aspect; 1357: aedeagus from the left; 1358: 1st–4th abdominal sterna from above. Scale: 0.1 mm.

Biology. On egg plant, potato, tomato, beet (Paoli, 1930). On *Trifolium alexandrinum*, kidney beans, lentils, barley, wheat, sugar-cane, maize, cotton, sweet potatoes, tomatoes, *Capparis* spec. in Egypt, and on various plants (vine, beans, *Helianthus*, *Ricinus*, tomato, potato, beet, oak) in Italy (Paoli, 1936). "Auch diese Art überwintert, offenbar aber nicht auf Koniferen" (Wagner, 1939). On *Cannabis sativa*, *Beta vulgaris*, *Amarantus hybridus*, *Potentilla grandiflora*, *Medicago sativa*, *Vicia faba*, *Phaseolus vulgaris*, *Lavatera arborea*, *Solanum tuberosum*, *S. melongena*, *S. nigrum*, *Lycopersicum esculentum*, *Nicotiana tabacum*, *Capsicum annuum*, *Chrysanthemum indicum* and *balsamita*, *Sambucus nigra*, *Clematis vitalba*, *Populus euro-americana*, *Salix alba*, *Alnus glutinosa*, *Corylus avellana*, *Ficus carica*, *Rubus fruticosus*, *Sophora japonica*, *Acer pseudoplatanus*, *Citrus aurantium*, *Ligustrum ovalifolium*, *Lonicera caprifolium*; on *Vitis vinifera* only in the adult stage. Not on conifers. Occasionally phloem feeder in the adult stage (Vidano, 1958). *E. decipiens* lays its eggs in petioles or the stems (of *Vicia faba*) (Koblet-Günthardt, 1975).

Economic importance. "In Italien, Spanien, Ägypten, Belutschistan an Eierfrucht, Capsicum, Kartoffeln, Tomaten, Rüben, Baumwolle und Reben schädlich (Ver- und Entfärbung der Blätter, Laubabfall besonders am Rebstock), in Mitteleuropa bisher harmlos" (Müller, 1956).

181. ***Empoasca (Empoasca) kontkaneni*** Ossiannilsson, 1949
Text-figs. 1359–1365.

Empoasca kontkaneni Ossiannilsson, 1949b: 71.

Body oblong, light green with whitish or yellow spots on head and prothorax. Fore wings extending considerably behind apex of abdomen, light green, apically indistinctly infuscate. Legs and veins of fore wings often with verdigris green streaks and spots. Male anal apparatus as in Text-fig. 1359, pygofer process as in Text-figs. 1360, 1361, genital style as in Text-fig. 1362, aedeagus as in Text-figs. 1363, 1364. 1st–3rd abdominal sterna in male as in Text-fig. 1365. Overall length (♂♀) 3.1–3.5 mm.

Distribution. So far not found in Denmark. - Distribution in Sweden imperfectly investigated, found in Gtl.: Östergarn, Katthammarsvik 15.VIII.1969 (Gyllensvärd); Upl.: Vänge, St. Kil 16.VI.1963 (Ossiannilsson), Uppsala, Norby 21.V. and 28.V. 1950 (Ossiannilsson), Uppsala, Lurbo 19.III.1950 (Ossiannilsson). - Norway: one male found in AK: Nesodden 5.VIII.1946 (Holgersen). - East Fennoscandia: distribution incompletely known; recorded from Ab: Pojo 30.VIII.1976 (Albrecht); N: Ekenäs lk.; Ingå; Sibbo; Strömfors (Albrecht); Ta: Hattula, Vuohiniemi (Nuorteva); Sa: Joutseno (Thuneberg); Kb: Hammaslahti 13.V.1948 (Kontkanen). - Poland, Latvia, Estonia, m. Russia, Altai Mts., Mongolia.

Biology. "On damp grassy ground, often in the shade of trees and in the undergrowth of sparse woods" (Kontkanen in litt.; Ossiannilsson, 1949b). In bushes, with e.g. *Rubus idaeus*, and in lush undergrowth of woods (Linnavuori, 1969). "Die Art überwintert auf *Picea excelsa* und besiedelt im Frühjahr die Laubbäume, in dem Masse wie sich diese belauben. Insbesondere auf *Prunus padus,* deren Blätter am aller frühesten hervorbrechen, findet man Individuen der Art reichlich vor. Nachdem die Blätter der Bäume ihre volle sommerliche Ausbildung erlangt haben, siedeln sich die Zikaden auf die Kräuter der Feldschicht über. Auf den Laubbäumen verweilt also die Art nur so lange, wie sich die Blätter nach ihrem Hervorbrechen im Wachstum befinden." (Nuorteva, 1952). - Most of my specimens from Norby (38 individuals) were collected by sweeping among *Caltha, Filipendula ulmaria,* and *Geum rivale* in a wood near its border.

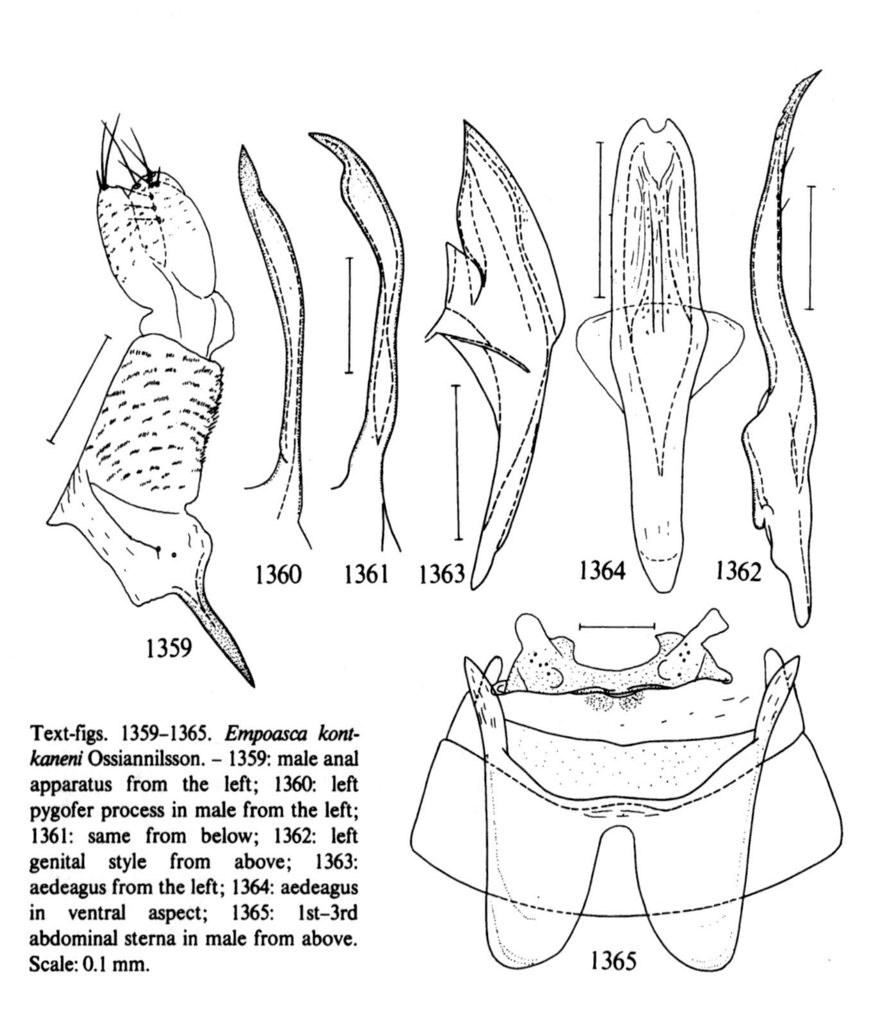

Text-figs. 1359–1365. *Empoasca kontkaneni* Ossiannilsson. - 1359: male anal apparatus from the left; 1360: left pygofer process in male from the left; 1361: same from below; 1362: left genital style from above; 1363: aedeagus from the left; 1364: aedeagus in ventral aspect; 1365: 1st–3rd abdominal sterna in male from above. Scale: 0.1 mm.

182. ***Empoasca (Empoasca) ossiannilssoni*** Nuorteva, 1948
Text-figs. 1366–1371.

Empoasca apicalis Ossiannilsson, 1943b: 16, 1946c: 118 (nec Flor, 1861).
Empoasca ossiannilssoni Nuorteva, 1948: 99 (n.n.).

Shape of body, pigmentation etc. as in *kontkaneni*. Apical part of fore wing in mature specimens distinctly infuscate but not conspicuously dark as in *apicalis*. Male anal apparatus as in Text-fig. 1366, pygofer appendage as in Text-fig. 1367, genital style as in Text-fig. 1368, aedeagus as in Text-figs. 1369, 1370, 1st–4th abdominal sterna in male as in Text-fig. 1371. Length (♂♀) 3.4–3.7 mm.

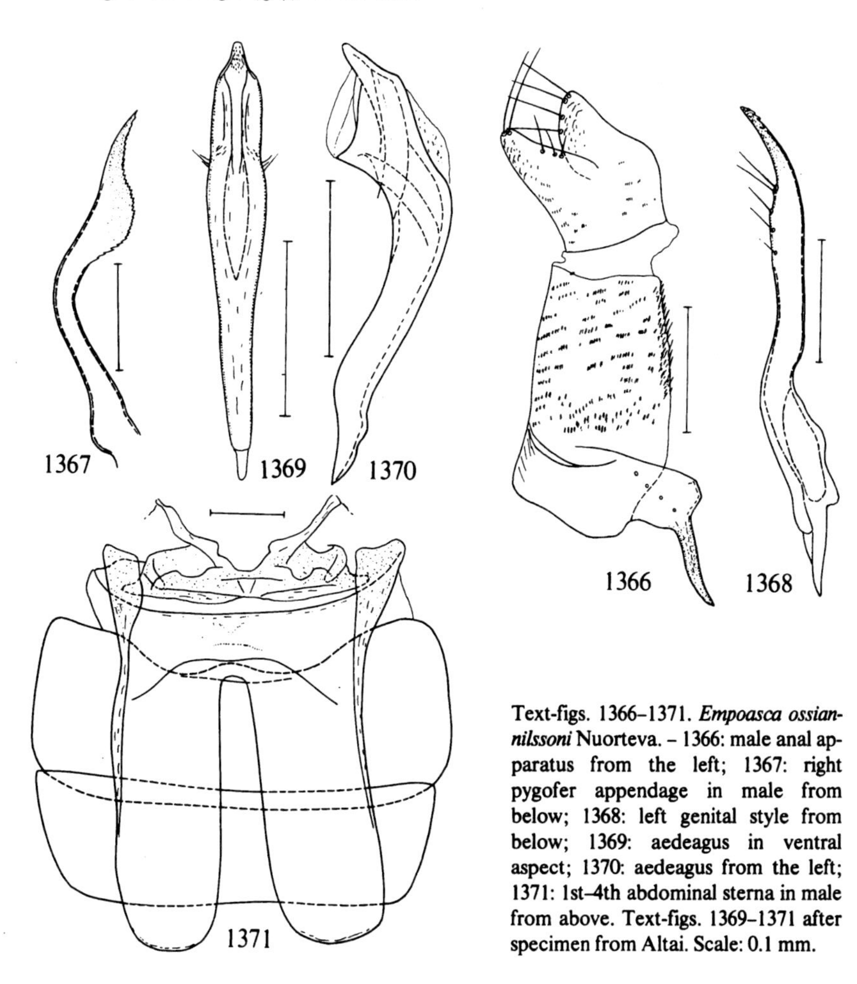

Text-figs. 1366–1371. *Empoasca ossiannilssoni* Nuorteva. – 1366: male anal apparatus from the left; 1367: right pygofer appendage in male from below; 1368: left genital style from below; 1369: aedeagus in ventral aspect; 1370: aedeagus from the left; 1371: 1st–4th abdominal sterna in male from above. Text-figs. 1369–1371 after specimen from Altai. Scale: 0.1 mm.

Distribution. Not found in Denmark and Norway. – Rare in Sweden, found only in Upl.: Solna, Bergshamra 27.IX.1941, 17.IX.1942, 20.X.1942, in all 4 specimens, and in Upl.: Vendel, Örbyhus 3.IV.1961, 1 ♂ (Ossiannilsson). – East Fennoscandia: distribution incompletely investigated; found in Al: Geta 31.VII.1942 (Håkan Lindberg); a few females from the vicinity of Åbo (Ab) probably also belong to the present species (Linnavuori, 1969). – Latvia, Russia (Gorky district), Altai Mts., Maritime Territory.

Biology. One of my males was found on *Prunus padus*, another male and two females on *Picea abies*. The specimens recorded from Latvia were found in a birch forest and in a park (Vilbaste, 1974).

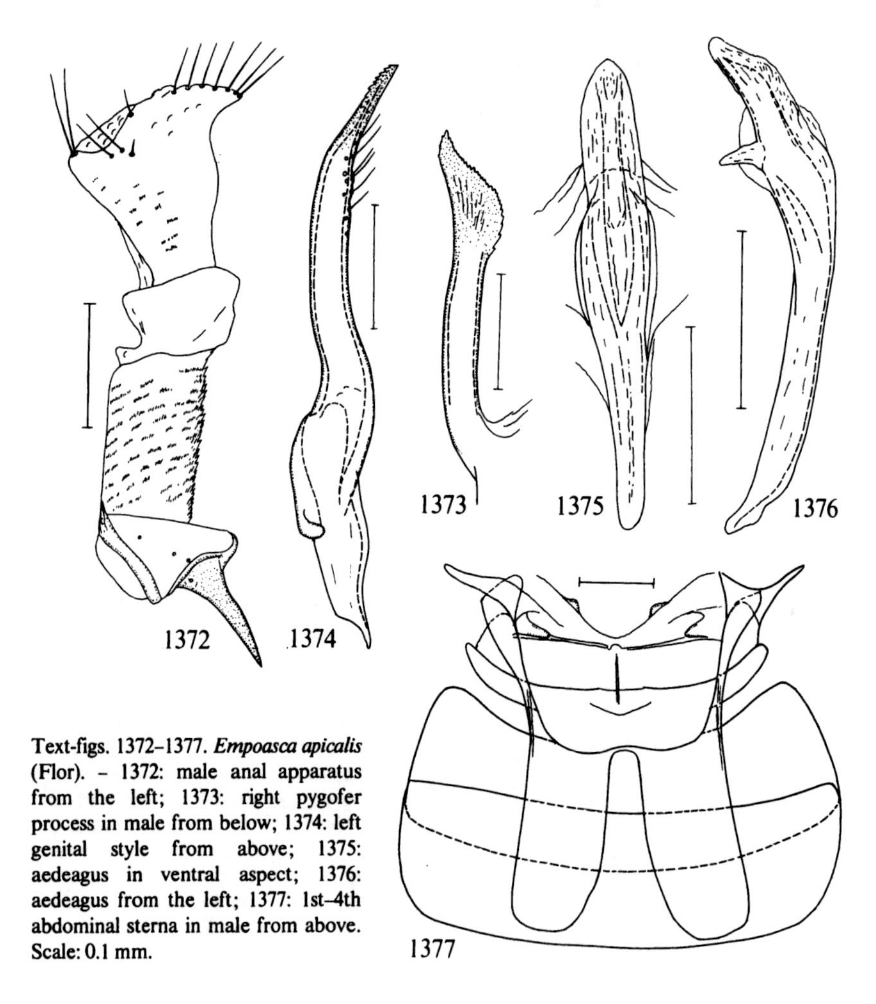

Text-figs. 1372–1377. *Empoasca apicalis* (Flor). – 1372: male anal apparatus from the left; 1373: right pygofer process in male from below; 1374: left genital style from above; 1375: aedeagus in ventral aspect; 1376: aedeagus from the left; 1377: 1st–4th abdominal sterna in male from above. Scale: 0.1 mm.

183. ***Empoasca (Empoasca) apicalis*** (Flor, 1861)
Plate-fig. 109, text-figs. 1372–1377.

Typhlocyba apicalis Flor, 1861: 396.
Empoasca apicalis Nast, 1938: 161.

Pale yellow, fore wings slightly paler, veins yellowish. (Fore wings in live specimens more greenish, according to Lindberg, 1960). Apical cells of fore wing and distal ends of median and cubital cells markedly smoke-coloured. Anal apparatus of male as in Text-fig. 1372, pygofer process as in Text-fig. 1373, genital style as in Text-fig. 1374, aedeagus as in Text-figs. 1375, 1376, 1st–4th abdominal sterna in male as in Text-fig. 1377. Overall length about 3.8 mm.

Distribution. Not in Denmark, Sweden and Norway. - East Fennoscandia: found in Ab: Lojo 27.VIII.1956 (Håkan Lindberg); Ta: Lammi 11.VIII.1948 (Linnavuori), Koski 13.VIII.1948 (Linnavuori), Renko, Renkajoki, Hiitta 29.V.1948 (Nuorteva), Hattula 2.–30.VII.1948 (Nuorteva), Urjala, Matku 26.V.1960 (Håkan Lindberg); also found in LkW; Kr: Solomino 28.VII.1869 (J. Sahlberg), Jalguba 1.IX.1969 (J. Sahlberg). - Austria, Belgium, Netherlands, Romania, Latvia, Estonia, n. and m. Russia, Ukraine, Kazakhstan, Maritime Territory, Korean Peninsula.

Biology. "Des Sommers monophag auf *Lonicera xylosteum,* nach dem Vergilben der Blätter dieses Strauches zahlreich auf *Alnus incana, Juniperus communis* und *Picea excelsa* angetroffen. Überwintert möglicherweise - - - auf Nadelbäumen. Imagines 29.V.–13.VI. und dann wieder 2.VII.–12.IX. - - - Nahrungsaufnahmestelle: Palissadenparenchym". (Nuorteva, 1952).

184. ***Empoasca (Kybos) smaragdula*** (Fallén, 1806)
Plate-fig. 89, text-figs. 1378–1386.

Cicada smaragdula Fallén, 1806: 37.

Strongly built, greenish, shining. Head, pronotum and scutellum often with whitish spots and bands. Pronotum and fore wings usually with a brownish tinge. Corioclaval suture and commissural border brownish. Veins of hind wings partly fuscous. Abdominal tergum often dark-spotted. Legs green, partly intensively verdigris green, tibiae with strong macrosetae. Genital plates (see Text-fig. 1378) bent upwards, with strong black macrosetae, long fine hairs, and short setae. Processes of anal collar (Text-fig. 1379) short, stout, with or without a thinner apical nib. Pygofer process as in Text-fig. 1380, genital style as in Text-fig. 1381, aedeagus as in Text-figs. 1382, 1383. Apodemes of 2nd abdominal sternum in male short, inconspicuous (Text-fig. 1384), phragma lobes of 3rd abdominal tergum in male as in Text-fig. 1385. 7th abdominal sternum in female as in Text-fig. 1386. Overall length of males 3.8–4.3 mm, of females 4.1–4.6 mm.

Distribution. Common and widespread in Denmark and Sweden (Sk.–Lu. Lpm.). -

Norway: so far found in AK, Os, Bø, TEi, AAi, Ry, Ri, HOi, MRi, Nnø. - Common in southern and central East Fennoscandia, found up to Ok and Ks; Kr. - Widespread in Europe, also in Armenia, Altai Mts., Kazakhstan, Tadzhikistan, Uzbekistan, m. Siberia, Sakhalin.

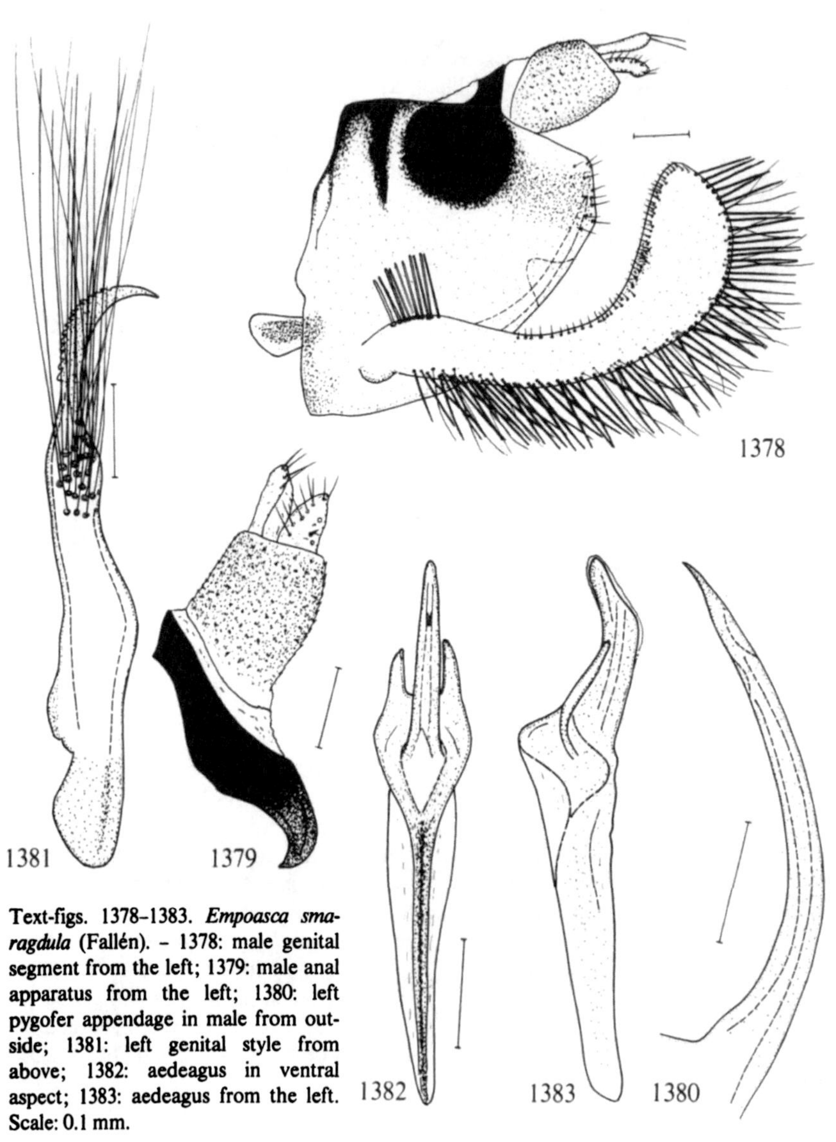

Text-figs. 1378–1383. *Empoasca smaragdula* (Fallén). - 1378: male genital segment from the left; 1379: male anal apparatus from the left; 1380: left pygofer appendage in male from outside; 1381: left genital style from above; 1382: aedeagus in ventral aspect; 1383: aedeagus from the left. Scale: 0.1 mm.

Biology. On *Alnus* (Nuorteva, 1952; Günthart, 1974; Claridge & Wilson, 1976). Hibernation in the egg stage (Müller, 1957). On *Alnus glutinosa* and *A. incana.* Adults in June–September.

185. ***Empoasca (Kybos) lindbergi*** Linnavuori, 1951
Text-figs. 1387–1393.

Empoasca lindbergi Linnavuori, 1951: 60.
Empoasca borealis Lindberg, 1952: 144.
Empoasca austriaca Ossiannilsson, 1955: 131 (nec. W. Wagner, 1949).

Resembling *E. smaragdula.* Male anal apparatus as in Text-figs. 1387, 1388, pygofer appendage as in Text-fig. 1389, aedeagus as in Text-figs. 1390, 1391, 2nd and 3rd abdominal sterna in male as in Text-fig. 1392, 2nd and 3rd abdominal terga in male as in Text-fig. 1393. 7th abdominal sternum in female as in *smaragdula.* Length (♂♀) 4.5–4.6 mm.

Distribution. So far not recorded from Denmark. - Not uncommon in central and northern Sweden, found in Upl., Nrk., Dlr., Hls., Ång., Vb., Nb. - Norway: found in AK: Ås, Kaja 25.IX.1973 (Baeschlin); Os: Biri 3.VII.1969 (Taksdal); On: Vågå, Vågåmo

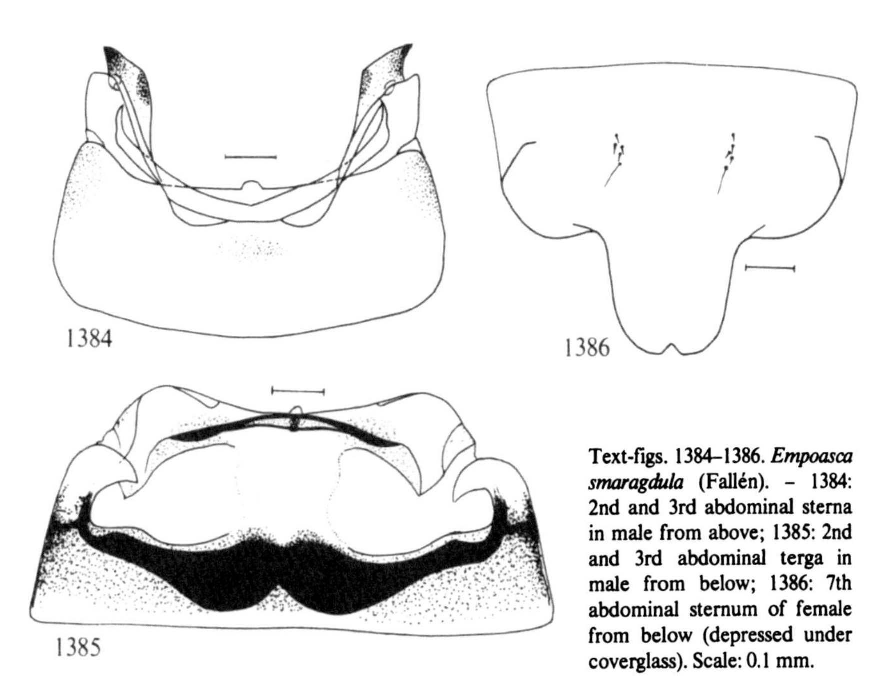

Text-figs. 1384–1386. *Empoasca smaragdula* (Fallén). - 1384: 2nd and 3rd abdominal sterna in male from above; 1385: 2nd and 3rd abdominal terga in male from below; 1386: 7th abdominal sternum of female from below (depressed under coverglass). Scale: 0.1 mm.

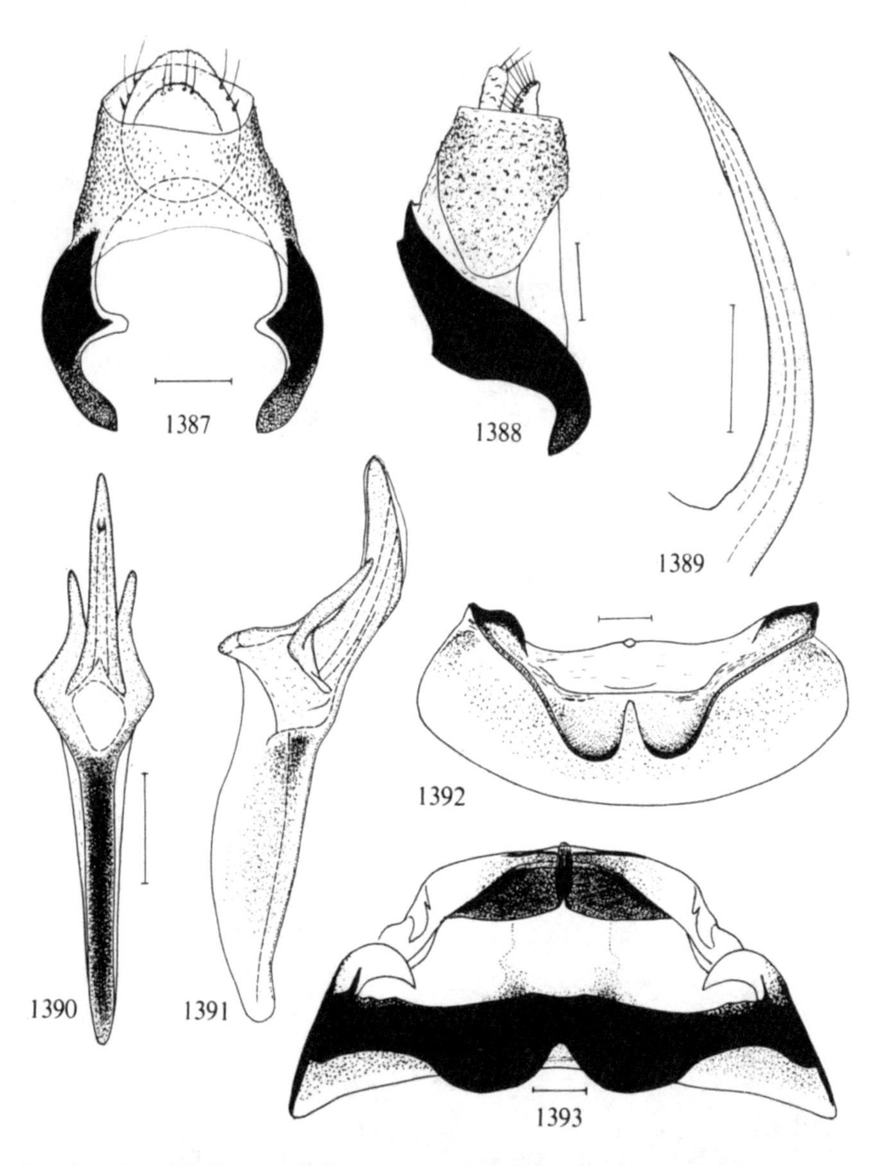

Text-figs. 1387–1393. *Empoasca lindbergi* Linnavuori. - 1387: male anal apparatus from behind; 1388: same from the left; 1389: left pygofer appendage of male from outside; 1390: aedeagus in ventral aspect; 1391: aedeagus from the left; 1392: 2nd and 3rd abdominal sterna in male from above; 1393: 2nd and 3rd abdominal terga in male from below. Scale: 0.1 mm.

12.VII.1953 (A. Løken); Bø: Modum, Bingen (Holgersen); STi: Eggen 24.VIII.1961 (Holgersen). - Fairly common in southern and central East Fennoscandia, found in Ab, Ta, Tb, Ok, ObN. - Bohemia, German D.R., Estonia, Latvia, m. Russia, Kazakhstan.

Biology. On *Betula verrucosa, B. pubescens*, and *B. nana* (Linnavuori, 1953a). Adults from end of June to September.

186. ***Empoasca (Kybos) betulicola*** W. Wagner, 1955
Text-figs. 1394–1401.

Empoasca betulicola W. Wagner, 1955: 178.

Resembling *E. smaragdula.* Appendage of anal collar in male as in Text-fig. 1394, appendage of male pygofer as in Text-fig. 1395, aedeagus as in Text-figs. 1396–1398, 2nd abdominal sternum in male as in Text-fig. 1399, 2nd abdominal tergum in male as in Text-fig. 1400, 3rd abdominal tergum in male as in Text-fig. 1401. 7th abdominal sternum in female as in *smaragdula.* Overall length ♂ 4.2–4.4 mm, ♀ 4.6–4.8 mm.

Distribution. Not found in Denmark. - Scarce in Sweden, found in Sk., Boh., Upl., Dlr., and Nb. - Norway: found in Ry: Hinna 2.VIII.1942 (Holgersen); HOi: Etne 24.VII.1952 (Holgersen), Voss, Bulken-Evanger 29.VII.1942 (Tambs-Lyche), Vindal, Granvin 21.VIII.1968 (J. Rönhovde), Ulvik, Osastøl, altitude 620 m, 2.IX.1970 (A. Fjellberg); SFi; Sogndal in Sogn 28.VII., 2.IX., and 4.IX.1967 (S. Hågvar). - So far not found in East Fennoscandia. - England, Scotland, Netherlands, German D.R. and F.R.; Mongolia.

Biology. On *Betula* (W. Wagner, 1955). On *Betula pubescens* (Claridge & Wilson, 1976; S. Hågvar in litt.). I found *E. betulicola* also on *Betula verrucosa.* Adults in July–September.

187. ***Empoasca (Kybos) strigilifera*** Ossiannilsson, 1941
Text-figs. 1402–1407.

Empoasca strigilifera Ossiannilsson, 1941b: 198.

Resembling *E. smaragdula.* Anal collar of male as in Text-fig. 1402, appendage of male pygofer as in Text-fig. 1403, aedeagus as in Text-figs. 1404, 1405, 2nd and 3rd abdominal sterna in male as in Text-fig. 1406, 3rd abdominal tergum in male as in Text-fig. 1407. 7th abdominal sternum in female as in *smaragdula.* Length (♂) 4.3 mm.

Distribution. Denmark: so far found in SJ, WJ, and B. - Not uncommon in Sweden, found in Bl., Hall., Sm., Upl., Vstm., Dlr., Hls., Ång., and Jmt. - Norway: found in Os: Ringebu 22–24.VII.1943, and in TEi: Rjukan, Dal, by Holgersen. - Scarce in East Fennoscandia, found in Al, Ab, N, Ta, Sb, Kb; Kr. - Austria, Bohemia, Moravia, Slovakia, England, German D.R. and F.R., Switzerland, Poland, Estonia, Latvia, n. Russia, Ukraine.

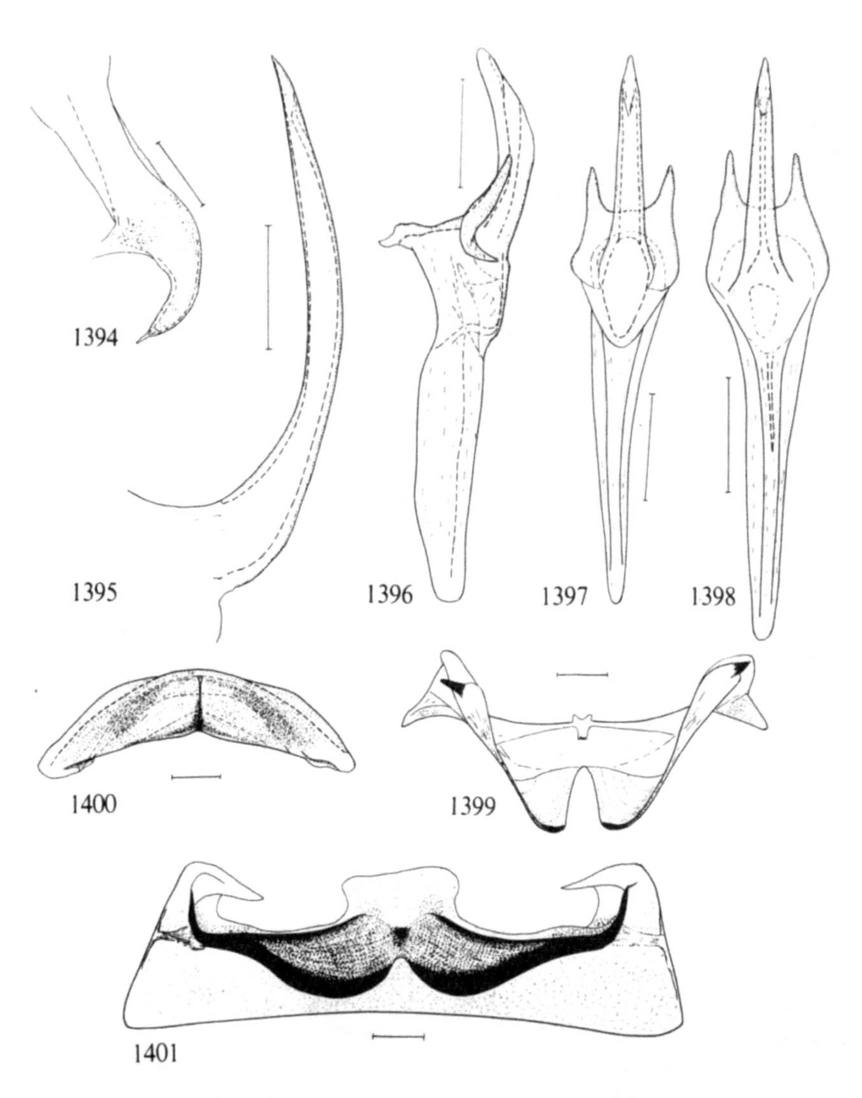

Text-figs. 1394–1401. ***Empoasca betulicola*** **W. Wagner. - 1394: left anal collar appendage of male from the left; 1395: left pygofer appendage of male from outside; 1396: aedeagus from the left; 1397: aedeagus in ventral aspect; 1398: aedeagus in ventral aspect; 1399: 2nd abdominal sternum in male from above; 1400: 2nd abdominal tergum in male from below; 1401: 3rd abdominal tergum in male from below. Text-figs. 1394–1397 after a paratype from Hamburg, the rest after a Swedish specimen. Scale: 0.1 mm.**

Biology. "Auf *Salix*-Arten, besonders *S. caprea*" (Wagner & Franz, 1961). On *Salix fragilis* (Gyllensvärd, 1965). On *Salix caprea* and *cinerea* (Linnavuori, 1969). I collected *E. strigilifera* on *Salix cinerea, caprea,* and *myrsinifolia.* Adults in June–September.

188. ***Empoasca (Kybos) virgator*** Ribaut, 1933
Text-figs. 1408–1413.

Empoasca virgator Ribaut, 1933: 152.

Resembling *E. smaragdula.* Anal collar of male as in Text-fig. 1408, pygofer appendage

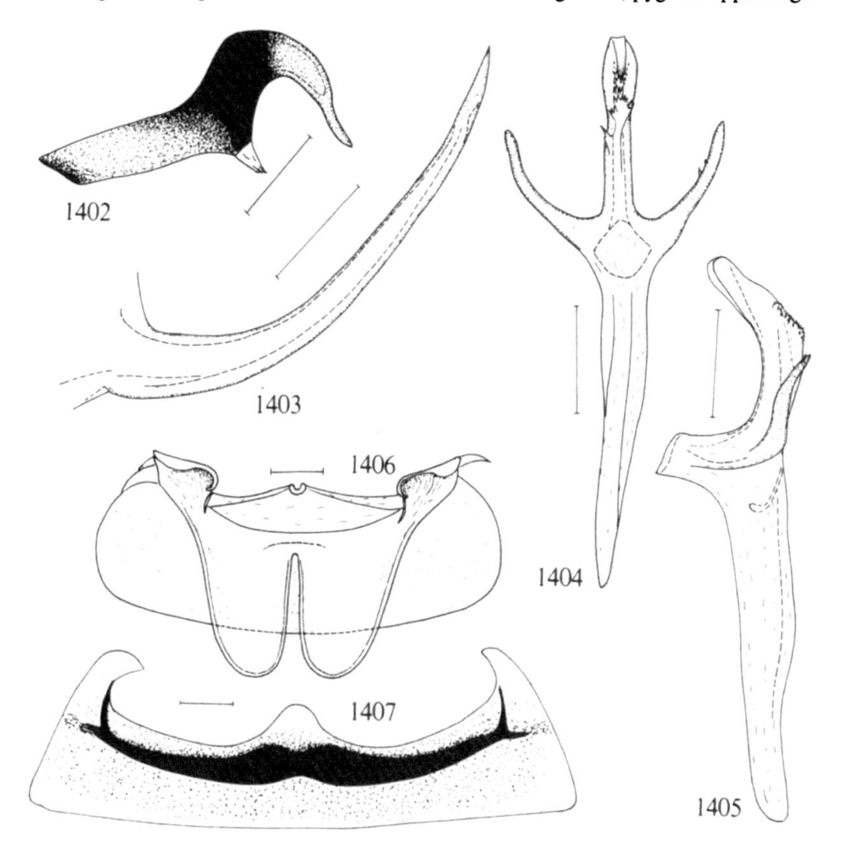

Text-figs. 1402–1407. *Empoasca strigilifera* Ossiannilsson. - 1402: male anal collar from the left; 1403: left pygofer appendage of male from outside; 1404: aedeagus in ventral aspect; 1405: aedeagus from the left; 1406: 2nd and 3rd abdominal sterna in male from above; 1407: 3rd abdominal tergum from below. Scale: 0.1 mm.

in male as in Text-fig. 1409, aedeagus as in Text-figs. 1410, 1411, 2nd and 3rd abdominal sterna in male as in Text-fig. 1412, 3rd abdominal tergum in male as in Text-fig. 1413. 7th abdominal sternum in female as in *smaragdula.* Overall length of male 4–4.5 mm, of female 4.2–4.65 mm.

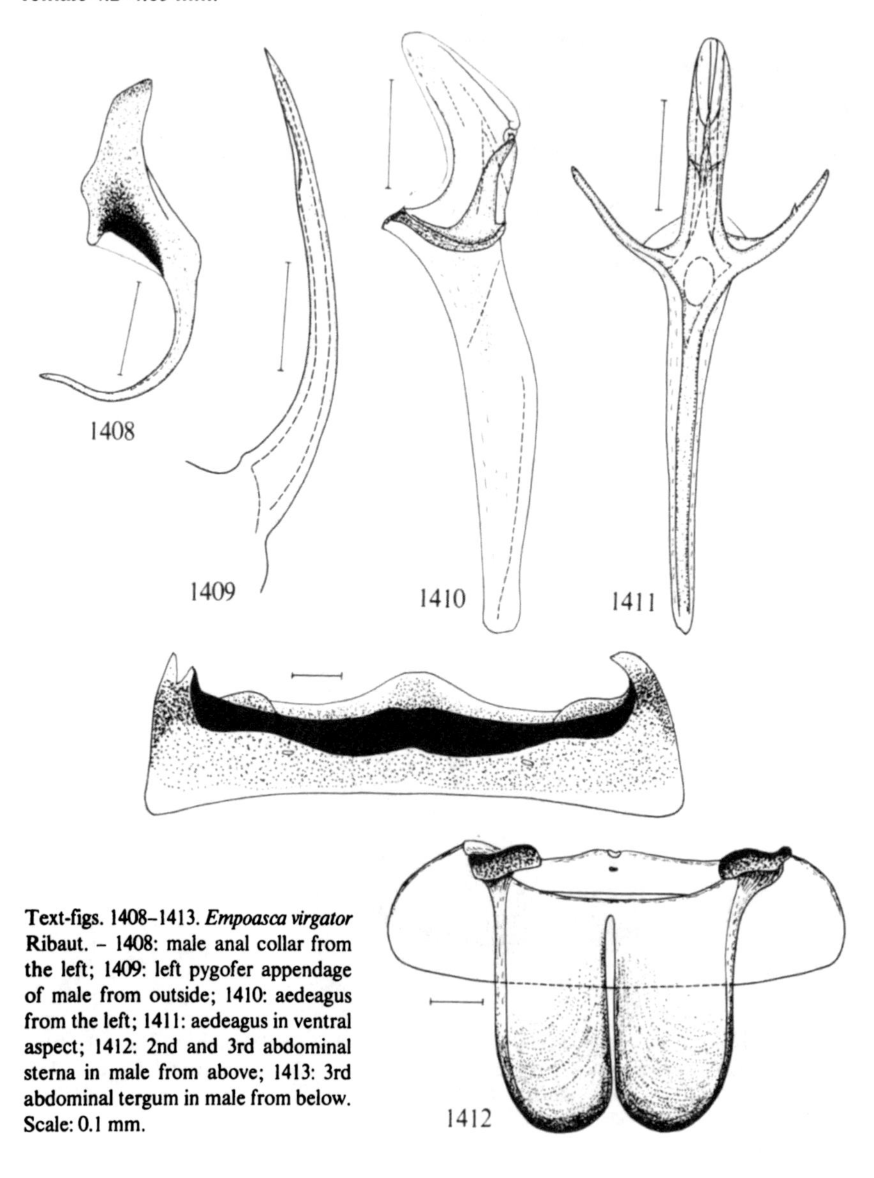

Text-figs. 1408–1413. *Empoasca virgator* Ribaut. – 1408: male anal collar from the left; 1409: left pygofer appendage of male from outside; 1410: aedeagus from the left; 1411: aedeagus in ventral aspect; 1412: 2nd and 3rd abdominal sterna in male from above; 1413: 3rd abdominal tergum in male from below. Scale: 0.1 mm.

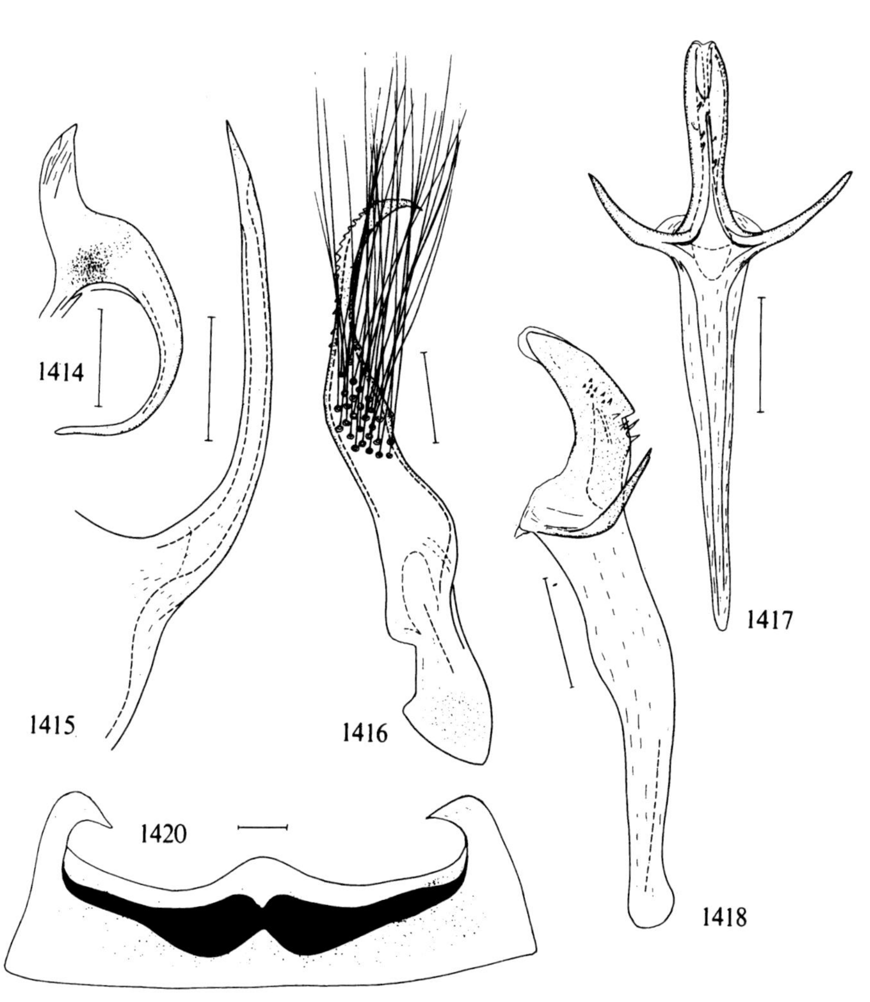

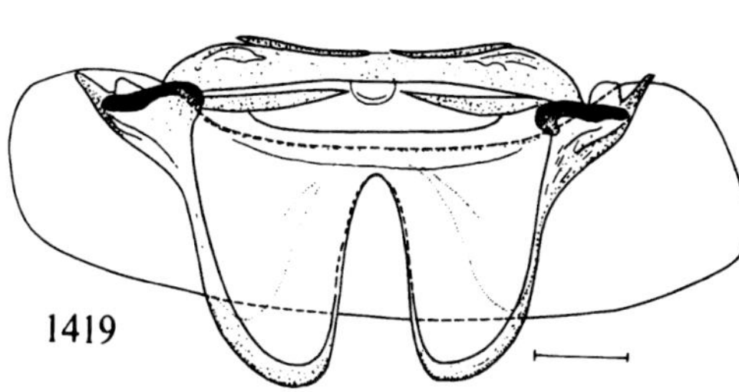

Text-figs. 1414–1420. *Empoasca vol-gensis* (Vilbaste). – 1414: male anal collar from the left; 1415: left pygofer appendage of male from the left; 1416: left genital style from above; 1417: aedeagus in ventral aspect; 1418: aedeagus from the left; 1419: 1st–3rd abdominal sterna in male from above; 1420: 3rd abdominal tergum of male from below. Scale: 0.1 mm.

Distribution. Denmark: found in several places in NEZ and B. – Scarce but widespread in Sweden, found in Sk., Bl., Sm., Gtl., Upl., Dlr., Hls., Jmt. – I have seen one male from Norway, BØ: Drammen 9.VII.1925, leg. Warloe. – Scarce in southern and central East Fennoscandia, found in Al: Finström; Ab: Raisio; Åbo; Lojo; N: Helsingfors; ObN: Kemi. – Widespread in Europe, also in Kazakhstan and Mongolia.

Biology. "In Flachmoor an *Salix viminalis*" (Kuntze, 1937). "*Auf Salix alba*" (Wagner & Franz, 1961). On *Salix alba* and *S. pentandra* (Le Quesne, 1961; Linnavuori, 1969). In Poland two generations (Dworakowska, 1976). I found *E. virgator* on *Salix pentandra*. Adults in June–September.

189. ***Empoasca (Kybos) volgensis*** (Vilbaste, 1961)
Text-figs. 1414–1420.

Kybos volgensis Vilbaste, 1961: 318.

Resembling *E. smaragdula*. Male anal collar as in Text-fig. 1414, appendage of male pygofer as in Text-fig. 1415, aedeagus as in Text-figs. 1417, 1418, 1st–3rd abdominal sterna in male as in Text-fig. 1419, 3rd abdominal tergum in male as in Text-fig. 1420. 7th abdominal sternum in female as in *smaragdula*. Overall length of males 3.75–3.97 mm, of females 4.0–4.2 mm (according to Vilbaste, l.c.). Swedish specimens are larger; males 4.1–4.45 mm, females 4.35–4.85 mm.

Distribution. Not in Denmark, Norway and East Fennoscandia. – Sweden: only found by me in Upl.: Uppsala, Kungsängen 10.IX.1970 (1 ♂), 12.VII.1979 (5 ♂♂, 6 ♀♀), and 6.IX.1979 (17 ♂♂, 27 ♀♀, 4 larvae). – S. Russia.

Biology. On (cultivated) *Salix pentandra*. Two generations.

190. ***Empoasca (Kybos) rufescens*** (Melichar, 1896)
Plate-fig. 107, text-figs. 1421–1426.

Kybos smaragdula rufescens Melichar, 1896b: 180.

Resembling *smaragdula* but claval suture not infuscate. On the contrary, the commissural border is more broadly brownish in *rufescens* than in *smaragdula*. Scutellum usually brownish with a light median longitudinal band. Male anal collar as in Text-fig. 1421, male pygofer appendage as in Text-fig. 1422, aedeagus as in Text-figs. 1423, 1424, 1st–4th abdominal sterna in male as in Text-fig. 1425, 3rd abdominal tergum in male as in Text-fig. 1426. 7th abdominal sternum in female as in *smaragdula*, median lobe wide. Overall length of males 3.9–4.5 mm, of females 4.15–4.6 mm.

Distribution. Denmark: so far recorded only once: LFM: Bremersvold 20.IX.1915 (Schlick). – Scarce but locally abundant in Sweden, found in Sk., Bl., Upl., Vrm. – So far not recorded from Norway, nor from East Fennoscandia. – Austria, Hungary, Bohemia, Moravia, Slovakia, Bulgaria, Romania, Yugoslavia, France, German D.R and F.R., England, Netherlands, Italy, Switzerland, Poland; Kazakhstan, Mongolia, Maritime Territory.

Biology. On *Salix purpurea* (Wagner, 1955a). "Two generations in Poland, the first appears in the beginning of June and the second at the end of August. *Salix purpurea* L. is mainly the food plant but in Pieniny Mts. this species breeds also on *Alnus glutinosa* (L.)" (Dworakowska, 1976). In Sweden adults have been found in June–August.

191. ***Empoasca (Kybos) butleri*** Edwards, 1908
Text-figs. 1427–1432.

Empoasca butleri Edwards, 1908b: 82.

Resembling *E. rufescens*. Anal apparatus in male as in Text-fig. 1427, male pygofer appendage as in Text-fig. 1428, aedeagus as in Text-figs. 1429, 1430, 2nd–4th abdominal

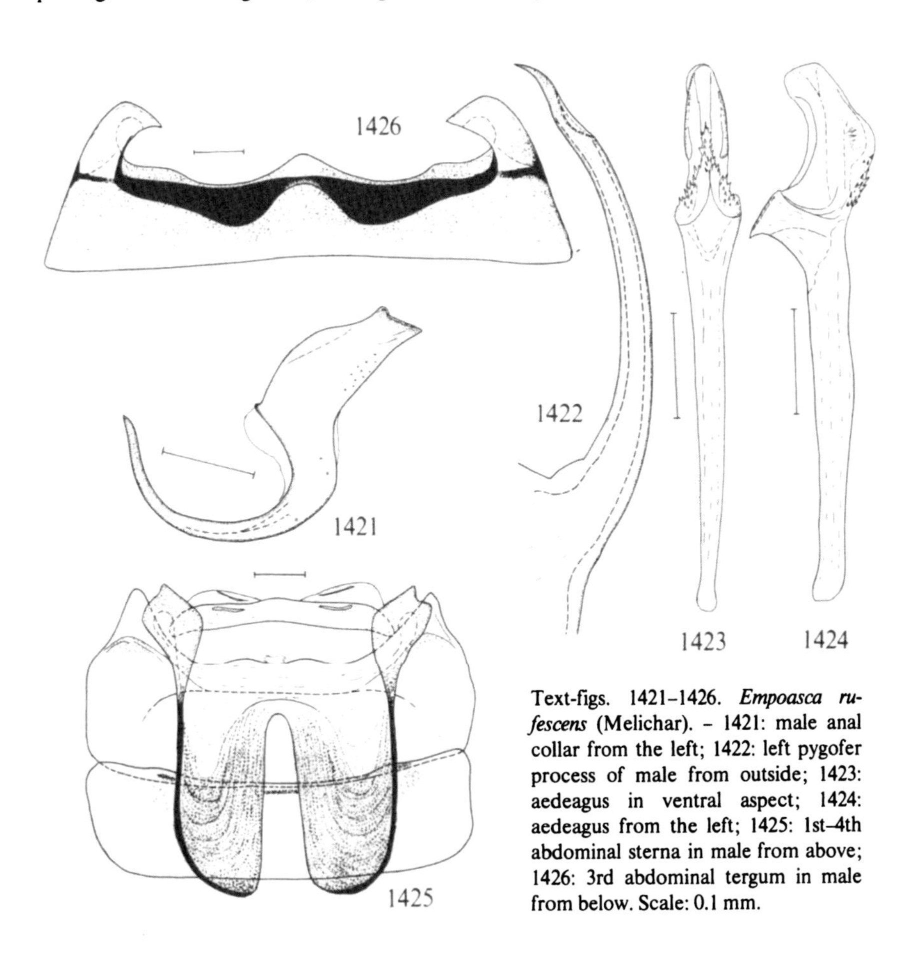

Text-figs. 1421–1426. *Empoasca rufescens* (Melichar). - 1421: male anal collar from the left; 1422: left pygofer process of male from outside; 1423: aedeagus in ventral aspect; 1424: aedeagus from the left; 1425: 1st–4th abdominal sterna in male from above; 1426: 3rd abdominal tergum in male from below. Scale: 0.1 mm.

sterna in male as in Text-fig. 1431, 3rd and 4th abdominal terga in male as in Text-fig. 1432. 7th abdominal sternum in female as in *rufescens,* median lobe slightly shorter. Overall length of males 3.9–4.3 mm, of females 4.2–4.7 mm.

Distribution. Denmark: so far found along the west coast of Jutland (WJ, NWJ). – Common in Sweden, found in Bl., Hall., Ög., Vg., Upl., Vstm., Dlr., Hls., Med., Ång., Jmt., Vb. – Norway: HEn: Åmot, 1♂ (Siebke). – East Fennoscandia: fairly common in the southern and central parts, found in Al, Ab, N, Sa, Kb, ObN, Ks, Le; Kr. – England, Netherlands, German F.R., Hungary, Yugoslavia, Poland, Estonia, Latvia, Lithuania, Mongolia.

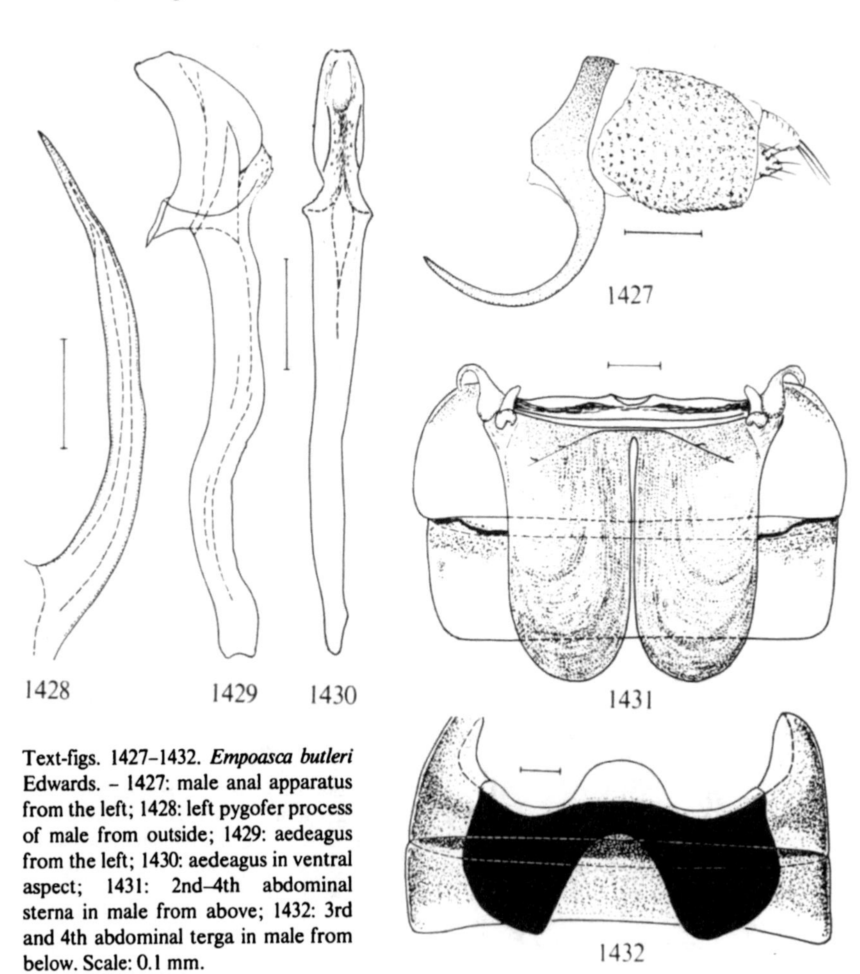

Text-figs. 1427–1432. *Empoasca butleri* Edwards. – 1427: male anal apparatus from the left; 1428: left pygofer process of male from outside; 1429: aedeagus from the left; 1430: aedeagus in ventral aspect; 1431: 2nd–4th abdominal sterna in male from above; 1432: 3rd and 4th abdominal terga in male from below. Scale: 0.1 mm.

Biology. On *Salix repens, cinerea, caprea* (Le Quesne, 1961; Linnavuori, 1969). "Develops two generations in Poland, adult forms of the first of them appear in June and the second one in August. Willows related to *Salix aurita* L. are food plants in Poland" (Dworakowska, 1976). In Sweden also *Salix myrsinifolia* serves as a food plant. Adults in June–September.

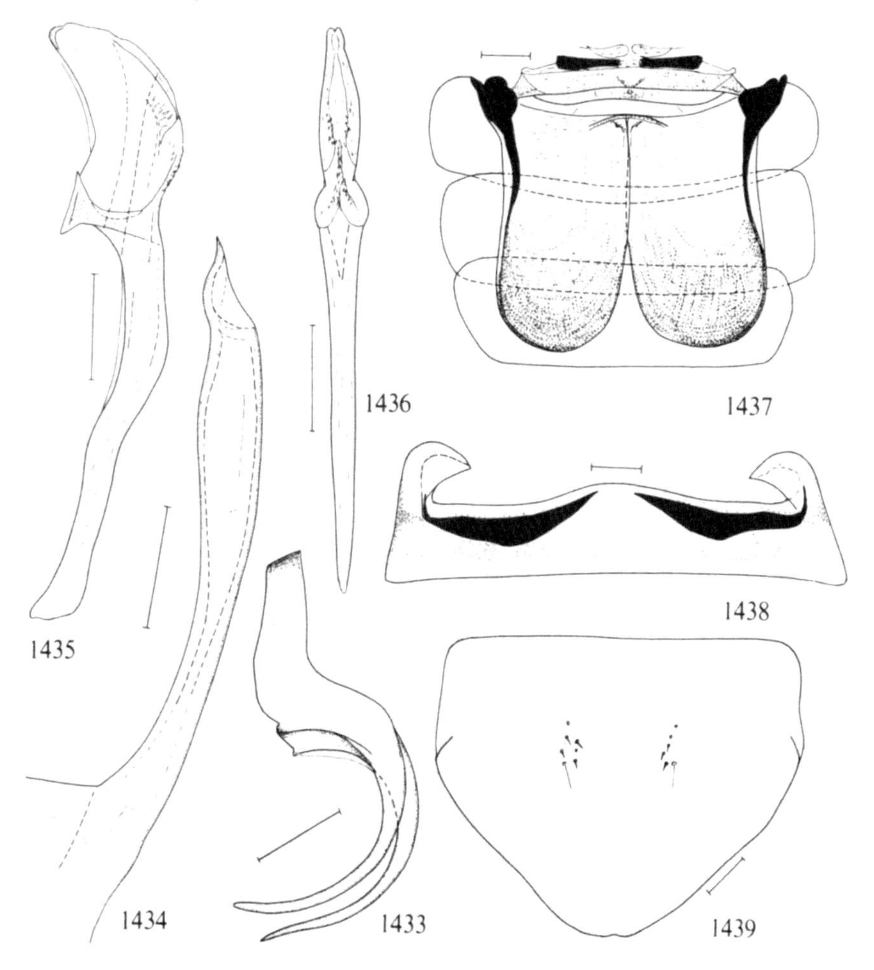

Text-figs. 1433–1439. *Empoasca populi* Edwards. - 1433: male anal collar from the left; 1434: left pygofer appendage of male from outside; 1435: aedeagus from the left; 1436: aedeagus in ventral aspect; 1437: 1st–5th abdominal sterna from above; 1438: 3rd abdominal tergum in male from below; 1439: 7th abdominal sternum of female from below (depressed under coverglass). Scale: 0.1 mm.

192. ***Empoasca (Kybos) populi*** Edwards, 1908
Text-figs. 1433–1439.

Empoasca populi Edwards, 1908b: 81.

Corioclaval suture light. Commissural border light or sometimes slightly darker than adjacent part of fore wing. Male anal collar as in Text-fig. 1433, appendage of male pygofer as in Text-fig. 1434, aedeagus as in Text-figs. 1435, 1436, 1st–5th abdominal sterna in male as in Text-fig. 1437, 3rd abdominal tergum in male as in Text-fig. 1438. 7th abdominal sternum in female convex, side margins not concave (Text-fig. 1439). Overall length of males 3.8–4.45 mm, of females 4.0–4.75 mm.

Distribution. Fairly common in Denmark as well as in Sweden up to Nb. – I have seen two males collected in Norway, Os: Biri 3.VII.1969, by Taksdal. – Scarce in East Fennoscandia, found in Al, Ab, N, Ta, Sb, ObN. – Belgium, Bulgaria, Bohemia, Moravia, France, England, German D.R. and F.R., Netherlands, Switzerland, Hungary, Poland, Estonia, Latvia, Moldavia, Ukraine, Kazakhstan, m. Siberia.

Biology. "Ich sammelte sie nur an *Populus tremula*" (Kuntze, 1937). "On poplars *(P. tremula, P. canescens, P. serotina)*" (Edwards, 1908). "In Poland *E. (K.) populi* Edw. breeds chiefly on *Populus nigra* L. and on cultivated related species and on *P. alba* L. Two generations a year. The first generation appears in June and adult forms of the second one occur in August–September." (Dworakowska, 1976). In Sweden adults have been found in June, July, August, and September.

193. ***Empoasca (Kybos) sordidula*** Ossiannilsson, 1941
Text-figs. 1440–1445.

Empoasca sordidula Ossiannilsson, 1941d: 69. (♂).
Empoasca sordidula Ossiannilsson, 1942b: 114. (♀).

The male differs from *smaragdula* by the green colour of the upper side being more or less mixed with or covered by diffusely limited fumose shades. Face of male usually largely fuscous. Pigmentation of fore wing as in *smaragdula.* Veins of hind wings partly black. Abdomen with dorsum largely fuscous, venter transversely black-striped. The female resembles the same sex in *populi.* Male anal collar as in Text-fig. 1440, appendage of male pygofer as in Text-fig. 1441, aedeagus as in Text-figs. 1442, 1443, 2nd and 3rd abdominal sterna in male as in Text-fig. 1444, 3rd abdominal tergum in male as in Text-fig. 1445. 7th abdominal sternum in female as in *populi.* Overall length (♂♀) 4–4.7 mm.

Distribution. Not found in Denmark. – Not uncommon in central and northern Sweden, found in Upl., Nrk., Vstm., Vrm., Dlr., Hls., Med., Ång., and Jmt. – Norway: I collected 4 specimens in Bø: vicinity of Drammen, 23.VII.1960. – East Fennoscandia: comparatively common in the southern and central parts, found in Al, Ab, N, Ta, Sa, Tb, Sb, Kb, Ks. – Central Russia, Canada, Alaska.

Biology. On *Salix caprea* (Linnavuori, 1969). On *Salix phylicifolia* (Linnavuori, 1949a). I have found one specimen on *Salix purpurea* but the normal host-plant of *E. sordidula* in Sweden seems to be *Salix myrsinifolia* on which I found it repeatedly. Since *Salix phylicifolia* is closely related to *myrsinifolia,* the former may be a suitable host-plant too. Possibly two generations in south-western Finland (Linnavuori, in Dworakowska, 1976). In Sweden adults have been found in June, July, and August.

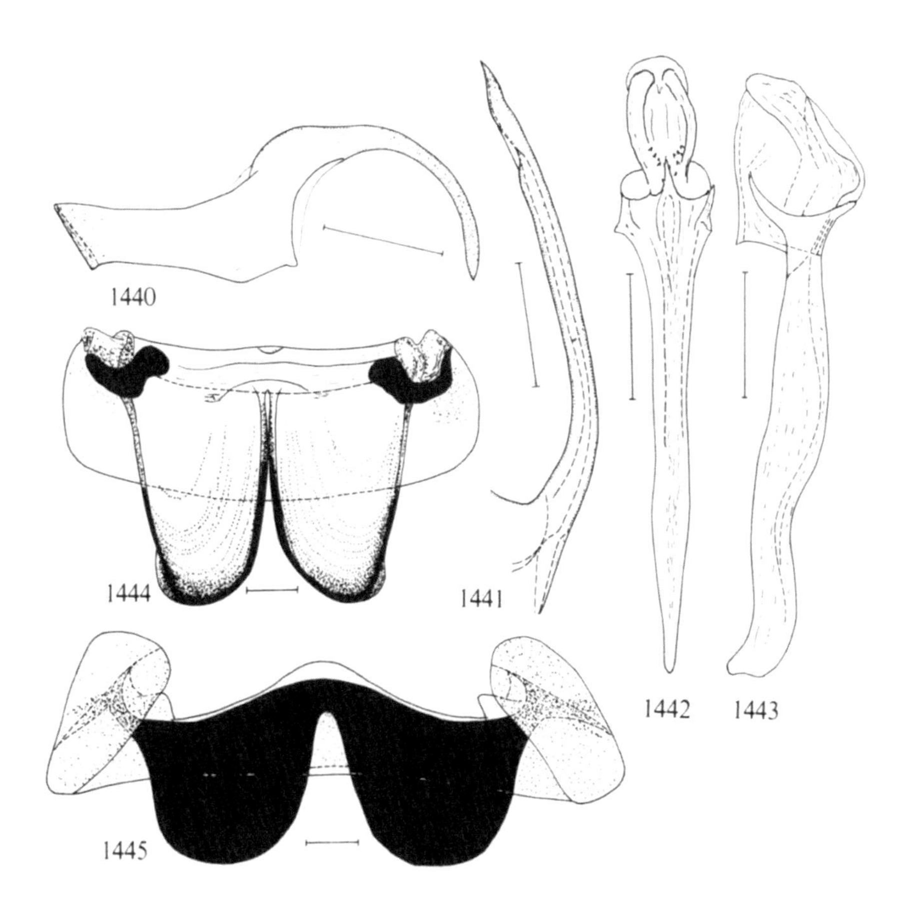

Text-figs. 1440–1445. *Empoasca sordidula* Ossiannilsson. – 1440: male anal collar from the left; 1441: left pygofer appendage of male from outside; 1442: aedeagus in ventral aspect; 1443: aedeagus from the left; 1444: 2nd and 3rd abdominal sterna in male from above; 1445: 3rd abdominal tergum in male from below. Scale: 0.1 mm.

194. ***Empoasca (Kybos) abstrusa*** Linnavuori, 1949
Text-figs. 1446–1451.

Empoasca abstrusa Linnavuori, 1949b: 148.
Empoasca taunica W. Wagner, 1955a: 177.

Above light green or yellowish green. Fore wings light green, costal and commissural borders darker green, the latter sometimes olivaceous. Claval suture usually not distinctly darker than fore wing membrane. Veins of hind wings usually light. Pronotum and scutellum medially with a light longitudinal band or with light spots. Pronotum often also laterally with a pale spot near fore border. Venter light green. Anal collar of male as in Text-fig. 1446, appendages of male pygofer as in Text-fig. 1447, aedeagus as

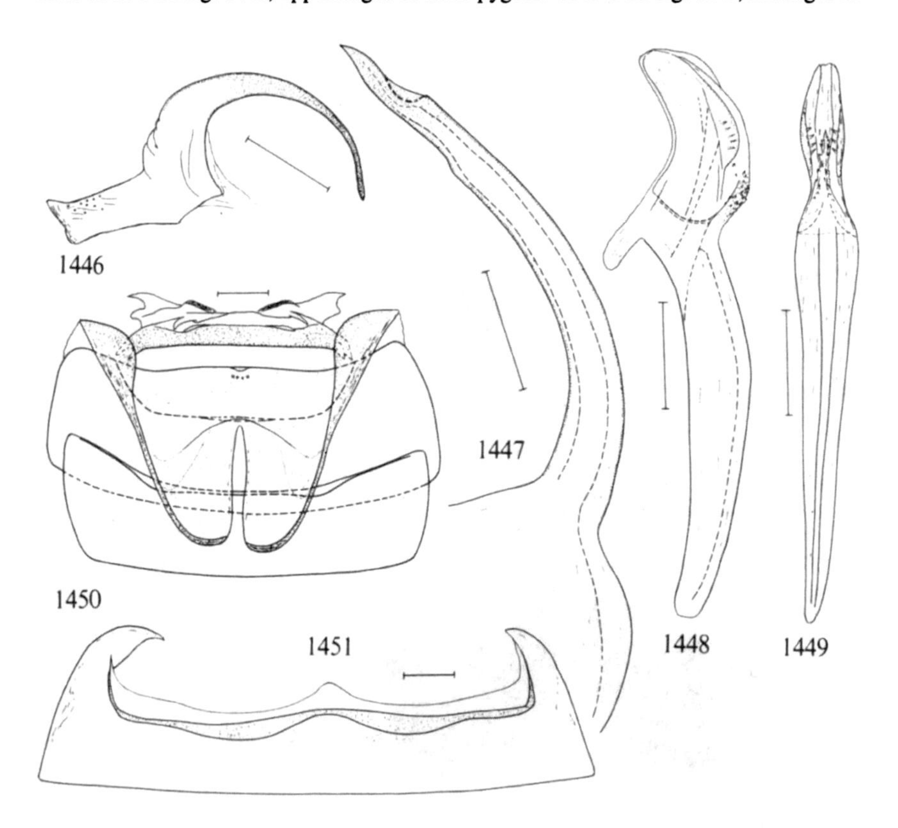

Text-figs. 1446–1451. *Empoasca abstrusa* Linnavuori. – 1446: male anal collar from the left; 1447: left pygofer appendage of male from outside; 1448: aedeagus from the left; 1449: aedeagus in ventral aspect; 1450: 1st–4th abdominal sterna in male from above; 1451: 3rd abdominal tergum in male from below. Scale: 0.1 mm.

in Text-figs. 1448, 1449, 1st–4th abdominal sterna as in Text-fig. 1450, 3rd abdominal tergum in male as in Text-fig. 1451. 7th abdominal sternum in female about as in *smaragdula*. Overall length of males 4.2–4.7 mm, of females 4.55–4.95 mm.

Distribution. So far not found in Denmark, nor in Norway. - Sweden: Bl.: Karlskrona 12.IX.1971 (Gyllensvärd); Upl.: Stockholm, Bergian Garden 8.IX.1942 (Ossiannilsson), Uppsala, Kungsängen 8.IX.1970 (Ossiannilsson). - East Fennoscandia: Ab: Åbo 25.VIII.1948 and later (Linnavuori, 1949, 1969). - German F.R., Poland, Hungary, Austria, Bulgaria, Romania, Yugoslavia, Lithuania, Central Russia.

Biology. On *Populus nigra, P. nigra italica,* and related species. In Central Europe two generations (Dworakowska, 1976).

Genus *Kyboasca* Zachvatkin, 1953

Kyboasca Zachvatkin, 1953c: 228.
Type-species: *Chloria bipunctata* Oshanin, 1871, by original designation.

Resembling *Empoasca* s.str. but more robust. Head anteriorly rounded, medially only slightly longer than near eyes. Pronotum a little broader than head with eyes. Subcostal cell in fore wing not longer than cubital cell. Apical veins of fore wing all arising from median cell. Male anal collar with appendages, male pygofer also with a pair of appendages. Genital style straight, without very long hairs. Aedeagus simple, without appendages. In Denmark and Fennoscandia one species.

195. ***Kyboasca bipunctata*** (Oshanin, 1871)
Text-figs. 1452–1459.

Chloria bipunctata Oshanin, 1871: 212.
Empoasca punctum Haupt, 1912: 189 (nec *Chlorita punctum* Lethierry, 1884).
Empoasca haupti Ribaut, 1933: 161 (n.n.).

Light green to yellowish green. Vertex, pronotum, and scutellum medially with a broad milky-white longitudinal band. Head and pronotum irregularly white spotted. Scutellum laterally of the median band largely yellowish. Fore wings light green or yellowish green, apically fumose, with small black or fuscous spot in distal end of cubital cell (Text-fig. 1452). Venter and abdomen light, legs partly verdigris green. Male genital plate as in Text-fig. 1453, process of male anal collar as in Text-fig. 1454, appendage of male pygofer as in Text-fig. 1455, genital style as in Text-fig. 1456, aedeagus as in Text-figs. 1457, 1458, 1st–3rd abdominal sterna in male as in Text-fig. 1459. 7th abdominal sternum of female strongly convex, caudal border medially with an incision varying in shape. Overall length (♂♀) 3.3–3.6 mm.

Distribution. Not in Denmark, Sweden and Norway. - East Fennoscandia: repeatedly found in Ab: Åbo by Linnavuori (27.VII.1949 and later). - Austria,

Bohemia, Moravia, Hungary, Romania, Yugoslavia, Italy, German D.R. and F.R., Poland, s. Russia, Ukraine, Moldavia, Azerbaijan, Georgia, Kazakhstan, Uzbekistan, Kirghizia, Manchuria, Mongolia, Maritime Territory. Introduced into North America.

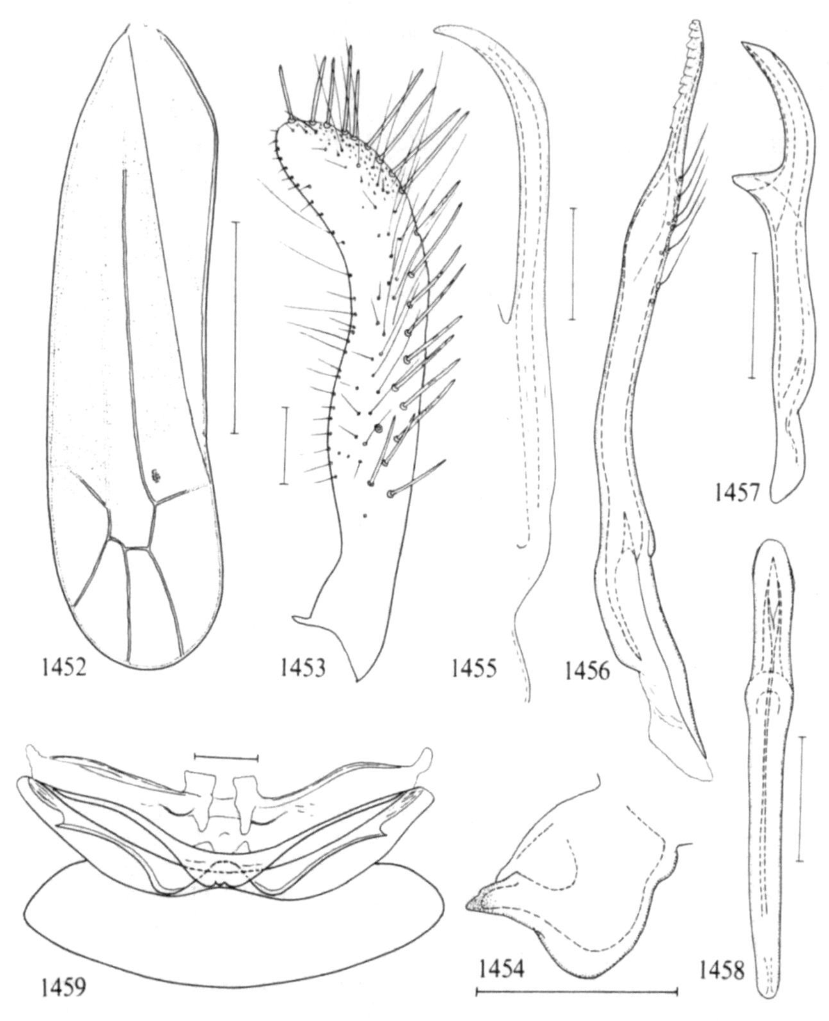

Text-figs. 1452–1459. *Kyboasca bipunctata* (Oshanin). – 1452: left fore wing of male; 1453: genital plate from outside; 1454: left anal collar process of male from outside; 1455: left pygofer process of male from the left; 1456: left genital style from above; 1457: aedeagus from the left; 1458: aedeagus in ventral aspect; 1459: 1st–3rd abdominal sterna in male from above. Scale: 1 mm for 1452, 0.1 mm for the rest.

Biology. On *Cannabis* (Paoli, 1936); on *Ulmus* (Linnavuori, 1969); on *Ulmus pumila* (Müller, 1956); on *Ulmus laevis, Ulmus campestris,* and *Cannabis sativa* (Dworakowska, 1973).

Genus *Chlorita* Fieber, 1872

Chloria Fieber, 1866: 508 (nec Schiner, 1862).
Chlorita Fieber, 1872: 14 (n.n.).
Type-species: *Cicada viridula* Fallén, 1806, by subsequent designation.

Small, short and broad leafhoppers. Subcostal cell in fore wing much shorter than cubital cell. Only one apical vein emanating from median cell. Pygofer lobes of male without appendages. Anal collar of male with long appendages. Genital styles apically faintly curved, without a group of very long hairs. 7th abdominal sternum in female caudally convex. In Denmark and Fennoscandia three species.

Key to species of *Chlorita*

1 Aedeagus with a pair of long processes arising from base of shaft. Abdominal tergum black spotted or striped 2
\- Aedeagus on each side with two or three pointed processes arising from base of shaft (Text-figs. 1476, 1477). Abdominal tergum entirely light .... 198. *dumosa* (Ribaut)
2 (1) Processes of aedeagus stout, compressed, each with a tooth on inside (Text-figs. 1470, 1471) .... 197. *paolii* (Ossiannilsson)
\- Processes of aedeagus more or less long, thin, normally unarmed (Text-figs. 1462–1465) .... 196. *viridula* (Fallén)

196. ***Chlorita viridula*** (Fallén, 1806)
Plate-fig. 111, text-figs. 1460–1467.

Cicada viridula Fallén, 1806: 37.
Empoasca subulata Ribaut, 1933: 155.

Head as seen from above somewhat angularly produced, medially only slightly longer than near eyes. Green, fore body with yellowish or whitish spots and stripes. Abdomen in both sexes dorsally, often also ventrally, transversely black striped. Appendages of male anal collar as in Text-fig. 1460, genital style as in Text-fig. 1461, aedeagus as in Text-figs. 1462–1465, 1st abdominal sternum in male as in Text-fig. 1466, 2nd–4th abdominal sterna in male as in Text-fig. 1467. Overall length (♂♀) 2.6–3 mm.

Distribution. Not very common in Denmark, found in EJ, F, and NEZ. - Fairly common in southern and central Sweden, found Sk. - Ång. - Norway: found in Ø, AK, Bø, Bv, TEi, AAy. - Rare in East Fennoscandia, found in Oa: Laikia; Mustasaari; Ylistaro. - Austria, Bulgaria, Hungary, Romania, Yugoslavia, England, France, Italy, Spain,

Switzerland, German F.R., Poland, Latvia, Estonia, n. and m. Russia, Moldavia, Ukraine, Algeria, Kazakhstan, Kirghizia, m. Siberia.

Biology. *On Artemisia vulgaris* (Günthart, 1974). Usually living on *Chrysanthemum*

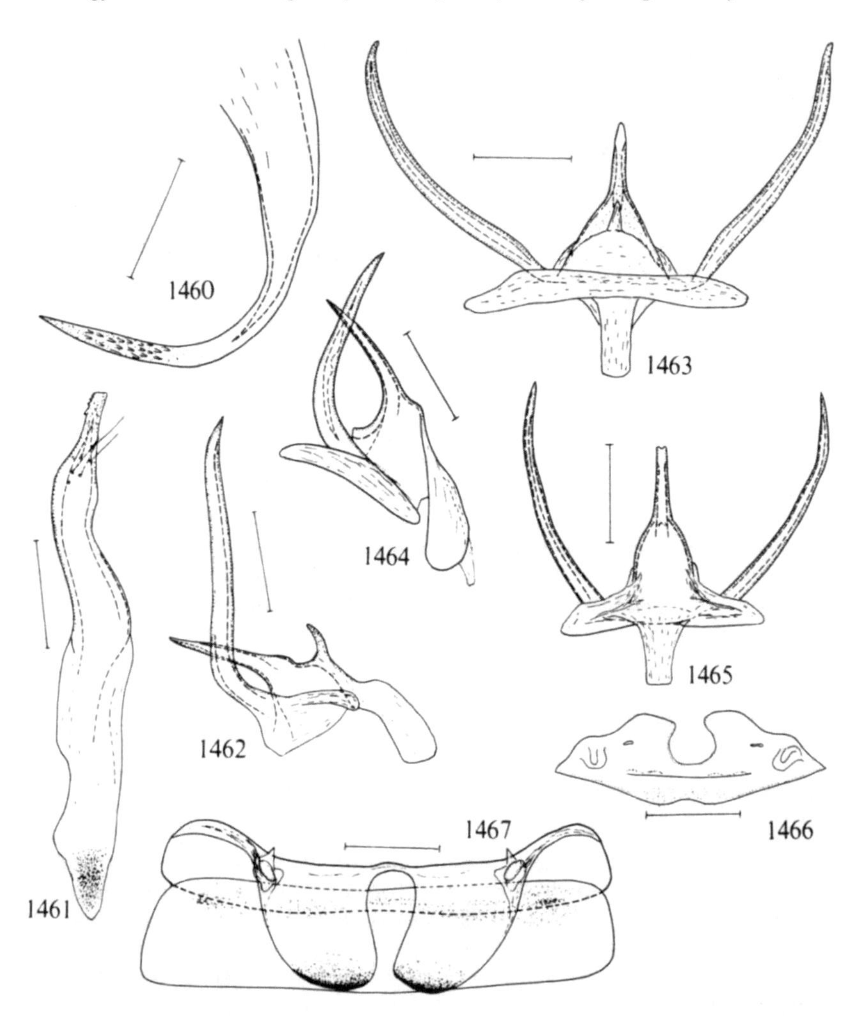

Text-figs. 1460–1467. *Chlorita viridula* (Fallén). – 1460: right anal collar process of male from the right; 1461: left genital style from above; 1462: aedeagus from the left; 1463: aedeagus in ventral aspect; 1464: aedeagus from the left; 1465: aedeagus in ventral aspect; 1466: 1st abdominal sternum in male from behind; 1467: 2nd–4th abdominal sterna in male from above. Text-figs. 1460–1463 after a specimen from Norway, the rest after a specimen from Öland. Scale: 0.1 mm.

vulgare, Artemisia campestris, and *Achillea millefolium,* also found on *Artemisia absinthium* and *Santolina chamaecyparissus* (Vidano & Arzone, 1978). Two generations in southern Sweden, adults in June–October.

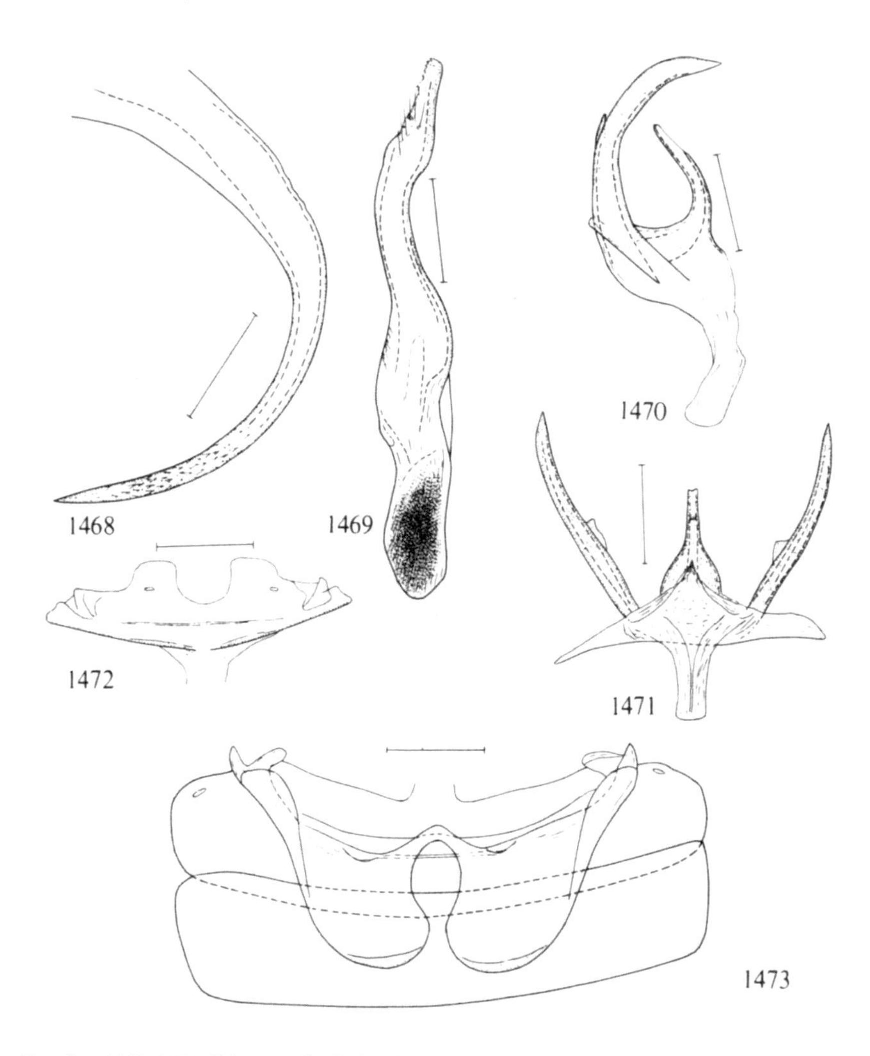

Text-figs. 1468–1473. *Chlorita paolii* (Ossiannilsson). – 1468: right anal collar process of male from the right; 1469: left genital style from above; 1470: aedeagus from the left; 1471: aedeagus in ventral aspect; 1472: 1st abdominal sternum in male from behind; 1473: 2nd–4th abdominal sterna in male from above. Scale: 0.1 mm.

197. ***Chlorita paolii*** (Ossiannilsson, 1939)
Text-figs. 1468–1473.

Empoasca viridula Ribaut, 1933: 155 (nec Fallén, 1806).
Empoasca paolii Ossiannilsson, 1939: 25 (n.n.).

Resembling *viridula*, abdomen dorsally black spotted. Appendage of male anal collar as in Text-fig. 1468, genital style as in Text-fig. 1469, aedeagus as in Text-figs. 1470, 1471,

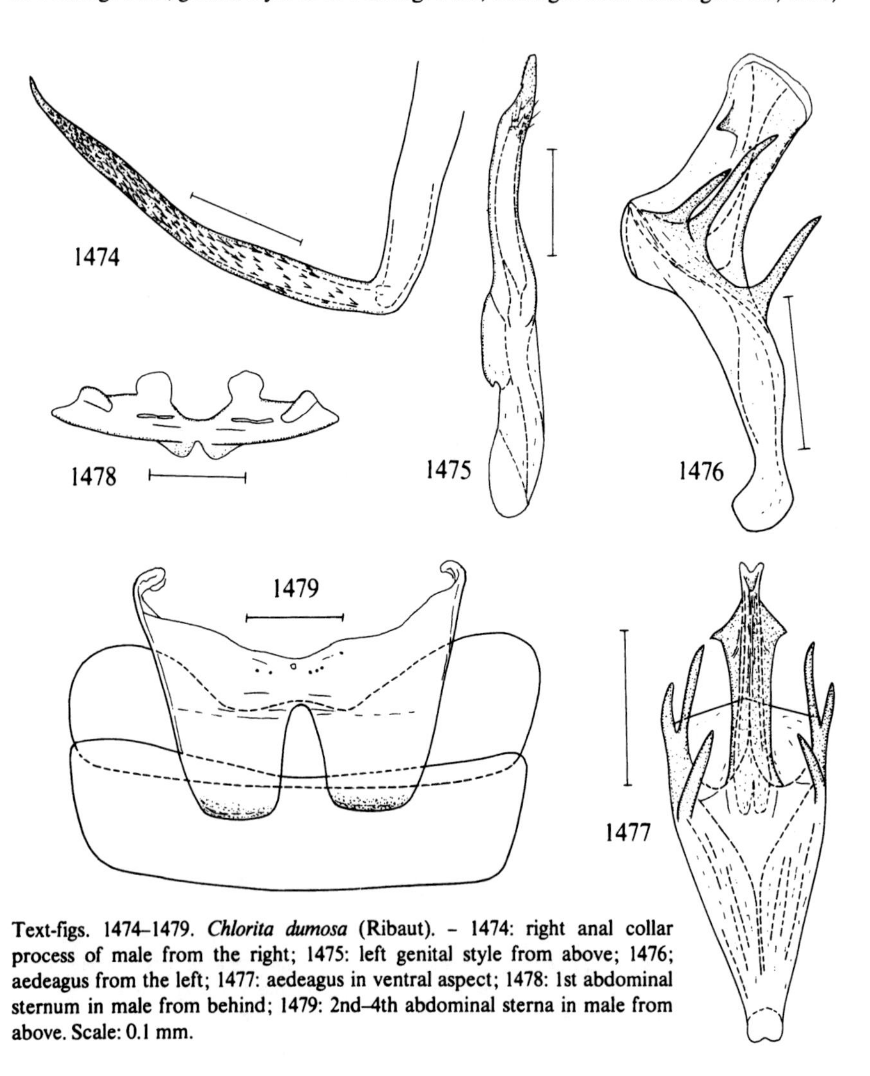

Text-figs. 1474–1479. *Chlorita dumosa* (Ribaut). – 1474: right anal collar process of male from the right; 1475: left genital style from above; 1476; aedeagus from the left; 1477: aedeagus in ventral aspect; 1478: 1st abdominal sternum in male from behind; 1479: 2nd–4th abdominal sterna in male from above. Scale: 0.1 mm.

1st abdominal sternum in male as in Text-fig. 1472, male abdominal sterna 2–4 as in Text-fig. 1473. Overall length in males 2.7–2.8 mm, in females 2.8–3.0 mm.

Distribution. So far not found in Denmark, Sweden and Norway. – East Fennoscandia: common in the southern and central parts, found in Al, Ab, N, Kb; Vib, Kr. – Austria, Bulgaria, Bohemia, Moravia, Slovakia, Hungary, Yugoslavia, Italy, Netherlands, France, German D.R. and F.R., Poland, Estonia, Latvia, Lithuania, Ukraine, Moldavia, Altai Mts., Kazakhstan.

Biology. On *Artemisia* species and *Achillea millefolium* (Lindberg, 1947). Two generations (Kontkanen, 1948). In dryish fields, moist sloping meadows, cultivated fields (Linnavuori, 1952a). Hibernation in the egg stage; in Central Europe 3 generations (Schiemenz, 1969b). "Frequent in the oat-growing region in dry meadows, nowadays almost exclusively in man-made meadows, pastures and fields" (Raatikainen & Vasarainen, 1976).

198. ***Chlorita dumosa*** (Ribaut, 1933)
Text-figs. 1474–1479.

Empoasca dumosa Ribaut, 1933: 158.

Resembling *viridula*, abdomen entirely light. Appendage of anal collar in male as in Text-fig. 1474, genital style as in Text-fig. 1475, aedeagus as in Text-figs. 1476, 1477, 1st abdominal sternum in male as in Text-fig. 1478, 2nd–4th abdominal sterna in male as in Text-fig. 1479. Overall length (♂♀) 2.3–2.6 mm.

Distribution. Not found in Denmark and Norway. – Rare in Sweden, found in Sk.: Löderup, on the hills, 21.VII.1950 (Bo Tjeder); Öl.: Vickleby 20.VI.1936 (Ossiannilsson); Möckelmossen 17.IX.1946 (Håkan Lindberg); Fårskyddet, Karlevi 1.VIII.-1967 (H. Bornfeldt). – Very rare in East Fennoscandia, found in Sa: Joutseno in 1950 (Linnavuori). – France, Italy, German D.R. and F.R., Austria, Bulgaria, Bohemia, Slovakia, Romania, Poland, Estonia, Latvia, Lithuania, Kazakhstan.

Biology. On *Thymus* (Wagner, 1941). "Stenotope Art der Trockenrasen" (Schiemenz, 1969b). "Ei-Überwinterer, 2 Generationen (?)" (Schiemenz, l.c.).

Genus *Fagocyba* Dlabola, 1958

Fagocyba Dlabola, 1958: 54.
Type-species: *Typhlocyba cruenta* Herrich-Schäffer, 1838, by original designation.

"Vertex longer than 1/2 of pronotum, fore margin rounded. Head narrower than pronotum. Pygofore without spinulation, roundly prolonged, with spine near the base of the anal tube, which is longer than the pygofore. Genital plates narrow on the base, broadening towards the tip, – – – and with a triangular prolongation near the dorsal margin, nearer to the base. Aedeagus simple, orificium apical" (Dlabola, l.c.). In Denmark and Fennoscandia three species.

Key to species of *Fagocyba*

1 Dorsum of fore body, clavus and adjacent part of corium reddish brown .... 199. *cruenta* (Herrich-Schäffer)
– Colour of fore body and fore wings different 2
2 (1) Fore wings usually largely lively yellow, often almost orange-coloured. Genital style of male as in text-fig. 1489 .... 200. *douglasi* (Edwards)
– Fore wings light yellow. Genital styles of male as in Text-fig. 1494 .... 201. *carri* (Edwards)

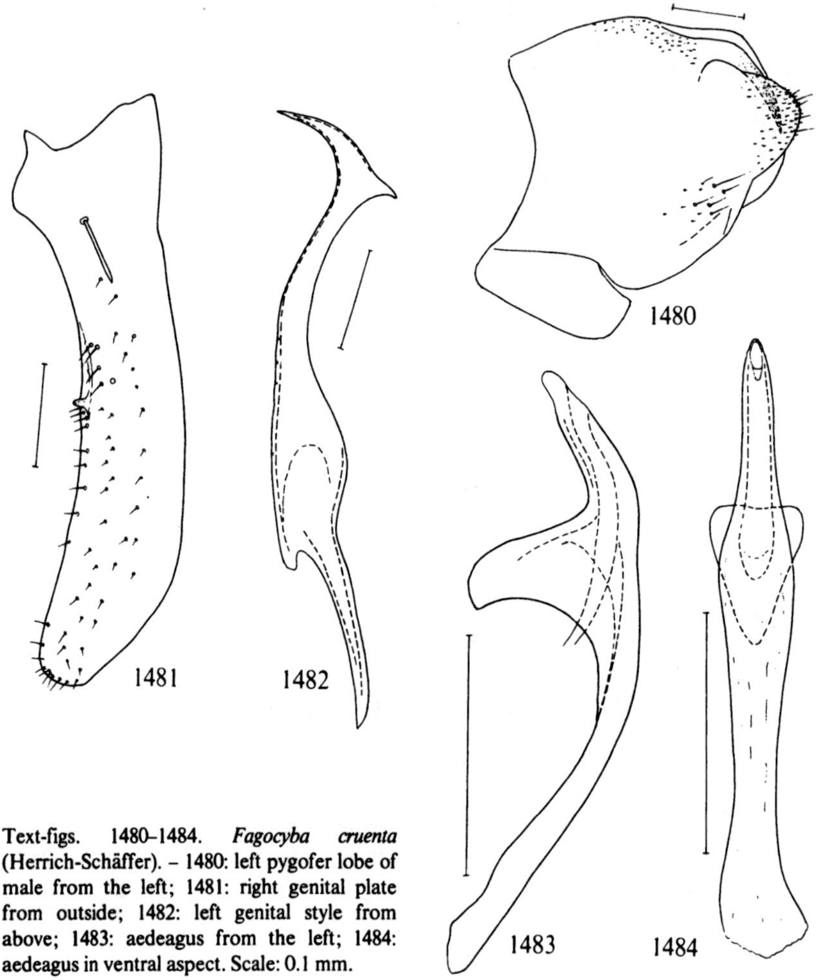

1480

1481

1482

1483

1484

Text-figs. 1480–1484. *Fagocyba cruenta* (Herrich-Schäffer). – 1480: left pygofer lobe of male from the left; 1481: right genital plate from outside; 1482: left genital style from above; 1483: aedeagus from the left; 1484: aedeagus in ventral aspect. Scale: 0.1 mm.

199. ***Fagocyba cruenta*** (Herrich-Schäffer, 1838)
Plate-fig. 112, text-figs. 1480–1486.

Typhlocyba cruenta Herrich-Schäffer, 1838: 15.

Elongate. Above reddish brown except for anterior border, apex of radial cell, and apical part of fore wing, these parts being colourless or very faintly brownish yellow. Apical veins of fore wing reddish brown. Vertex, pronotum and scutellum may be lighter or darker than fore wings. Vertex sometimes with a pair of diffusely limited dark spots. Venter yellowish or brownish yellow, legs light. Male pygofer lobes as in Text-fig. 1480, genital plate as in Text-fig. 1481, genital style as in Text-fig. 1482, aedeagus as in Text-figs. 1483, 1484, 2nd–6th abdominal sterna as in Text-fig. 1485. Ventral aspect of caudal part of female abdomen as in Text-fig. 1486. Overall length (♂♀) 3.4–4 mm. – Last instar larva dorsally dark red, with pronotum and wing pads lacking spines. Anterior margin of vertex with two pairs of widely spaced spines, median pair usually shorter han lateral pair. Abdomen with one pair of median spines on segments 3–8 and in addition lateral spines on segments 5–8.

Distribution. Quite rare in Denmark, found in EJ, LFM, NEZ. – Rare in southern Sweden, found in Sk. (C. G. Thomson); Sk.: Stehag 21.VII.1939 (Ossiannilsson), Brunnby, Kullen 1.VIII.1962 and 24.VIII.1964 (Ossiannilsson); Bl.: Lyckeby, Afvelsgärde 3.IX.1963, 2.IX.1964 (Gyllensvärd); Vg.: Göteborg 7.VII.1953 (Ossiannilsson). – Not found in Norway, nor in East Fennoscandia. – England, France, Netherlands, Italy, Switzerland, Austria, Bohemia, Moravia, Yugoslavia, German D.R. and F.R., Poland.

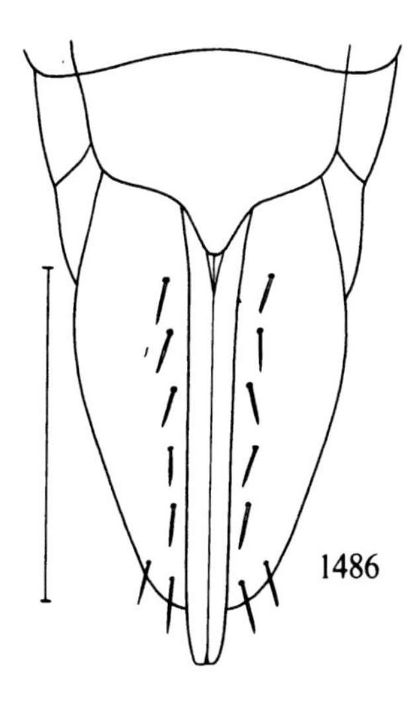

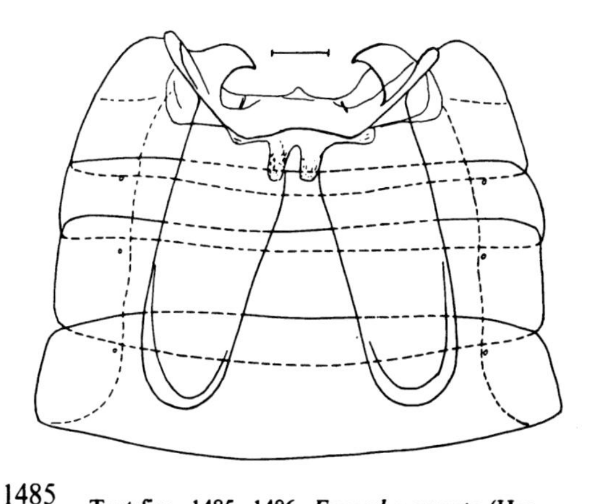

Text-figs. 1485, 1486. *Fagocyba cruenta* (Herrich-Schäffer). – 1485: 2nd–6th abdominal sterna in male from above; 1486: apex of female abdomen from below. Scale: 0.1 mm for 1485, 0.5 mm for 1486.

Biology. "An Buchen" (Kuntze, 1937). On *Fagus* (Dlabola, 1954). "Auf *Fagus silvatica, Carpinus betulus* und anderen Laubhölzern" (Wagner & Franz, 1961).

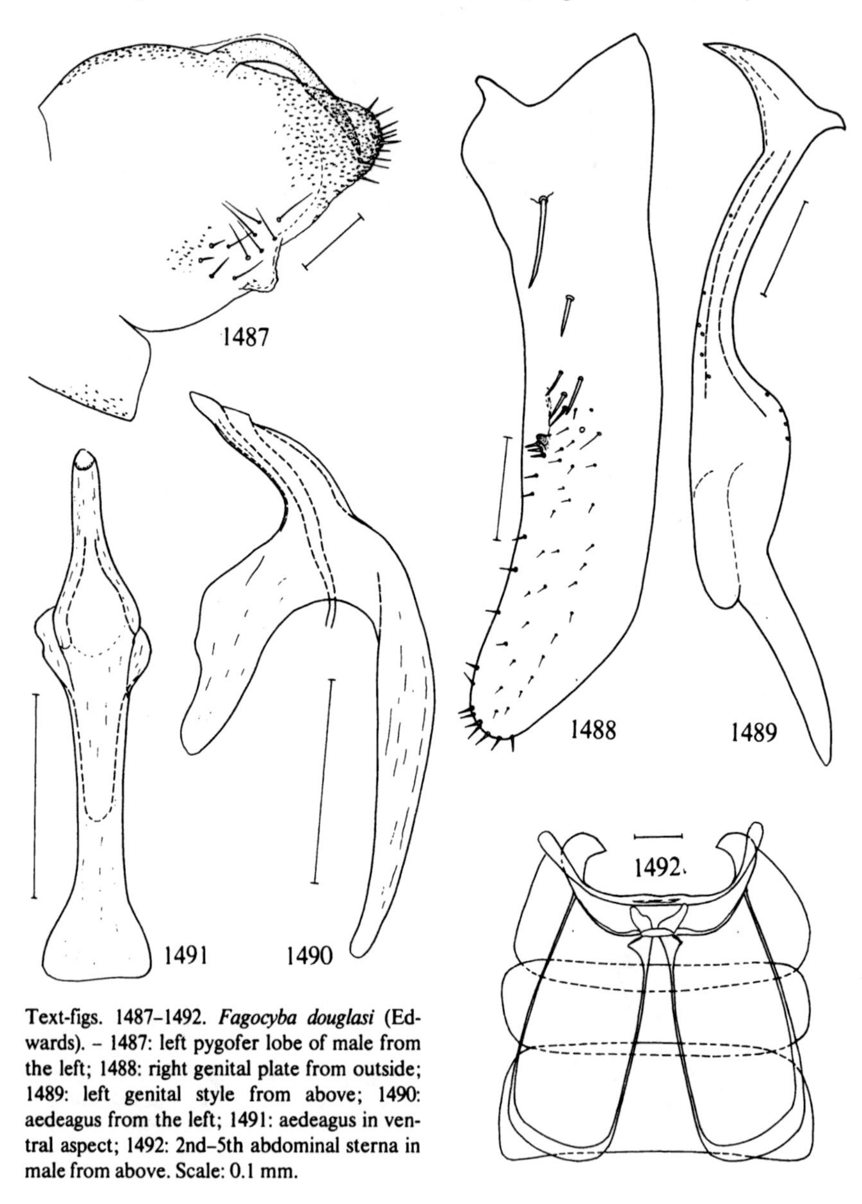

Text-figs. 1487–1492. *Fagocyba douglasi* (Edwards). – 1487: left pygofer lobe of male from the left; 1488: right genital plate from outside; 1489: left genital style from above; 1490: aedeagus from the left; 1491: aedeagus in ventral aspect; 1492: 2nd–5th abdominal sterna in male from above. Scale: 0.1 mm.

200. ***Fagocyba douglasi*** (Edwards, 1878)
Text-figs. 1487–1492.

Typhlocyba douglasi Edwards, 1878: 248.
Typhlocyba inquinata Ribaut, 1936b: 88.

Elongate, yellowish white, shining. Fore wings sometimes yellowish white, usually lively yellow or almost orange-coloured; apical cells and distal ends of radial, median and cubital cells more or less distinctly fumose. In f. *inquinata* (Ribaut) also clavus is entirely fumose. Male pygofer lobes as in Text-fig. 1487, genital plate as in Text-fig. 1488, genital style as in Text-fig. 1489, aedeagus as in Text-figs. 1490, 1491, 2nd–5th abdominal sterna in male as in Text-fig. 1492. (These text-figures refer to *douglasi* s.str.; I have not seen males of f. *inquinata*). Overall length 3.15–3.9 mm. – Last instar larvae entirely yellow, otherwise as in *cruenta*.

Distribution. Common and widespread in Denmark as well as in Sweden up to Vb. – Norway: found in AK, VE, Ry, and MRi. I have seen two females of f. *inquinata* from VE: Larvik (S. Hågvar leg.). – Probably fairly common in the southern part of East Fennoscandia, found in Al, Ab, N, St, Sa. – Widespread in Europe.

Biology. On *Fagus, Carpinus, Ulmus* and other deciduous trees (Ossiannilsson, 1946c). On *Quercus robur, Alnus incana, Sorbus suecica* (Linnavuori, 1969). Also found on *Quercus petraea, Alnus glutinosa* and *Corylus avellana* in Sweden, and on *Prunus domestica* in Norway. Adults in June–September.

201. ***Fagocyba carri*** (Edwards, 1914)
Text-figs. 1493–1497.

Typhlocyba carri Edwards, 1914: 170.

Yellowish white, shining. Apical part of fore wing faintly fumose. Male pygofer lobe as in Text-fig. 1493, genital style as in Text-fig. 1494, aedeagus as in Text-figs. 1495, 1496, abdominal sterna 2–6 in male as in Text-fig. 1497. Overall length (♂) about 4 mm. – 5th instar larva entirely pale yellow, head, pronotum, and wing pads without long spines, abdominal segments 5–8 with prominent lateral spines.

Distribution. Denmark: found in several places in EJ. – Scarce in southern Sweden, found in Sk., Bl., Sm., Öl., Ög., Boh. – Norway: found in AK: Nesodden 5.VIII.1946 (Holgersen); VE: Horten (Holgersen); AAy: Lillesand 3.VIII. (Holgersen). – So far not found in East Fennoscandia. – England, France, German D.R. and F.R., Netherlands, Switzerland, Bohemia, Moravia, Poland, Latvia, Lithuania.

Biology. On *Quercus sessilis* (Wagner, 1941). Monophagous on *Quercus* (Claridge & Wilson, 1976). Adults in July–September.

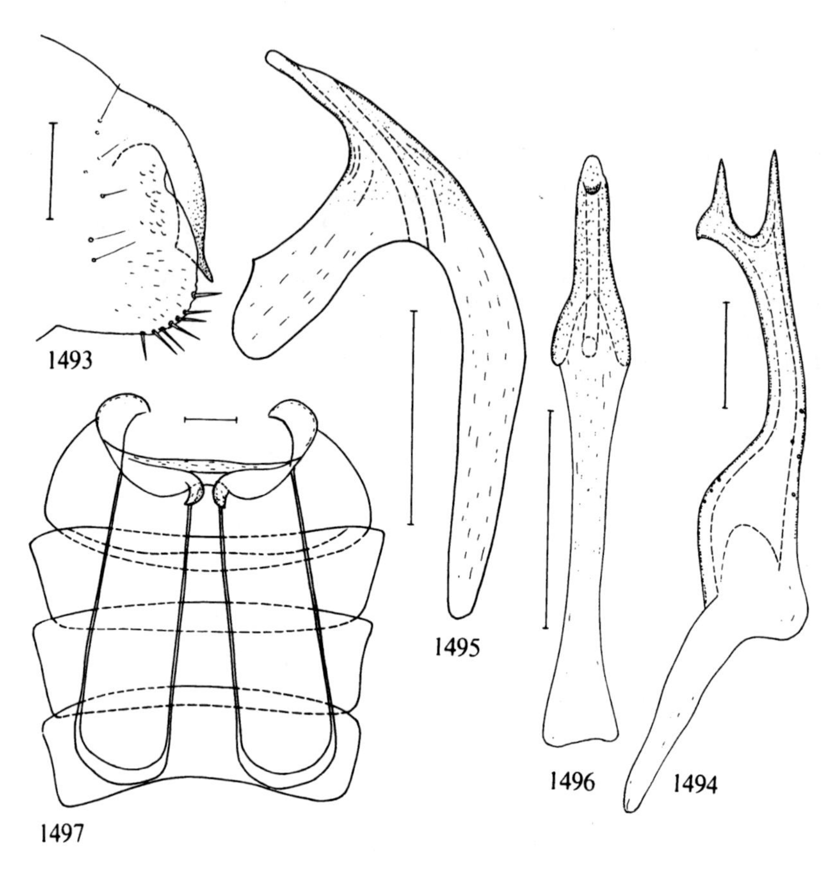

Text-figs. 1493–1497. *Fagocyba carri* (Edwards). – 1493: left pygofer lobe of male from the left; 1494: right genital style from above; 1495: aedeagus from the left; 1496: aedeagus in ventral aspect; 1497: 2nd–6th abdominal sterna in male from above. Scale: 0.1 mm.

Genus *Ossiannilssonola* Christian, 1953

Ossiannilssonia Young & Christian, 1952: 97 (nec Hille Ris Lambers, 1952).
Type-species: *Typhlocyba berenice* Mc Atee, 1926, by original designation.
Ossiannilssonola Christian, 1953: 1132 (n.n.).

Vertex medially much shorter than pronotum. Ocelli present or absent. Male genital plate without a basal macroseta. Shaft of aedeagus reduced to a flattened membranous structure. Many species in the Nearctic Region, in Europe only one species.

202. ***Ossiannilssonola callosa*** (Then, 1886)
Text-figs. 1498–1501.

Typhlocyba (Anomia) callosa Then, 1886: 56.
Typhlocyba distincta Edwards, 1914: 170.

Strongly built. Pale yellow. Commissural and scutellar borders of clavus more or less broadly brownish (f. *typica*) or concolorous, pale (f. *distincta*). Apical part of fore wing including distal ends of cubital and median cells, often also of radial cell, fumose. These parts are pale in f. *distincta*. Male genital style as in Text-fig. 1498, aedeagus as in Text-figs. 1499, 1500, 2nd and 3rd abdominal sterna in male as in Text-fig. 1501. Overall length of males 4.0–4.4 mm, of females 4.25–4.8 mm. – Last instar nymph pale yellow, anterior margin of vertex rounded with 2 pairs of spines with bases widely spaced, median pair tending to be as long as lateral pair. Face on upper part with some shorter setae. Chaetotaxy of abdomen as in *Fagocyba cruenta*.

Distribution. Quite rare in Denmark, found in EJ, F, NEZ, SZ, and LFM. – Sweden: only found in Sk.: Lund, Botanical Garden 19.VII.1916 by Tullgren (*f. distincta*). – Not found in Norway and East Fennoscandia. – Austria, Bulgaria, Bohemia, Moravia, Slovakia, Romania, Yugoslavia, Belgium, France, England, German D.R. and F.R., Netherlands, Poland.

Biology. "Auf Alnus" (Then, 1886). On *Acer pseudoplatanus* (Wagner & Franz, 1961; Claridge & Wilson, 1976). Monophagous (Claridge & Wilson, l.c.). The Swedish specimens were found on *Acer platanoides*.

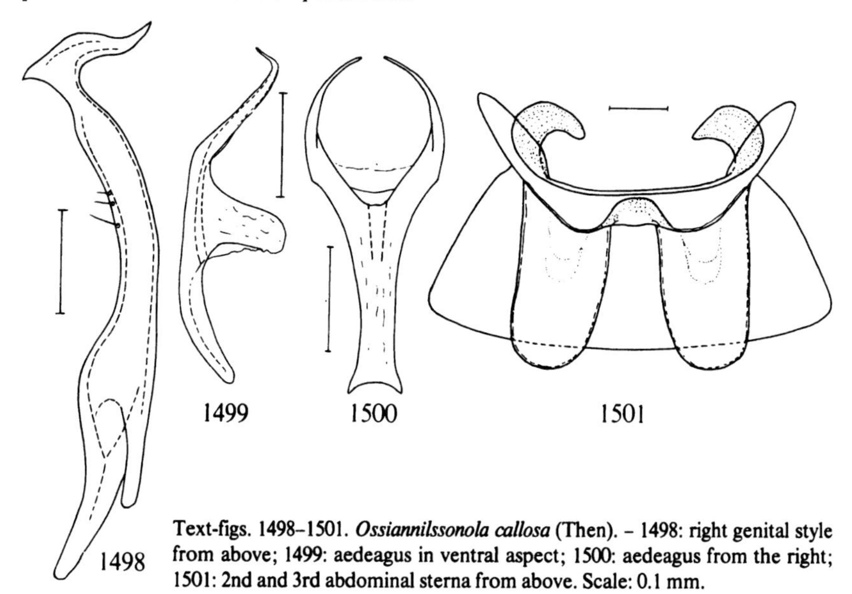

Text-figs. 1498–1501. *Ossiannilssonola callosa* (Then). – 1498: right genital style from above; 1499: aedeagus in ventral aspect; 1500: aedeagus from the right; 1501: 2nd and 3rd abdominal sterna from above. Scale: 0.1 mm.

Genus *Edwardsiana* Zachvatkin, 1929

Edwardsiana Zachvatkin, 1929: 262.
Type-species: *Cicada rosae* Linné, 1758, by original designation.

Vertex medially much shorter than pronotum. 2nd apical cell in fore wing triangular, usually stalked. Hind wings with 2 apical cells. Shaft of aedeagus strongly sclerotized, with or (rarely) without apical processes, without processes arising from base. Genital style of male without an angular subapical extension. Elongate leafhoppers with elongate fore wings. Usually only males can be identified with certainty. – Nymphs of species so far studied pale yellow, some with dark bases to dorsal macrosetae. Anterior margin of vertex rounded, with two pairs of strong macrosetae, sides in front of eyes slightly concave. Face with some shorter macrosetae and short hairs. Thorax, wing pads and abdomen with long macrosetae. In Denmark and Fennoscandia 24 species.

Key to species of *Edwardsiana*

1 Fore wings and scutellum with distinct dark brown markings 2
– Fore wings without well-defined dark markings 3
2 (1) Fore wings with a broad dark brown longitudinal band occupying cubital cell, adjacent part of clavus, and 3rd and 4th apical cells (Plate-fig. 115) .... 221. *geometrica* (Schrank)
– Distal half of clavus fuscous or brownish, scutellar border and proximal part of commissural border bordered with the same colour (Plate-fig. 114) .... 220. *gratiosa* (Boheman)
3 (1) Aedeagus simple, without appendages, only apex shallowly incised (Text-figs. 1519, 1520) .... 206. *stehliki* Lauterer
– Aedeagus apically with two or more appendages 4
4 (3) Aedeagus with one pair of apical appendages (Text-fig. 1511) 204. *avellanae* (Edwards)
– Aedeagus with two pairs of apical appendages 5
5 (4) Aedeagus more or less as in *avellanae* but with an additional subapical pair of very thin and short appendages (Text-figs. 1515, 1516) .... 205. *staminata* (Ribaut)
– Aedeagus different 6
6 (5) Aedeagus apically with two pairs of simple appendages, a "lateral" pair (l in Text-fig. 1508) and an "anterior" pair (a in the same Text-fig.) 7
– Aedeagus with two pairs of appendages, at least the anterior pair ramified 10
7 (6) Shaft of aedeagus with dorsal margin dilated and compressed laterally. Appendages flattened like knife-blades, in lateral aspect situated in approximately the same level in

about at a right angle to shaft (Text-figs. 1506–1508) .... 203. *rosae* (Linné)
- Aedeagus different 8

8 (7) Caudal-ventral angle of male pygofer black (Text-fig. 1530). Aedeagus as in Text-fig. 1531 .... 208. *nigriloba* (Edwards)
- Caudal-ventral angle of male pygofer not black, aedeagus different 9

9 (8) Anterior appendages of aedeagus with a common median stem, in lateral aspect sickle-shaped, apices directed upwards (Text-fig. 1541). Shaft of aedeagus not laterally compressed .... 210. *alnicola* (Edwards)
- Anterior appendages each arising directly from apex of aedeagus without a common stalk. Shaft of aedeagus laterally compressed (Text-figs. 1535, 1536) .... 209. *salicicola* (Edwards)

10 (6) Anterior appendages basally fused, emitting a common median branch, apex of aedeagus thus with 5 processes (Text-figs. 1525, 1528) .... 207. *crataegi* (Douglas)
- Anterior appendages forked 11

11 (10) Lateral appendages simple 12
- Lateral appendages ramified 22

12 (11) Branches of each anterior appendage with apices converging, in side view shaped as a lobster's pincer (Text-fig. 1552) 212. *frustrator* (Edwards)
- Aedeagus different 13

13 (12) Upper branches of anterior appendages situated at the same level as lateral pair (Text-fig. 1563) .... 214. *prunicola* (Edwards)
- Lateral appendages and branches of anterior appendages diverging 14

14 (13) Appendages short, only 1/4–1/5 of length of aedeagal shaft (Text-figs. 1578, 1579) .... 217. *plebeia* (Edwards)
- Longest appendage 1/3–1/2 length of aedeagal shaft 15

15 (14) Anterior appendages much longer than lateral appendages, forked considerably distally of middle (Text-fig. 1620) 223. *soror* (Linnavuori)
- Anterior appendages not much longer than lateral appendages, forked near middle or proximally of middle 16

16 (15) Anterior appendages forked near base, without a distinct stalk, in lateral aspect strongly diverging. Shaft of aedeagus basally on dorsal side with (Text-fig. 1546) or without a laminate carina .... 211. *sociabilis* (Ossiannilsson)
- Anterior appendages either with a distinct stalk or branches not strongly diverging 17

17 (16) Anterior appendages much shorter than the lateral pair (Text-figs. 1571, 1573, 1574) .... 216. *flavescens* (Fabricius)
- Anterior appendages approximately as long as lateral pair 18

18 (17) Shaft of aedeagus basally with a laminate carina on dorsal side (Text-fig. 1614). Hind wings broad, fore and hind margins practically parallel (Text-fig. 1612) .... 222. *tersa* (Edwards)

– Shaft of aedeagus without a basal laminate carina on dorsal side 19

19 (18) Outer branches of anterior appendages in terminal aspect arched, converging (Text-figs. 1559, 1596) 20

– Outer branches of anterior appendages in terminal aspect not converging 21

20 (19) Anterior appendages forked near middle, branches with a distinct stalk (Text-fig. 1559), in lateral aspect curved upwards (Text-fig. 1557) .... 213. *ishidae* (Matsumura)

– Anterior appendages forked near base (Text-fig. 1596), branches in lateral aspect not or only faintly curved upwards (Text-figs. 1594, 1595) .... 219. *candidula* (Kirschbaum)

21 (19) Branches of anterior appendages curved upwards (Text-fig. 1567) .... 215. *menzbieri* Zachvatkin

– Branches of anterior appendages almost straight (Text-fig.

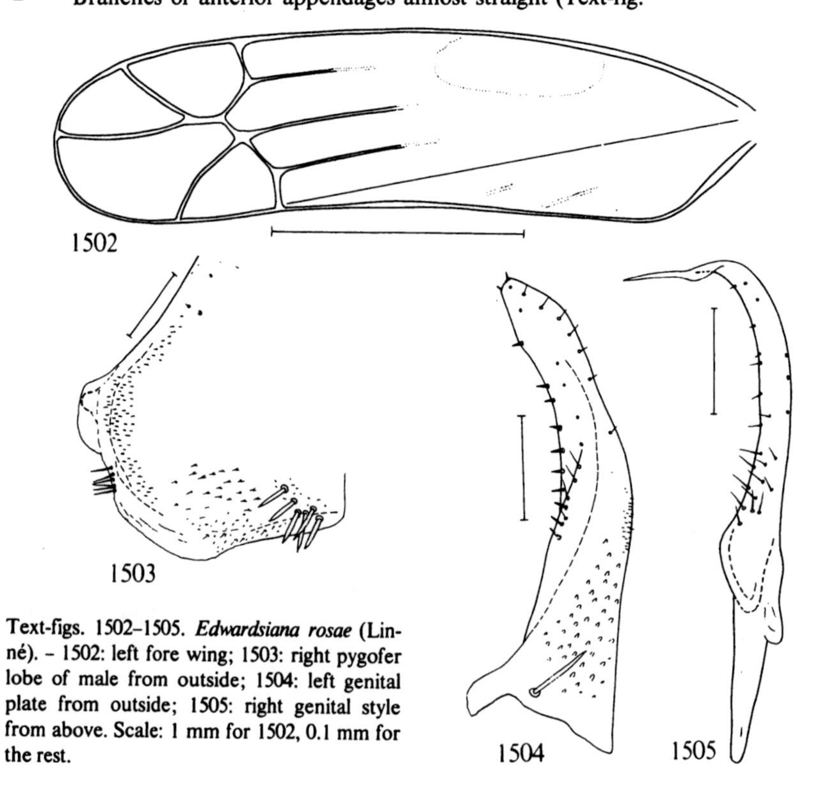

Text-figs. 1502–1505. *Edwardsiana rosae* (Linné). – 1502: left fore wing; 1503: right pygofer lobe of male from outside; 1504: left genital plate from outside; 1505: right genital style from above. Scale: 1 mm for 1502, 0.1 mm for the rest.

1586) .... 218. *kemneri* (Ossiannilsson)

22 (11) Anterior appendages forked near middle (Text-fig. 1626)
224. *bergmani* (Tullgren)

– Anterior appendages forked near base 23

23 (22) Upper branch of anterior appendage in lateral aspect with apex more or less distinctly curved upwards (Text-figs. 1634, 1635), in terminal aspect less strongly diverging (Text-fig. 1637) .... 225. *hippocastani* (Edwards)

– Branches of anterior appendages in lateral aspect converging (Text-figs. 1641, 1643), in terminal aspect more strongly diverging (Text-fig. 1642) .... 226. *lethierryi* (Edwards)

203. ***Edwardsiana rosae*** (Linné, 1758)
Plate-fig. 113, text-figs. 1269, 1502–1509.

Cicada rosae Linné, 1758: 439.
Typhlocyba rosae manca Ribaut, 1936b: 93.

White or yellowish white, shining. Apical part of fore wing indistinctly fumose. Venation of fore wing as in Text-fig. 1502, of hind wing as in Text-fig. 1269. Male pygofer lobe as in Text-fig. 1503, genital plate as in Text-fig. 1504, aedeagus as in Text-figs. 1506–1508, genital style as in Text-fig. 1505, 1st–5th abdomianl sterna as in Text-fig. 1509. Males with aedeagus more or less deformed by parasitization (anterior and lateral appendages fused) have been described by Ribaut as "aberr. *manca*". – Overall length (♂♀) 3.4–3.9 mm. – Last instar nymphs with dorsal macrosetae partly arising from black dots.

Distribution. Common and widespread in Denmark as well as in southern and central Sweden (Sk.–Vstm.). – I have seen specimens from Norway: AK: Sem, Asker 2.IX.–1.XI.1973 (Taksdal), and 17.IX, 17.X.1973 (Baeschlin). – East Fennoscandia: Al: Sund 20.VIII.1975 (Albrecht); N: Helsingfors 25.IX.1955 (Nuorteva). – Widespread in Europe, also in Azerbaijan, Kazakhstan, Tadzhikistan, Uzbekistan, Kirghizia, Japan; Nearctic and Oriental regions.

Biology. On *Rosa* spp., *Sorbus, Pyrus malus, Rubus, Fragaria.* In Sweden two generations. Hibernation takes place in the egg stage (Tullgren, 1916b). "Taken predominantly from rowan and hawthorn. It was hand-collected abundantly from rose, which appears to be its main host plant. It has been bred from rose and rowan" (Claridge & Wilson, 1976). "We have found *E. rosae* nymphs of the first generation hatching only on *Rosa* species, often in large numbers. In the second generation, a wider range of hosts in attacked. It is clear that *E. rosae* significantly expands its host plant range during this generation" (Claridge & Wilson, 1978). Adults 20.VI.–1.XI.

Economic importance. Injurious mainly on roses. "Die befallenen Blätter werden durch die Saugstiche zunächst fein weissfleckig, bleichen schliesslich völlig und fallen in extremen Fällen ab, junge kräuseln sich" (Müller, 1956).

204. ***Edwardsiana avellanae*** (Edwards, 1888)
Text-figs. 1510–1513.

Typhlocyba avellanae Edwards, 1888b: 157.

Light yellow or whitish yellow, shining. Apical part of fore wing indistinctly fumose. Male pygofer lobe as in Text-fig. 1510, aedeagus as in Text-figs. 1511, 1512, 2nd–5th abdominal sterna as in Text-fig. 1513. Overall length 3.35–3.8 mm.

Distribution. Denmark: only found once: – ♂ in NEZ: Bøllemosen 10.IX.1915 (Oluf Jacobsen). – Sweden: Sk.: Åkarp, in light traps 28–29.VII.1951, 12.VIII and 14.VIII.1952 (E. Sylvén); Bl.: Förkärla, Kvalmsö 14.VIII.1960; Lyckeby 21.VII.1963 (Gyllensvärd). – So far not found in Norway. – East Fennoscandia: only found in St: Yläne, by J. Sahlberg. – Austria, Bulgaria, Bohemia, Slovakia, Hungary, England, Scotland, Wales, France, Netherlands, German D.R. and F.R., Switzerland, Poland, Latvia, Lithuania, Estonia, Moldavia, m. Russia, Ukraine.

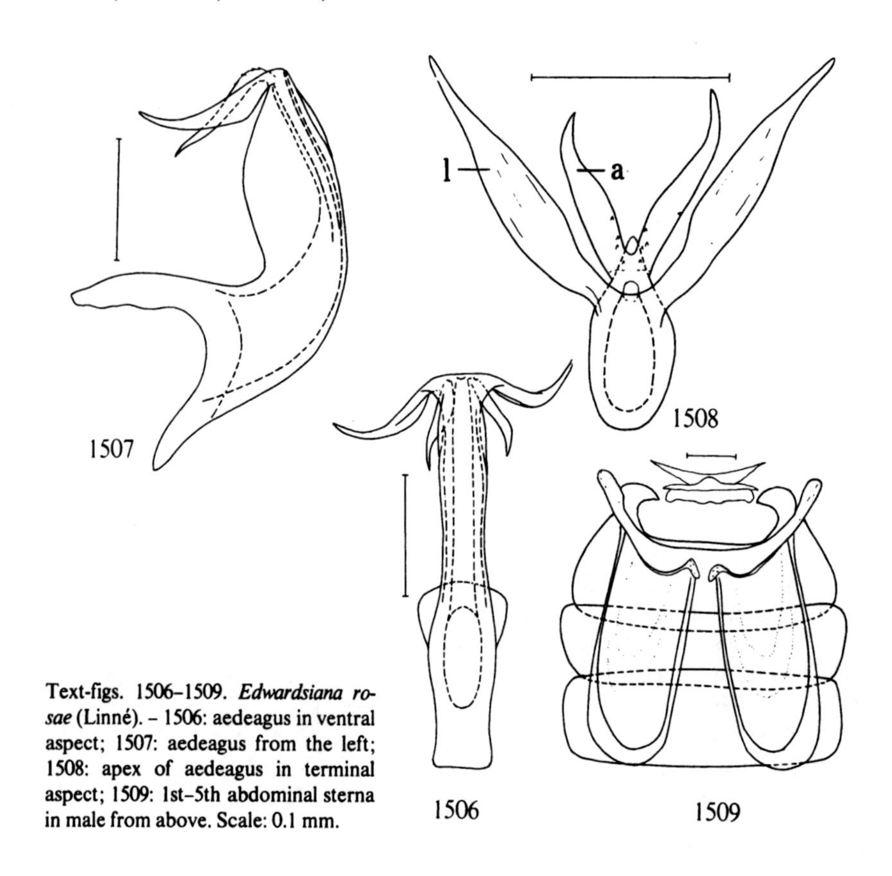

Text-figs. 1506–1509. *Edwardsiana rosae* (Linné). – 1506: aedeagus in ventral aspect; 1507: aedeagus from the left; 1508: apex of aedeagus in terminal aspect; 1509: 1st–5th abdominal sterna in male from above. Scale: 0.1 mm.

Biology. On *Corylus avellana*. "*E. avellanae* (Edwards) was dominant on hazel - - - but also occurred on elm, alder and beech. Previous sampling records - - - and breeding records suggest hazel as the most important host plant" (Claridge & Wilson, 1976).

205. ***Edwardsiana staminata*** (Ribaut, 1931)
Text-figs. 1514–1517.

Typhlocyba staminata Ribaut, 1931a: 334.

Yellowish white or light yellow, shining. Apical part of fore wing indistinctly fumose. Male pygofer lobe as in Text-fig. 1514, aedeagus as in Text-figs. 1515, 1516, 1st–5th abdominal sterna in male as in Text-fig. 1517. Overall length 3.5–3.8 mm.

Distribution. Not found in Denmark. – Probably not rare in southern Sweden, found

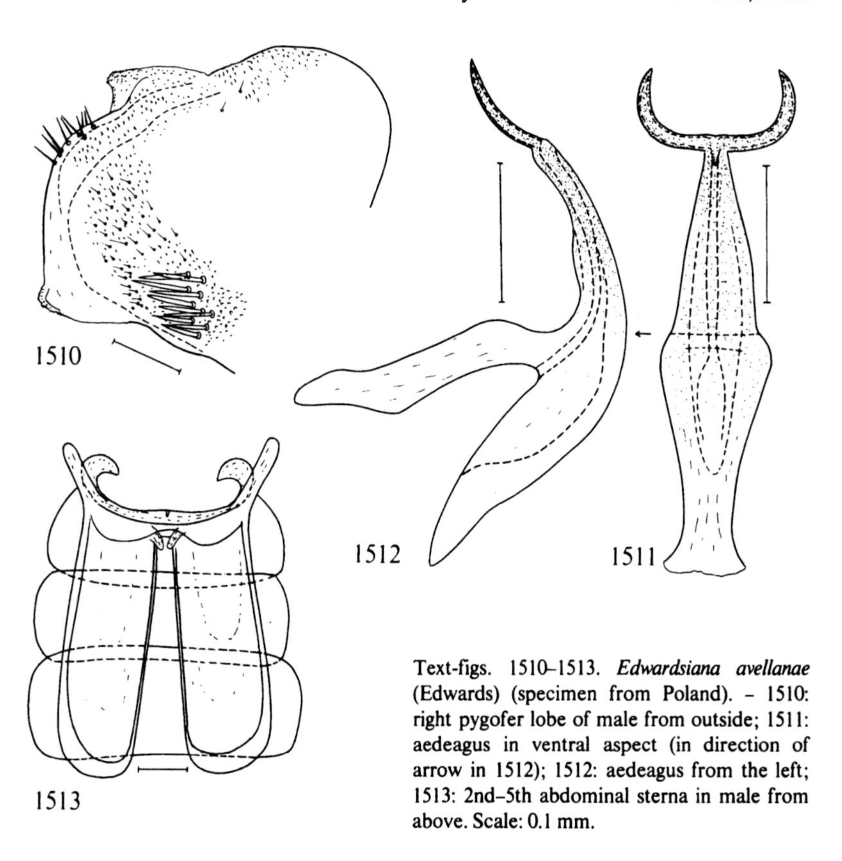

Text-figs. 1510–1513. *Edwardsiana avellanae* (Edwards) (specimen from Poland). – 1510: right pygofer lobe of male from outside; 1511: aedeagus in ventral aspect (in direction of arrow in 1512); 1512: aedeagus from the left; 1513: 2nd–5th abdominal sterna in male from above. Scale: 0.1 mm.

in Sk.: Lund 19.VII.1916 (Tullgren); Nyhus, Börringe 29.VII.1934 (Kemner); Brunnby, Kullen 2.VIII.1961 and 28.VIII.1964 (Ossiannilsson); Bl.: Förkärla 6.VIII.1959, and Förkärla, Kvalmsö 14.VIII.1960 (Gyllensvärd); Ög.: Vårdsberg 4.VII.1933; Bankekind 13.VII.1935; Törnevalla 7.VII.1935; Kimstad 11.VII.1935 (Ossiannilsson). - Norway: found in AK: Ås, Sem 7.IX.1974 by Baeschlin. - East Fennoscandia: only found in Vib: Valkjärvi, by J. Sahlberg. - France, England, Netherlands, German F.R. and D.R., Austria, Bohemia, Slovakia, Lithuania, Estonia, Ukraine.

Biology. On *Corylus avellana* (Ribaut, 1936b; Kuntze, 1937; China, 1943; Dlabola, 1954). Adults in July–September.

206. ***Edwarsiana stehliki*** Lauterer, 1958
Text-figs. 1518–1521.

Edwardsiana stehliki Lauterer, 1958: 130, 135.

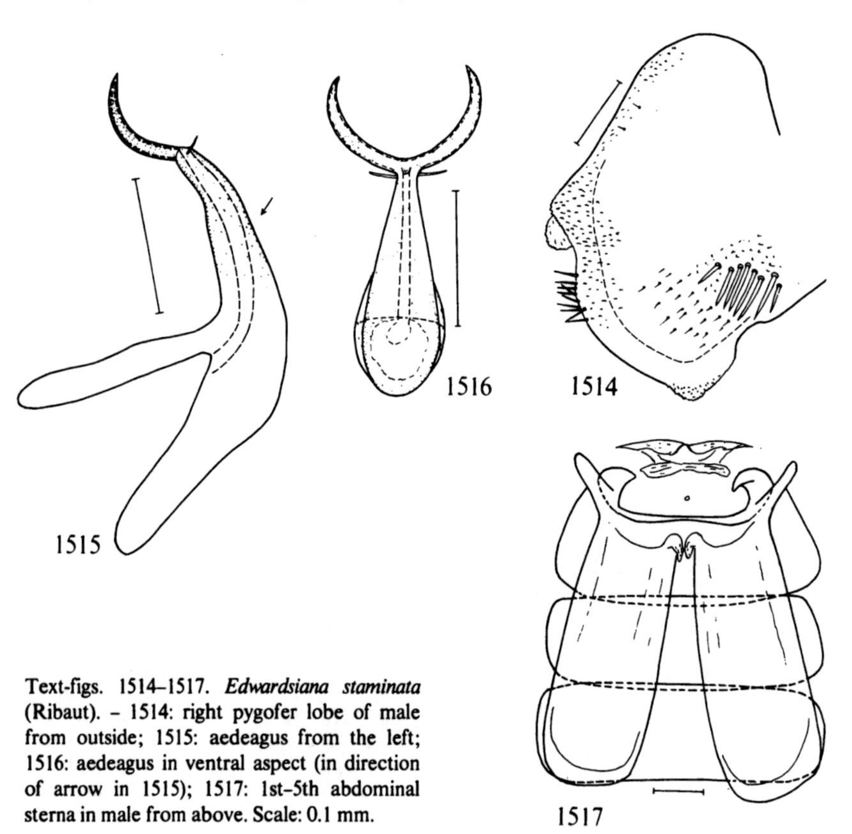

Text-figs. 1514–1517. *Edwardsiana staminata* (Ribaut). - 1514: right pygofer lobe of male from outside; 1515: aedeagus from the left; 1516: aedeagus in ventral aspect (in direction of arrow in 1515); 1517: 1st-5th abdominal sterna in male from above. Scale: 0.1 mm.

"Male: Length 3.5–3.8 mm. Vertex, thorax and abdomen isabelline, fore-wings deeper yellow in colour or in some gradation of it. Claws and arolia of the fore- and middle legs black, last tarsite yellow or brownish in its end. Hind tarsus totally light, praetarsus brownish till black" (Lauterer, l.c.). Male pygofer lobes as in Text-fig. 1518, aedeagus as in Text-figs. 1519, 1520, 2nd–6th abdominal sterna in male as in Text-fig. 1521.

Distribution. Not found in Denmark and Norway. – Sweden: one male and five females found in Jmt.: Åre, Aug. 3–5, 1926, by W. Siefke (Dworakowska, 1971). – Also recorded from East Fennoscandia, N: Helsingfors, Sept.1, 1927, W. Siefke leg. (Dworakowska, l.c.), but according to Dr. Dworakowska (personal communication) the specimen in question (1 ♀) may belong to some other species. – Moravia, Poland.

Biology. On *Corylus avellana* and *Corylus maxima*, adults in June–October (Lauterer, l.c.).

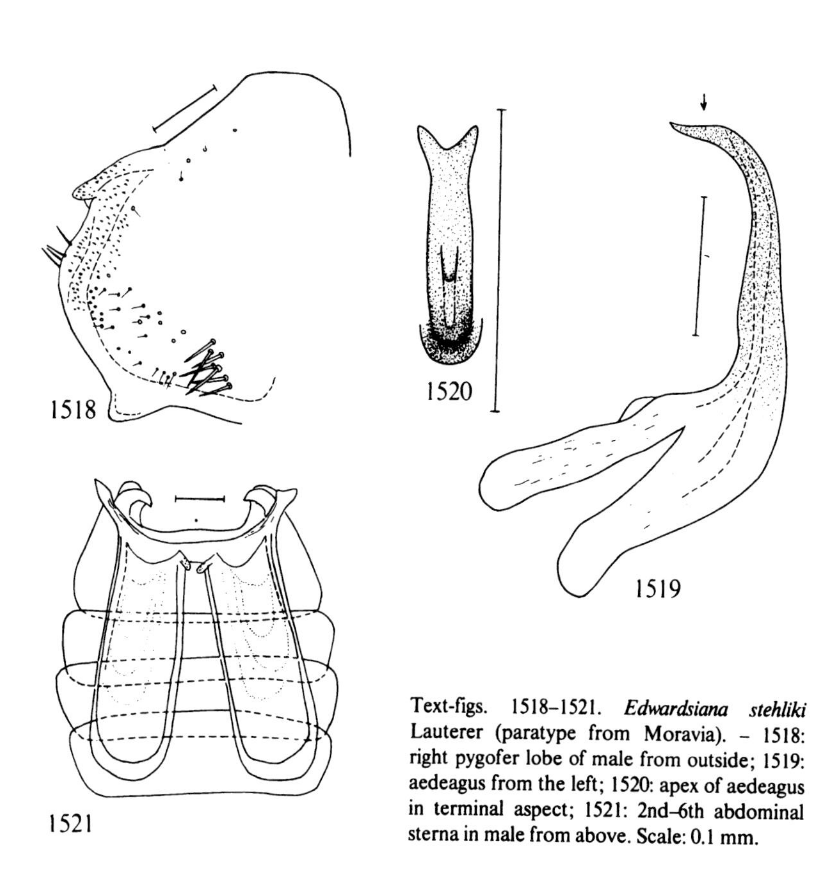

Text-figs. 1518–1521. *Edwardsiana stehliki* Lauterer (paratype from Moravia). – 1518: right pygofer lobe of male from outside; 1519: aedeagus from the left; 1520: apex of aedeagus in terminal aspect; 1521: 2nd–6th abdominal sterna in male from above. Scale: 0.1 mm.

207. ***Edwardsiana crataegi*** (Douglas, 1876)
Text-figs. 1522–1529.

Typhlocyba crataegi Douglas, 1876: 203.
Empoasca australis Froggatt, 1918: 568 (nec Walsh, 1862).
Typhlocyba froggatti Baker, 1925: 537 (n.n.).
Typhlocyba oxyacanthae Ribaut, 1931a: 334.

Yellowish, apical cells and distal ends of cells in corium often strongly fumose. A diffusely limited fumose longitudinal band along claval suture is usually present, sometimes more or less reduced, sometimes covering the entire surface of clavus. Male pygofer lobe as in Text-fig. 1522. Males are dimorphous in length of lateral appendages of aedeagus; aedeagus in f. *typica* as in Text-figs. 1523–1525, of f. *froggatti* as in Text-figs. 1526–1528. 1st–6th abdominal sterna in male as in Text-fig. 1529. Overall length 3.1–3.6 mm.

Distribution. Denmark: known from EJ, WJ, LFM, and B. – Sweden: found in Sk., Bl., Öl., Ög., Boh., Upl., Nrk. – Baeschlin collected *E. crataegi* in Norway, AK: Ås,

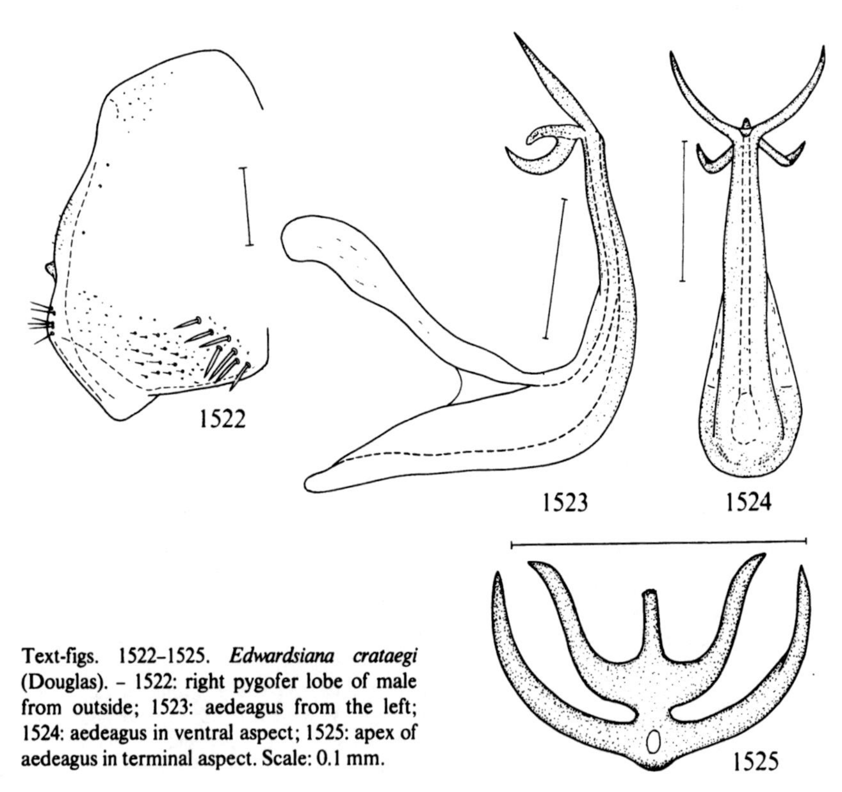

Text-figs. 1522–1525. *Edwardsiana crataegi* (Douglas). – 1522: right pygofer lobe of male from outside; 1523: aedeagus from the left; 1524: aedeagus in ventral aspect; 1525: apex of aedeagus in terminal aspect. Scale: 0.1 mm.

Kaja, Ås, Sem, and Asker, Sem in 1973, 1974, and 1975. - East Fennoscandia: Al (Reuter); N: Helsingfors (Håkan Lindberg); Lovisa (Albrecht); Ab: Åbo; Raisio (Linnavuori). - England, Scotland, Wales, France, Netherlands, German D.R. and F.R., Switzerland, Austria, Bulgaria, Bohemia, Moravia, Slovakia, Hungary, Yugoslavia, Poland, Moldavia, Kazakhstan, Uzbekistan, Kirghizia; Nearctic and Australian regions.

Biology. Hibernation in the egg stage (Müller, 1957). "Auf *Pirus malus, Prunus domestica, Crataegus*" (Wagner & Franz, 1961). Oligophagous, "dominant on hawthorn

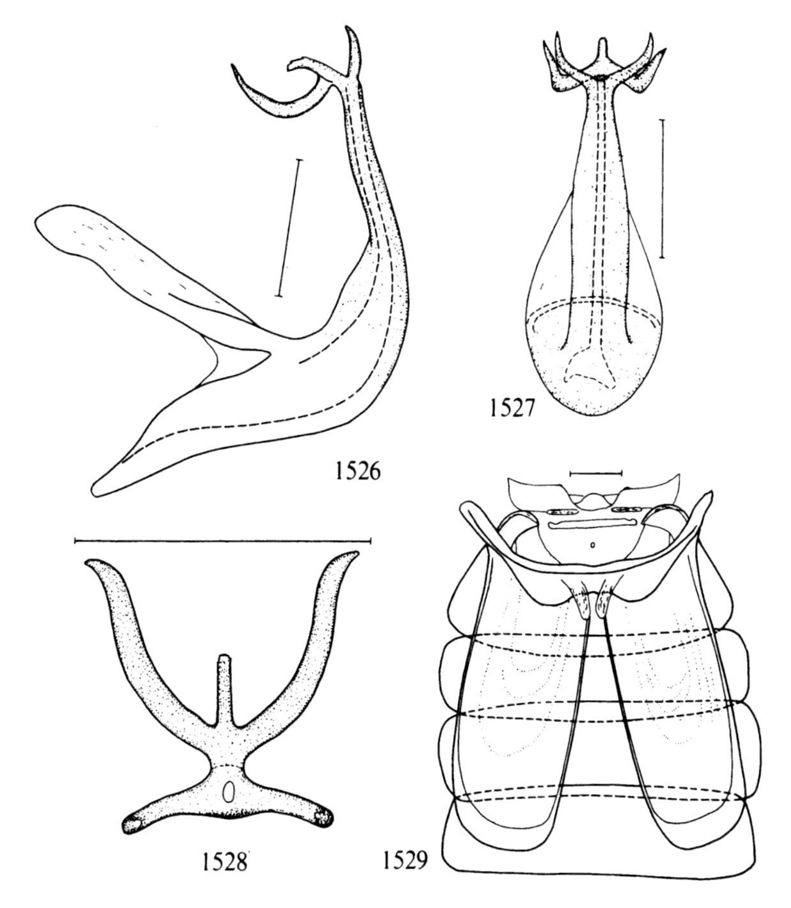

Text-figs. 1526–1529. *Edwardsiana crataegi* (Douglas). - 1526: aedeagus from the left (f. "*froggatti*"); 1527: same in ventral aspect; 1528: same in terminal aspect; 1529: 1st–6th abdominal sterna in male from above. Scale: 0.1 mm.

– – – with fewer on rowan and alder" (Claridge & Wilson, 1976). Günthart (1971) found *Edwardsiana australis* on "Apfel, Kirsche, Mespilus", *E. crataegi* on "Apfel, Crataegus, Mespilus". She also studied *crataegi* and *australis* in rearing experiments. "Wir entnahmen je ein einzelnes befruchtetes Weibchen aus der Mischpopulation und setzten es auf eine neue Apfelpflanze. In drei Nachkommenschaften erhielten wir nur *E. australis*, in drei weiteren aber *E. australis* und *E. crataegi* je gemischt, und zwar in der F_1- und der F_2-Generation; Uebergangsformen sind keine beobachtet worden" (Günthart, l.c.). Of course this demonstrates that the difference between *crataegi* and *australis* is intraspecific.

Economic importance. *Edwardsiana crataegi* is recorded as a pest on various fruit-frees in Australia, America, and Europe. "Vorwiegend an Apfelbäumen durch Besaugen der Blätter, die vorzeitig welken und abfallen, erheblich schädlich, ferner durch Verunreinigung der Früchte durch die zunächst weisslichen, bei Regen braun werdenden Kottropfen (Honigtau), die schwer zu entfernen sind und die Ernte bis zur Unverkäuflichkeit entwerten können. An Pflaumen und anderen Obstarten ist der Schaden meist gering" (Müller, 1956). Damage on cultivated plants in Denmark and Fennoscandia has not been observed so far.

208. ***Edwardsiana nigriloba*** (Edwards, 1924)
Text-figs. 1530, 1531.

Typhlocyba nigriloba Edwards, 1924: 55.

Yellowish white or light yellow, sometimes with an orange tinge. Apical part of fore wing more or less fumose. Male pygofer lobe as in Text-fig. 530, aedeagus as in Text-fig. 531. Overall length 3.5–4.25 mm.

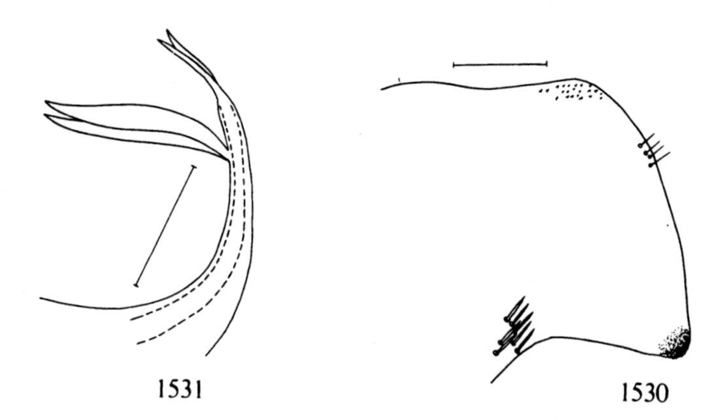

Text-figs. 1530, 1531. *Edwardsiana nigriloba* (Edwards). – 1530: left pygofer lobe of male from outside; 1531: aedeagus from the left. Scale: 0.1 mm.

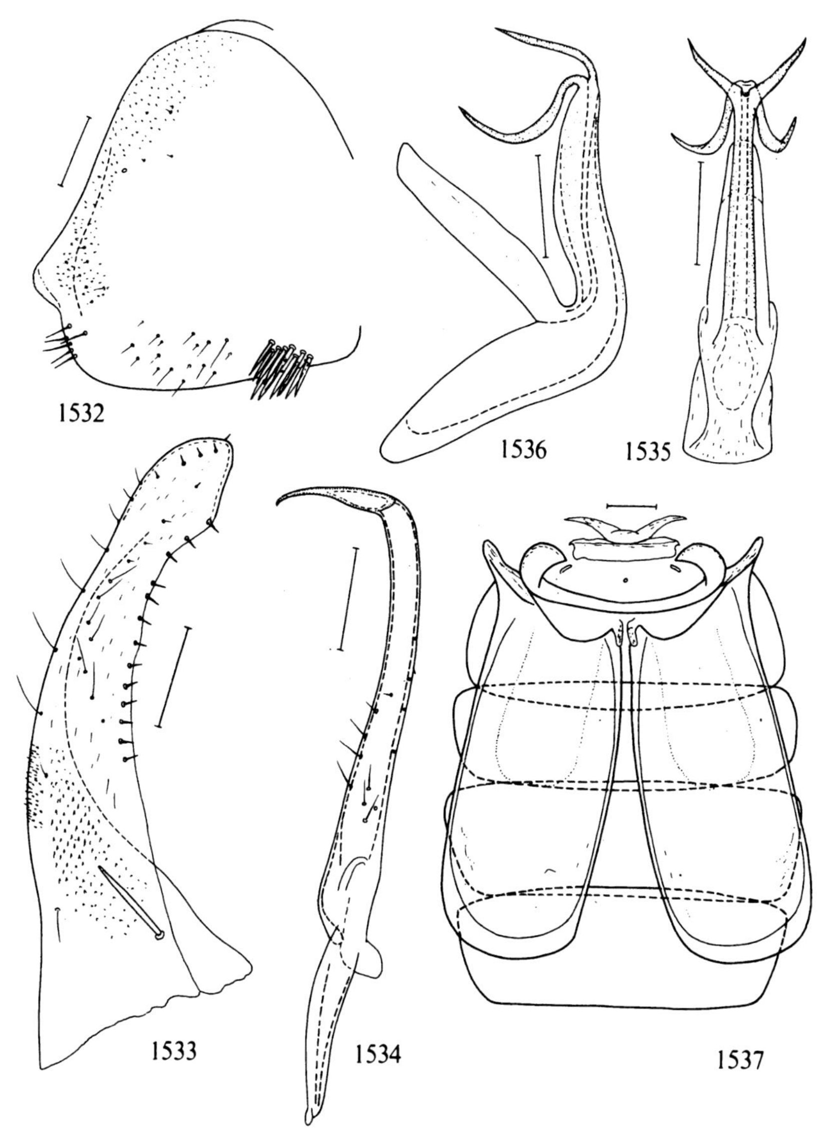

Text-figs. 1532–1537. *Edwardsiana salicicola* (Edwards). – 1532: right pygofer lobe of male from outside; 1533: right genital plate from outside; 1534: right genital style from above; 1535: aedeagus in ventral aspect; 1536: aedeagus from the left; 1537: 1st–6th abdominal sterna in male from above. Scale: 0.1 mm.

Distribution. Not found in Denmark, Norway, and East Fennoscandia. – Rare in Sweden, only found in Sk.: Lund 19.VII.1916 by Tullgren, and in Boh.: Uddevalla 23.VII.1943 by Ossiannilsson. – England, France, Belgium, German D.R. and F.R., Switzerland, Bulgaria, Bohemia, Moravia, Poland.

Biology. "Vit sur le Sycomore" (Ribaut, 1936b). "On sycamore" (China, 1943). On *Acer* (Günthart, 1974). Monophagous, on *Acer pseudoplatanus* (Claridge & Wilson, 1976). Tullgren found his specimen on *Acer platanoides*.

209. ***Edwardsiana salicicola*** (Edwards, 1885)

Text-figs. 1532–1537.

Typhlocyba salicicola Edwards, 1885: 230.

Yellowish white, shining. Apical part of fore wing faintly fumose. Male pygofer lobe as in Text-fig. 1532, genital plate as in Text-fig. 1533, genital style as in Text-fig. 1534, aedeagus as in Text-figs. 1535, 1536, 1st–6th abdominal sterna in male as in Text-fig. 1537. Overall length 4–4.2 mm.

Distribution. Denmark: found in several places in EJ, WJ, and B. – Sweden: apparently not uncommon, found in Sk., Bl., Sm., Gtl., Ög., Sdm., Upl., Nb. – Norway: one male found in Ry: Madla 12.VIII.1952 by Holgersen. – East Fennoscandia: found in Ab: Raisio (Linnavuori); Ta: Lammi (Linnavuori), Hattula (Nuorteva). – England, France, Netherlands, German D.R. and F.R., Switzerland, Austria, Hungary, Bohemia, Moravia, Slovakia, Romania, Poland, Latvia, Lithuania, Estonia, Altai Mts., Kazakhstan, Iran, Mongolia.

Biology. "Auf *Salix*-Arten, besonders *Salix aurita*" (Wagner & Franz, 1961). "Monophag auf *Salix aurita*. Imagines 23.VIII. – 4.IX." (Nuorteva, 1952).

210. ***Edwardsiana alnicola*** (Edwards, 1924)

Text-figs. 1538–1543.

Typhlocyba alnicola Edwards, 1924: 54.

Yellowish white, shining. Apical part of fore wing faintly fumose. Male pygofer lobe as in Text-fig. 1538, genital style as in Text-fig. 1539, aedeagus as in Text-figs. 1540–1542, 2nd–5th abdominal sterna in male as in Text-fig. 1543. Overall length 4–4.3 mm.

Distribution. Denmark: so far known from F: Staurby Skov 6.VII.1968 (Trolle), and B: Hundsemyre 25.VII.1979 (Trolle). – Not uncommon in Sweden, found in Sk., Bl., Hall., Ög., Dlsl., Sdm., Dlr., Ång., Vb. – Norway: found in Os, TEi, AAi, MRi. – East Fennoscandia: found in Ab: Lojo (Håkan Lindberg), Raisio (Linnavuori); Ta: Hattula (Nuorteva); Sb: Kiuruvesi (Linnavuori). – England, German F.R., Austria, Romania, Poland, Estonia, Latvia.

Biology. On *Alnus* (Wagner & Franz, 1961). "Monophag auf *Alnus incana,* auf welcher Zuchtversuchen gemäss auch die Larven leben. - - - Imagines 6.VI.-13.VIII" (Nuorteva, 1952). On *Alnus glutinosa* (Linnavuori, 1969). "*E. alnicola* (Edwards) is almost restricted to Alder, from which we have reared it" (Claridge & Wilson, 1976).

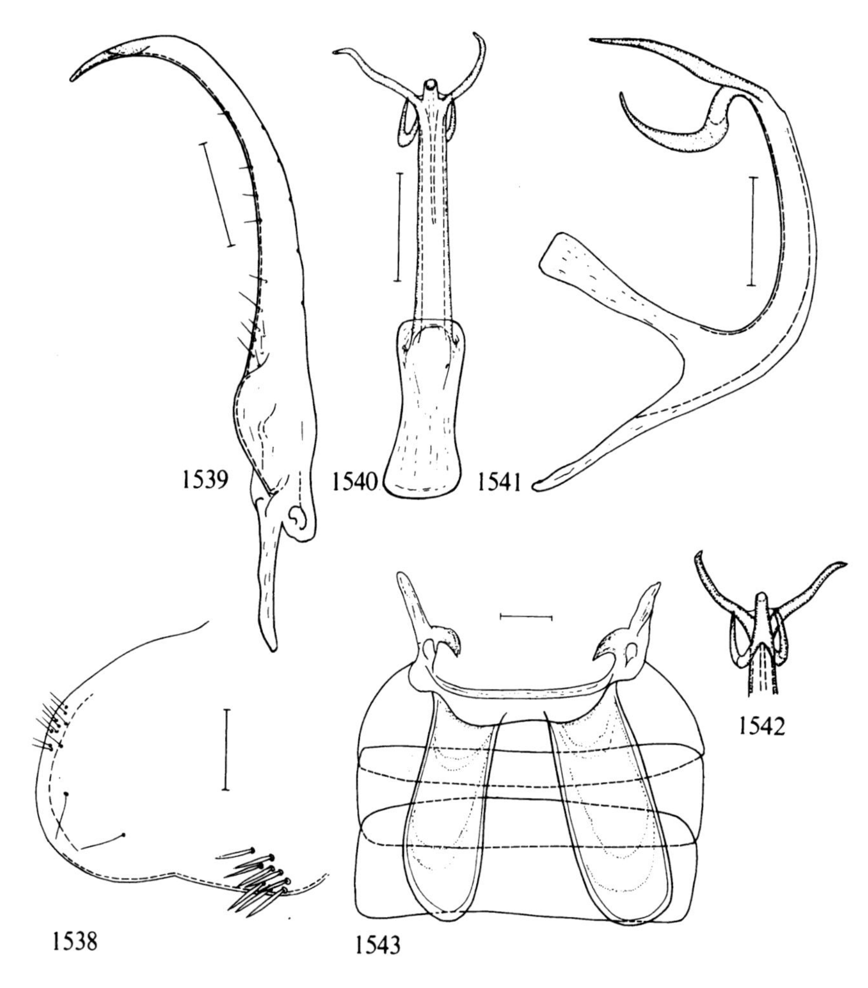

Text-figs. 1538–1543. *Edwardsiana alnicola* (Edwards). – 1538: right pygofer lobe of male from outside; 1539: right genital style from above; 1540: aedeagus in ventral aspect; 1541: aedeagus from the left; 1542: apex of aedeagus in dorsal aspect; 1543: 2nd–5th abdominal sterna in male from above. Scale: 0.1 mm.

211. ***Edwardsiana sociabilis*** (Ossiannilsson, 1936)
Text-figs. 1544–1549.

Typhlocyba sociabilis Ossiannilsson, 1936a: 257.

Yellowish white, shining. Apical part of fore wing faintly fumose. Male pygofer lobe as in Text-fig. 1544, genital style as in Text-fig. 1545, aedeagus as in Text-figs. 1546–1548, 1st–6th abdominal sterna in male as in Text-fig. 1549. Overall length 3.4–3.6 mm. – Macrosetae on thorax and wing pads of last instar larvae with fuscous bases.

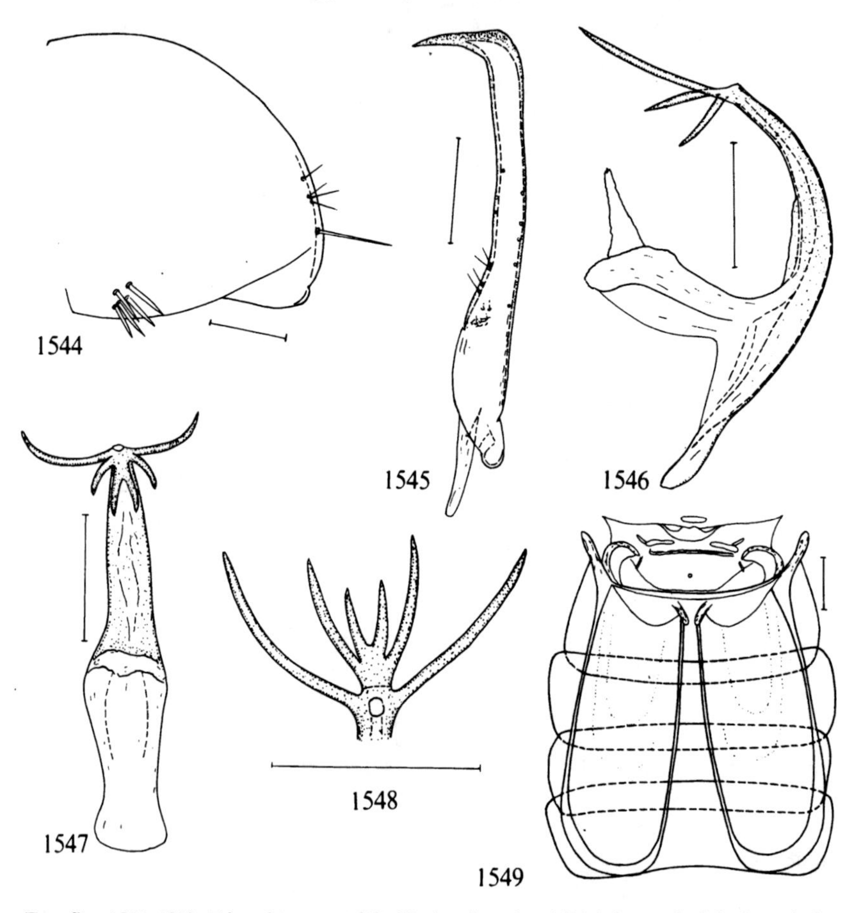

Text-figs. 1544–1549. *Edwardsiana sociabilis* (Ossiannilsson). – 1544: left pygofer lobe in male from outside; 1545: right genital style from above; 1546: aedeagus from the left; 1547: aedeagus in dorsal aspect; 1548: apex of aedeagus in terminal aspect; 1549: 1st–6th abdominal sterna in male from above. Scale: 0.1 mm.

Distribution. Denmark: only found in EJ: Silkeborg 1917 (Jensen-Haarup), and in B: Østermarie 28.IX.1972 (Trolle). – Not uncommon in southern and central Sweden, found in Sk., Bl., Öl., Ög., Sdm., Upl., Nrk. – Norway: found in AK: Ås 8.VII.1970 (Stenseth); Sem, Asker 1.X.1974 (Taksdal); Bø: Bingen, Modum (Holgersen). – Locally common in southern and central East Fennoscandia, found in Al, Ab, N, Sa, Sb. – Bohemia, Moravia, German F.R., Poland, Georgia.

Biology. "Common on *Rosa* and *Rubus idaeus* in gardens" (Linnavuori, 1952a). Breeding also on *Filipendula ulmaria*. Adults in July–October.

Economic importance. "*T. sociabilis* Oss. also occurs as a pest on roses in Finland" (Nuorteva, 1955).

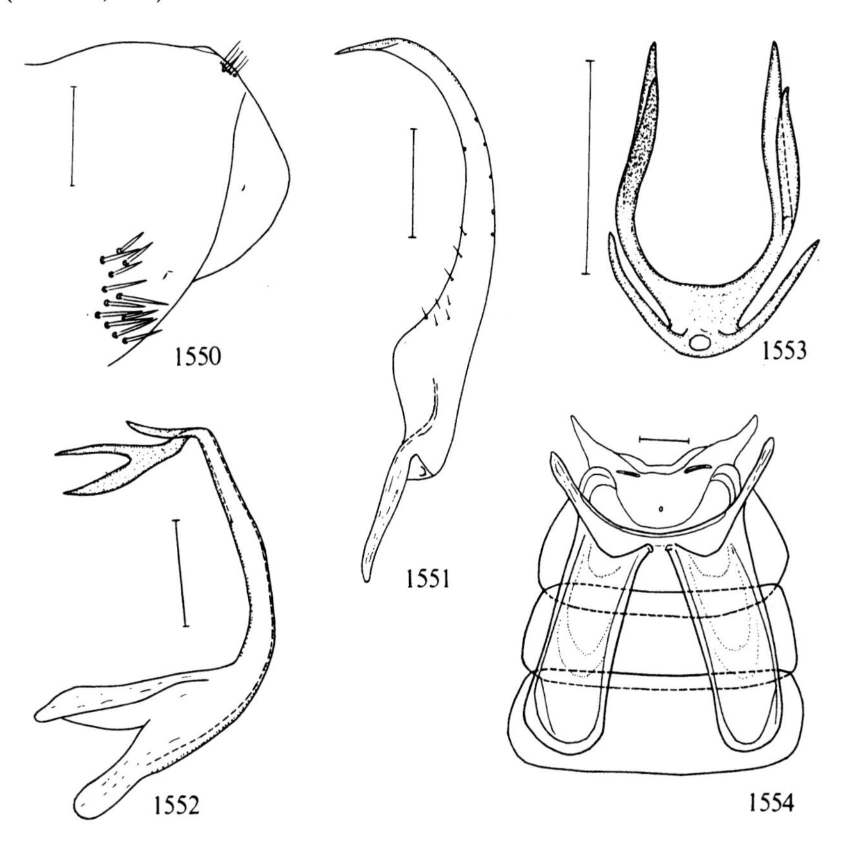

Text-figs. 1550–1554. *Edwardsiana frustrator* (Edwards). – 1550: left pygofer lobe in male from the left; 1551: right genital style from above; 1552: aedeagus from the left; 1553: apex of aedeagus in terminal aspect; 1554: 1st–5th abdominal sterna in male from above. Scale: 0.1 mm.

212. ***Edwardsiana frustrator*** (Edwards, 1908)
Text-figs. 1550–1554.

Typhlocyba frustrator Edwards, 1908b: 84.
Typhlocyba solearis Ribaut, 1931a: 339.

Yellowish white, shining, fore wings light yellow, apical part faintly fumose. Male pygofer lobes as in Text-fig. 1550, genital style as in Text-fig. 1551, aedeagus as in Text-figs. 1552, 1553, 1st–5th abdominal sterna in male as in Text-fig. 1554. Overall length 3.1–3.4 mm.

Distribution. Denmark: found in several places in EJ and B. – Sweden: not uncommon in the southern part, found in Sk., Bl., Öl., Gtl., Ög. – Norway: found in AK: Sem, Asker 7.IX.1974 (Taksdal), and in MRy: Bolsøy (Holgersen). – East Fennoscandia: found in Ab: Pargas (J. Sahlberg), Lojo (Håkan Lindberg), Åbo (Linnavuori); N: Helsingfors (Håkan Lindberg). – England, France, Netherlands, German D.R. and F.R., Switzerland, Bohemia, Poland, Ukraine, Georgia; Nearctic region.

Biology. Found on *Castanea vesca, Acer platanoides, Tilia cordata, Aesculus hippocastanum* (Ossiannilsson, 1946c). Host-plant *Aesculus,* also found on *Acer, Tilia, Crataegus* (Günthart, 1974). "Occurred widely on most trees in small numbers, with no obvious dominant. It was reared in small numbers from several hosts" (Claridge & Wilson, 1976). Adults in July–October.

213. ***Edwardsiana ishidae*** (Matsumura, 1932)
Text-figs. 1555–1560.

Typhlocyba ishidae Matsumura, 1932: 98.
Typhlocyba lanternae Ossiannilsson, 1946c: 139 (nec W. Wagner, 1937).

Yellowish white, shining, apical part of fore wings faintly fumose. Male pygofer lobes as in Text-fig. 1555, genital style as in Text-fig. 1556, aedeagus as in Text-figs. 1557–1559, 2nd–5th abdominal sterna in male as in Text-fig. 1560. Overall length (♂♀) 3.7–4.3 mm.

Distribution. Denmark: so far found in several places in EJ and F. – Sweden: Upl.: Solna 17.–18.IX.1942 (Ossiannilsson), Uppsala Botanical Garden 10.VIII.1943 (Ossiannilsson), Uppsala, Ultuna 1943 (Kemner); Dlr.: Falun 22.IX.1936 and 14.IX.1943 (Bo Tjeder); Jmt.: Frösön 21.VII.1941 and 16.VIII.1942 (Ossiannilsson). – Norway: Bø: Bingen, Modum (Holgersen). – East Fennoscandia: Ab: Raisio (Linnavuori); Perniö 21.VI.1948 (Linnavuori); Ta: Lammi (Linnavuori), and Tb: Jyväskylä (Linnavuori). – Japan.

Biology. On *Ulmus glabra* (Ossiannilsson, 1946c; Linnavuori, 1952a). Adults in July–October.

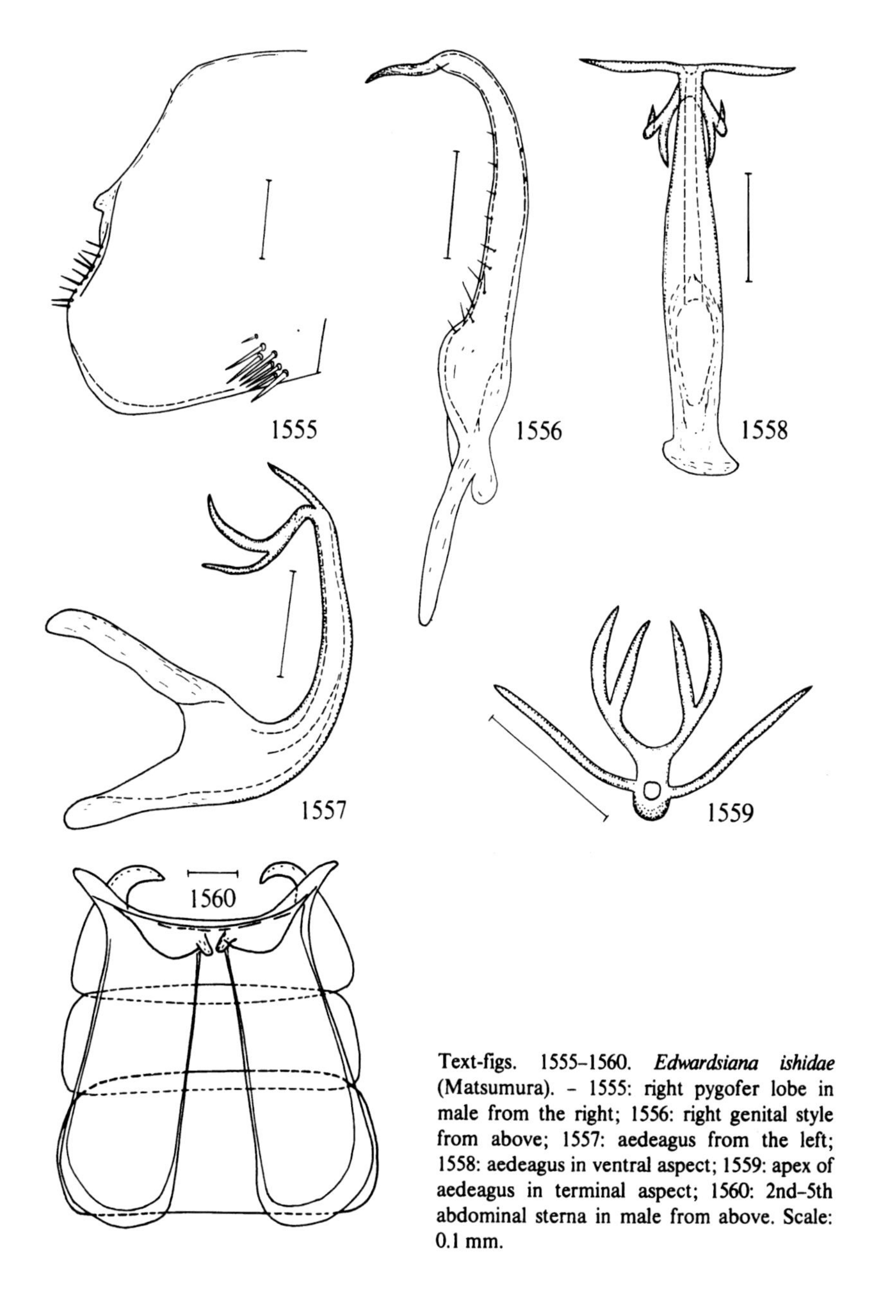

Text-figs. 1555–1560. *Edwardsiana ishidae* (Matsumura). – 1555: right pygofer lobe in male from the right; 1556: right genital style from above; 1557: aedeagus from the left; 1558: aedeagus in ventral aspect; 1559: apex of aedeagus in terminal aspect; 1560: 2nd–5th abdominal sterna in male from above. Scale: 0.1 mm.

214. ***Edwardsiana prunicola*** (Edwards, 1914)
Text-figs. 1561–1565.

Typhlocyba prunicola Edwards, 1914: 168.
Typhlocyba barbata Ribaut, 1931a: 338.

Yellowish white or light yellow, shining, apical part of fore wing fumose. Male genital plate as in Text-fig. 1561, genital style as in Text-fig. 1562, aedeagus as in Text-figs. 1563, 1564, 2nd–5th abdominal sterna in male as in Text-fig. 1565. Length about 4 mm.

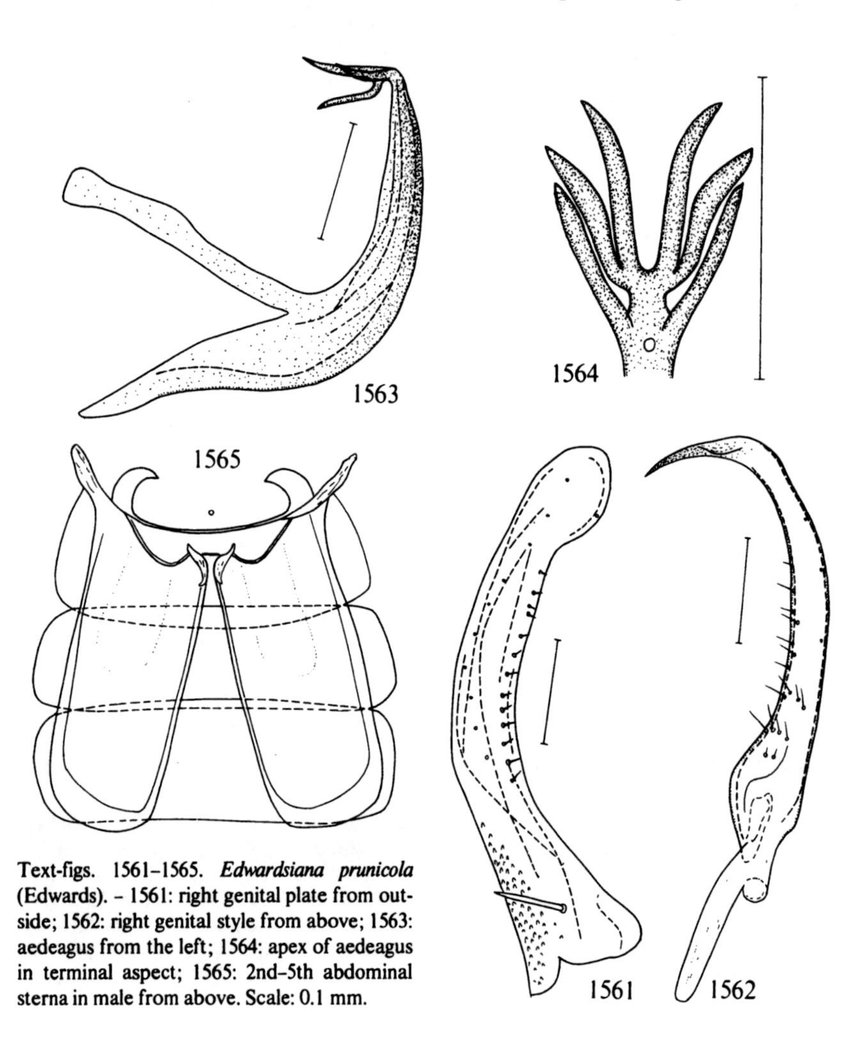

Text-figs. 1561–1565. *Edwardsiana prunicola* (Edwards). – 1561: right genital plate from outside; 1562: right genital style from above; 1563: aedeagus from the left; 1564: apex of aedeagus in terminal aspect; 1565: 2nd–5th abdominal sterna in male from above. Scale: 0.1 mm.

Distribution. Denmark: found in EJ: Lisbjerg Skov 15.IX.1965 (Trolle) and Kalø 21.VII.1966 (Trolle). - Sweden: found in Sk.: Brunnby, Kullen 25.VII.1961, 30.VII. and 31.VII.1962 (Ossiannilsson); Bl.: Nättraby 18.VIII.1959 (Gyllensvärd), Bräkne-Hoby 14.VI.1960 (Gyllensvärd), Förkärla 19.IX.1960, 30.VII.1962 (Gyllensvärd); Gtl.: Östergarn, Katthammarsvik 23.VIII.1961 (Gyllensvärd); Ög.: Rystad, Fröstad 14.VIII.-1935 (Ossiannilsson). - Norway: found in AK: Sem, Asker 2.X.1973 (T. Edland), Ås, Sem 7.IX.1974, 9.VII.1975 (Baeschlin). - East Fennoscandia: Ab: Ispoinen near Åbo 21.VIII.1948 (Linnavuori); Ta: Lammi 1948 (Linnavuori). Linnavuori also recorded a male found on *Salix* in Ab: Raisio 27.VII.1948 as *barbata* Ribaut. - England, France, Netherlands, German D.R. and F.R., Switzerland, Austria, Bulgaria, Bohemia, Moravia, Slovakia, Hungary, Romania, n. Yugoslavia, Poland, Estonia, Lithuania, Moldavia, s. Russia; introduced into the Nearctic region.

Biology. On *Prunus spinosa (prunicola)*; on *Salix aurita (barbata)* (Kuntze, 1937). "Auf *Salix*" (Wagner & Franz, 1961). Günthart (1971) collected *E. prunicola* on "Pflaume, Schwarzdorn, Weide"; by rearing experiments she established "Pflaume" as "Wirtspflanze (= Brutpflanze)". I found this species on *Crataegus monogyna* and *Salix aurita* in Scania. Adults in June–October.

215. ***Edwardsiana menzbieri*** Zachvatkin, 1948
Text-figs. 1566–1570.

Edwardsiana menzbieri Zachvatkin, 1948a: 184.
Typhlocyba sundholmi Gyllensvärd, 1964: 170.

"Body length, ♂♂, 2.8–3.0 mm., ♀♀ 3.0–3.4 mm. Total length including fore wings, ♂♂ 3.8–4.2 mm (mean 4.0 mm.), ♀♀ 4.0–4.4 mm. Colour, yellowish-white to pale yellow. Anterior border of head seen from the dorsal aspect, rounded. Length of head rather greater in the midline than at the compound eyes. Vertex convex. Angle between vertex and frontoclypeus markedly rounded. Fore-wings long, narrow, clear to slightly hyaline, apical part not smoke-coloured. Veins pale" (Gyllensvärd, l.c.). Male genital plate as in Text-fig. 1566, aedeagus as in Text-figs. 1567–1569, 1st–5th abdominal sterna in male as in Text-fig. 1570.

Distribution. So far not found in Denmark. - Sweden: Jmt.: Åre 3.-5.VIII.1927, 4 ♂♂ (W. Siefke); Lu. Lpm.: Kamajokk 1–2 km N. Kvickjokk 4.IX.1962, 1 ♂ (P. Brinck & al.); on a small island in the Farrajokk delta near Kvickjokk 1. - 2.VIII.1963, and at Kvarnsjön, close to Mount Nammates c. 2 km south of Tarrajokk 3.VIII.1963 (Gyllensvärd). - Norway: found in Nsi: Stormo 18.VIII.1961, 1 ♂ (Holgersen); TRi: Skibotn 4.VIII.1961, 4 ♂♂ (Holgersen), Kvesmenes 5.VIII.1961, 1 ♂ (Holgersen). - East Fennoscandia: only found in Ab: Bjärnå 6.VII.1976 by Albrecht. - Latvia, Lithuania, m. Russia, Altai Mts., Maritime Territory, Mongolia, Japan.

Biology. On *Salix* sp. (Gyllensvärd, l.c., Albrecht, 1977).

216. ***Edwardsiana flavescens*** (Fabricius, 1794)
Text-figs. 1571-1574.

Cicada flavescens Fabricius, 1794: 46.
Typhlocyba fratercula Edwards, 1908b: 84.
Typhlocyba sororcula Ossiannilsson, 1936a: 259.

Yellowish white, shining, fore wings light yellow, apical part not obviously fumose. Aedeagus as in Text-figs. 1571–1574. Overall length 3.5–3.7 mm.

Distribution. Denmark: found in NEZ: Damhussøen 3.X.1913 (Oluf Jacobsen) and Tisvilde 13.IX.1914 (Oluf Jacobsen). – Sweden: one male found in Sk.: Lund 19.VII.-

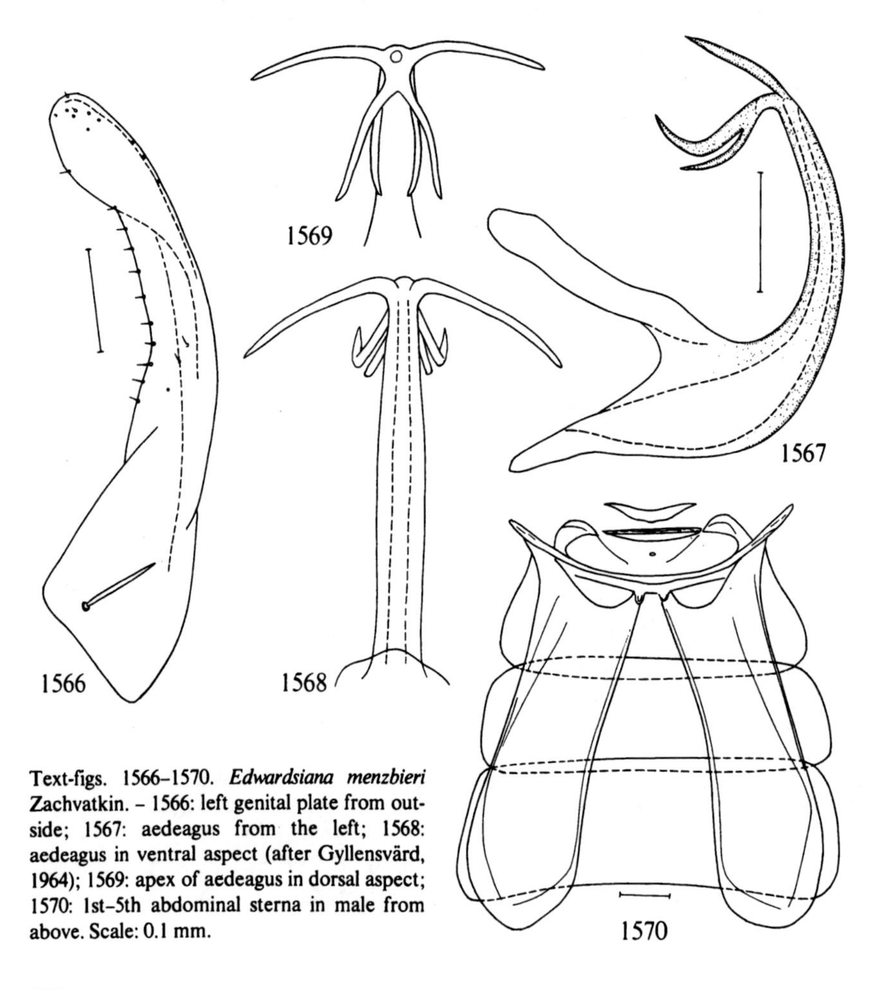

Text-figs. 1566–1570. *Edwardsiana menzbieri* Zachvatkin. – 1566: left genital plate from outside; 1567: aedeagus from the left; 1568: aedeagus in ventral aspect (after Gyllensvärd, 1964); 1569: apex of aedeagus in dorsal aspect; 1570: 1st–5th abdominal sterna in male from above. Scale: 0.1 mm.

1916 (A. Tullgren); two males found in Öl.: Halltorps hage 19.-23.VII. 1976 by H. Andersson & R. Danielsson. - So far not found in Norway and East Fennoscandia. - England, France, Netherlands, German D.R. and F.R., Austria, Bulgaria, Bohemia, Moravia, Slovakia, Romania, Hungary, Poland, Lithuania, s. Russia, Ukraine, Moldavia, Georgia.

Biology. "Sur le Charme" (Ribaut, 1936b). "Auf *Carpinus betulus,* aber sicher auch auf anderen Laubhölzern, da die Art in den Alpen weit über das Verbreitungsgebiet der Weissbuche hinausreicht" (Wagner & Franz, 1961). Found on *Juglans, Fagus silvatica, Populus,* host-plant: *Fagus silvatica* (Günthart, 1974). "Abundant on hornbeam - - - and beech - - -, from which it was also reared. It also occurred on sycamore, maple, elm and oak" (Claridge & Wilson, 1976).

217. ***Edwardsiana plebeja*** (Edwards, 1914)
Text-figs. 1575–1581.

Typhlocyba plebeja Edwards, 1914: 169.
Typhlocyba divergens Ribaut, 1931a: 339.

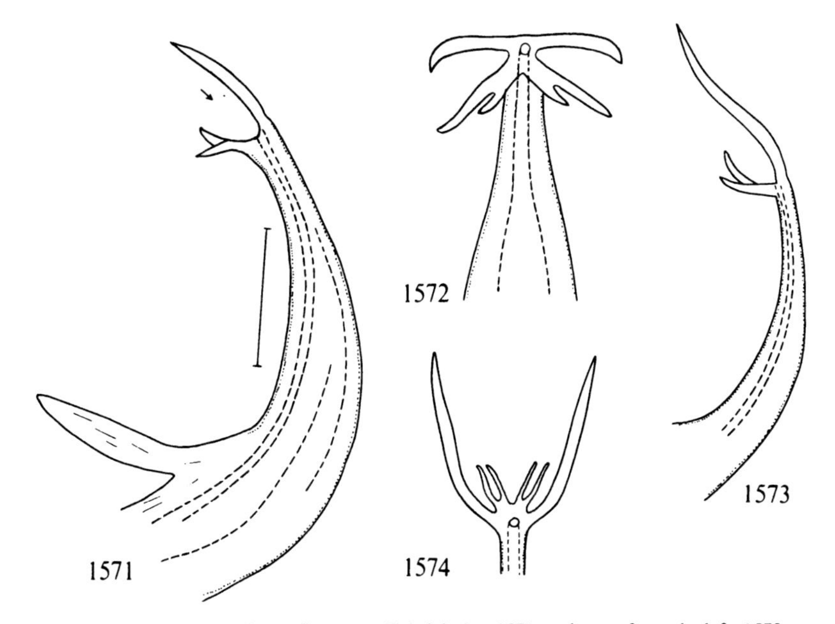

Text-figs. 1571–1574. *Edwardsiana flavescens* (Fabricius). - 1571: aedeagus from the left; 1572: apex of aedeagus in dorsal aspect (as seen in direction of arrow in 1571); 1573: aedeagus from the left; 1574: apex of aedeagus in ventral aspect (1573 and 1574 after type of *sororcula* Ossiannilsson). Scale: 0.1 mm.

Yellowish white, shining, fore wings light yellow, apical part faintly fumose. Male pygofer lobe as in Text-fig. 1575, genital plate as in Text-fig. 1576, genital style as in Text-fig. 1577, genital plate as in Text-fig. 1576, genital style as in Text-fig. 1577, aedeagus as in Text-figs. 1578–1581. Overall length 3.5–3.8 mm.

Distribution. Denmark: found in SJ: Sønderborg 26.VII.1881 (Wüstnei), and in EJ: Pinds Mølle 28.VIII.1965 (Trolle). – Not uncommon in southern Sweden, found in Sk.: Lund 19.VII.1916 (Tullgren), Svalöv 7.VII.1936 (Ossiannilsson), Ven 2.VIII.1936 (Ossiannilsson), and Alnarp 1.VIII.1944 (Ossiannilsson). E. Sylvén caught in all 21 males in light traps in Sk.: Åkarp 10.VII.–19.VIII.1952. Found in Hall.: Halmstad by Wieslander, and in Hall.: Ö. Karup in 1939 by Bo Tjeder. One male found in Gtl.: Visby

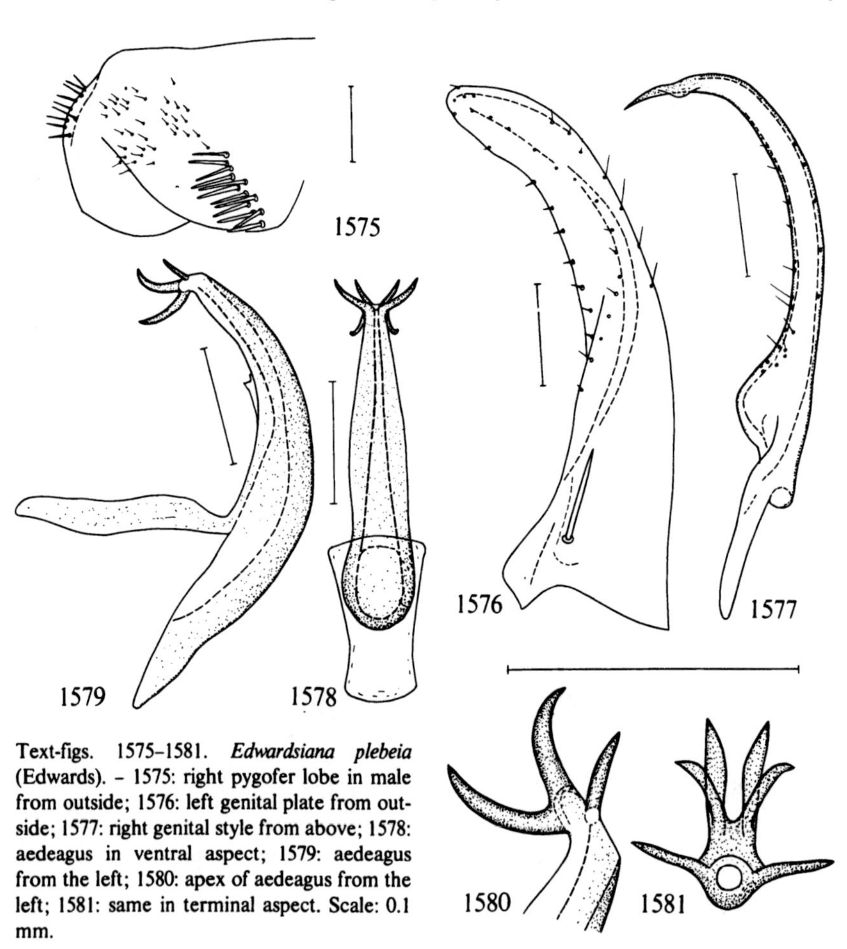

Text-figs. 1575–1581. *Edwardsiana plebeia* (Edwards). – 1575: right pygofer lobe in male from outside; 1576: left genital plate from outside; 1577: right genital style from above; 1578: aedeagus in ventral aspect; 1579: aedeagus from the left; 1580: apex of aedeagus from the left; 1581: same in terminal aspect. Scale: 0.1 mm.

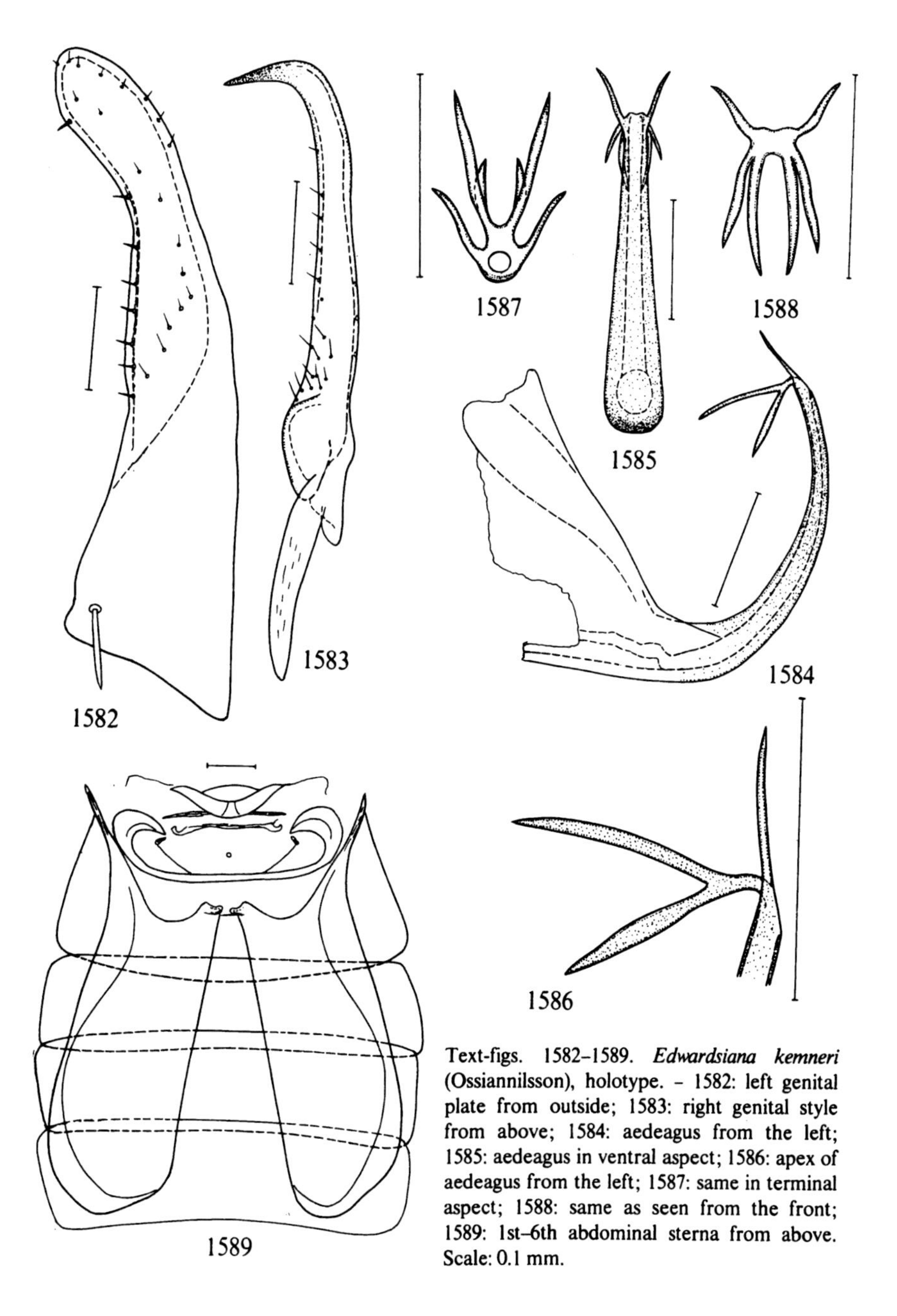

Text-figs. 1582–1589. *Edwardsiana kemneri* (Ossiannilsson), holotype. – 1582: left genital plate from outside; 1583: right genital style from above; 1584: aedeagus from the left; 1585: aedeagus in ventral aspect; 1586: apex of aedeagus from the left; 1587: same in terminal aspect; 1588: same as seen from the front; 1589: 1st–6th abdominal sterna from above. Scale: 0.1 mm.

1.VIII.1935 (Ossiannilsson). – Norway: so far only found in Ry: Kleppe by Holgersen (1♂). – East Fennoscandia: found in Ab: Lojo 18.VIII.1923 by Håkan Lindberg, and in Åbo by Linnavuori. – England, France, German D.R. and F.R., Switzerland, Austria, Bulgaria, Bohemia, Slovakia, Yugoslavia, Moravia, Poland, Latvia, Lithuania, m. and s. Russia, Kazakhstan, Uzbekistan.

Biology. "Sur l'Orme" (Ribaut, 1936b). "Taken in small numbers only from elm" (Claridge & Wilson, 1976).

218. ***Edwardsiana kemneri*** (Ossiannilsson, 1942)
Text-figs. 1582–1589.

Typhlocyba kemneri Ossiannilsson, 1942b: 113.

Whitish, shining. I have seen only one specimen of this species, viz. the holotype. The apical part of the fore wing of this specimen is not distinctly fumose but there is a fumose spot in the distal end of the cubital cell. Male genital plate as in Text-fig. 1582, genital style as in Text-fig. 1583, aedeagus as in Text-figs. 1584–1588, 1st–6th abdominal sterna in male as in Text-fig. 1589. Length of holotype 4.1 mm.

Distribution. Not found in Denmark, Norway and East Fennoscandia. – Sweden Jmt.: Frösön 16.VIII.1942, 1♂ (Ossiannilsson). – German F.R., Bohemia, Romania, Cyprus, Mongolia, Canada.

Biology. The holotype was found on *Salix* sp. Schwoerbel (1975) listed the present species among "Arten die im Gebiet des Calluneto-Genistetum bzw. Querceto-Betuletum gefunden wurden".

219. ***Edwardsiana candidula*** (Kirschbaum, 1868)
Text-figs. 1590–1597.

Typhlocyba candidula Kirschbaum, 1868b: 185.

White, not particularly shining, apical part of fore wing faintly fumose. Male pygofer lobes as in Text-fig. 1590, genital plate as in Text-fig. 1591, genital style as in Text-fig. 1592, aedeagus as in Text-figs. 1593–1596, 1st–5th abdominal sterna in male as in Text-fig. 1597. Overall length about 3.7 mm.

Distribution. Not yet recorded from Denmark, nor from Norway. – Sweden: so far only found in Bl.: Lyckeby 29.VII.1944, 2 ♂♂ (Ossiannilsson). – East Fennoscandia: Al: Jomala 20.VIII.1975 (Albrecht). – England, France, Netherlands, German D.R. and F.R., Belgium, Austria, n. Italy, Bohemia, Moravia, Slovakia, Hungary, Yugoslavia, Poland, Latvia, Lithuania, s. Russia, Ukraine; Nearctic region.

Biology. On *Populus alba* (Kuntze, 1937; Ossiannilsson, 1946c; Wagner & Franz, 1961; Albrecht, 1977).

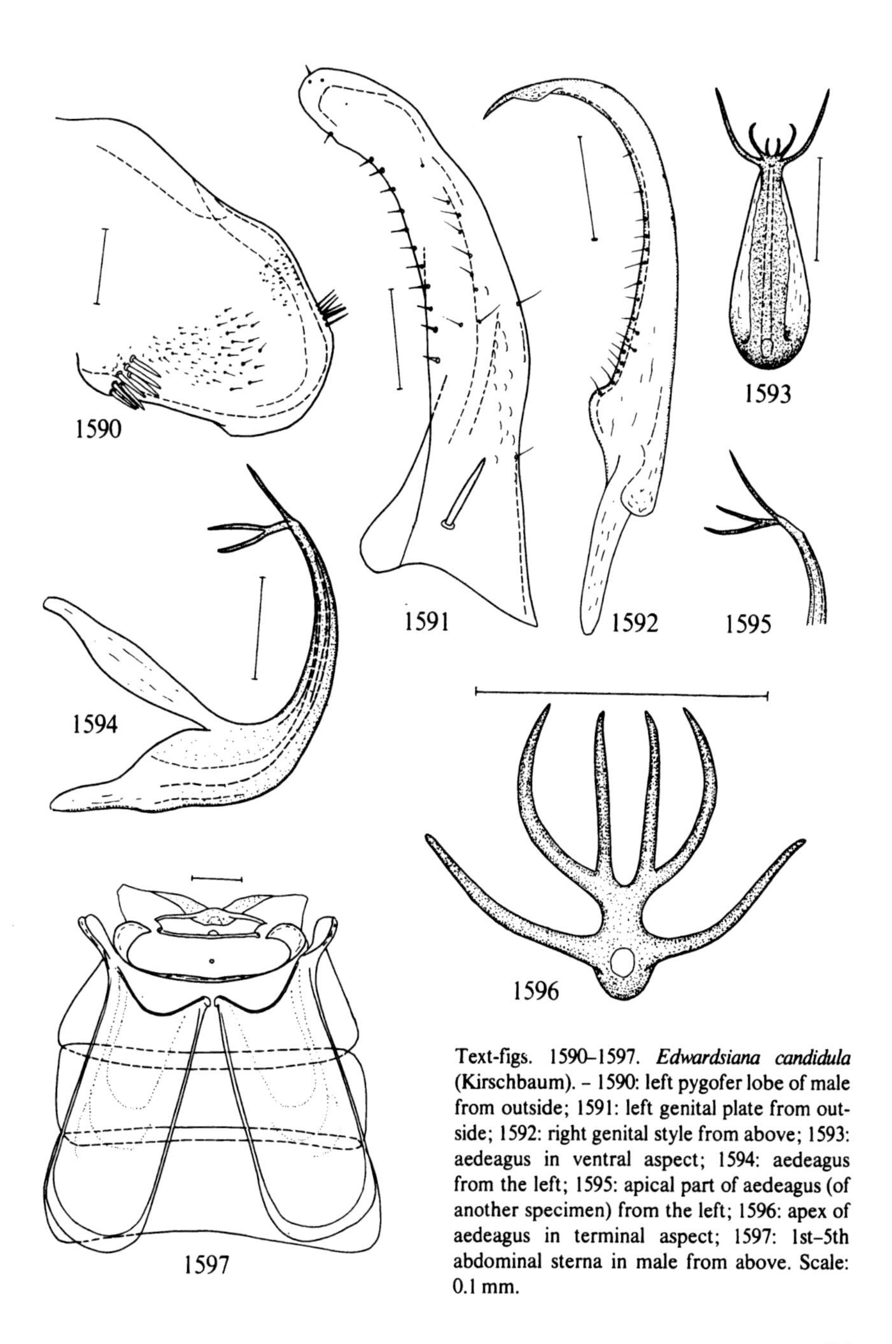

Text-figs. 1590–1597. *Edwardsiana candidula* (Kirschbaum). – 1590: left pygofer lobe of male from outside; 1591: left genital plate from outside; 1592: right genital style from above; 1593: aedeagus in ventral aspect; 1594: aedeagus from the left; 1595: apical part of aedeagus (of another specimen) from the left; 1596: apex of aedeagus in terminal aspect; 1597: 1st–5th abdominal sterna in male from above. Scale: 0.1 mm.

220. ***Edwardsiana gratiosa*** (Boheman, 1852)
Plate-fig. 114, text-figs. 1598–1604.

Typhlocyba gratiosa Boheman, 1852b: 121.
Typhlocyba suturalis Flor, 1861b: 634.

Yellowish white, shining. Scutellum with two triangular black-brown spots usually covering the entire visible part of scutum except the narrow lateral margins. In the fore wing there is a black-brown longitudinal band along scutellar border of clavus tapering towards apex. Distal half of clavus entirely black-brown. A dark transverse band immediately proximally of the transverse veinlets, veins being light also within this band. 4th apical cell brownish, remaining apical cells fumose, also wax area sometimes fumose. Male pygofer lobes as in Text-fig. 1598, genital plate as in Text-fig. 1599, genital style as in Text-fig. 1600, aedeagus as in Text-figs. 1601–1603, 1st–5th abdominal sterna in male as in Text-fig. 1604. Overall length 3.15–3.7 mm.

Distribution. Scarce in Denmark, found in EJ, F, SZ, and NEZ. – Rare in Sweden, found in Sk.: Brunnby, Kullen (Boheman), Ivetofta 11.VIII.1939 (Ossiannilsson), Sövde 27.VIII.1939 (Ossiannilsson); Bl: Augerum, Säljö 21.VIII.1971 (Gyllensvärd); Öl.: Färjestaden 5.VIII.1935 (Ossiannilsson). – Not found in Norway and East Fennoscandia. – France, Netherlands, Belgium, Spain, Italy, Switzerland, Austria, Hungary, Bohemia, Moravia, Slovakia, Yugoslavia, German F.R. and D.R., Poland, Estonia, Latvia, Lithuania, n. and m. Russia, Ukraine.

Biology. "Auf *Alnus glutinosa*" (Wagner & Franz, 1961). "Wirtspflanze: *Alnus*" (Günthart, 1974). Adults in August and September.

221. ***Edwardsiana geometrica*** (Schrank, 1801)
Plate-fig. 115, text-figs. 1605–1611.

Cercopis geometrica Schrank, 1801: 57.
Cicada lineatella Fallén, 1806: 36.

Yellowish white or light yellow, shining. Scutellum entirely or partly black-brown. Fore wings each with a black-brown longitudinal band of nearly equal width starting on scutellar border and running along both sides the claval suture. 1st and 2nd apical cells fumose, 3rd and 4th apical cells black-brownish. Male pygofer lobes as in Text-fig. 1605, genital plate as in Text-fig. 1606, genital style as in Text-fig. 1607, aedeagus as in Text-figs. 1608–1610, 1st–6th abdominal sterna in male as in Text-fig. 1611. Overall length (♂♀) 3.9–4.4 mm.

Distribution. Widespread though not abundant in Denmark. – Common and widespread in Sweden (Sk.–Nb.). – Norway: found in AK, Os, On, Bø, TEi, VAy, Ri, SFi, MRi, STi, Nsy. – East Fennoscandia: common in the southern and central parts, found in Al, Ab, N, Ta, Sa, Sb, Kb, Om, ObS, ObN; Vib, Kr. – England, Scotland, Wales, Ireland, Netherlands, Belgium, France, Italy, Switzerland, Austria, Bulgaria,

Bohemia, Moravia, Slovakia, German D.R. and F.R., Romania, Poland, Estonia, Latvia, Lithuania, n. and m. Russia, Ukraine.

Biology. On *Alnus glutinosa* (Kontkanen, 1938). "Auf Alnus, in den Nordostalpen vorwiegend auf *Alnus incana*" (Wagner & Franz, 1961). "Wirtspflanze: *Alnus*" (Günthart, 1974). "Dominant on alder --- from which it was also reared" (Claridge & Wilson, 1976). Adults in July–October.

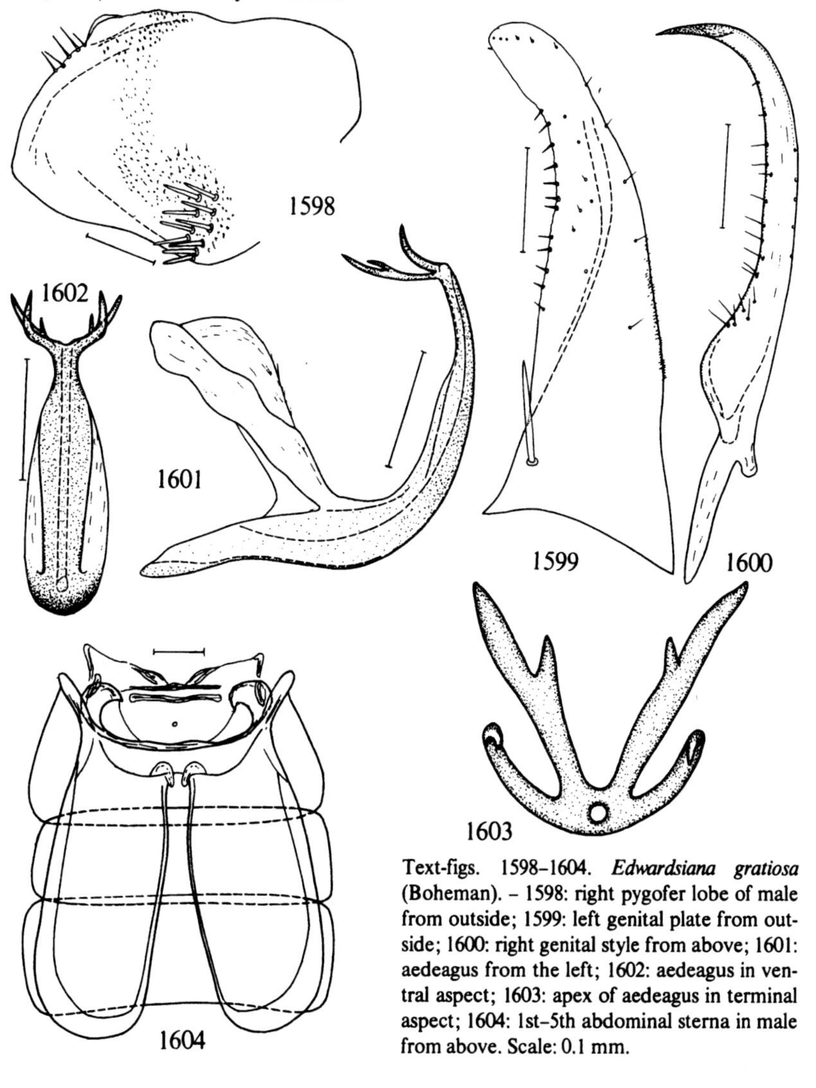

Text-figs. 1598–1604. *Edwardsiana gratiosa* (Boheman). – 1598: right pygofer lobe of male from outside; 1599: left genital plate from outside; 1600: right genital style from above; 1601: aedeagus from the left; 1602: aedeagus in ventral aspect; 1603: apex of aedeagus in terminal aspect; 1604: 1st–5th abdominal sterna in male from above. Scale: 0.1 mm.

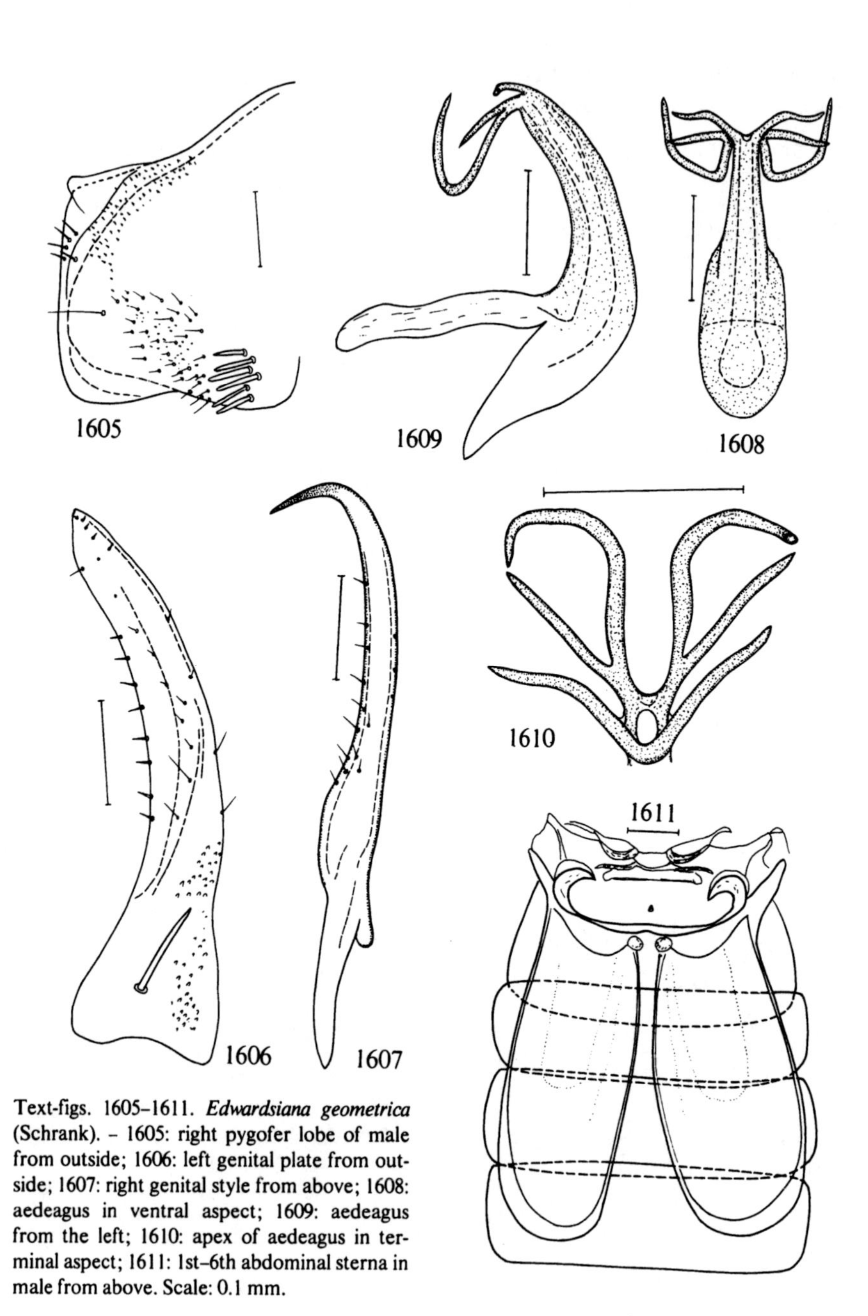

Text-figs. 1605–1611. *Edwardsiana geometrica* (Schrank). – 1605: right pygofer lobe of male from outside; 1606: left genital plate from outside; 1607: right genital style from above; 1608: aedeagus in ventral aspect; 1609: aedeagus from the left; 1610: apex of aedeagus in terminal aspect; 1611: 1st–6th abdominal sterna in male from above. Scale: 0.1 mm.

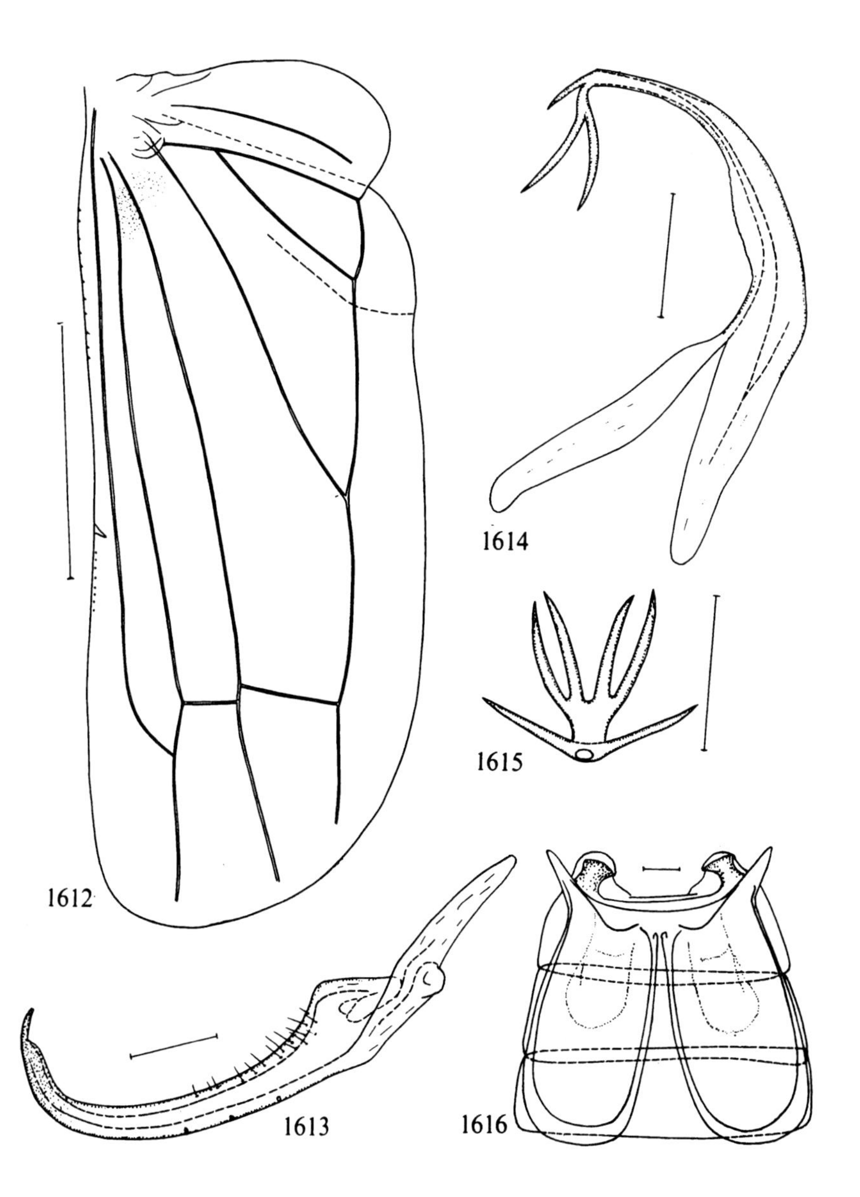

Text-figs. 1612–1616. *Edwardsiana tersa* (Edwards). – 1612: left hind wing of female; 1613: left genital style from above; 1614: aedeagus from the left, 1615: apex of aedeagus in terminal aspect; 1616: 2nd–5th abdominal sterna in male from above. Scale: 1 mm for 1612, 0.1 mm for the rest.

222. ***Edwardsiana tersa*** (Edwards, 1914)
Text-figs. 1612–1616.

Typhlocyba tersa Edwards, 1914: 169.

Whitish, shining. Fore wing with a fumose spot in distal end of cubital cell, also apical cells more or less strongly fumose. Hind wings broad (Text-fig. 1612). Male genital style as in Text-fig. 1613, aedeagus as in Text-figs. 1614, 1615. 2nd–5th abdominal sterna in male as in Text-fig. 1616. Overall length (♂♀) 3.9–4.4 mm.

Distribution. Denmark: known from several places in EJ and NEZ. – Not uncommon in Sweden, found in Sk., Uppl., Vrm., Hls., Ång., Vb., Lu.Lpm. – Norway: found in STi, Nsi, Nnø, TRi. – So far not recorded from East Fennoscandia. – England, France, Netherlands, German D.R. and F.R., Moravia, Poland, Lithuania, Anatolia, Altai Mts.

Biology. On *Salix aurita* (Schwoerbel, 1957). The species is abundant on *Salix viminalis* in Uppsala; in northern Sweden I found it on *Salix lapponum*. Adults in July–September.

223. ***Edwardsiana soror*** (Linnavuori, 1950)
Text-figs. 1617–1623.

Typhlocyba soror Linnavuori, 1950: 185.

White to light yellow, fore wings apically fairly dull, very faintly fumose. Male pygofer lobe as in Text-fig. 1617, genital plate as in Text-fig. 1618, genital style as in Text-fig. 1619, aedeagus as in Text-figs. 1620–1622, 1st–5th abdominal sterna in male as in Text-fig. 1623. Length 3.5–4 mm.

Distribution. Not found in Denmark. – Sweden: found in Jmt.: Åre 3.–5.VIII.1927, 3 ♂♂ (Siefke). – Norway: found in Nsi: Krutå 2.VIII.1961, 1 ♂ (Holgersen); Nnø: Øyjord 10.VIII.1961, 2 ♂♂ (Holgersen). – East Fennoscandia: Sa: Joutseno 20.VII.1950 (Linnavuori); Ks: Kuusamo 1949 (Linnavuori). – Austria, German D.R., Poland, Latvia, Lithuania, Estonia, Mongolia.

Biology. Linnavuori (1950) found the type material on *Alnus incana*. Later he collected some specimens on *Prunus padus*.

224. ***Edwardsiana bergmani*** (Tullgren, 1916)
Text-figs. 1624–1629.

Typhlocyba bergmani Tullgren, 1916a: 65.

Yellowish white, shining. Fore wings light yellow or yellowish white, apical part very faintly fumose, cubital cell with a fumose spot in distal end. Saw-case of female apically black. Male pygofer lobe as in Text-fig. 1624, genital style as in Text-fig. 1625, aedeagus as in Text-figs. 1626–1628, 1st–6th abdominal sterna in male as in Text-fig. 1629. Overall length 3.9–4.5 mm.

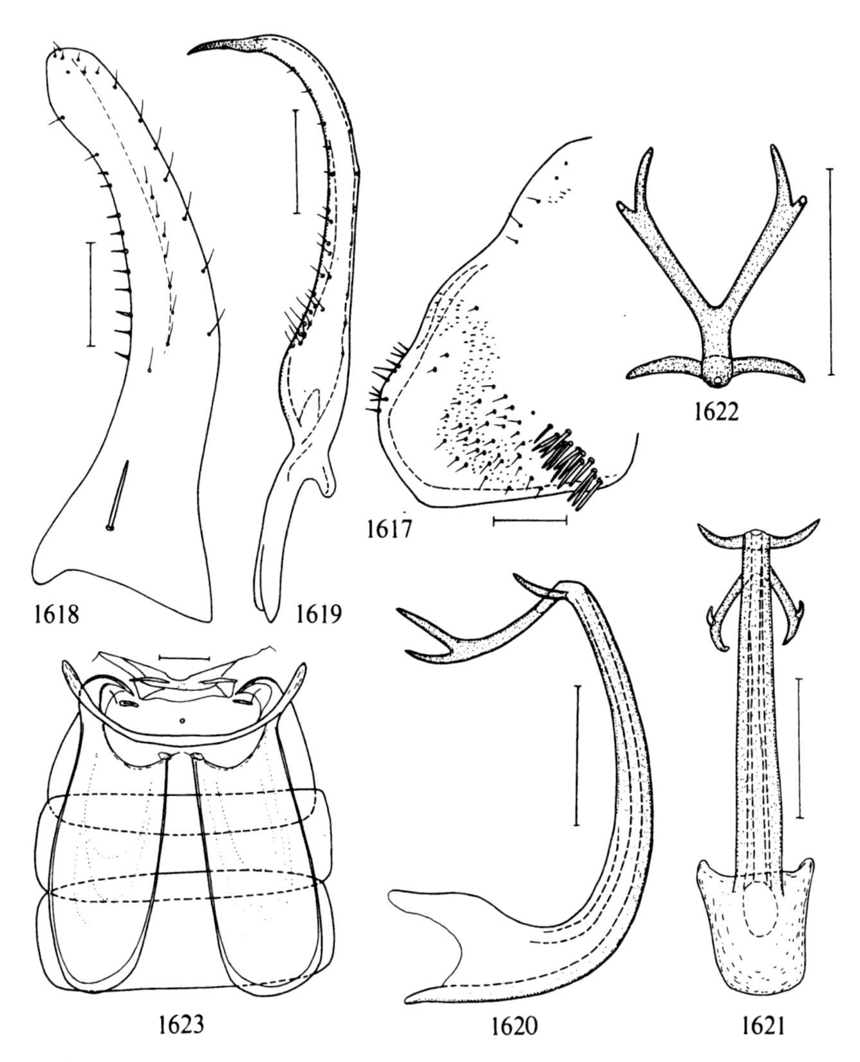

Text-figs. 1617–1623. *Edwardsiana soror* (Linnavuori). – 1617: right pygofer lobe of male from outside; 1618: left genital plate from outside; 1619: right genital style from above; 1620: aedeagus from the left; 1621: aedeagus in ventral aspect; 1622: apex of aedeagus in terminal aspect; 1623: 1st–5th abdominal sterna in male from above. Scale: 0.1 mm.

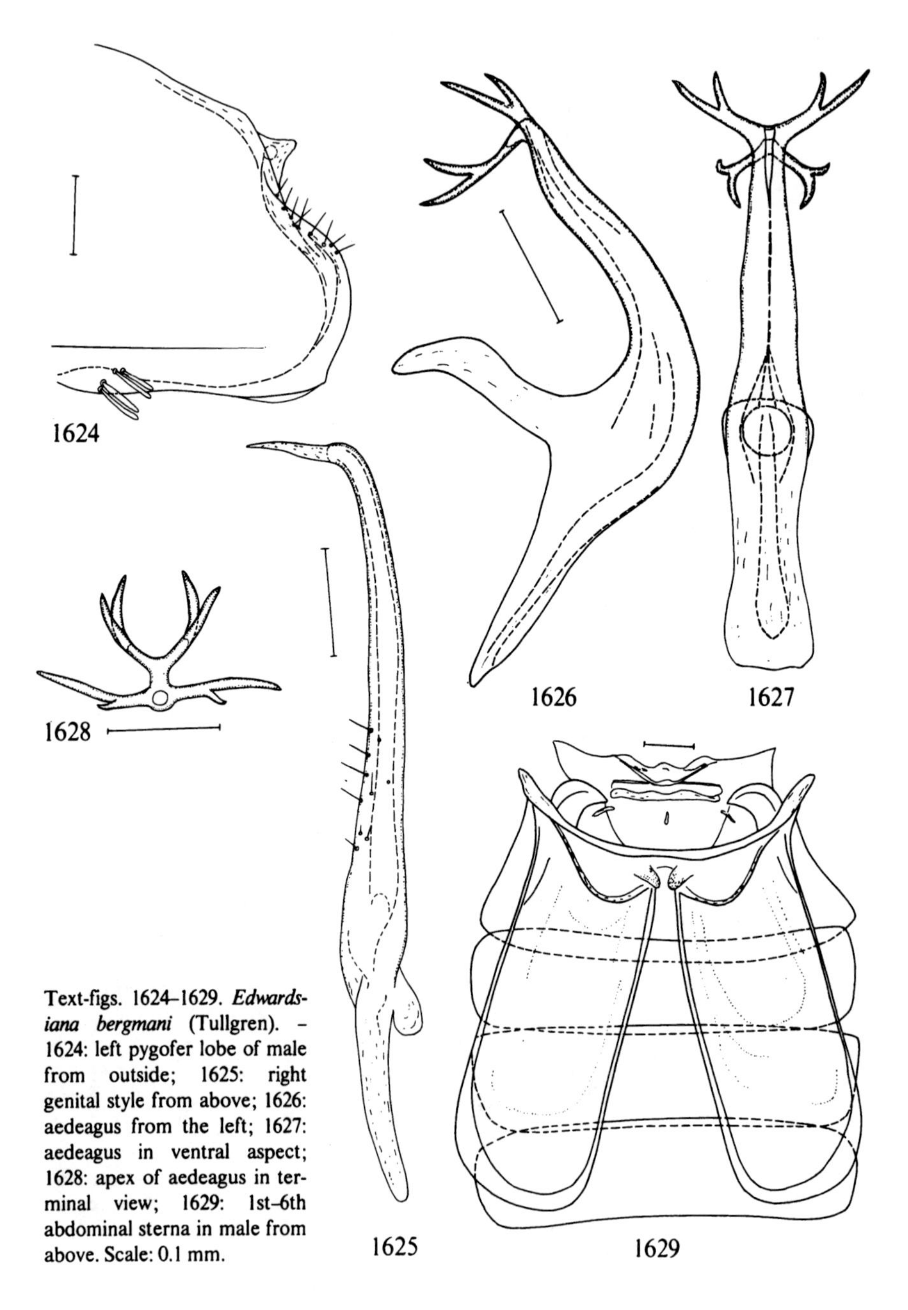

Text-figs. 1624–1629. *Edwardsiana bergmani* (Tullgren). – 1624: left pygofer lobe of male from outside; 1625: right genital style from above; 1626: aedeagus from the left; 1627: aedeagus in ventral aspect; 1628: apex of aedeagus in terminal view; 1629: 1st–6th abdominal sterna in male from above. Scale: 0.1 mm.

Distribution. Denmark: so far found only in EJ: Pøt Mølle 1.VII.1965 (Trolle) and Moesgaard 4.VIII.1966 (Trolle). – Rare in southern Sweden, more common in the north, found in Bl., Nrk., Hls., Hrj., Jmt., Vb., P.Lpm., Lu. Lpm., T.Lpm. – Norway: first found in tremendous masses ("milliards", "litres") in TRi: Målselv, between Rundhaug and Nordmo, in the first week of August, 1915, by A. Bergman. Later found

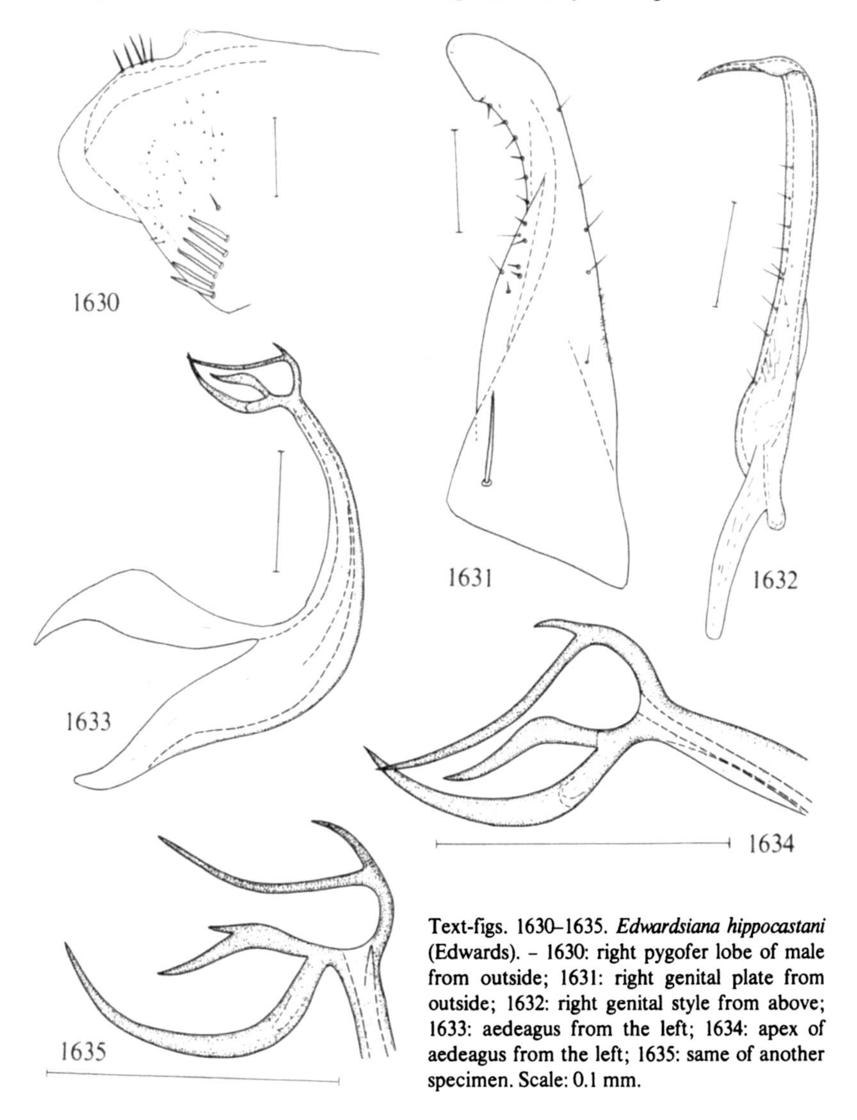

Text-figs. 1630–1635. *Edwardsiana hippocastani* (Edwards). – 1630: right pygofer lobe of male from outside; 1631: right genital plate from outside; 1632: right genital style from above; 1633: aedeagus from the left; 1634: apex of aedeagus from the left; 1635: same of another specimen. Scale: 0.1 mm.

in AK, Os, SFi, STi, and Nnø. – East Fennoscandia: widespread, found in Al, Ab, N, Ta, Sa, Oa, Sb, Kb, Om, LkW. – England, France, German D.R. and F.R., Switzerland, Austria, Bulgaria, Bohemia, Poland, Latvia, Estonia, Mongolia. – In the Nearctic region represented by subspecies *ariadne* (Mc Atee, 1926).

Biology. "Auf *Alnus*-Arten; in den Alpen an *Alnus viridis* bis in hochsubalpine Lagen emporsteigend" (Wagner & Franz, 1961). On birch and alder (Linnavuori, 1969). "Taken in relatively small numbers from birch and, to a lesser extent, from alder. Breeding records and previous sampling suggests a close association with birch" (Claridge & Wilson, 1976). Adults in June–September.

225. ***Edwardsiana hippocastani*** (Edwards, 1888)
Text-figs. 1630–1638.

Typhlocyba hippocastani Edwards, 1888b: 157.

Yellowish white to light yellow, shining. Fore wings often fairly lively yellow with an

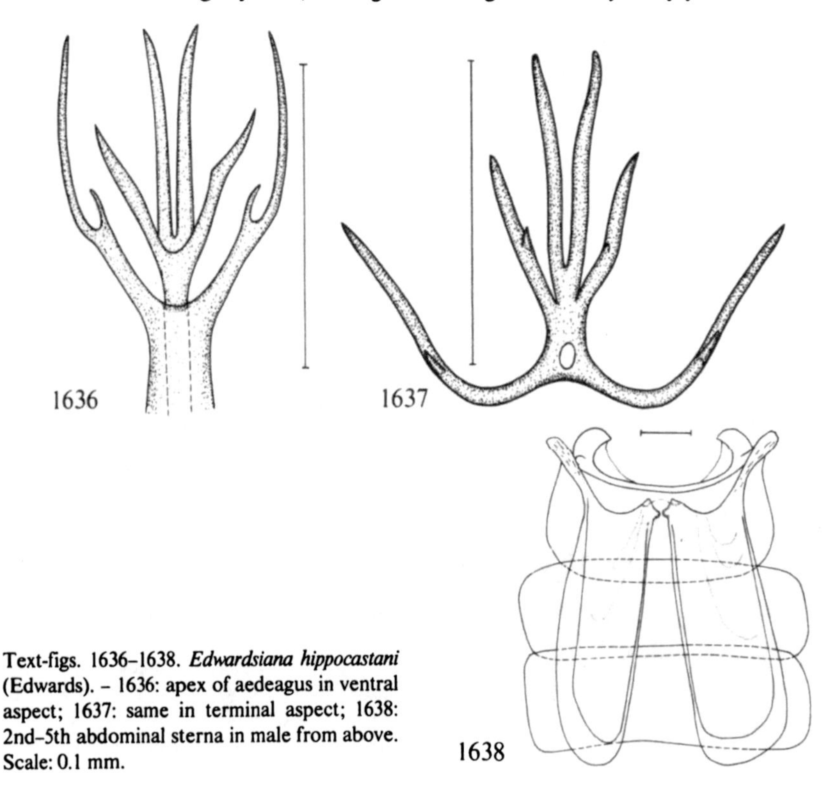

Text-figs. 1636–1638. *Edwardsiana hippocastani* (Edwards). – 1636: apex of aedeagus in ventral aspect; 1637: same in terminal aspect; 1638: 2nd–5th abdominal sterna in male from above. Scale: 0.1 mm.

orange tinge, apical part at most very faintly fumose. Male pygofer lobe as in Text-fig. 1630, genital plate as in Text-fig. 1631, genital style as in Text-fig. 1632, aedeagus as in Text-figs. 1633–1637, 2nd–5th abdominal sterna in male as in Text-fig. 1638. Length 3.4–3.9 mm.

Distribution. Common and widespread in Denmark. – Common in southern Sweden, Sk.–Dlr. – Norway: found in AK: Sem, Asker 2.X., 17.X.1973 (Taksdal), and in AAy: Risør 27.VIII.1905 (Warloe). – East Fennoscandia: Ab: Åbo (Linnavuori), Pojo

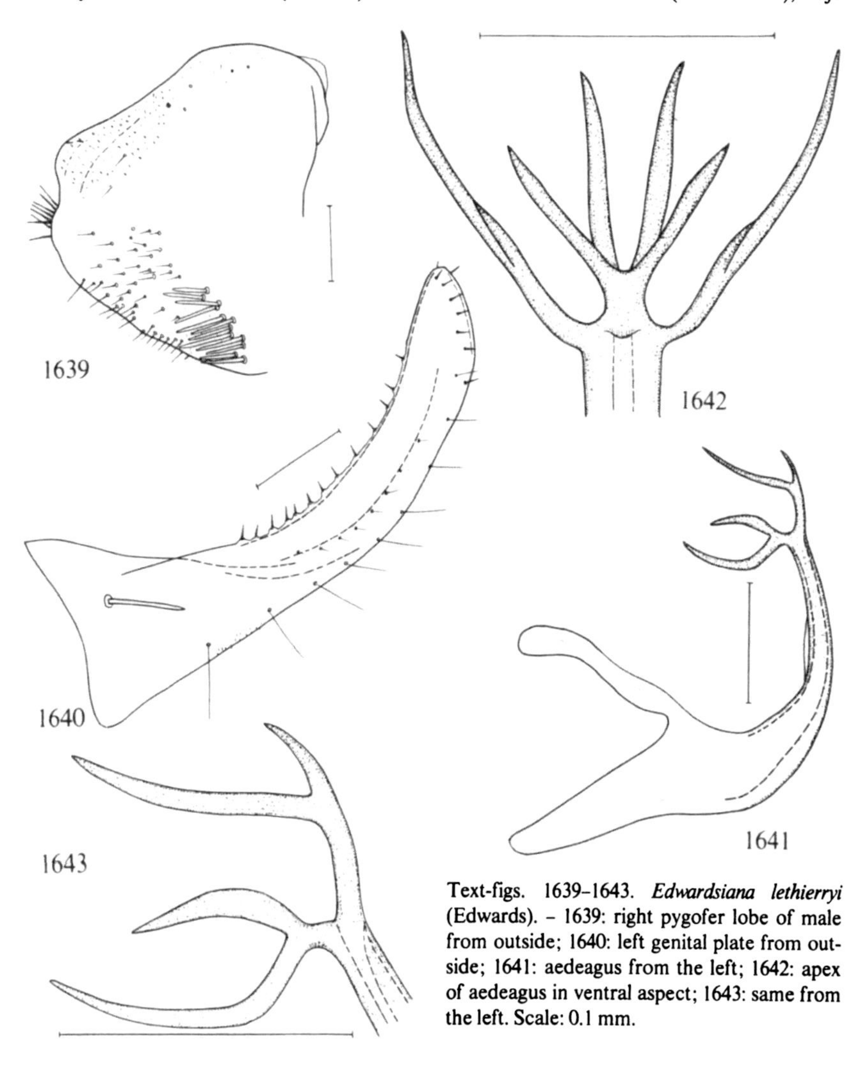

Text-figs. 1639–1643. *Edwardsiana lethierryi* (Edwards). – 1639: right pygofer lobe of male from outside; 1640: left genital plate from outside; 1641: aedeagus from the left; 1642: apex of aedeagus in ventral aspect; 1643: same from the left. Scale: 0.1 mm.

22.VI.1976 (Albrecht); N: Helsingfors 18.IX.1975 (Albrecht). – England, France, Belgium, Netherlands, Switzerland, Austria, German D.R. and F.R., Bohemia, Moravia, Slovakia, Poland, Lithuania, Moldavia. – Nearctic region.

Biology. "Auf verschiedenen Laubhölzern" (Wagner & Franz, 1961). "Sur le Marronier, le Charme, le Noisetier et surtout l'Orme" (Ribaut, 1936b). Reared on *Ulmus* and *Corylus*, also found on *Juglans* and *Alnus* (Günthart, 1974). "The most abundant species of *Edwardsiana* in our samples, with large numbers on elm – – –, alder – – – and hazel – – – (from which it has also been reared), and fewer on sycamore, rowan, birch and oak" (Claridge & Wilson, 1976). Adults in June–October, probably two generations.

226. ***Edwardsiana lethierryi*** (Edwards, 1881)
Text-figs. 1639–1643.

Typhlocyba lethierryi Edwards, 1881: 224.
Typhlocyba lethierryi plurispinosa W. Wagner, 1935a: 32.

Usually pale yellow, sometimes light yellow. Apical part of fore wing, sometimes also distal ends of disc cells faintly fumose. Male pygofer lobe as in Text-fig. 1639, genital plate as in Text-fig. 1640, aedeagus as in Text-figs. 1641–1643. Upper branch of anterior appendage of aedeagus sometimes with an additional branch (f. *plurispinosa* W. Wagner). Length 3.3–4.05 mm.

Distribution. So far not found in Denmark, nor in East Fennoscandia. – Sweden: only found in Gtl.: Lojsta 17.VII.1956 (one male, f. *plurispinosa*) by O. Lundblad. – Norway: one male found in AK: Asker, Sem 7.XI.1975 by Baeschlin. – Widespread in Europe, also in Algeria, Morocco, and Georgia.

Biology. "An Ahorn und Ulmen" (Kuntze, 1937). Reared on *Acer campestre* by Günthart (1971). "Predominantly found on maple" (Claridge & Wilson, 1976).

Genus *Eupterycyba* Dlabola, 1958

Typhlocyba, subgenus *Eupterycyba* Dlabola, 1958: 55.
Type-species: *Typhlocyba jucunda* Herrich-Schäffer, 1837, by original designation.

"Very remarkable for the black pattern on fore body. Head broad, a little narrower than pronotum. Vertex in the middle only slightly prolonged, shorter than 1/2 of pronotum, fore margin rounded as in *Eupteryx* or *Ribautiana*, hind margin nearly parallel, vertex much higher than pronotum, difference about 1 mm. Style curved regularly, in two thirds with a torsion, there with triangular broadening, as in *Eupteryx*, *Ribautiana* etc." (Dlabola, l.c.). One species.

227. ***Eupterycyba jucunda*** (Herrich-Schäffer, 1837)
Plate-fig. 90, text-figs. 1644–1651.

Typhlocyba jucunda Herrich-Schäffer, 1837: 16.
Typhlocyba zetterstedti Boheman, 1845a: 47.

Yellowish green, shining. Vertex broad, twice as wide as long, with two large rounded black spots, and with a small pointed black spot on middle of hind border. Pronotum with three large black spots, viz. one median longitudinal oval, and a lateral rounded or

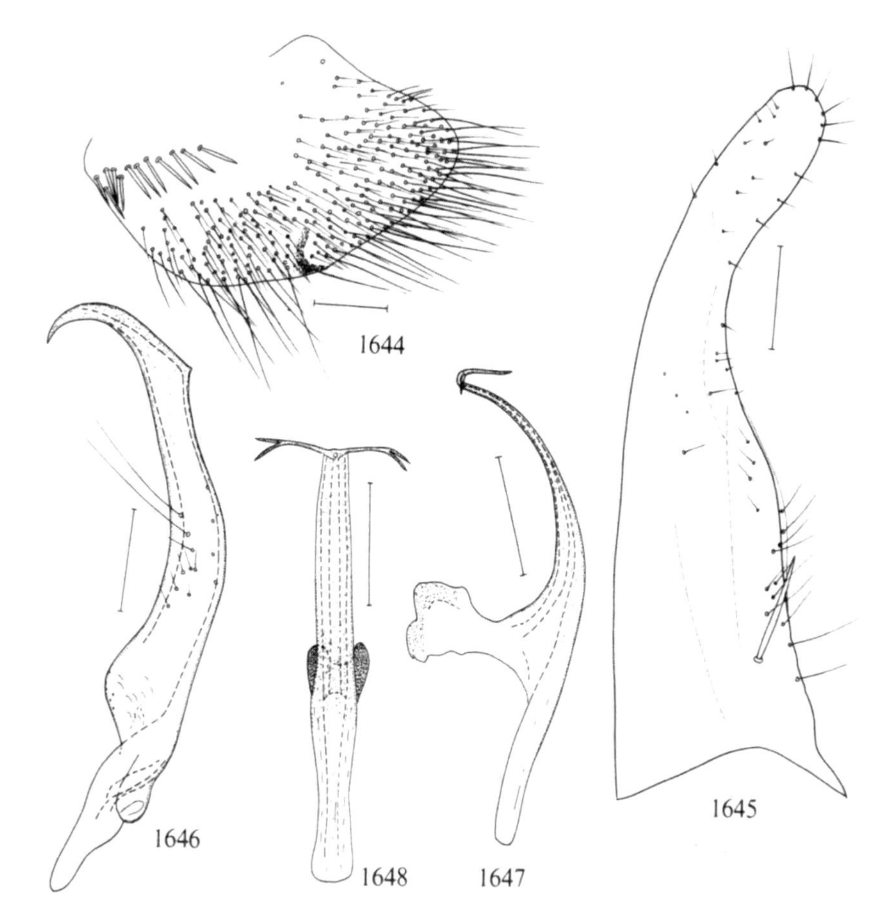

Text-figs. 1644–1648. *Eupterycyba jucunda* (Herrich-Schäffer). - 1644: left pygofer lobe of male from outside; 1645: right genital plate from outside; 1646: right genital style from above; 1647: aedeagus from the left; 1648: aedeagus in ventral aspect. Scale: 0.1 mm.

transversely oval pair, usually also with some smaller black spots near fore border. Scutellum with a black spot near each lateral corner, between these with a clamp-shaped black spot. Major part of clavus, a longitudinal band in median cell, and apical parts of cubital and radial cells black-brown, apical cells strongly fumose. Veins of fore wing yellow. Legs light, abdomen largely black, ventrally transversely striped. Male pygofer lobe as in Text-fig. 1644, genital plate as in Text-fig. 1645, genital style as in Text-fig. 1646, aedeagus as in Text-figs. 1647–1650. 1st–4th abdominal sterna in male as in Text-fig. 1651. Overall length (♂♀) 4–4.5 mm. – Nymphs yellow with varying dark brown markings; anterior margin of vertex with one pair of widely separated macrosetae, between these smoothly concave; notum, wing pads, and abdomen without long macrosetae, abdomen laterally with small spines.

Distribution. Scarce in Denmark, so far found in several places in EJ and NEZ. – Scarce in Sweden, found in Sk., Bl., Öl., Ög., Boh., Sdm., Upl. – Not found in Norway and East Fennoscandia. – England, Wales, France, Belgium, Netherlands, German D.R. and F.R., Austria, Switzerland, Spain, Italy, Bulgaria, Bohemia, Moravia, Slovakia, Hungary, Yugoslavia, Romania, Poland, Estonia, Latvia, Lithuania, Ukraine.

Biology. "An Erlen" (Kuntze, 1937). On *Alnus* (Wagner & Franz, 1961). "Wirtspflanze: *Alnus*" (Günthart, 1974). "Predominantly found on alder – – –, from which it was also reared" (Claridge & Wilson, 1976). Adults in July–September.

Genus *Linnavuoriana* Dlabola, 1958

Linnavuoriana Dlabola, 1958: 54.

Type-species: *Cicada decempunctata* Fallén, 1806, by original designation.

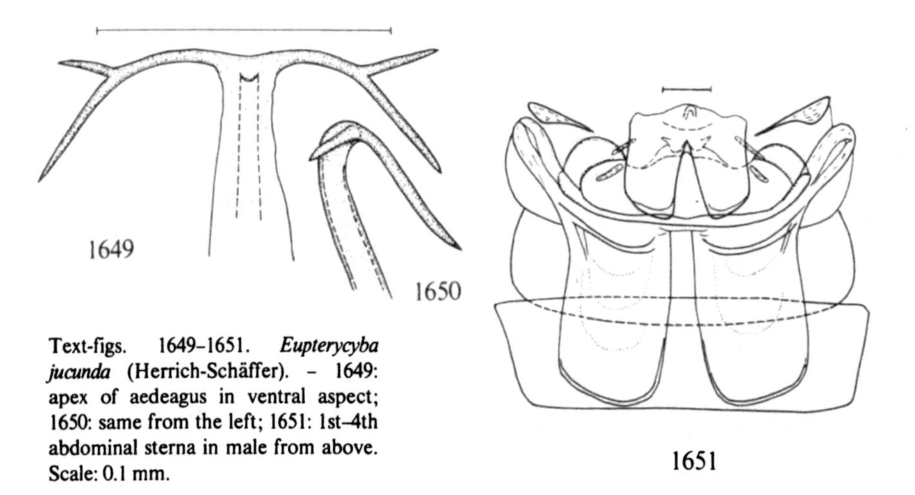

Text-figs. 1649–1651. *Eupterycyba jucunda* (Herrich-Schäffer). – 1649: apex of aedeagus in ventral aspect; 1650: same from the left; 1651: 1st–4th abdominal sterna in male from above. Scale: 0.1 mm.

"Vertex about 1/3 of pronotum-length, fore margin angularly rounded, hind margin rounded. Head narrower than the pronotum. Pygofore rounded, with varying pilosity. Genital plate subapically strongly emarginated, obliquely curved and rounded on the tip. Anal tube slightly longer than the pygofore. Style scarcely curved, sharpened to the apex and with a long spine under an angle of 90°, subapically situated. Aedeagus with developed shaft, lacking apical processes and other appendices on the dorsal apodeme and preatrium. Orificium apical. Hind wings: both cross veins widely separated. Differences from other genera: genital plates without macrosetae, pilosity on pygofore, distinctness of style and aedeagus" (Dlabola, l.c.). In Europe two species.

Key to species and subspecies of *Linnavuoriana*

1 Face entirely light. Pronotum with four black spots (Plate-fig. 93). Aedeagus dorsally near base of shaft on each side with a low laminate lobe (Text-fig. 1657) .... 228. *sexmaculata* (Hardy)
– Face laterally darker. Pronotum with six black spots (Plate-fig.92). Aedeagus different 2
2 (1) Shaft of aedeagus dorsally near base on each side with an elongate lobe (Text-figs. 1660, 1661). On *Betula*. 229. *decempunctata* (Fallén), s.str.
– Shaft of aedeagus dorsally near base on each side with a more or less triangular lobe (Text-fig. 1663). On *Alnus incana*
229a. *decempunctata* (Fallén), ssp. *intercedens* (Linnavuori)

228. ***Linnavuoriana sexmaculata*** (Hardy, 1850)
Plate-fig. 93, text-figs. 1652–1658.

Cicada sexpunctata Fallén, 1826: 51 (Primary homonym).
Typhlocyba sexmaculata Hardy, 1850: 421.

Yellowish white or almost white, often with a greenish tinge. Face unicolorous light, vertex frontally with two rounded black spots. Pronotum with 4 black spots, the aterior pair usually smaller, situated near fore border behind the spots of vertex, the posterior pair situated near posterior side-corners of pronotum. In strongly pigmented specimens these spots may coalesce into two oblique longitudinal bands. Scutellum with a triangular black spot near each lateral corner. Fore wing usually with two black-brown transverse bands, one proximally and one distally of middle. These bands may be disintegrated into spots. Often with a dark spot in the proximal part of clavus. Apical cells of fore wing fumose, distal ends of apical veins fuscous or enclosed in fuscous spots. Venter largely black-brown, abdomen transversely striped. Male pygofer lobe as in Text-fig. 1652, genital plate as in Text-fig. 1653, genital style as in Text-fig. 1654, aedeagus as in Text-figs. 1655–1657, 1st–5th abdominal sterna in male as in Text-fig. 1658. Length (♂♀) 3.3–4 mm. – Last instar larvae whitish yellow with dark brown markings, head, notum, wing pads and abdomen with many macrosetae varying in length; the setae may be black when arising in dark spots.

Distribution. Scarce in Denmark, found in EJ, NEZ, and B. - Common and widespread in Sweden, found Sk.-Lu. Lpm. - Fairly common in southern and central Norway, found in Ø, AK, HEs, HEn, VE, TEi, AAy, VAy, Ry, HOy, HOi, STi. - Common in the major part of East Fennoscandia, found in Al, Ab, N, Ta, Oa, Tb, Sb, Kb;

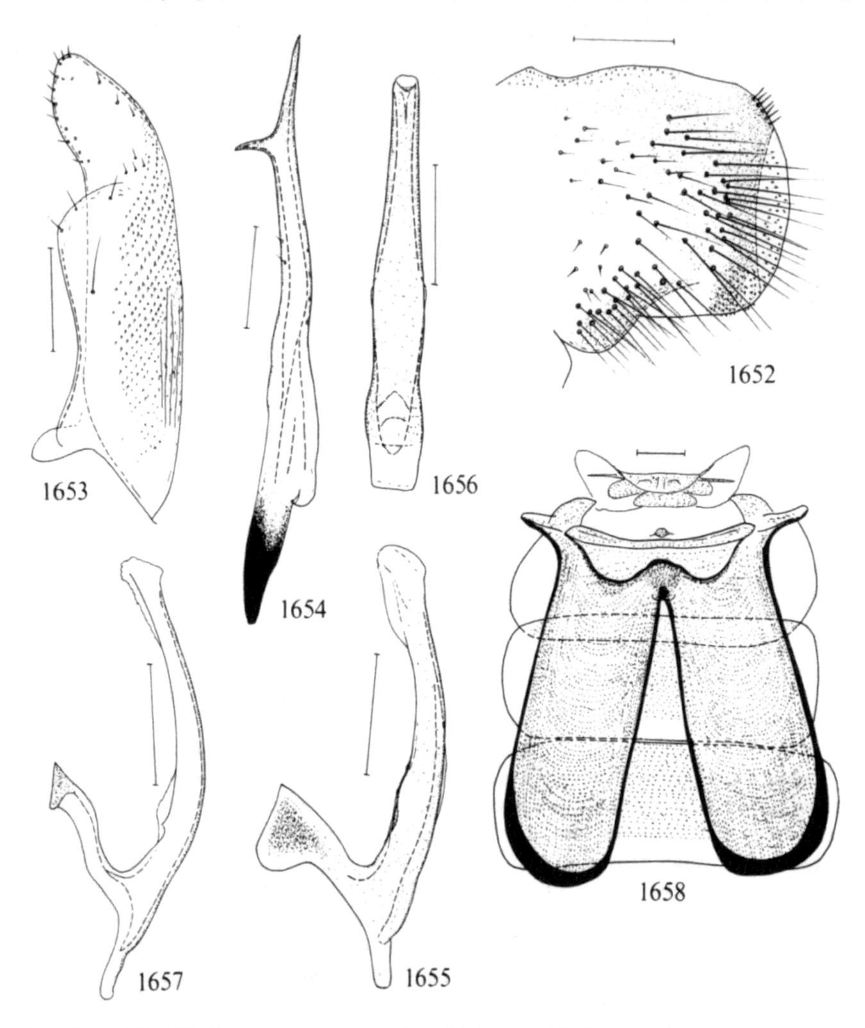

Text-figs. 1652–1658. *Linnavuoriana sexmaculata* (Hardy). - 1652: left pygofer lobe of male from outside; 1653: left genital plate from outside; 1654: right genital style from above; 1655: aedeagus from the left; 1656: aedeagus in ventral aspect; 1657: aedeagus of another specimen from the left; 1658: 1st–5th abdominal sterna in male from above. Scale: 0.1 mm.

Vib, Kr, Lr. – Widespread in Europe, also found in Anatolia, Altai Mts., Kazakhstan, Kurile Isl., m. Siberia, Maritime Territory, Japan, Nearctic region.

Biology. "An Weiden, meist *Salix caprea*" (Kuntze, 1937). "Auf *Salix aurita*" (Schwoerbel, 1957). "Wirtspflanze: *Salix*" (Günthart, 1974). I collected *L. sexmaculata* on *Salix caprea, aurita, viminalis,* and *myrsinifolia.* Also found on conifers in April and September, so hibernation probably takes place in the adult stage.

229. ***Linnavuoriana decempunctata*** (Fallén, 1806)
Plate-fig. 92, text-figs. 1659–1663.

Cicada decempunctata Fallén, 1806: 41.
Typhlocyba betulicola Edwards, 1925: 64.
Typhlocyba decempunctata intercedens Linnavuori, 1949b: 152.

Resembling *sexmaculata,* often with a reddish tinge. Face usually with more or less extended brownish areas. Pronotum with a pair of small dark spots at fore border behind

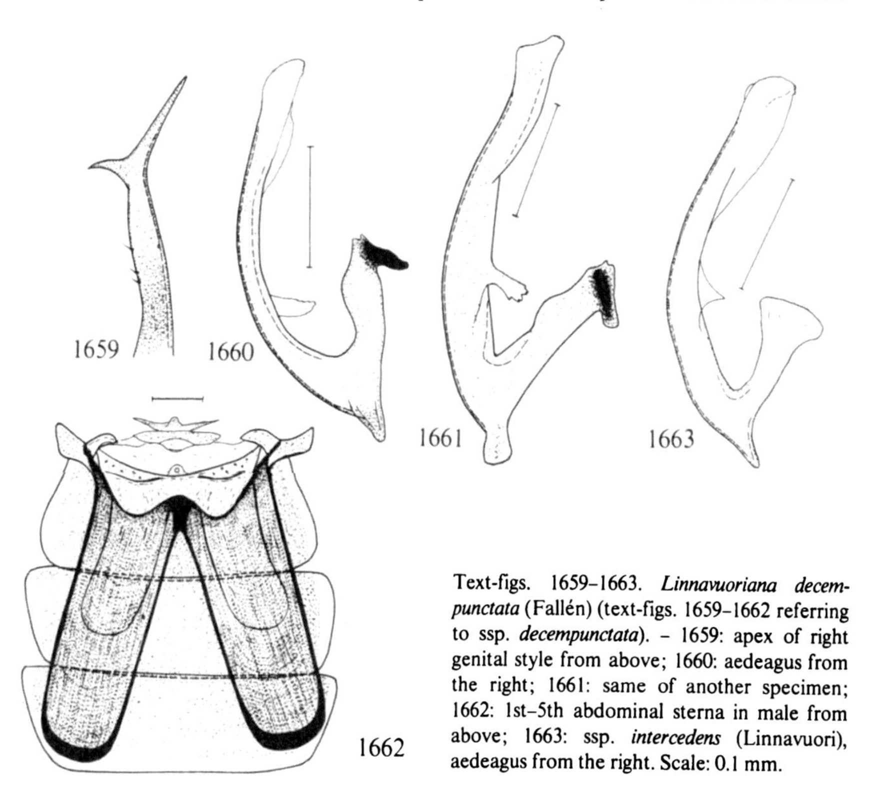

Text-figs. 1659–1663. *Linnavuoriana decempunctata* (Fallén) (text-figs. 1659–1662 referring to ssp. *decempunctata*). – 1659: apex of right genital style from above; 1660: aedeagus from the right; 1661: same of another specimen; 1662: 1st–5th abdominal sterna in male from above; 1663: ssp. *intercedens* (Linnavuori), aedeagus from the right. Scale: 0.1 mm.

eyes; an addition to those present in *sexmaculata.* In strongly pigmented specimens the spots of each side may coalesce forming a pair of angular longitudinal bands. Sometimes the pronotum is largely dark. Scutellum with a black spot on each lateral corner. Markings of fore wings as in *sexmaculata,* more often with a tendency of coalescing. Legs sometimes partly dark. Apex of male genital style as in Text-fig. 1659, aedeagus of *decempunctata* s.str. as in Text-figs. 1660, 1661, aedeagus of ssp. *intercedens* as in Text-fig. 1663, 1st–5th abdominal sterna in male as in Text-fig. 1662. Overall length (♂♀) 3.4–3.7 mm. - Last instar larva as in *sexmaculata.*

Distribution. Scarce in Denmark, found in EJ and NEZ. - Not found in Scania, for the rest common in Sweden, Bl.–T. Lpm. - Norway: found in HEs, HEn, HOi, SFi, STi, Fn. - East Fennoscandia: comparatively scarce and sporadic in southern and central parts, found in Ab, N, Ta, Sa, Oa, Kb. - England, Scotland, France, Netherlands, German D.R. and F.R., Switzerland, Austria, Bohemia, Slovakia, Romania, n. Italy, Poland, Estonia, Latvia, m. Russia, Mongolia, Maritime Territory.

Distribution of ssp. *intercedens.* Not found in Denmark. - Sweden: so far found in Dlr.: Leksand, Tällberg 19.VI.1949, 2 ♂♂ (Tord Tjeder); Hls.: Edsbyn, Räken 2.IX.-1975, 1 ♂ (Bo Henriksson); Ång.: Gudmunrå 23.VIII.1944, 1 ♂ (E. Runquist). - Norway: HEs: Flisa 30.VIII.1961, 1 ♂ (Holgersen); HEn: Snippen, Rena 28.VIII.1961, 1 ♂, and Åsk. 27.VIII.1961 (Holgersen); VAy: Øye, Kvinesdal, 1 ♂ (Holgersen). - East Fennoscandia: comparatively common in the southern parts, found in Al, Ab, N, Ta, Sa; Kr. - Latvia, Moravia.

Biology. On *Betula* spp. (*decempunctata* s.str.) or *Alnus incana* (ssp. *intercedens*). Adults in March–December, hibernation takes place in the adult stage.

Genus *Ribautiana* Zachvatkin, 1947

Ribautiana Zachvatkin, 1947: 113.

Type-species: *Cicada ulmi* Linné, 1758, by original designation.

Fore wings with dark markings at apices of apical veins or at least of 1st and 3rd apical vein. Male genital plate with one long macroseta near base and two shorter macrosetae near middle. Male pygofer without processes. Genital style on inner margin with an angular protuberance. Aedeagus with a pair of processes arising from socle and often with preapical or apical processes on shaft. Last instar larva with long macrosetae on head, notum, wing pads and abdomen. In Denmark and Fennoscandia three species.

Key to species of *Ribautiana*

Adults

1 Caudal apex of scutellum with a black spot .... 230. *ulmi* (Linné)
– Caudal apex of scutellum concolorous 2
2 (1) Overall length > 3.5 mm. Basal processes of aedeagus shorter than shaft (Text-fig. 1667). On *Ulmus* .... 230. *ulmi* (Linné)

- Overall length < 3.5 mm. Basal processes of aedeagus longer than shaft (Text-figs. 1674, 1679) 3

3 (2) Aedeagus as in Text-figs. 1674, 1675. On *Rubus* spp. .... 231. *tenerrima* (Herrich-Schäffer)

- Aedeagus as in Text-figs. 1679, 1680. On *Quercus* spp. .... 232. *scalaris* (Ribaut).

Nymphs (From Wilson, 1978)

1 Pronotum with anterior median spines present. Abdominal segments 5–8 with laterally directed spines arising medially in addition to median spines. On *Rubus* and *Quercus* spp. 2

- Pronotum with anterior median spines absent. Abdominal segments 5–8 with only one pair of medially arising macrosetae. On *Ulmus* .... *230. ulmi* (Linné)

2 (1) On *Quercus* species .... 232. *scalaris* (Ribaut)

- On *Rubus* species. Also on trees in second generation .... 321. *tenerrima* (Herrich-Schäffer).

230. ***Ribautiana ulmi*** (Linné, 1758)
Plate-fig. 91, text-figs. 1664–1670.

Cicada ulmi Linné, 1758: 439.

Yellowish, shining. Fore body light with or without some black-brown spots not always coexistent: two spots on fore margin of vertex, one single spot on fore border of pronotum, and one on caudal apex of scutellum. Fore wing proximally of transverse veins with a transverse band interrupted by the light longitudinal veins; distal ends of apical veins black; apical part of fore wing more or less strongly fumose. Abdomen above largely black, ventrally largely light. Male pygofer lobe as in Text-fig. 1664, genital plate as in Text-fig. 1665, genital style as in Text-fig. 1666, aedeagus as in Text-figs. 1667, 1668, 1st–6th abdominal sterna as in Text-fig. 1669, 2nd abdominal sternum in frontal aspect as in Text-fig. 1670. Overall length 3.7–4 mm. Last instar larvae entirely pale whitish-yellow or with pigmentation much varying in extent, macrosetae often brownish.

Distribution. Common and widespread in Denmark as well as in southern and central Sweden (Sk.–Ång.). – Norway: found in several places in AK, also found in HEn and HOi. – Common in southern East Fennoscandia, found in Al, Ab, N, Ka, Ta, Sa, Oa; Vib. – Widespread in Europe, recently found even in Iceland; introduced into North America.

Biology. Collected and reared on *Ulmus* (Günthart, 1974; Claridge & Wilson, 1976). Two generations in Sweden, hibernation takes place in the egg stage (Ossiannilsson, 1946c). Females of the 2nd generation deposit their eggs in the bark of young twigs of the host-plant where they hibernate. Eggs of first generation females are placed in leaf-stalks and thicker leaf-veins.

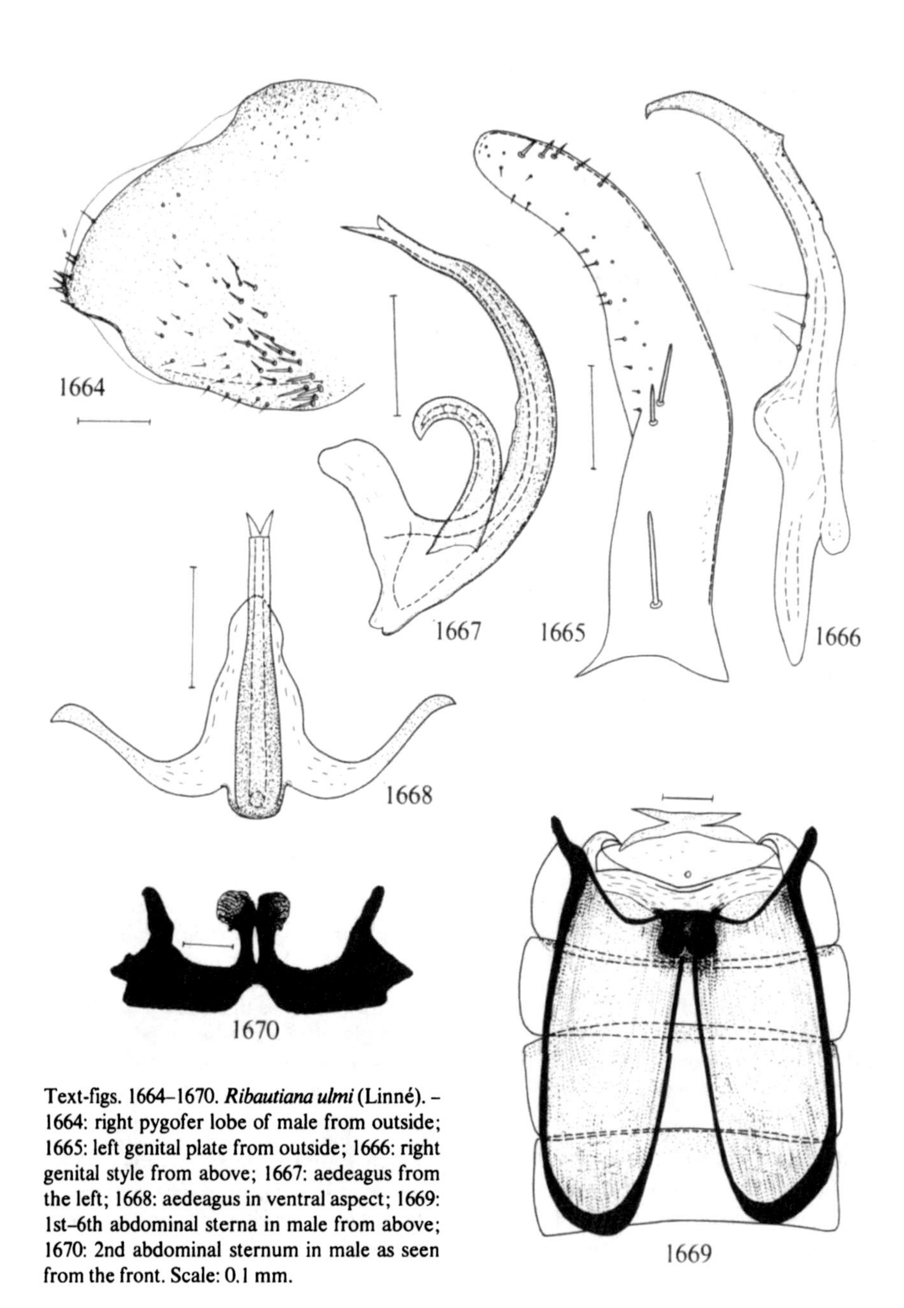

Text-figs. 1664–1670. *Ribautiana ulmi* (Linné). – 1664: right pygofer lobe of male from outside; 1665: left genital plate from outside; 1666: right genital style from above; 1667: aedeagus from the left; 1668: aedeagus in ventral aspect; 1669: 1st–6th abdominal sterna in male from above; 1670: 2nd abdominal sternum in male as seen from the front. Scale: 0.1 mm.

231. ***Ribautiana tenerrima*** (Herrich-Schäffer, 1834)
Plate-fig. 95, text-figs. 1671–1676.

Typhlocyba tenerrima Herrich-Schäffer, 1834b: 10.
Typhlocyba misella Boheman, 1852b: 122.

Smaller and more delicate than *ulmi.* Head more pointed. Fore body usually without dark markings. Fore wings (Plate-fig. 95) with a hyaline streak along claval suture, for

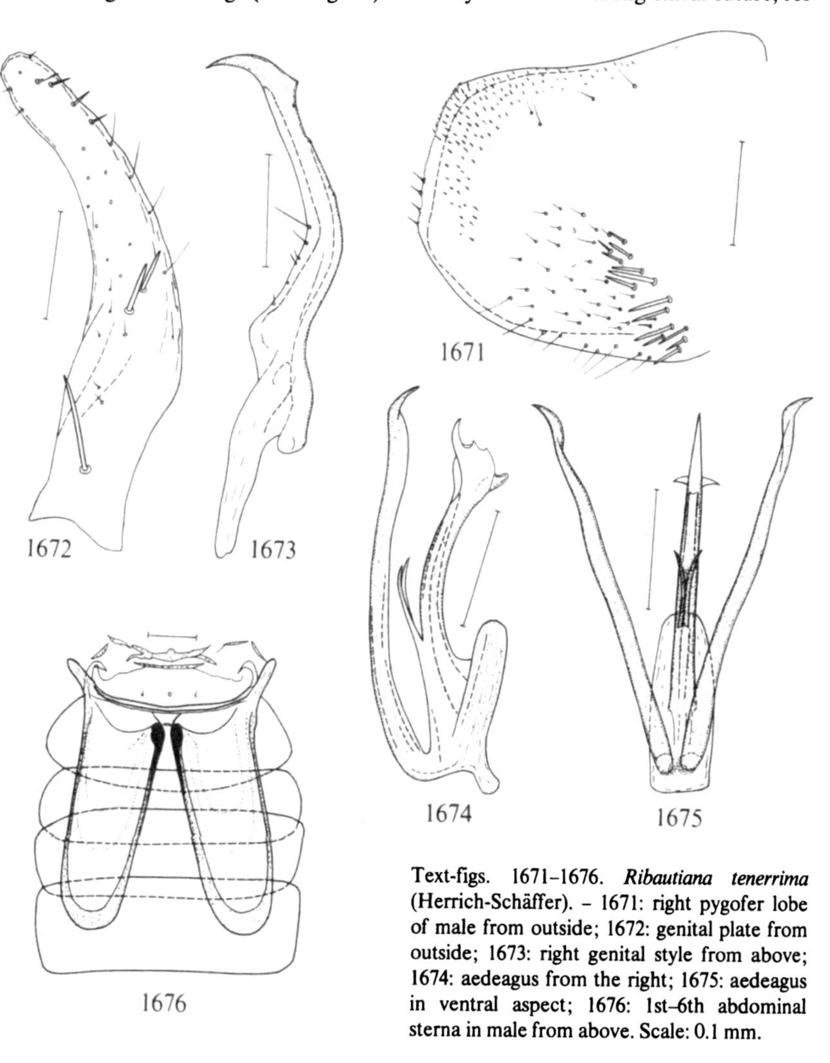

Text-figs. 1671–1676. *Ribautiana tenerrima* (Herrich-Schäffer). – 1671: right pygofer lobe of male from outside; 1672: genital plate from outside; 1673: right genital style from above; 1674: aedeagus from the right; 1675: aedeagus in ventral aspect; 1676: 1st–6th abdominal sterna in male from above. Scale: 0.1 mm.

the rest more or less as in *ulmi*. Venter light, abdomen dorsally with or without a black spot varying in size. Male pygofer lobe as in Text-fig. 1671, genital plate as in Text-fig. 1672, genital style as in Text-fig. 1673, aedeagus as in Text-figs. 1674, 1675, 1st–6th abdominal sterna in male as in Text-fig. 1676. Overall length 2.7–3.5 mm. – Last instar larvae usually pale whitish yellow with brown bases to dorsal macrosetae.

Distribution. Probably quite common in Denmark, so far found in EJ, F, and B. – Scarce in southern Sweden, found in Sk., Bl., Öl., Gtl. – Not found in Norway, nor in East Fennoscandia. – Widespread in Europe, also found in Israel, Iran, Georgia, Azerbaijan; introduced into North America and N. Zealand.

Biology. "An Himbeeren und Brombeeren" (Kuntze, 1937). On various trees and bushes, preferably oak (Ribaut, 1936b). "Auf *Salix* und *Rubus*" (Wagner & Franz, 1961). "Collected in numbers from alder, hazel and oak. It was hand-collected very commonly on brambles (*Rubus* species), form which it was also reared and which are undoubtedly its major host plants" (Claridge & Wilson, 1976). Hibernation takes place in the egg stage (Müller, 1957). Adults in July–October.

Economic importance. Recorded as a pest on loganberries in Canada and N. Zealand; for details consult Müller (1956).

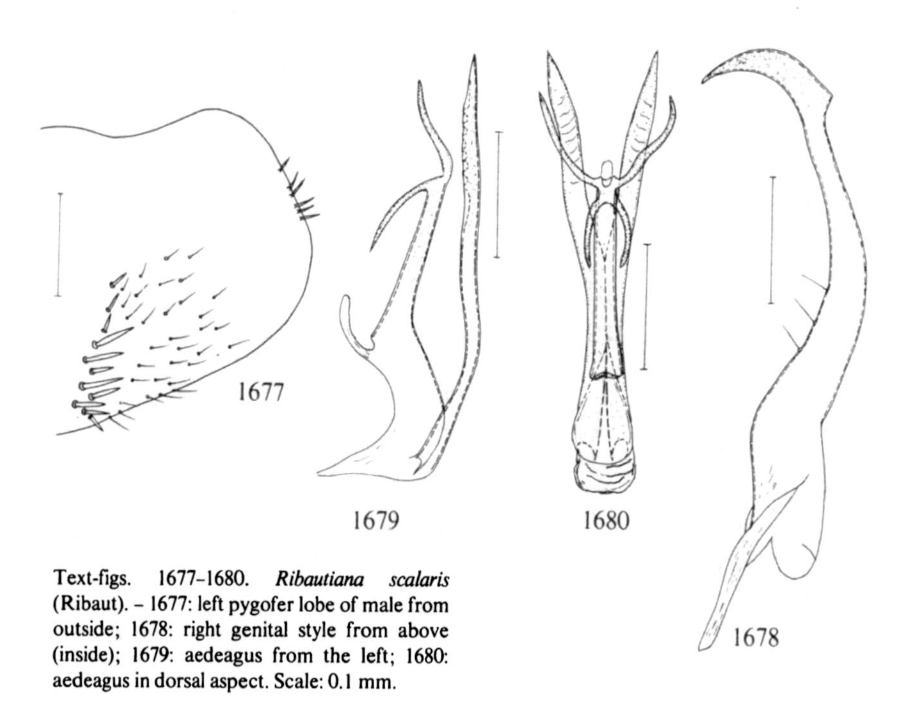

Text-figs. 1677–1680. *Ribautiana scalaris* (Ribaut). – 1677: left pygofer lobe of male from outside; 1678: right genital style from above (inside); 1679: aedeagus from the left; 1680: aedeagus in dorsal aspect. Scale: 0.1 mm.

232. ***Ribautiana scalaris*** (Ribaut, 1931)
Text-figs. 1677–1680.

Typhlocyba scalaris Ribaut, 1931b: 281.

Resembling *tenerrima,* head less pointed. Macrosetae of upper lateral row on hind tibiae arising from black dots. Male pygofer lobe as in Text-fig. 1677, genital style as in Text-fig. 1678, aedeagus as in Text-figs. 1679, 1680. Overall length (♂♀) 3.25–3.65 mm. Last instar larvae as in *tenerrima.*

Distribution. Denmark: so far only found in EJ: Funder Egekrat 10.VII.1968 (Trolle). – Sweden: found in Sk.: Brunnby, Kullen 18.VII. and 29.VII.1961 (Ossiannilsson). – Norway: found in AAy: Lillesand, and in Ry: Høyland by Holgersen. – Not found in East Fennoscandia. – England, France, German F.R., Switzerland, Austria, Bulgaria, Bohemia, Slovakia, Hungary, Romania, Ukraine, Azerbaijan.

Biology. On *Quercus* (Ribaut, 1936b). "Auf *Quercus* in warmen Lagen" (Wagner & Franz, 1961). "Taken only from oak, from which it was also reared" (Claridge & Wilson, 1976).

Genus *Typhlocyba* Germar, 1833

Typhlocyba Germar, 1833: 180.
Type-species: *Cicada quercus* Fabricius, 1777, by subsequent designation.
Anomia Fieber, 1866: 509.
Type-species: *Cicada quercus* Fabricius, 1777, by subsequent designation.

Small species. Genital plate of male with only one macroseta arising near base. Genital style without a subapical angle. – The genus *Typhlocyba* in the sense here adopted comprises only two European species, but they are by no means nearly related. The characters of the genus *Typhlocyba* s.str. were precised by Anufriev (1973). A new genus should be erected for *bifasciata* Boheman and its allies but a revision should be based also on non-European species at present included in *Typhlocyba* s.lat.

Key to species of *Typhlocyba*

1 Fore wings whitish with large red or orange-coloured spots (Plate-fig. 141) .... 233. *quercus* (Fabricius)
– Fore wings light yellow with two broad dark brown transverse bands (Plate-fig. 140) .... 234. *bifasciata* Boheman.

233. ***Typhlocyba quercus*** (Fabricius, 1777)
Plate-fig. 141, text-figs. 1681–1688.

Cicada quercus Fabricius, 1777: 298.

Whitish, shining. Head produced in front, angular. Vertex usually with an orange-coloured band parallel with fore border, or with a pair of orange spots near eyes. Pronotum with fore border fairly broadly orange-coloured and with a central orange spot. Scutellum with one spot on each lateral corner and one spot on caudal apex orange-coloured. Alternatively the entire dorsum of fore body may be whitish or yellowish. Fore wings adorned with the following spots usually being orange in the proximal, blood-red in the distal part of cells: three in clavus (one large C-shaped near base, one smaller on middle and one still smaller in apex of clavus, the last-mentioned spot often extending over adjacent part of corium), two in corium at level of the colourless parts of clavus, these spots often united by red longitudinal streaks. At costal border distally of wax area there is an oblique black-brown streak. Veins in apical part of fore wing bordered with fuscous, apical cells partly infuscate. Face and venter whitish, black pigmentation on dorsum of abdomen much varying in extension. Male pygofer lobe as in Text-fig. 1681, genital plate as in Text-figs. 1682, 1683, genital style as in Text-fig. 1684, aedeagus as in Text-figs. 1685, 1686, 1st–6th abdominal sterna in male as in Text-fig. 1687, 2nd and 3rd abdominal terga in male as in Text-fig. 1688. Overall length (♂♀) 3–3.4 mm. – Last instar larvae pale whitish yellow with red-orange coloration on mesothoracic wing pads frequently present. Vertex extending anteriorly bearing one pair of widely separated short setae directed laterally and slightly caudally. Dorsum of fore body including wing pads bare, abdominal segments 5–8 with short lateral setae, median abdominal setae very small.

Distribution. Common and widespread in Denmark and in southern Sweden (Sk.–Upl.). – Norway: found in AK, VE, AAy, VAy, Ry. – Not found in East Fennoscandia. – Widespread in Europe, also recorded from Kazakhstan, Mongolia, and North America.

Biology. Hibernation in the egg stage (Müller, 1957). "Auf Apfel, Kirsche, Prunus und Rhamnus" (Günthart, 1971). "Dominant on oak – – –. Smaller but significant numbers were also taken from sycamore, alder and hornbeam. – – – We have reared it from oak and hornbeam" (Claridge & Wilson, 1976). I found *T. quercus* abundantly on *Sorbus suecica.* Adults in July–September.

234. ***Typhlocyba bifasciata*** Boheman, 1851
Plate-fig. 140, text-figs. 1689–1694.

Cicada nitidula Fabricius, 1794: 46 (nec Fabricius, 1787).
Typhlocyba bifasciata Boheman, 1851: 212.
Anomia norgueti Lethierry, 1874: 277.
Typhlocyba nitidula fenestrata Melichar, 1896: 345.
Typhlocyba nitidula atrata Melichar, 1896: 345.
Typhlocyba bifasciata perrieri Ribaut, 1936b: 122.

Head frontally obtuse rounded. Body and fore wings comparatively short and broad. Light yellow, shining. Scutellum and two broad transverse bands across fore wings black-brown. In f. *norgueti* (Lethierry) (= *atrata* Melichar) the fore wings are entirely

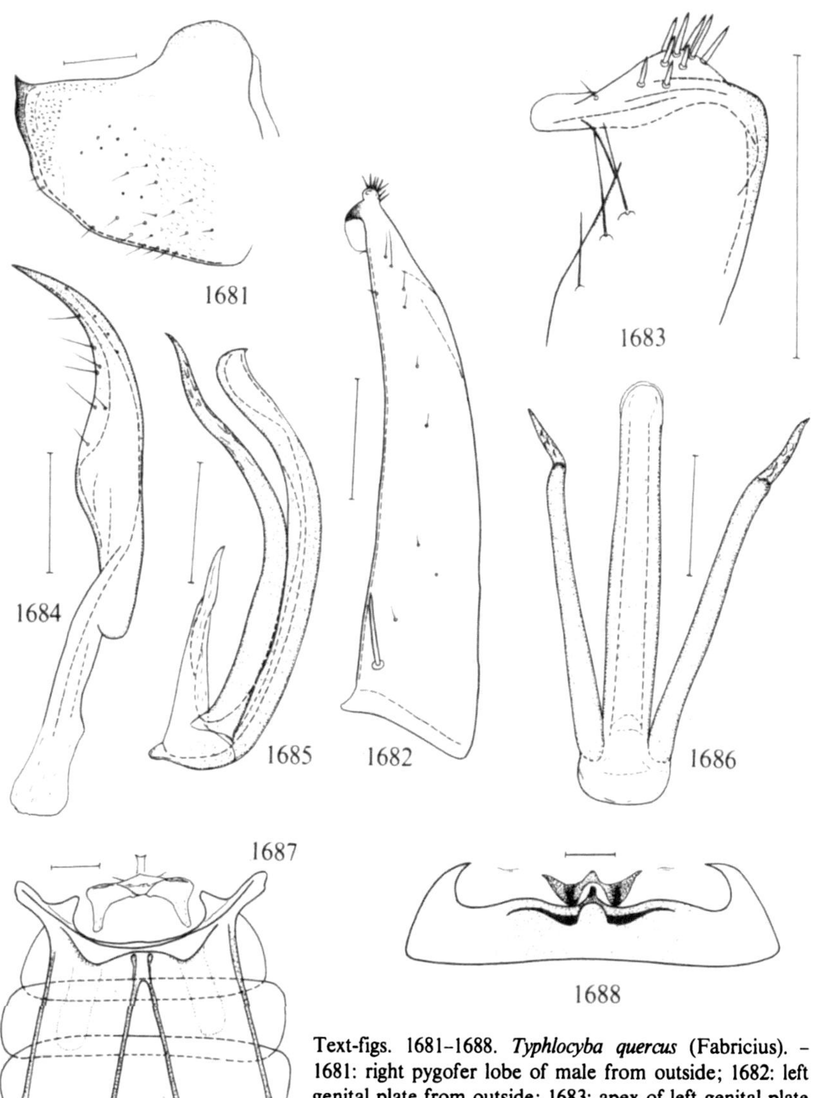

Text-figs. 1681–1688. *Typhlocyba quercus* (Fabricius). – 1681: right pygofer lobe of male from outside; 1682: left genital plate from outside; 1683: apex of left genital plate from below; 1684: right genital style from above; 1685: aedeagus from the left; 1686: aedeagus in ventral aspect; 1687: 1st–6th abdominal sterna in male from above; 1688: 2nd and 3rd abdominal terga from below: Scale: 0.1 mm.

dark except apical part. In f. *perrieri* Ribaut they are dark with apical part, a basal spot and wax area light. F. *fenestrata* Melichar differs from f. *perrieri* only by the presence of a small light spot in apex of clavus. Male pygofer lobe as in Text-fig. 1689, genital plate as in Text-fig. 1690, genital style as in Text-fig. 1691, aedeagus as in Text-figs. 1692, 1693, 1st–6th abdominal sterna as in Text-fig. 1694. Overall length 3.3–3.6 mm. – Last instar larva pale whitish-yellow, dorsally with dark brown markings, setae mostly dark pigmented. Vertex extended anteriorly between eyes, margin truncate with two pairs of long setae, almost equal in length. Dorsal setae moderately long, few in number on thorax and wing pads, abdominal segments 3–8 each with 1–2 pairs of setae.

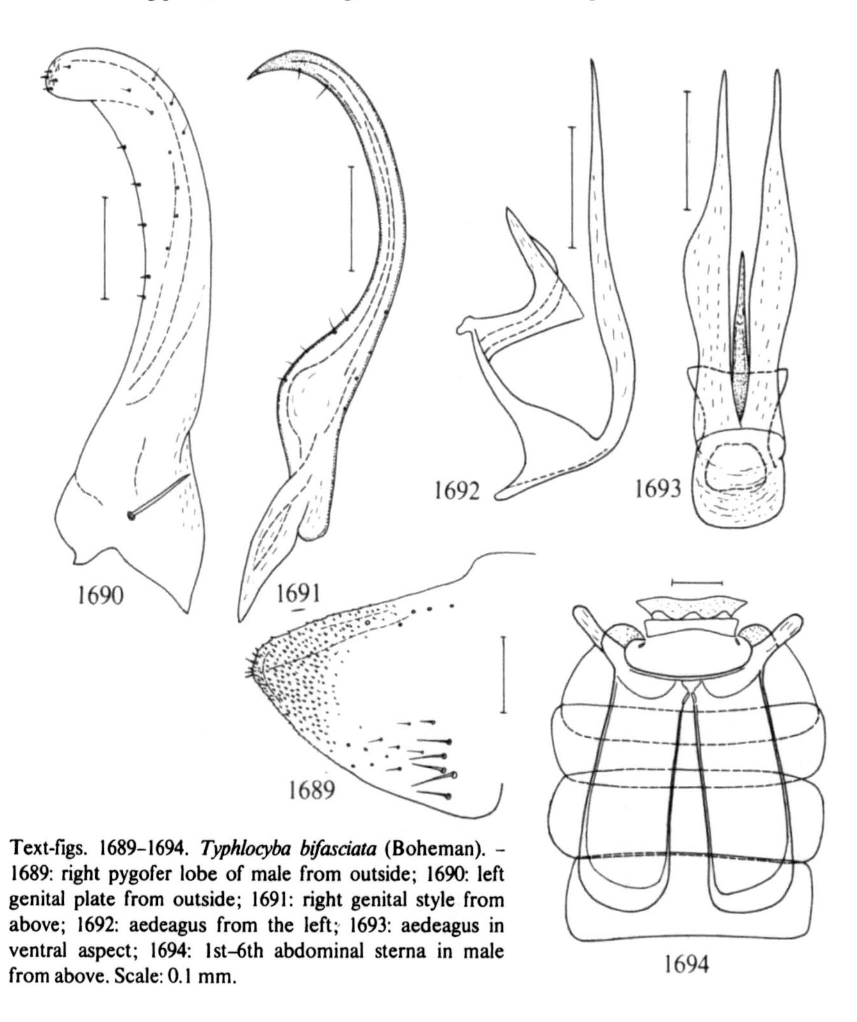

Text-figs. 1689–1694. *Typhlocyba bifasciata* (Boheman). – 1689: right pygofer lobe of male from outside; 1690: left genital plate from outside; 1691: right genital style from above; 1692: aedeagus from the left; 1693: aedeagus in ventral aspect; 1694: 1st–6th abdominal sterna in male from above. Scale: 0.1 mm.

Distribution. Widespread but rather scarce in Denmark, also in southern Sweden (Sk.–Vstm.). – Not found in Norway. – East Fennoscandia: found in Ab: Pojo, Spakanäs 30.VIII.1976 (A. Albrecht). – Widespread in Europe, also recorded from Tadzhikistan.

Biology. On *Carpinus betulus* (Kuntze, 1937; Schwoerbel, 1957; Wagner & Franz, 1961). "Wirtspflanze: *Carpinus*" (Günthart, 1974). "Recorded from several trees, but most prominently from elm – – – and hornbeam – – –, the only species from which we have reared it" (Claridge & Wilson, 1976). Found on *Ulmus* in Sweden by Kemner and Gyllensvärd, in Finland by Albrecht. Adults in July–September.

Genus *Eurhadina* Haupt, 1929

Eurhadina Haupt, 1929: 1075.

Type-species: *Cicada pulchella* Fallén, 1806, by original designation.

Outline of face in lateral aspect obtuse angular (in *Eurhadina* s.str.). Fore wing much narrower near apex than at middle. Male genital plate with one single macroseta on basal half, on lateral margin with a triangular preapical projection. Pygofer lobe without a distinct group of macrosetae near base. Genital style elongate, slender, without a preapical angular protuberance. Aedeagus with two pairs of apical processes. 2nd abdominal sternum in male (in our species) on each side laterally of apodeme with a clavate process. Last instar larvae dorso-ventrally flattened, anterior margin of vertex with two pairs of macrosetae arising from conical projections. In Denmark and Fennoscandia five species, all belonging to subgenus *Eurhadina* Haupt.

Key to species of *Eurhadina*

1 First apical vein in fore wing crossing a rounded black spot (Plate-fig. 96) 2

– First apical vein in fore wing not crossing a rounded black spot, at most with a black streak on the corresponding place (Plate-fig. 97) 3

2 (1) General colour yellow or orange. Anterior appendages of aedeagus forked (Text-figs. 1696–1698) .... 235. *pulchella* (Fallén)

– General colour white. Anterior appendages of aedeagus simple (Text-figs. 1711–1713) .... 238. *kirschbaumi* W. Wagner

3 (1) Lateral appendages of aedeagus simple (Text-figs. 1717, 1718). Large species, length > 4.1 mm. On *Acer pseudoplatanus* 239. *untica* Dlabola

– Lateral appendages of aedeagus forked. Smaller, length < 4.1 mm. 4

4 (3) Median apical cell with a fairly large semivitreous spot near the black streak on first apical vein. Lateral appendages of aedeagus shorter than anterior appendages, two-branched (Text-figs. 1701–1703) .... 236. *concinna* (Germar)

– Median apical cell almost entirely fumose. Lateral appendages of aedeagus longer than anterior appendages, three-branched (Text-figs. 1708, 1709) .... 237. *ribauti* W. Wagner.

235. ***Eurhadina pulchella*** (Fallén, 1806)
Plate-figs. 96, 142, 143, text-figs. 1695–1699.

Cicada pulchella Fallén, 1806: 36.
Eupteryx ornatipennis Curtis, 1837b: pl.640.
Typhlocyba pulchella thoracica Fieber, 1884: 99.

Face whitish. Vertex, pronotum, scutellum and fore wings except apical part light yellow or golden yellow with black markings (f. typica). Apical part of fore wing fumose

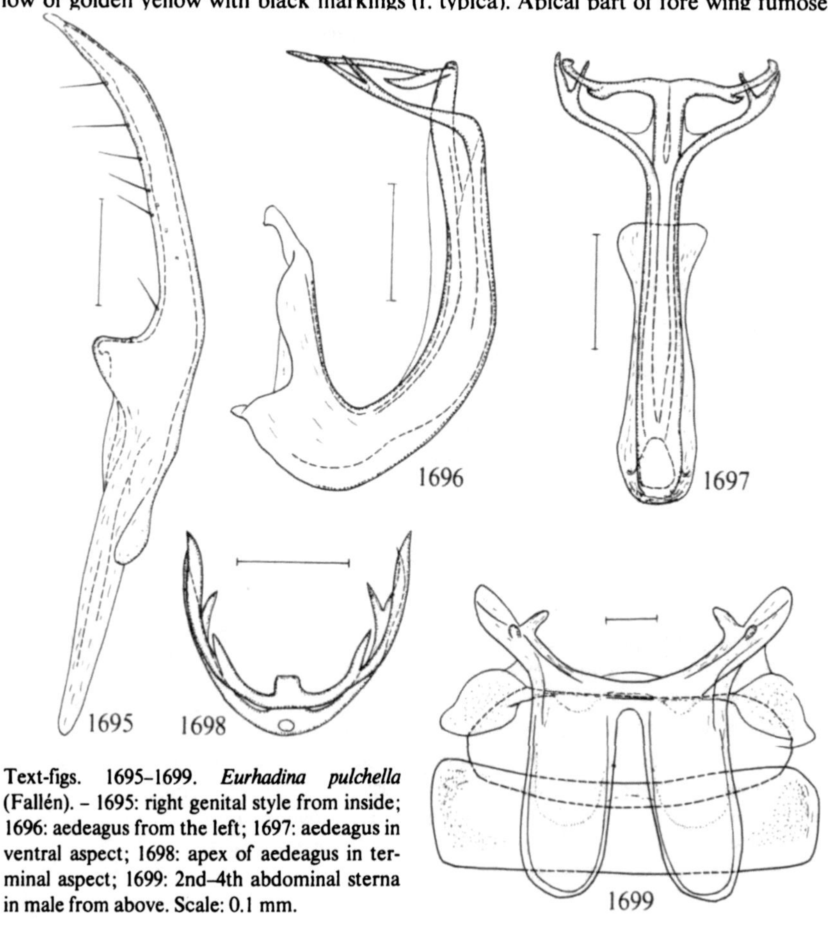

Text-figs. 1695–1699. *Eurhadina pulchella* (Fallén). – 1695: right genital style from inside; 1696: aedeagus from the left; 1697: aedeagus in ventral aspect; 1698: apex of aedeagus in terminal aspect; 1699: 2nd–4th abdominal sterna in male from above. Scale: 0.1 mm.

or light brown, with a large whitish spot in first apical cell. First apical vein passing through a rounded black spot. In f. *ornatipennis* (Curtis) (Plate-fig. 143), clavus, cubital, median and radial cells are largely reddish with an orange tinge. In f. *thoracica* (Fieber) (Plate-fig. 142), more or less large parts of fore body and fore wings are black-brownish, apical part of fore wing lighter. Male genital style as in Text-fig. 1695, aedeagus as in Text-figs. 1696–1698, 2nd–4th abdominal sterna in male as in Text-fig. 1699. Overall length (♂♀) 3.6–4.2 mm. – Last instar larvae varying in colour from entirely pale yellow to almost entirely dark brown dorsally. Setae on pronotum and wing pads absent or present only on lateral margins of the latter.

Distribution. Common and widespread in Denmark as well as in southern Sweden (Sk.–Vstm.). – Norway: found in Ø, AK, VAy, Ry. – Scarce in East Fennoscandia, found in Al, Ab, N, Ka, Ta, Sa, Kb; Vib, Kr. – Widespread in Europe, also recorded from Algeria, Kazakhstan, Maritime Territory, and Japan.

Biology. "An Eichen" (Kuntze, 1937). "Auf *Quercus*" (Wagner & Franz, 1961). "Elle vit exclusivement sur le Chêne" (Ribaut, 1936b). "Centred completely on oak, and we have no regular nymphal records from other plants" (Claridge & Wilson, 1976). On *Quercus robur* and *Quercus petraea*. Univoltine (Claridge & Wilson, 1978). Adults in July–September.

236. ***Eurhadina concinna*** (Germar, 1831)
Plate-fig. 97, text-figs. 1700–1704.

Tettigonia concinna Germar, 1831: pl. 12.

Resembling *pulchella* but general colour whitish yellow or white. A rounded black spot on first apical vein absent, often replaced by a short black streak. Abdominal tergum with a large black patch. Male genital style as in Text-fig. 1700, aedeagus as in Text-figs. 1701–1703, 2nd–4th abdominal sterna in male as in Text-fig. 1704. Overall length (♂♀ 3.3–4 mm. – Median setae on eighth abdominal tergum of last instar larvae with bases very close or touching; setae on pronotum and wing pads as in *pulchella*. Dorsal pigmentation often arranged in an 8-shaped pattern, vertex usually unmarked.

Distribution. In Denmark there are reliable records from EJ, LFM and B. – Not uncommon in southern Sweden, found in Sk., Bl., Sm., Öl., Gtl., G. Sand., Ög., Vstm. – Norway: found in Ø, AK, AAy, VAy, Ry. – East Fennoscandia: scarce in the southern part, found in Ab, N, St. – England, Scotland, Wales, France, Belgium, Netherlands, German D.R. and F.R., Spain, Italy, Greece, Austria, Hungary, Yugoslavia, Bohemia, Moravia, Slovakia, Bulgaria, Poland, Estonia, Latvia, Lithuania, m. Russia, Moldavia, Ukraine, Algeria, Switzerland.

Biology. "Sur le Chêne" (Ribaut, 1936b). "An Eichen und Buchen" (Kuntze, 1937). On oak (Linnavuori, 1952a). "Auf *Fagus silvatica*" (Wagner & Franz, 1961). "Wirtspflanze: *Castanea*" (Günthart, 1974). "Dominant on oak – – – but present significantly also on alder – – – and beech. – – – This fits very well with breeding records – nymphs are common on oak and rarer on alder and beech" (Claridge & Wilson, 1976). On *Quercus petraea* and *Q. robur*, also found on *Alnus*, *Betula* and *Fagus*. Univoltine (Claridge & Wilson, 1978). Adults in July–August.

237. ***Eurhadina ribauti*** W. Wagner, 1935
Plate-fig. 98, text-figs. 1705–1709.

Eurhadina ribauti W. Wagner, 1935: 34.
Eupteryx ribauti umbrata Ribaut, 1936b: 136.

Resembling *concinna*, median apical cell without a distinct semi-hyaline spot. Fore wings in f. *umbrata* (Ribaut) with an elongate, ill-defined fuscous patch along claval

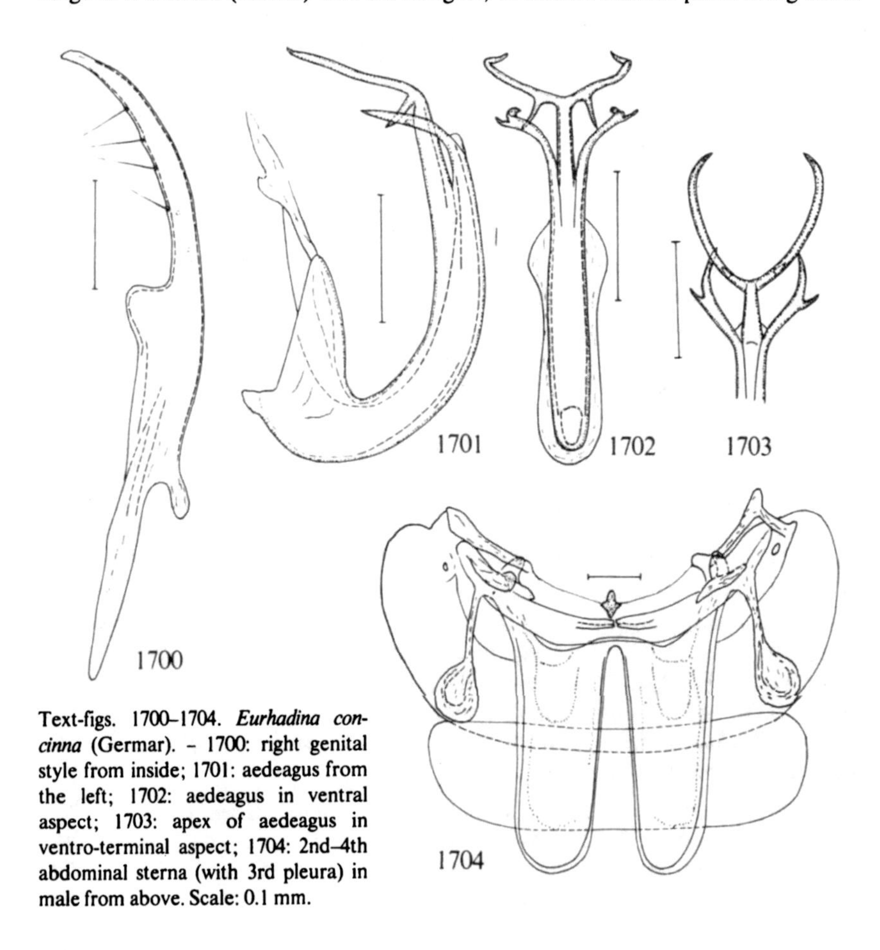

Text-figs. 1700–1704. *Eurhadina concinna* (Germar). – 1700: right genital style from inside; 1701: aedeagus from the left; 1702: aedeagus in ventral aspect; 1703: apex of aedeagus in ventro-terminal aspect; 1704: 2nd–4th abdominal sterna (with 3rd pleura) in male from above. Scale: 0.1 mm.

suture (Plate-fig. 98). Male pygofer lobe as in Text-fig. 1705, genital plate as in Text-fig. 1706, genital style as in Text-fig. 1707, aedeagus as in Text-figs. 1708, 1709. Overall length (♂♀) 3.3–4 mm. – Last instar larvae much as those of *concinna.*

Distribution. Denmark: so far reliable records from EJ and B. – Not uncommon in southern Sweden, found in Sk., Bl., Hall., Sm., Ög., Upl. – Norway: one female of f. *umbrata* found in VE: Larvik by Holgersen. – East Fennoscandia: so far only found in N: Helsingfors (Håkan Lindberg). – German D.R. and F.R., France, Wales, Netherlands, Cyprus.

Biology. On *Quercus, Tilia, Alnus,* f. *umbrata* on *Alnus* (Ribaut, 1936b). "An Eichen" (Kuntze, 1937). "Wirtspflanze: *Quercus*" (Günthart, 1974). "We have bred it from oak" (Claridge & Wilson, 1976). On *Quercus petraea* and *Q. robur;* univoltine (Claridge & Wilson, 1978). Adults in July–September.

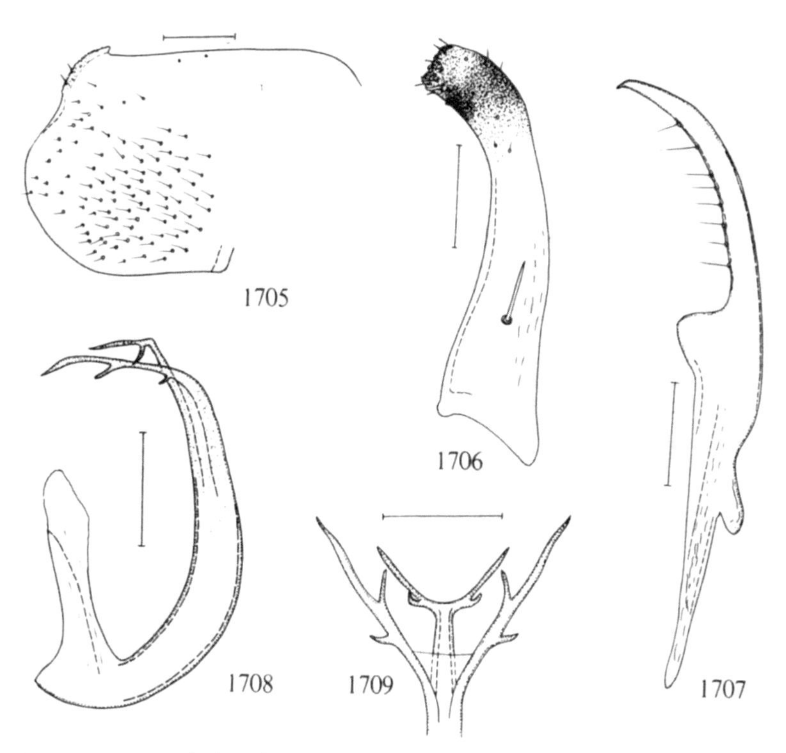

Text-figs. 1705–1709. *Eurhadina ribauti* W. Wagner. – 1705: right pygofer lobe of male from outside; 1706: left genital plate from outside; 1707: right genital style from inside; 1708: aedeagus from the left; 1709: apex of aedeagus in ventro-terminal aspect. Scale: 0.1 mm.

238. ***Eurhadina kirschbaumi*** W. Wagner, 1937
Text-figs. 1710–1713.

Eurhadina kirschbaumi W. Wagner, 1937a: 69.

Resembling *pulchella*, general colour whitish. Male genital style as in Text-fig. 1710, aedeagus as in Text-figs. 1711–1713. Length 3.3–4 mm. – Last instar larva pale yellow with scattered raised spots of dark brown pigment on dorsal surface and also light brown markings on pronotum, wing pads and occasionally abdominal segments. Median spines on eighth abdominal segment with bases widely spaced. Prominent setae on pronotum and wing pads (Wilson, 1978).

Distribution. So far not recorded from Denmark. – Sweden: found in Sk.: Brunnby, Kullen 31.VII.1961, 21.VIII.1964 and 26.VIII.1964 (Ossiannilsson). – Norway: found in VAy: Kristiansand 4.IX.1930, 1 ♀ (Warloe), and in AK: Nesodden 5.VIII.1946, 1 ♂, 4 ♀♀ (Holgersen). – East Fennoscandia: found in Ab: Åbo and Raisio by Linnavuori, and in Ab: Pojo by Albrecht. – German F.R., Wales, France, Netherlands, Austria, Hungary, Moravia, Bulgaria, Poland, Latvia, Lithuania, m. Russia.

Biology. On *Quercus sessilis* (= *petraea*) (Wagner, 1937a). "Several specimens from

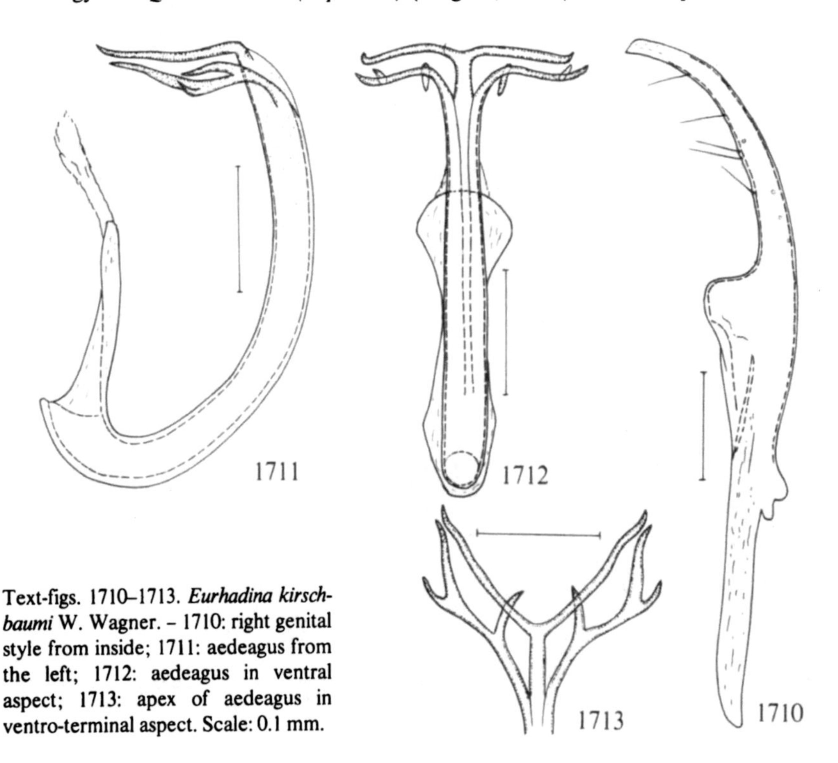

Text-figs. 1710–1713. *Eurhadina kirschbaumi* W. Wagner. – 1710: right genital style from inside; 1711: aedeagus from the left; 1712: aedeagus in ventral aspect; 1713: apex of aedeagus in ventro-terminal aspect. Scale: 0.1 mm.

large oaks in a woodland" (Linnavuori, 1952a). "*E. kirschbaumi* appears to be confined to *Q. petraea* - - -. Occurs later than other species on Oak. Univoltine" (Claridge & Wilson, 1978). But *Quercus petraea* is absent in Finland where Linnavuori (l.c.) found the present species.

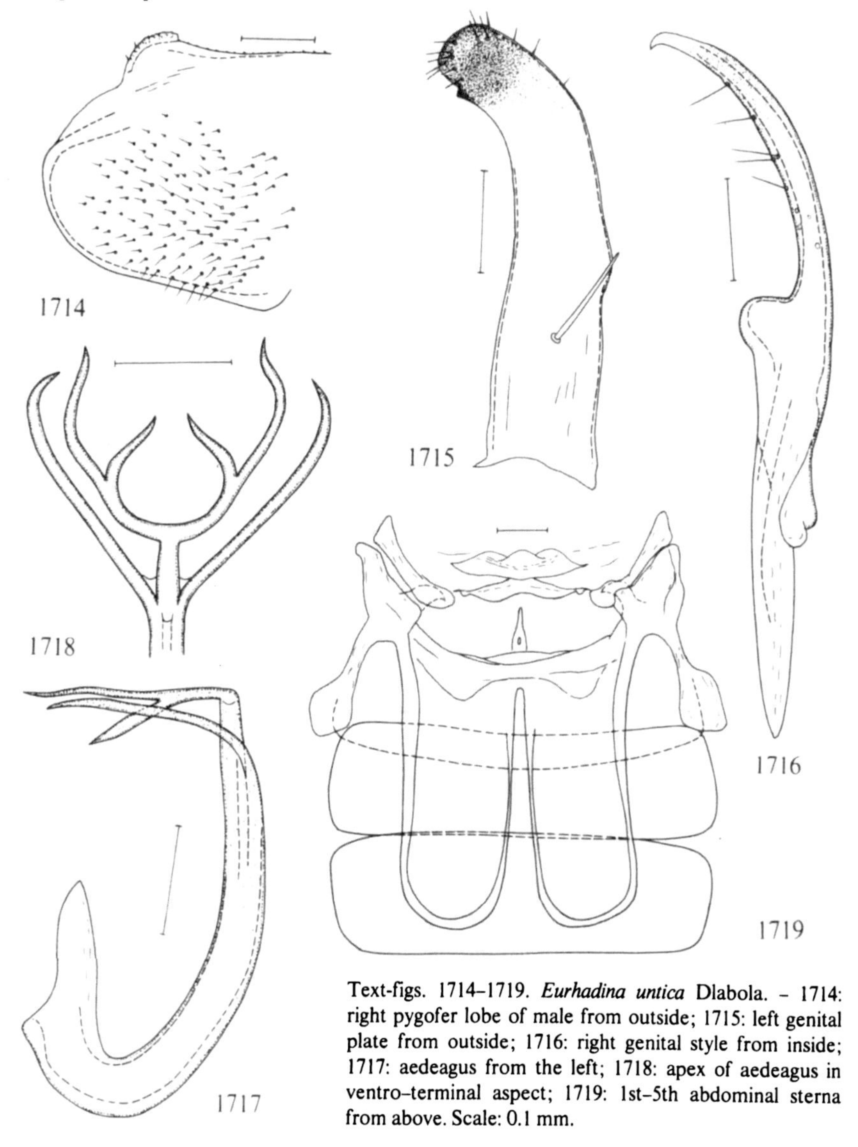

Text-figs. 1714–1719. *Eurhadina untica* Dlabola. - 1714: right pygofer lobe of male from outside; 1715: left genital plate from outside; 1716: right genital style from inside; 1717: aedeagus from the left; 1718: apex of aedeagus in ventro-terminal aspect; 1719: 1st–5th abdominal sterna from above. Scale: 0.1 mm.

239. ***Eurhadina untica*** Dlabola, 1967
Text-figs. 1714–1719.

Eurhadina löwi W. Wagner, 1935: 34 (nec Then, 1886).
Eupteryx loewi Ribaut, 1936b: 136 (p.p.).
Eurhadina untica Dlabola, 1967: 24.
Eurhadina loewi Claridge & Wilson, 1976: 245; Wilson, 1978: 87.

Resembling *concinna* and *ribauti* but larger (4.12–4.5 mm). Fore wings in some individuals with a more or less distinct broad black-brown of fuscous longitudinal band on both sides of claval suture, caudally curving towards costal border distally of wax area, as in *ribauti* f. *umbrata.* Male pygofer lobe as in Text-fig. 1714, genital plate as in Text-fig. 1715, genital style as in Text-fig. 1716, aedeagus as in Text-figs. 1717, 1718, 1st-5th abdominal sterna as in Text-fig. 1719. – Tergum of 5th instar larvae usually with an 8-shaped dark brown figure extending also on vertex. Chaetotaxy of dorsum as in *concinna.*

Distribution. Uncommon in Denmark, found in EJ, F, LFM and B. – Scarce in Sweden, only found in several places in Sk. – Not found in Norway, nor in East Fennoscandia. – England, Wales, German F.R., Poland, Mongolia.

Biology. In Mongolia on *Ulmus pumila* (Dlabola, l.c.). In Europe on *Acer pseudoplatanus* (Wagner, 1935; Ossiannilsson, 1946c; Claridge & Wilson, 1976). "Bred commonly on sycamore, with one record from field maple" (Claridge & Wilson, 1976). Confined to *Acer pseudoplatanus.* Univoltine (Claridge & Wilson, 1978). Adults in July and August.

Note. Le Quesne (1978) suggests that *Eurhadina untica* is only a geographical form of *loewii* (Then). This is by no means unlikely, but until definite proof has been produced I prefer treating them as separate taxa.

Genus *Eupteryx* Curtis, 1833

Eupteryx Curtis, 1833: 192.
Type-species: *Cicada picta* Fabricius, 1794, by original designation.

First and fourth apical cells in fore wing not reaching wing apex, second apical cell petiolate, third apical cell much broader at apex than at base. Fore wing not much narrower in apical part than at middle. Hind wing with three apical cells. Genital plate with a single macroseta near base or near middle or with a few macrosetae. Pygofer with a group of macrosetae near ventral part of base, distal margin strongly sclerotized, often with hooks or appendages. Genital style elongate, curved, with a preapical angular protuberance on median margin. Aedeagus usually with processes arising from or near apex of shaft, socle without processes. Species often brightly coloured. In Denmark and Fennoscandia 14 species, all associated with herbaceous plants.

Key to species of *Eupteryx*

1 Pronotum and scutellum entirely light without black markings 2
– Pronotum and/or scutellum with black markings or entirely dark 3
2 (1) General colour greenish yellow or greyish green. Small species, overall length not exceeding 3.05 mm. On *Artemisia* ..... 244. *artemisiae* (Kirschbaum)
– General colour light yellow or whitish yellow. Overall length not below 3.3 mm. On *Filipendula ulmaria* . 243. *signatipennis* (Boheman)
3 (1) Pronotum entirely or almost entirely black. Distal half of clavus entirely black. Apices of fore wings truncate 4
– Pronotum and distal half of clavus not entirely black. Apices of fore wings rounded 5
4 (3) Overall length at least 3 mm. Male anal tube with a pair of long appendages (Text-fig. 1783). Male pygofer on inside without a long erect appendage. Aedeagus as in Text-figs. 1786, 1787 ..... 251. *vittata* (Linné)
– Overall length at most 2.7 mm. Male anal tube without appendages. Male pygofer lobe on inside with a long and slender erect appendage (Text-fig. 1790). Aedeagus as in Text-figs. 1791, 1792 ..... 252. *notata* Curtis
5 (3) Vertex without black markings ..... 243. *signatipennis* (Boheman)
– Vertex with black spots or entirely black 6
6 (5) Fore wings with two dark longitudinal bands. Pronotum with a broad brownish transverse band but without black spots. Vertex with three black spots (Plate-fig. 116) ..... 253. *tenella* (Fallén)
– Markings of fore wings different. Pronotum black spotted 7
7 (6) Third apical vein in fore wing emanating from distal end of median cell (Text-fig. 18, plate-figs. 99, 131, 132) 8
– Apical veins in fore wing all emanating from radial cell (Plate-figs. 100, 101) 15
8 (7) Vertex with two black spots (sometimes confluent posteriorly) 9
– Vertex with three black spots, the odd (median) one situated at hind border. Sometimes these spots are confluent, forming an Y-shaped figure or a longitudinal band widening anteriorly 12
9 (8) Lateral spots on pronotum not extending to fore border ..... 243. *signatipennis* (Boheman)
– Lateral spots on pronotum extending to fore border 10
10 (9) Smaller, overall length 2.9–3.4 mm. Pronotum usually with only two black spots, these not much larger than spots on vertex. Appendage on inside of male pygofer lobe hook-

like, simple (Text-figs. 1731, 1732) or with a fine spine (Text-figs. 1733, 1734). Appendages of aedeagus in lateral aspect extending below ventral margin of shaft (Text-fig. 1736). On *Origanum* .... 241. *origani* Zachvatkin

– Larger, overall length 3.4–4.3 mm. Pronotum with two or four black spots, one smaller median pair (sometimes confluent) and one larger lateral pair, the latter often much larger than spots on vertex. Appendages on inside of male pygofer forked, shaped as in Text-figs. 1721, 1722, 1739 11

11 (10) Larger, overall length 3.5–4.3 mm. Main colour yellow, usually with an orange tinge. Black spots on vertex usually more or less elongate or irregularly shaped. Lateral black spots on pronotum often very much larger than spots on vertex. Black spots on scutellum often confluent. Female with 1st apical cell in fore wing largely dark (Plate-fig. 131). 242. *aurata* (Linné)

– Smaller, overall length 3.4–3.75 mm. Main colour whitish yellow, usually with a greenish tinge. Black spots on vertex more rounded. Lateral black spots on pronotum not much larger than black spots on vertex. First apical cell in fore wing largely pale in both sexes (Plate-fig. 132) .... 240. *atropunctata* (Goeze)

12 (8) Black spots on vertex confluent, forming an Y-shaped figure (Plate-fig. 126) or a longitudinal band widening anteriorly .... 246. *cyclops* Matsumura

– Vertex with three separate black spots 13

13 (12) Median spot on vertex basally wider than long (Plate-figs. 127, 128). Aedeagus with a pair of bifurcate appendages 14

– Median spot on vertex longer than basal width (Plate-fig. 125). Aedeagus with a pair of serrate appendages (Text-fig. 1767) .... 246. *cyclops* Matsumura

14 (13) Branches of appendages of aedeagus equal in thickness (Text-figs. 1770, 1771) .... 247. *calcarata* Ossiannilsson

– One branch of aedeagal appendage much thinner than the other, varying in length (Text-figs. 1761–1763) .... 245. *urticae* (Fabricius)

15 (7) Median spot on vertex at least twice as wide as long (Plate-fig. 129). Black spot distally of wax area completely or almost completely covering the part of R crossing this spot (Plate-fig. 100). Hind tarsi long, 0.7–0.8 mm .... 248. *stachydearum* (Hardy)

– Vein R not covered by the black spot present distally of wax area (Plate-fig. 101). Hind tarsi shorter, below 0.7 mm in length 16

16 (15) First segment of hind tarsus in male fuscous at least on its distal third. Aedeagus as in Text-figs. 1775–1777. Last segment of hind tarsus in female almost entirely fuscous .... 249. *collina* (Flor)

– First segment of hind tarsus in male entirely light. Aedea-

gus as in Text-figs. 1779, 1780. Last segment of hind tarsus
in female with a light brown tinge at most on distal half . 250. *thoulessi* Edwards.

240. ***Eupteryx atropunctata*** (Goeze, 1778)
Plate-figs. 132, 145, text-figs. 18, 19, 1720–1730.

Cicada atropunctata Goeze, 1778: 161.
Cicada carpini Fourcroy, 1785: 191.

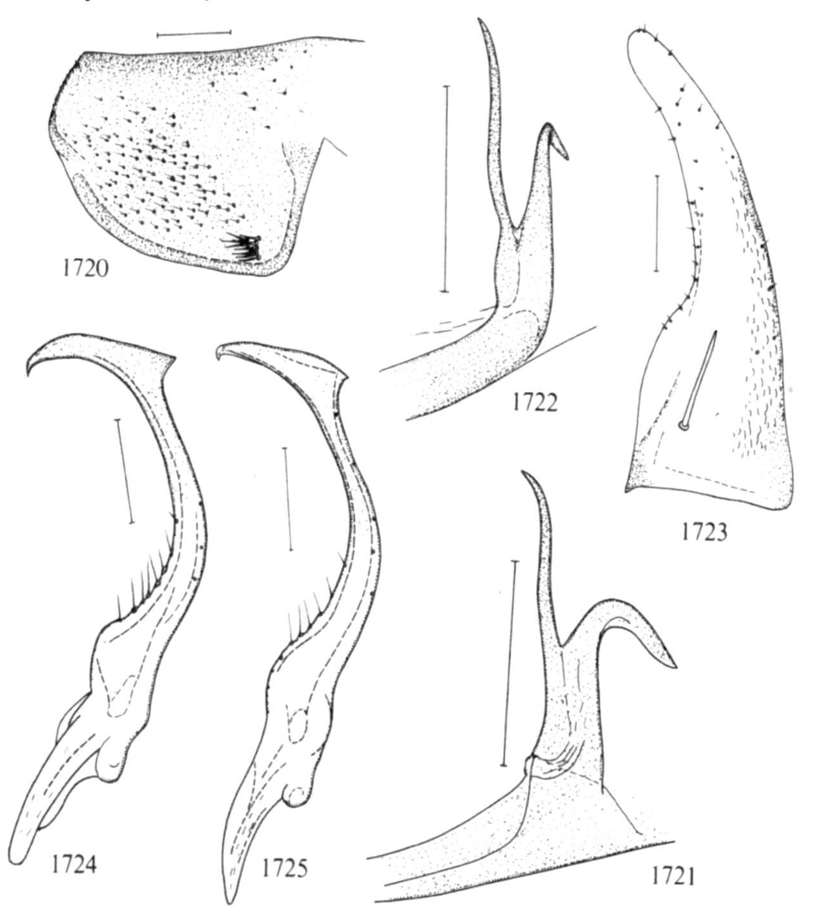

Text-figs. 1720–1725. *Eupteryx atropunctata* (Goeze). - 1720: right pygofer lobe in male from outside; 1721: process of right pygofer lobe of male from inside; 1722: same of another specimen; 1723: left genital plate from outside; 1724: right genital style from above; 1725: same of another specimen. Scale: 0.1 mm.

Cicada picta Fabricius, 1794: 32.
Typhlocyba aureola Boheman, 1845a: 49, 1845b: 161.
Swedish: potatisstrit.

Elongate, shining. Head anteriorly rounded, medially slightly longer than near eyes. Whitish yellow, usually with a greenish tinge. Vertex with two large rounded or approximately quadrangular black spots usually not extending to fore margin of head as seen

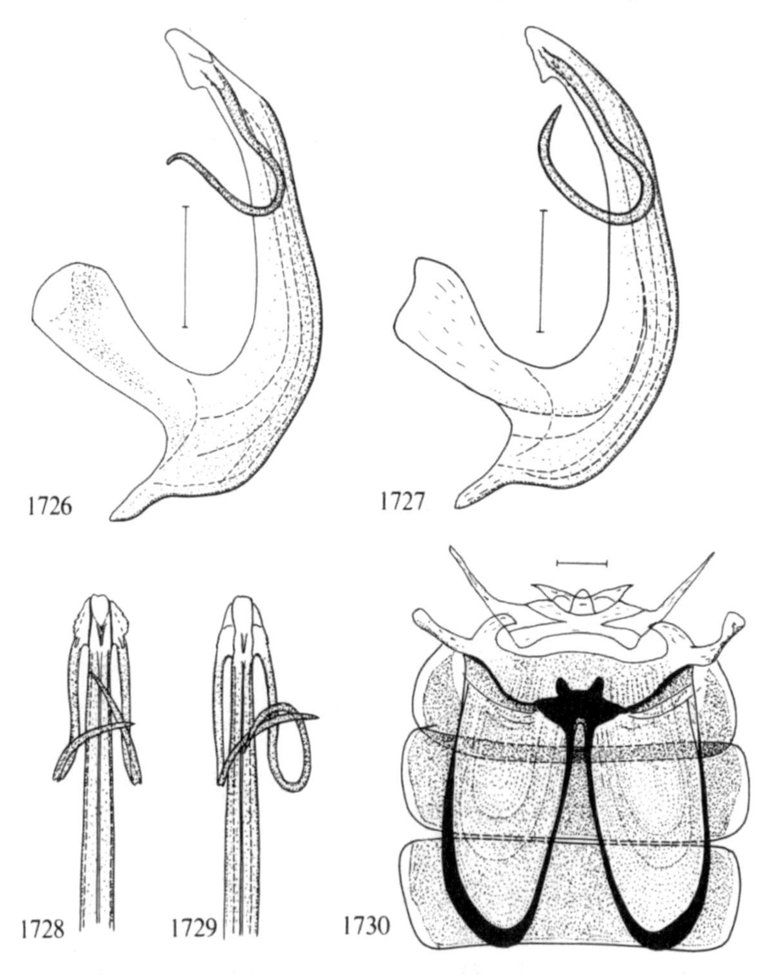

Text-figs. 1726–1730. *Eupteryx atropunctata* (Goeze). – 1726: aedeagus from the left; 1727: same of another specimen; 1728: shaft of aedeagus in dorsal aspect; 1729: same of another specimen; 1730: 1st–5th abdominal sterna in male from above. Scale: 0.1 mm.

from above. Anteclypeus broadly black-bordered or almost entirely black (♀, 2nd generation), frontoclypeus more or less strongly black-bordered. Pronotum usually with two pairs of black spots, the median smaller pair often confluent. Scutellum with two large black spots. Fore wing with two black spots, one on middle of clavus and another at costal border just distally of wax area (Plate-fig. 132). A more or less distinct oblique black or fuscous streak usually present in subcostal and radial cells proximally of wax area. Apex of clavus fuscous; a diffuse fuscous longitudinal band in cubital and median cells present or absent. First and fourth apical cells largely colourless in both sexes. 3rd apical cell dark, apex and a transverse spot proximally of middle colourless. Male pygofer lobe as in Text-fig. 1720, process of pygofer lobe as in Text-figs. 1721, 1722, genital plate as in Text-fig. 1723, genital style as in Text-figs. 1724, 1725, aedeagus as in Text-figs. 1726–1729, 1st–5th abdominal sterna in male as in Text-fig. 1730. Overall length 3.4–3.75 mm. - Larvae yellowish, dorsally with comparatively long setae. These are capitate in early instars, stump or pointed in older instars. The chaetotaxy of all the instars has been described by Ossiannilsson (1943a). In the last instar there are two pairs of setae on fore border of head, one pair on vertex, two lateral pairs on pronotum, two pairs on mesonotum, one pair on metanotum, one pair on 3rd abdominal tergum and two pairs on each of abdominal terga 4–8; fore wing pads each with 3 setae. Abdominal setae arising from black-pigmented papillae. Length of body 2.4–3.3 mm, of antennal flagellum 0.84–0.99 mm.

Distribution. Denmark: found in EJ, F, LFM, B. - Common in southern Sweden, found Sk.–Upl., Vstm. - I have seen specimens collected in Norway: AK: Ås 4.VIII.-1975 by T. Rygg. - Scarce in East Fennoscandia, found in Ab, N, Ta, Sa, Kb; Vib., Kr. - Widespread in Europe, also in Algeria, Anatolia, Armenia, Georgia.

Biology. Polyphytophagous. On *Solanum tuberosum* and *Lamium album*. Eggs are deposited in stalks and thicker leaf veins. On nettles only where these are growing together with *Lamium*. Eggs often parasitized by *Anagrus atomus* Haliday (Ossiannilsson, 1943a). On *Lamium album, Glechoma hederacea, Arctium*, beets; 2 generations in Sweden (Ossiannilsson, 1946c). Hibernation in the egg stage (Müller, 1957; Schiemenz, 1969b). Ubiquitous (Schiemenz, 1964). Host-plants: *Dahlia, Verbascum, Apium graveolens* (Günthart, 1974). On *Angelica archangelica, Melissa officinalis, Mentha piperita, Ocimum basilicum, Salvia officinalis* and *sclarea, Chrysanthemum balsamita* and *vulgare* (Vidano & Arzone, 1978). Koblet-Günthardt (1975) studied *E. atropunctata* with regard to life cycle, food-intake, energy requirements and detrimental effects to the host-plant.

Economic importance. "Im nördlichen und mittleren Europa u.a. an Kartoffeln häufig, aber wenig schädlich durch die Erzeugung von Blattrollungen. Auch an Rüben, Getreide usw." (Müller, 1956). See further Koblet-Günthardt (l.c.).

241. ***Eupteryx origani*** Zachvatkin, 1948
Text-figs. 1731–1738.

Eupteryx origani Zachvatkin, 1948b: 193.
Eupteryx origani f. *edwardsi* Le Quesne, 1974: 203.

Resembling *atropunctata*. Black markings on anteclypeus and frontoclypeus less exten-

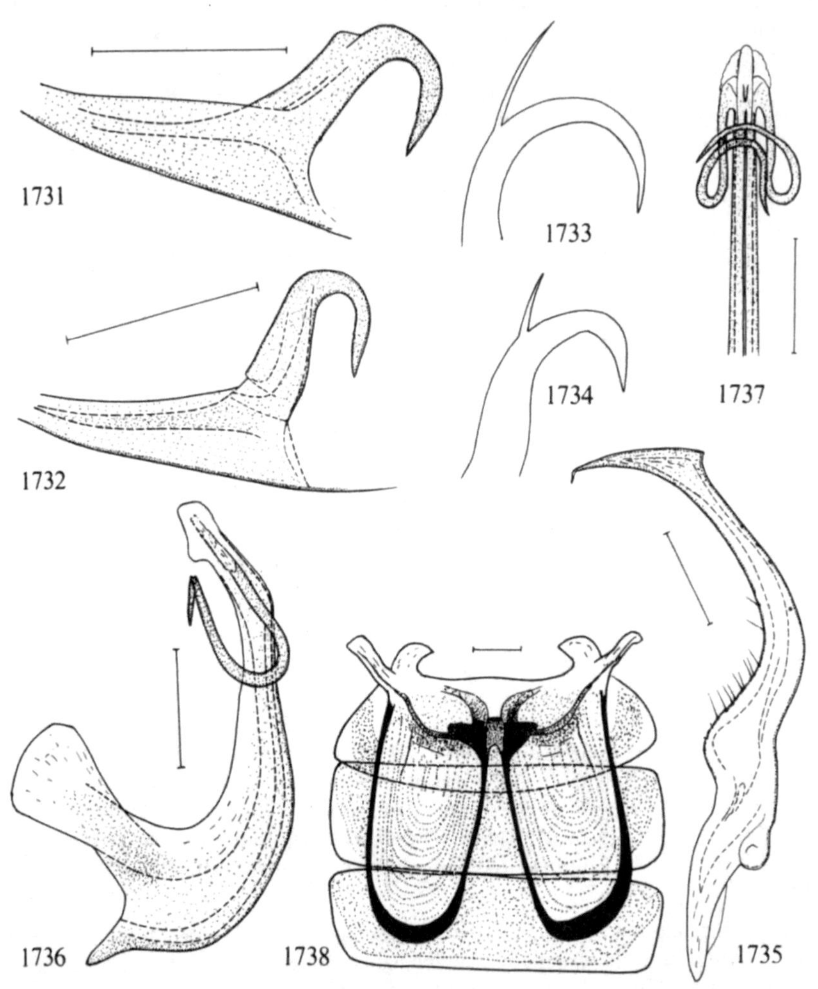

Text-figs. 1731–1738. *Eupteryx origani* Zachvatkin. – 1731: process of right pygofer lobe of male from inside; 1732: same of *f. edwardsi* Le Quesne; 1733, 1734: process of pygofer lobe in two males (after Le Quesne, 1974); 1735: right genital style from above; 1736: aedeagus from the left; 1737: apical part of shaft of aedeagus in dorsal aspect; 1738: 2nd–5th abdominal sterna in male from above. Scale: 0.1 mm.

sive. Pronotum usually without dark spots medially, lateral spots comparatively small. 3rd apical cell in fore wing dark, apex colourless, no colourless transversely oval spot present. Process on inside of pygofer lobe with (Text-figs. 1733, 1734) or without (Text-fig. 1732) a fine spine on convex side, or with a slight protuberance (Text-fig. 1731); specimens with process as in Text-fig. 1732 belong to f. *edwardsi* Le Quesne. Genital style as in Text-fig. 1735, aedeagus as in Text-figs. 1736, 1737, 2nd–5th abdominal sterna in male as in Text-fig. 1738. Overall length (according to Le Quesne, l.c.) 2.9–3.4 mm.

Distribution. So far not found in Denmark, nor in Norway. – Sweden: only found in Sm.: Annerstad, Skeen 15.VII.1934, 1 ♀, and 1.VIII.1934, 1 ♂, 1 ♀ (Ossiannilsson). – Scarce in southern and central East Fennoscandia, found in Ab, N, Sa, Oa, Tb, Sb. – England, Netherlands, German F.R., Bulgaria, Poland, Estonia, Latvia, Lithuania, m. and s. Russia.

Biology. On *Origanum vulgare* (Le Quesne, 1974). Adults in July–September.

242. ***Eupteryx aurata*** (Linné, 1758)
Plate-figs. 131, 144, text-figs. 1739–1744.

Cicada aurata Linné, 1758: 439.

Elongate, parallel-sided, shining. Head frontally rounded, medially slightly longer than near eyes. Light yellow, usually with an orange tinge. Black markings of face less extended than in *atropunctata*. On the other hand, the dark markings on dorsum and fore wings are more extended than in *atropunctata*. Vertex with a pair of large black spots, pronotum at fore border with or without a median pair of small spots sometimes confluent, and with a lateral pair of black spots sometimes occupying the major part of pronotum. Scutellum on lateral corners with a pair of triangular black spots sometimes extending on the entire frontal half of scutellum; caudal half of scutellum always light. Fore wings (Plate-fig. 131) with a broad black-brown longitudinal band proximally extending to scutellar angle of clavus and an opposite point on costal border and touching claval commissure at two spots; in light specimens this band may be divided into separate patches. A black-brown spot present at costal border just distally of wax area. Apical part of fore wing largely fumose; proximal major part of first apical cell in male colourless. Veins of hind wings dark. Mesosternum with a black patch; abdomen largely black. Process of male pygofer lobe as in Text-fig. 1739, genital plate as in Text-fig. 1740, genital style as in Text-figs. 1742, 1743, 1st–5th abdominal sterna in male as in Text-fig. 1744. Overall length (♂♀) 3.5–4.3 mm.

Distribution. Locally common in Denmark, found in EJ, NEJ, F and B. – Locally common in southern Sweden, found in Sk., Bl., Hall., Sm., Sdm., Upl. – I have seen a specimen from Norway, HOy: Bergen, Museihagen 1.–5.X.1936, collector unknown. – East Fennoscandia: only found in Kr. – Widespread in Europe, also recorded from Georgia.

Biology. "An *Urtica dioica* und vielen Kräutern (Labiaten) fast überall" (Kuntze,

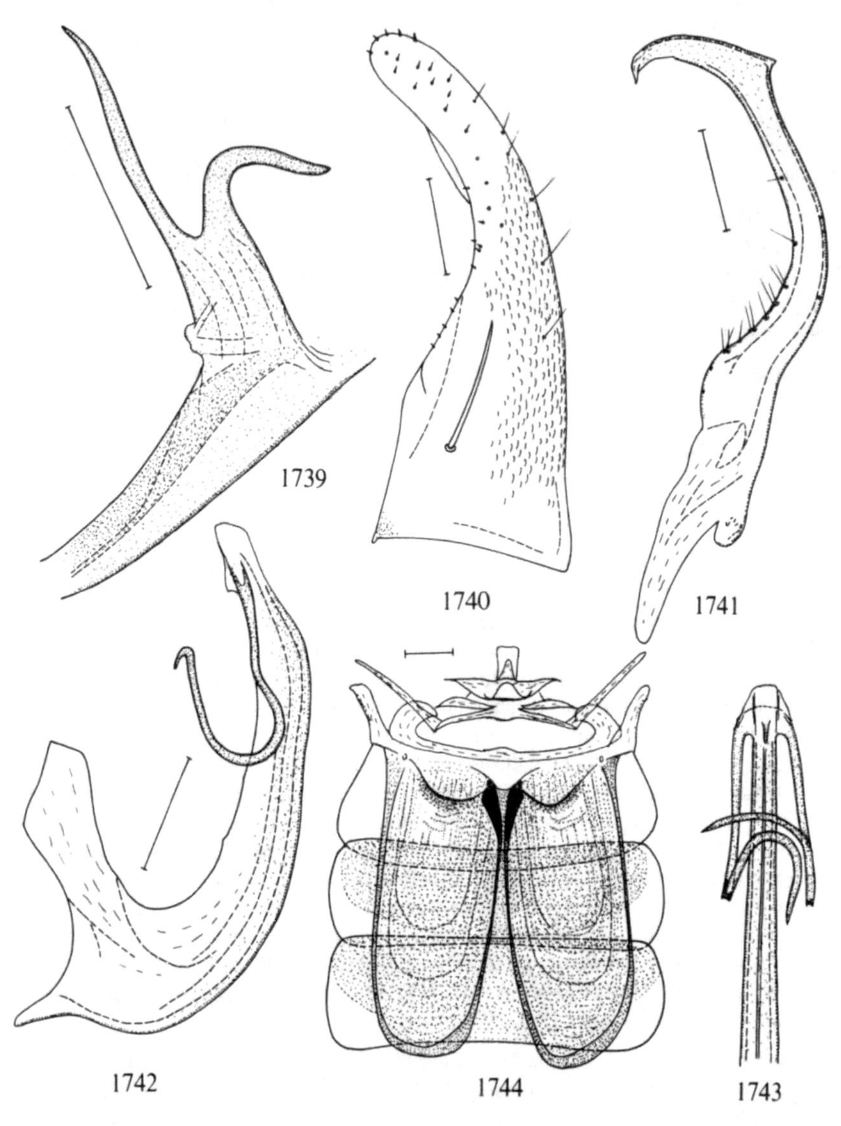

Text-figs. 1739–1744. *Eupteryx aurata* (Linné). – 1739: process of right pygofer lobe of male from inside; 1740: left genital plate from inside; 1741: right genital style from above; 1742: aedeagus from the left; 1743: shaft of aedeagus in dorsal aspect; 1744: 1st–5th abdominal sterna in male from above. Scale: 0.1 mm.

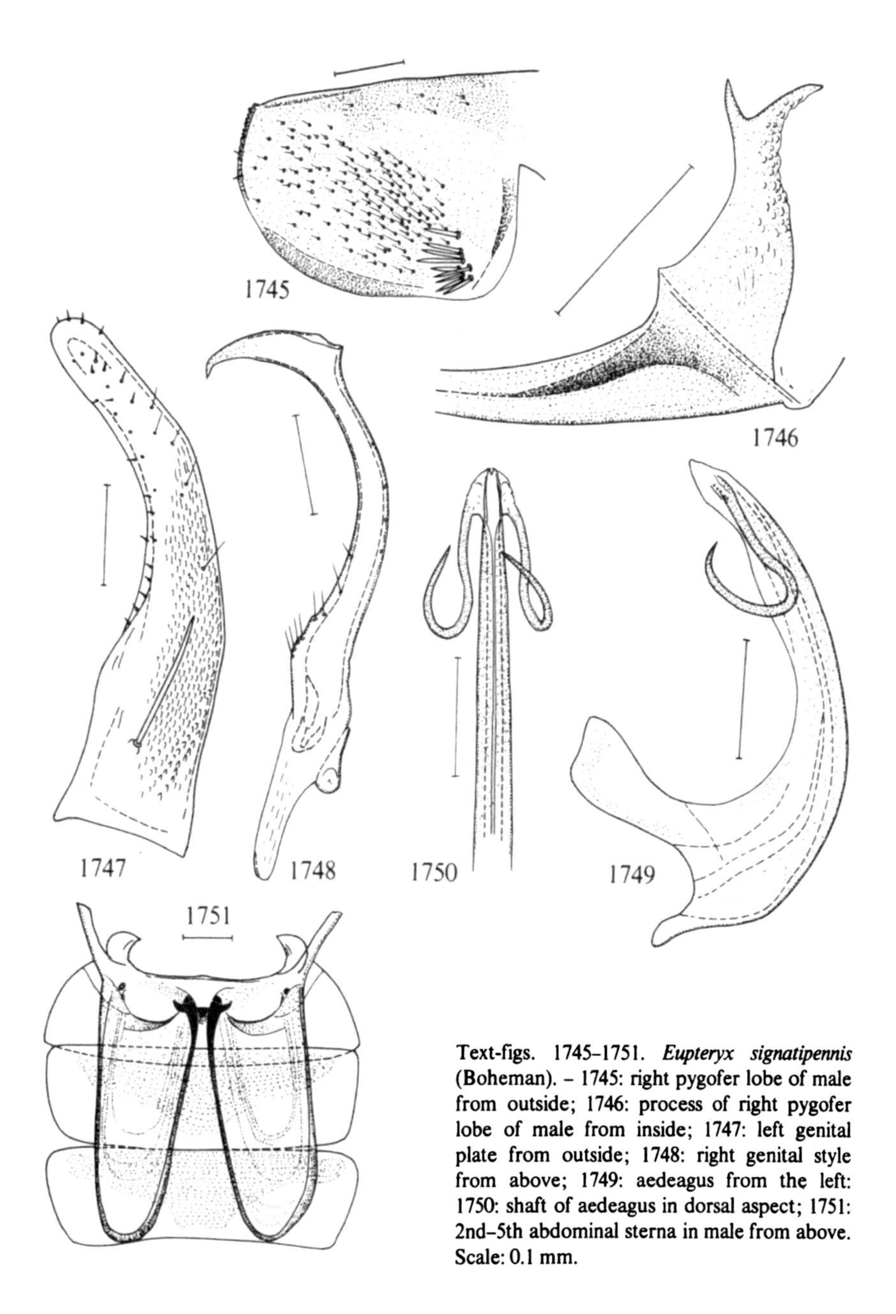

Text-figs. 1745–1751. *Eupteryx signatipennis* (Boheman). – 1745: right pygofer lobe of male from outside; 1746: process of right pygofer lobe of male from inside; 1747: left genital plate from outside; 1748: right genital style from above; 1749: aedeagus from the left: 1750: shaft of aedeagus in dorsal aspect; 1751: 2nd–5th abdominal sterna in male from above. Scale: 0.1 mm.

1937). On *Urtica* (Ossiannilsson, 1946c). On *Chaerophyllum spec., Urtica dioica, Carduus personatus* (Wagner & Franz, 1961). Host-plants: "*Senecio alp., Mentha*", also found on *Urtica* (Günthart, 1974). Adults in June–October.

243. ***Eupteryx signatipennis*** (Boheman, 1847)
Plate-fig. 118, text-figs. 1745–1751.

Typhlocyba signatipennis Boheman, 1847a: 36.

Related to *aurata* and *atropunctata* but lighter and with varying, usually only indistinct dark markings. Pale greenish yellow or whitish. Face in male unicolorous light, in female with or without a black spot in antennal pit. Vertex sometimes with 2 dark spots. Pronotum with or without a rounded dark spot on each side. A pair of dark spots on scutellum are also inconstant. Markings of fore wings consisting of same components as in *atropunctata* but always little prominent. The spot at scutellar corner of clavus present in *atropunctata* is always absent in *signatipennis*. Fore wings sometimes entirely light without markings. Dark markings of abdomen also much varying in extension. Male pygofer lobe as in Text-fig. 1745, process of pygofer lobe as in Text-fig. 1746, genital plate as in Text-fig. 1747, genital style as in Text-fig. 1748, aedeagus as in Text-figs. 1749, 1750, 2nd–5th abdominal sterna in male as in Text-fig. 1751. Overall length (♂♀) 3.4–3.8 mm.

Distribution. Common, widespread and abundant in Denmark. – Common in southern Sweden, found in Sk., Bl., Hall., Sm., Öl., Vg., Boh, Dlsl., Nrk. – Norway: found in AK, VE, AAy, Ry. – Comparatively scarce and sporadic in southern and central East Fennoscandia, found in Ab, N, St, Ta, Oa, Om; Kr. – England, Scotland, Wales, Ireland, France, Belgium, German D.R. and F.R., Netherlands, Switzerland, Estonia, Latvia, n. Russia.

Biology. On *Filipendula ulmaria* (Sahlberg, 1871; Ossiannilsson, 1946c). Host-plant: *Filipendula* (Günthart, 1974). Adults in June–September.

244. ***Eupteryx artemisiae*** (Kirschbaum, 1868)
Plate-fig. 117, text-figs. 1752–1757.

Typhlocyba artemisiae Kirschbaum, 1868b: 190.
Eupteryx abrotani Doulglas, 1874: 118.

Not very elongate, faintly shining, yellowish green to greyish green. Muscle traces on frontoclypeus often brownish. Markings of upper side strongly varying. A yellowish white median line on vertex extending on fore border of pronotum. A pair of dark spots often present laterally of the median line on vertex. Pronotum on each side with a pair of small dark spots. Fore wings with scattered small brownish spots or streaks especially in clavus and on vein Cu, often extending into longitudinal bands in the cells. Apical cells of fore wing partly fumose. Venter partly infuscate, abdomen above largely and ventrally partly greenish black. Male pygofer lobe as in Text-fig. 1752, genital plate as in Text-fig. 1753, genital style as in Text-fig. 1754, aedeagus as in Text-figs. 1755, 1756, 1st–5th abdominal sterna in male as in Text-fig. 1757. 2nd abdominal tergum on inside with two small black phragma lobes. Overall length 2.5–3.05 mm.

Distribution. Scarce but abundant in Denmark, found in EJ, F, LFM, NEZ. - Scarce in Sweden, found in Sk.: Vellinge (Tullgren), Landskrona (Håkan Lindberg),

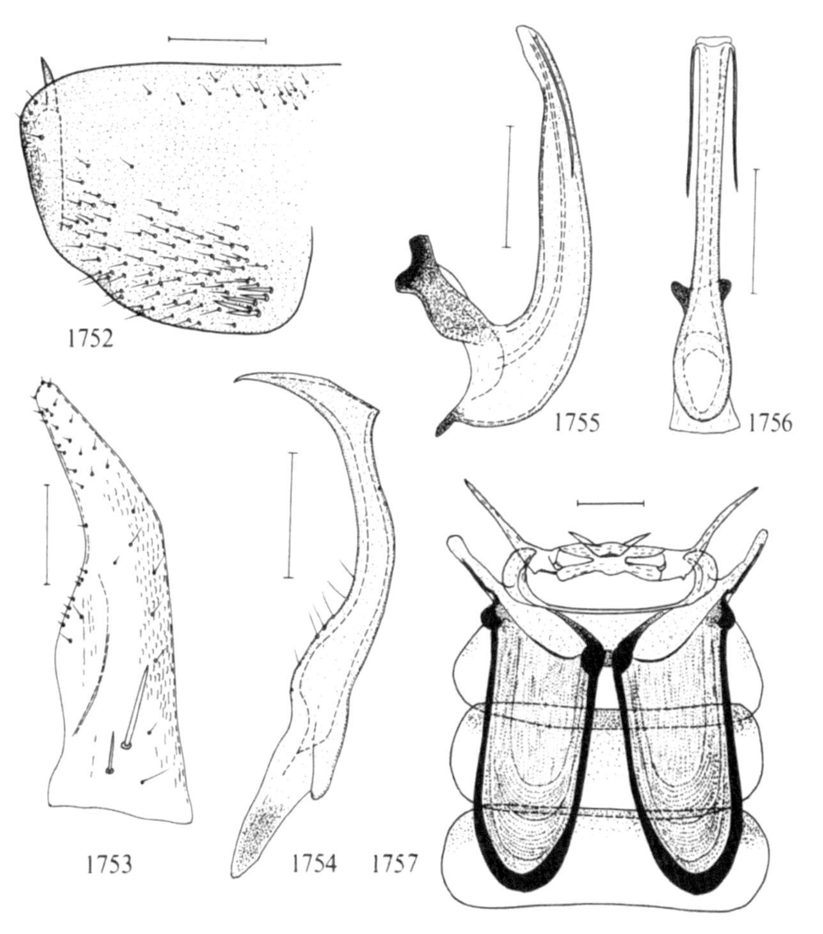

Text-figs. 1752–1757. *Eupteryx artemisiae* (Kirschbaum). - 1752: right pygofer lobe of male from outside; 1753: left genital plate from outside; 1754: right genital style from above; 1755: aedeagus from the left; 1756: aedeagus in ventral aspect; 1757: 1st–5th abdominal sterna from above. Scale: 0.1 mm.

Landskrona 22.VIII.1937 (Ossiannilsson), Eskilstorps holmar 7.VII.1935 (Kemner), Lund 15.VIII.1937 and 22.VII.1948 (Ossiannilsson); Gtl.: Visby 14.VII.1965 (Ossiannilsson); Upl.: Experimentalfältet 16.IX.1936 (Ossiannilsson). Abundant in most places. – So far not found in Norway and East Fennoscandia. – England, Scotland, France, Netherlands, German D.R. and F.R., Switzerland, Austria, Slovakia, Hungary, Romania, Poland. m. Russia; Nearctic region.

Biology. On *Artemisia abrotanum* and *maritima* (Ribaut, 1936b; Ossiannilsson, 1946c). Adults in July–September.

245. ***Eupteryx urticae*** (Fabricius, 1803)
Plate-figs. 99, 119, text-figs. 1758–1764.

Cicada urticae Fabricius, 1803: 77.
Eupteryx urticae conjuncta Rey, 1894: 77.
Eupteryx urticae haupti W. Wagner, 1935: 29.
Eupteryx urticae deficiens Ribaut, 1936b: 155.
Eupteryx urticae glomerata Ribaut, 1936b: 155.

Elongate, light yellow, shining. Head with black markings varying in extension. Frontoclypeus on each side with a brownish band extending to antennal pit. Frontoclypeus between eyes with a pair of rounded black spots (f. *typica*) absent in f. *deficiens* Ribaut, confluent in f. *haupti* W. Wagner. Vertex frontally with two large black spots, medially at caudal border with a somewhat smaller, broadly triangular black spot. These spots are confluent in f. *conjuncta* Rey, f. *glomerata* Ribaut combining the characters of f. *conjuncta* and f. *haupti*. Pronotum frontally with four, caudally with four black spots, an M-shaped greyish brown figure connecting the median caudal spots. Scutellum laterally with two large black spots. Thoracal venter with large black patches. Fore wings with yellowish brown markings (Plate-fig. 99). Hind tibiae laterally usually black. Last segment, often also distal half of 2nd segment of hind tarsus in male black-brown, in female faintly brownish. Abdomen largely black, genital plates of male (Text-fig. 1759) yellow. Male pygofer lobe as in Text-fig. 1758, genital style as in Text-fig. 1760. Aedeagus with two forked apical appendages, lower branches much varying in length and thickness, usually much thinner than upper branches (Text-figs. 1761–1763). 2nd–5th abdominal sterna in male as in Text-fig. 1764. Overall length (♂♀) 2.9–3.6 mm.

Distribution. Common and widespread in Denmark. – Not uncommon in southern Sweden, found in Sk., Bl., Sm., Öl., Ög. – Norway: found in VAy: Kristiansand (Warloe), Lyngdal and Loshavn (Holgersen), and in Ry: Madla and Hognestad (Holgersen). – East Fennoscandia: only found in Ab: Raisio 7.VIII.1975 (I. Dworakowska). – Widespread in Europe, also found in Armenia, Georgia, Kazakhstan, w. Siberia.

Biology. On *Urtica dioica* (Kuntze, 1937; Wagner & Franz, 1961). Host-plant: *Urtica* (Günthart, 1974). Adults in June–September.

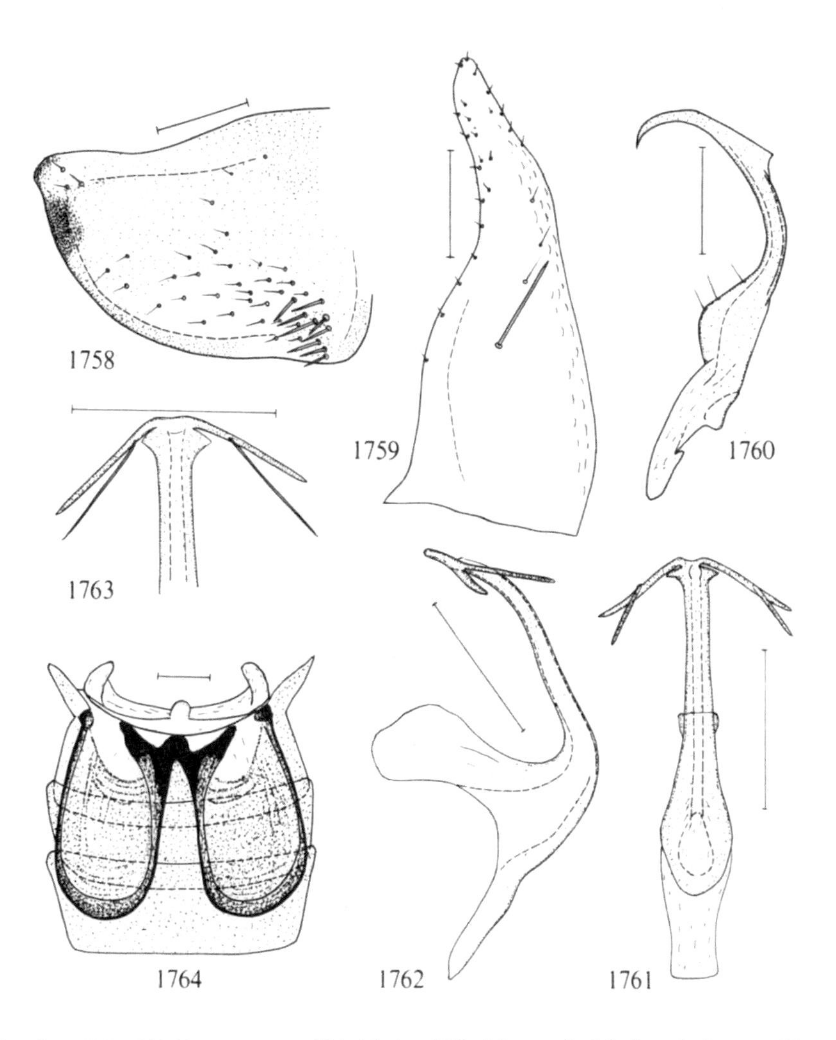

Text-figs. 1758–1764. *Eupteryx urticae* (Fabricius). – 1758: right pygofer lobe in male from outside; 1759: left genital plate from outside; 1760: right genital style from above; 1761: aedeagus in ventral aspect; 1762: aedeagus from the left; 1763: apex of aedeagus of another specimen in ventral aspect; 1764: 2nd–5th abdominal sterna in male from above. Scale: 0.1 mm.

246. ***Eupteryx cyclops*** Matsumura, 1906
Plate-figs. 124–126, 146, text-figs. 1765–1768.

Eupteryx cyclops Matsumura, 1906: 78.
Eupteryx urticae nigrifrons Haupt, 1912: 191.
Eupteryx britteni Edwards, 1924: 54.
Eupteryx affinis Ossiannilsson, 1936a: 255.
Eupteryx cyclops mendax Ribaut, 1936b: 157.
Eupteryx cyclops trilobata Ribaut, 1936b: 157.

Resembling *urticae.* In f. *typica* there is a large dark spot on middle of anteclypeus, lower half of frontoclypeus being entirely brownish, upper half medially with a rounded, transversely oval or pentagonal brown spot (Plate-fig. 124); vertex with the same three spots as in *urticae,* the median one longer and narrower than in *urticae* (Plate-fig. 125). In f. *nigrifrons* Haupt frontoclypeus is black up to level of antennae, a broad black longitudinal band extending to caudal border of vertex. In f. *trilobata* Ribaut the spots of vertex are confluent into an Y-shaped figure (Plate-fig. 126). Upper part of face in f. *mendax* Ribaut without black spots or with two separate spots as in *urticae* f. *typica.* Pronotum as in *urticae,* the lateral caudal spot usually being smaller than in that species. Fore wing with brown markings as in *urticae,* its very apex usually without a hyaline spot, the colourless spot in 2nd and 3rd cells usually being larger than in *urticae.* Colour of hind tibiae varying. Hind tarsi in both sexes usually entirely light, only extreme end of 3rd segment sometimes brownish. Male genital style and connective as in Text-fig.

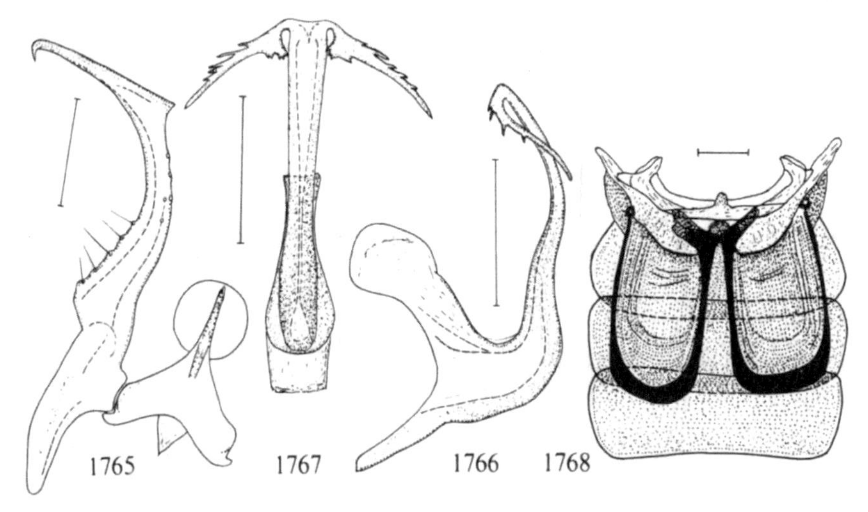

Text-figs. 1765–1768. *Eupteryx cyclops* Matsumura. – 1765: right genital style and connective from above; 1766: aedeagus from the left; 1767: aedeagus in ventral aspect; 1768: 2nd–5th abdominal sterna in male from above. Scale: 0.1 mm.

1765, aedeagus as in Text-figs. 1766, 1767, 2nd–5th abdominal sterna in male as in Text-fig. 1768. Overall length 3.0–3.7 mm.

Distribution. Fairly common in Denmark, found in EJ, LFM and NEZ. – Common and widespread in Sweden, Sk.–Lu.Lpm. – Widespread in Norway, found in AK, HEn, Bø, TEi, HOy, HOi, SFy, STi, Nsy, TRi. – Common in southern and central East Fennoscandia, Al–ObN. – England, Scotland, France, Netherlands, German D.R. and F.R., Switzerland, Italy, Austria, Bulgaria, Bohemia, Moravia, Slovakia, Romania, Poland, Latvia, Lithuania, m. Russia, Ukraine, Moldavia, Kazakhstan, m. Siberia.

Biology. On *Urtica* (Ribaut, 1936b; Ossiannilsson, 1946c; Linnavuori, 1952a). On *Urtica dioica* (Wagner & Franz, 1961; Schiemenz, 1965; Raatikainen & Vasarainen, 1976). Host-plant: *Urtica* (Günthart, 1974). Adults in June–October.

247. ***Eupteryx calcarata*** Ossiannilsson, 1936
Plate-figs. 127, 128, text-figs. 1769–1772.

Eupteryx calcarata Ossiannilsson, 1936a: 256.
Eupterix vallesiaca Cerutti, 1939: 90.

Resembling *urticae* f. *typica;* dark bands along sides of frontoclypeus usually very broad, lateral spots on vertex usually larger, the median one smaller than in *urticae* (Plate-fig. 127). Lateral spots sometimes almost confluent with median spot (Plate-fig. 128). Third

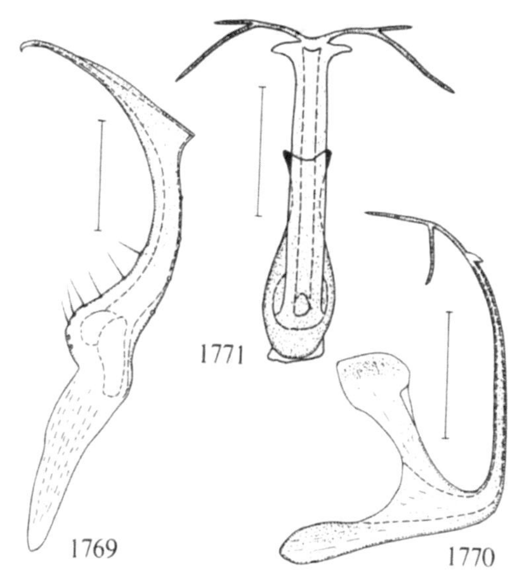

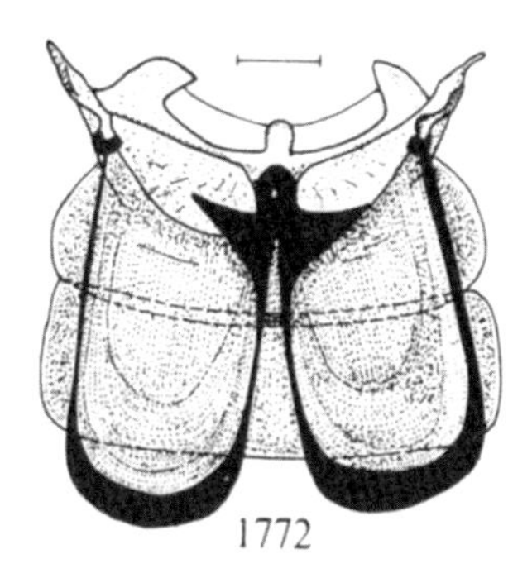

Text-figs. 1769: 1772. *Eupteryx calcarata* Ossiannilsson. – 1769: right genital style from above; 1770: aedeagus from the left; 1771: aedeagus in ventral aspect; 1772: 2nd–4th abdominal sterna in male from above. Scale: 0.1 mm.

segment and distal half of second segment of hind tarsus usually dark in males, often also in females. Male genital style as in Text-fig. 1769, aedeagus as in Text-figs. 1770, 1771, 2nd–4th abdominal sterna in male as in Text-fig. 1772. Overall length (♂♀) 3.3–3.6 mm.

Distribution. Denmark: one male found in NEZ: Dyrehaven 22.X.1917 by Oluf Jacobsen. - Sweden: not uncommon in the southern part, found in Sk., Öl., Ög., Boh., Sdm., Upl., Vstm. - So far not found in Norway. - East Fennoscandia: Al: Sund 20.VIII.1975 (Albrecht); Ab: Lojo 1918, 1920, 1922 (P. H. Lindberg, Håkan Lindberg), Pojo 30.VIII.1976 (Albrecht). - German F.R. and D.R., Switzerland, Austria, Bulgaria, Bohemia, Slovakia, Moravia, Hungary, Romania, Poland, Estonia, Latvia, Lithuania, Kazakhstan.

Biology. On nettles, 2 generations (Ossiannilsson, 1946c; Schwoerbel, 1957). Host-plant: *Urtica* (Günthart, 1974). First-instar larvae found on *Urtica* 19.V.(1944), adults in June–September.

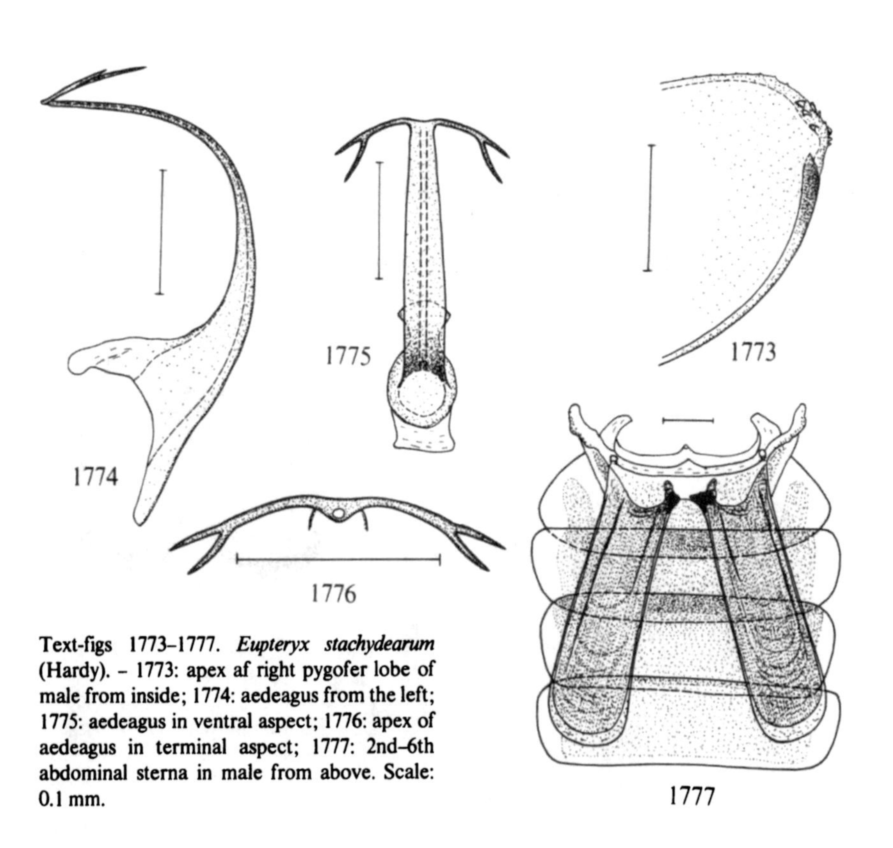

Text-figs 1773–1777. *Eupteryx stachydearum* (Hardy). - 1773: apex af right pygofer lobe of male from inside; 1774: aedeagus from the left; 1775: aedeagus in ventral aspect; 1776: apex of aedeagus in terminal aspect; 1777: 2nd–6th abdominal sterna in male from above. Scale: 0.1 mm.

248. ***Eupteryx stachydearum*** (Hardy, 1850)
Plate-figs. 100, 123, 129, 147, text-figs. 1773–1777.

Typhlocyba stachydearum Hardy, 1850: 422.
Typhlocyba curtisii Flor, 1861a: 431.

Resembling *calcarata*. Median spot on vertex transversely oval (Plate-fig. 129). Face in male usually with a pair of small black spots (sometimes missing) in upper part, and with a pair of black spots below antennae. Anteclypeus in female black-brown, frontoclypeus with a pair of brownish longitudinal bands, above these a pair of rounded black spots, laterally of these above antennae a black streak (Plate-fig. 123). Space below antennae black. Fore wings as in Plate-fig. 100. Hind tibiae in male apically narrowly black, in female narrowly brownish. 1st segment of hind tarsus in male apically broadly or narrowly black, third segment apically dark; these markings may be present or absent in females. Male pygofer lobe as in Text-fig. 1773, aedeagus as in Text-figs. 1774–1776, 2nd–6th abdominal sterna in male as in Text-fig. 1777. Overall length (♂♀) 2.8–3.6 mm.

Distribution. Uncommon in Denmark, found in EJ, F, NEZ and B. – Not uncommon, often abundant in southern Sweden, found in Sk., Bl., Öl., Gtl., Vg., Upl. – I have not seen unquestionable specimens from Norway. – Not recorded from East Fennoscandia. – Widespread in Europe, also in Algeria, Cyprus, Anatolia, Palestine, Armenia, Georgia, Azerbaijan, Kazakhstan.

Biology. On *Stachys sylvatica* (Kuntze, 1937). On *Stachys silvatica, Lamium album;* two generations (Ossiannilsson, 1946c). "An feuchten, schattigen Stellen auf Labiaten, z.B. *Stachys silvatica*" (Wagner & Franz, 1961). In xerophilous and mesophilous biotopes (Schiemenz, 1969b). Oligophagous species on Labiatae, infesting *Melissa officinalis* and *Mentha piperita* in Italy (Vidano & Arzone, 1978). – In the spring of 1944 I reared *E. stachydearum* on *Lamium album* in an unheated insectarium in Upland. The first adult (♂) appeared 14.VI, all specimens having turned into adults 20.VI, copulation observed 20.VI. Adults in June–October.

249. ***Eupteryx collina*** (Flor, 1861)
Text-figs. 1778–1781.

Typhlocyba collina Flor, 1861a: 433.
Eupteryx alticola Ribaut, 1936b: 165.

Genae black along inner border. Anteclypeus blackish in both sexes, only exceptionally light in males. Frontoclypeus with two dark brown longitudinal bands consisting of horizontal muscle traces, confluent towards anteclypeus. On upper part of frontoclypeus two black spots. Black streak above antenna parallel-sided. Vertex with three black spots, the median one much varying in shape and size. Dark spots on pronotum well marked, those on anterior border separate. Scutellum with two lateral black spots and two black points in front of the transverse furrow. Dark patches on fore wings with large light centra. Abdomen black, transversely yellow-striped. In male, apex of hind tibia, terminal 1/3 of 1st segment (except projecting ventral apical lobe), and at least terminal 3/4 of 3rd segment of hind tarsus blackish, 2nd segment light. In females, the 3rd tarsal segment is entirely black, apices of hind tibia and 1st hind tarsal segment often distinctly infuscate. Caudal part of male pygofer lobe as in Text-fig. 1778, aedeagus as in Text-figs. 1779–1781. Overall length (♂♀) 2.7–3.25 mm.

Distribution. Not found in Denmark, Sweden and Norway. – East Fennoscandia: only found in Kr: Parikkala by J. Sahlberg. – France, German F.R., Switzerland, Austria, n. Italy, Bohemia, Moravia, Slovakia, Romania, Latvia, Lithuania, n. Russia, Moldavia, Anatolia, Iraq, Afghanistan.

Biology. "Sur les Menthes des régions élevées et des bas-fonds" (Ribaut, 1936b). "An *Mentha* sp." (Wagner & Franz, 1961). Infesting *Melissa officinalis, Origanum majorana, Origanum onites, Origanum vulgare, Salvia officinalis, Salvia sclarea, Satureja hortensis* in Piedmont, Italy. Originates from mountainous territories (Vidano & Arzone, 1978).

250. ***Eupteryx thoulessi*** Edwards, 1926
Plate-figs. 101, 122, 130, 148, text-figs. 1782–1785.

Eupteryx thoulessi Edwards, 1926: 53.

Resembling *collina.* Anteclypeus in male always light, in female infuscate at most on sides. Black streaks above antennae usually club-shaped, well developed, often almost as large as spots between them. Light colour of vein R sometimes getting thinner within the black spot distally of wax area. Caudal part of male pygofer lobe as in Text-fig.

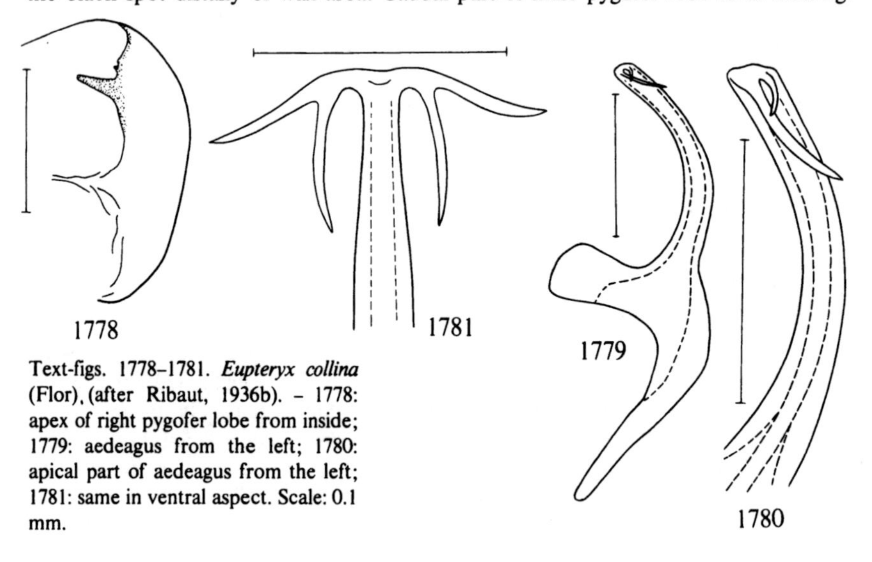

Text-figs. 1778–1781. *Eupteryx collina* (Flor), (after Ribaut, 1936b). – 1778: apex of right pygofer lobe from inside; 1779: aedeagus from the left; 1780: apical part of aedeagus from the left; 1781: same in ventral aspect. Scale: 0.1 mm.

1782, aedeagus as in Text-figs. 1783, 1784, 2nd-5th abdominal sterna in male as in Text-fig. 1785. Overall length (♂♀) 2.7-3.25 mm.

Distribution. Not found in Denmark, Norway and East Fennoscandia. - Rare in southern Sweden, found in Sk.: Revingehed 24.VII.1938 (Ossiannilsson), Brunnby, Kullen and Mölle 27.VIII.1960 (Håkan Lindberg); Bl.: Karlskrona, Gullberna 29.VI.-1972 (Gyllensvärd). - England, France, Netherlands, German D.R. and F.R., Italy, Switzerland, Austria, Hungary, Romania, Slovakia, Yugoslavia, Anatolia, Georgia, Ukraine.

Biology. "Sur les Menthes et le Lycope" (Ribaut, 1936b). "An *Mentha aquatica*" (Kuntze, 1937). Oligophagous on Labiatae, infesting *Mentha piperita* in Italy (Vidano & Arzone, 1978).

251. ***Eupteryx vittata*** (Linné, 1758)
Plate-figs. 133, 149, text-figs. 1786–1792.

Cicada vittata Linné, 1758: 438.

Fairly elongate, parallel-sided. Head frontally rounded obtuse angular. Whitish yellow to light yellow. Vertex with a pair of large, often confluent black spots at hind border. Pronotum brownish black, often with a light median streak and a few small light spots. Scutellum brownish black, usually with a narrow median light stripe towards caudal apex. Fore wing with a broad undulated black-brown longitudinal band (Plate-fig. 133). Abdomen largely blackish. Male pygofer lobe as in Text-fig. 1786, male anal apparatus as in Text-fig. 1787, genital plate as in Text-fig. 1788, genital style as in Text-fig. 1789,

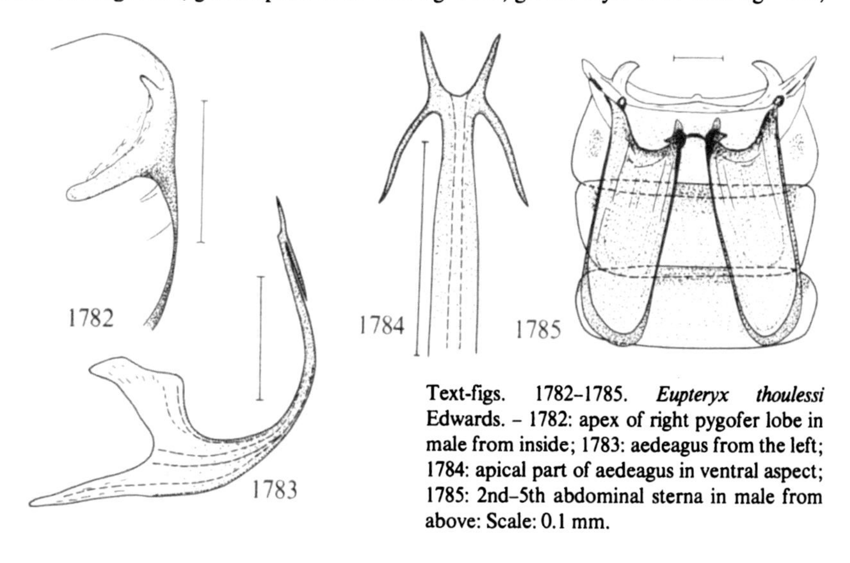

Text-figs. 1782–1785. *Eupteryx thoulessi* Edwards. - 1782: apex of right pygofer lobe in male from inside; 1783: aedeagus from the left; 1784: apical part of aedeagus in ventral aspect; 1785: 2nd–5th abdominal sterna in male from above: Scale: 0.1 mm.

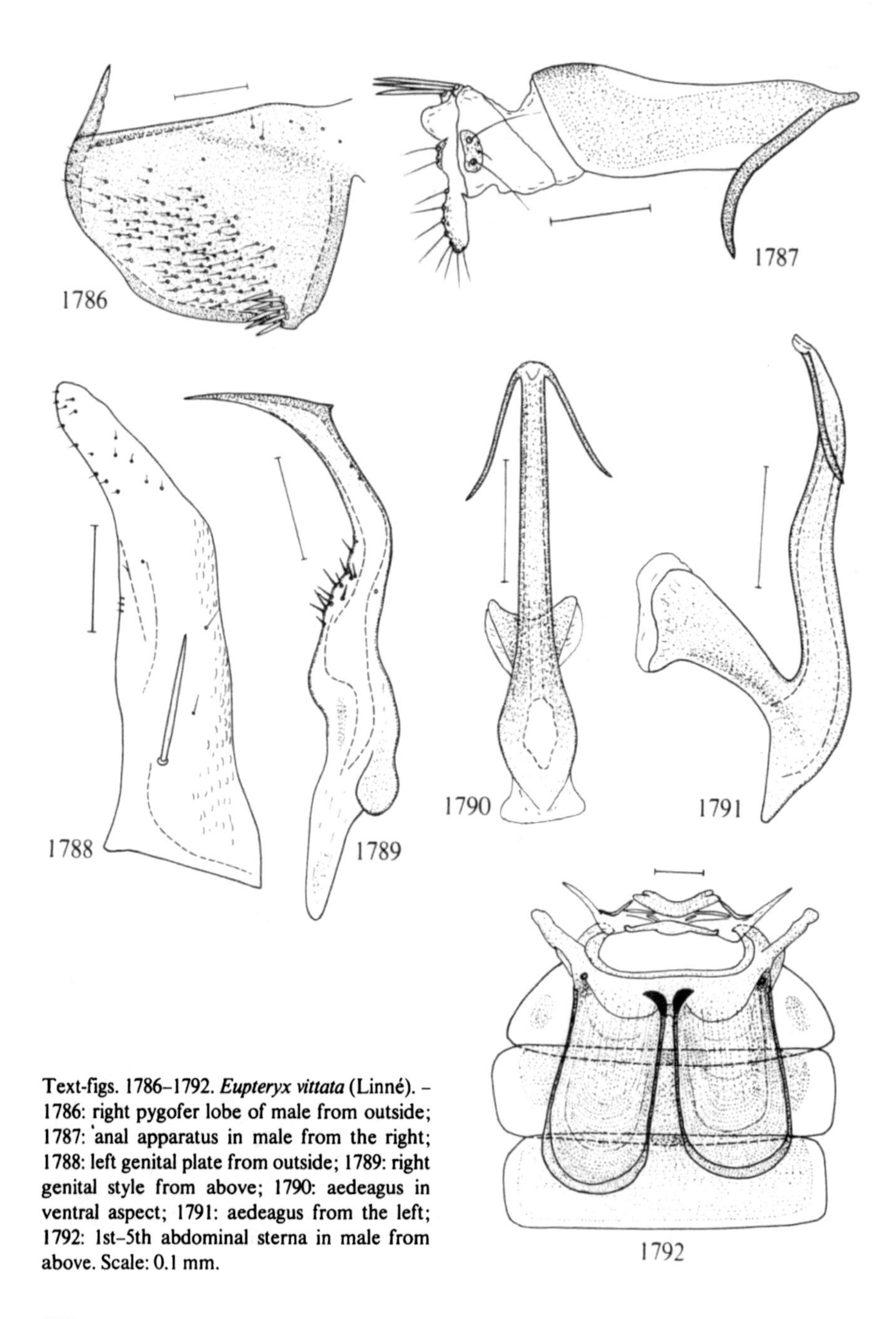

Text-figs. 1786–1792. *Eupteryx vittata* (Linné). – 1786: right pygofer lobe of male from outside; 1787: anal apparatus in male from the right; 1788: left genital plate from outside; 1789: right genital style from above; 1790: aedeagus in ventral aspect; 1791: aedeagus from the left; 1792: 1st–5th abdominal sterna in male from above. Scale: 0.1 mm.

aedeagus as in Text-figs. 1790, 1791, 1st–5th abdominal sterna in male as in Text-fig. 1792. Overall length (♂♀) 3.1–3.7 mm.

Distribution. Common and widespread in Denmark. – Fairly common in southern Sweden (Sk.–Vstm., Gstr., Jmt.). – Norway: found in AK: Drøbak (Siebke), and in Bv: Uvdal, Geitsjøen (in coll. Mus. Bergen). – Fairly common but sporadic in southern and central East Fennoscandia, found in Al, Ab, N, Ka, St, Ta, Oa; Kr, Lr. – Widespread in Europe, also in Algeria.

Biology. "Elle vit dans les endroits humides, ou simplement frais et ombragés, et peut y être récoltée en fauchant les Orties et les Menthes" (Ribaut, 1936b). "An Kräutern (vor allem Compositen)" (Kuntze, 1937. Larvae living on *Ranunculus repens* (Dlabola, 1954). Hibernation in the egg stage (Remane, 1958). Adults in June–October.

252. ***Eupteryx notata*** Curtis, 1937
Plate-fig. 134, text-figs. 1793–1797.

Eupteryx notata Curtis, 1837b: pl. 640.
Typhlocyba wallengreni Stål, 1853: 177.

Resembling *vittata* but much smaller, also comparatively less elongate. Spots of vertex often reduced. Fore wing as in Plate-fig. 134, male pygofer lobe as in Text-figs. 1793, 1794, aedeagus as in Text-figs. 1795, 1796, 2nd–5th abdominal sterna in male as in Text-fig. 1797. Overall length 2.1–2.7 mm. – Morcos (1953) reproduced the 5th instar larva.

Distribution. Common and widespread in Denmark as well as in Sweden (Sk.–P.Lpm.). – Norway: AK: vicinity of Oslo (Reuter); Ø: Tveiten, Trøgstad (Holgersen); HOy: Herdla, July 1937 (Knaben?); SFy: Brekke 23.VIII.1945 (S. Johnsen). – Common in major part of East Fennoscandia, found as far to the north as in Petsamo. – Widespread in Europe, also found in Altai Mts., Kazakhstan, m. Siberia, Mongolia.

Biology. "An *Hieracium pilosella*" (Kuntze, 1937). In dry localities, on *Hieracium pilosella, Prunella vulgaris, Thymus* (Ossiannilsson, 1946c). Two generations (Kontkanen, 1948). On *Hypochoeris radicata.* Three generations in England, hibernation in the egg stage (Morcos, 1953). "Mehr in trockenen oder wenigstens zeitweilig trockenen Wiesen" (Marchand, 1953). Two generations, hibernation in the egg stage (Remane, 1958; Schiemenz, 1969b). Host-plants: *Crepis, Crepis aurea* (Günthart, 1974). Adults in May–October.

253. ***Eupteryx tenella*** (Fallén, 1806)
Plate-fig. 116, text-figs. 1798–1804.

Cicada tenella Fallén, 1806: 43.

Fairly elongate, shining, whitish yellow to dirty yellow. Anteclypeus usually dark, frontoclypeus brown-bordered. Vertex with three large black markings: two rounded or polygonal spots on transition to frontoclypeus, and one transversely oval spot at hind border. Pronotum entirely light or with a large brownish patch sometimes occupying its entire surface except a transverse spot behind each eye, sometimes leaving anterior and posterior borders. Scutellum with a large triangular black spot on each lateral corner, caudal apex often orange coloured. Fore wings with 3 or 4 brownish longitudinal bands: a narrow one along claval commissure, a broad one on both sides of claval suture, and a narrow one on each side of vein M, the posterior one of the latter often indistinct, only fumose. Apical part of fore wing fumose. Hind wings with dark veins. Abdomen largely black. Male pygofer and anal apparatus as in Text-figs. 1798, 1799, genital plate as in Text-fig. 1800, genital style as in Text-fig. 1801, aedeagus as in Text-figs. 1802, 1803, 1st–5th abdominal sterna in male as in Text-fig. 1804. Overall length (♂♀) 2.7–3 mm.

Distribution. Not found in Denmark. – Scarce in Sweden, found in several places in Sk., Ög., and Upl. – I have not seen specimens from Norway. – Rare in southern East Fennoscandia, found in Ab, N, Ka, Sa, Kb; Vib, Kr. – England, Scotland, Netherlands, German D.R. and F.R., Switzerland, Italy, Austria, Hungary, Romania, Moravia, Bohemia, Slovakia, Poland, Estonia, Lithuania, Latvia, n. and m. Russia, Ukraine, Moldavia, Anatolia.

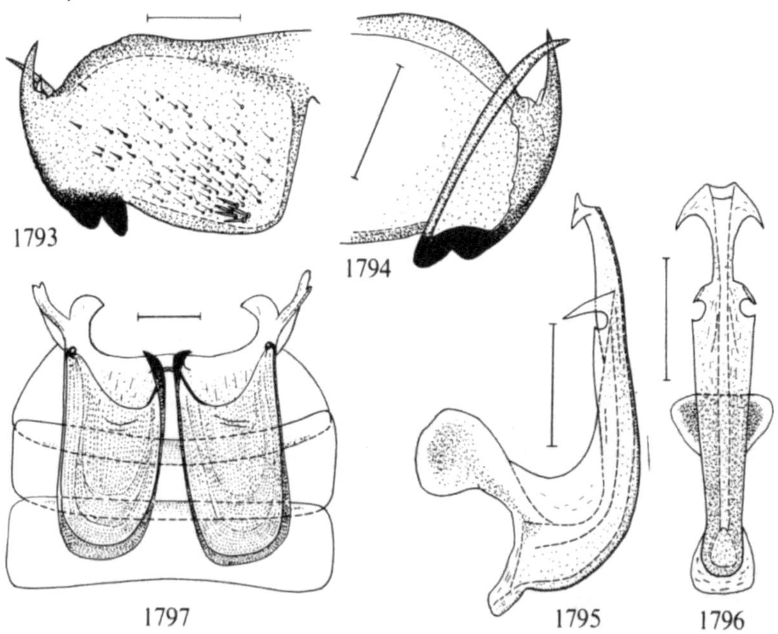

Text-figs. 1793–1797. *Eupteryx notata* Curtis. – 1793: right pygofer lobe of male from outside; 1794: apical part of right pygofer lobe of male from inside; 1795: aedeagus from the left; 1796: aedeagus in ventral aspect; 1797: 2nd–5th abdominal sterna in male from above. Scale: 0.1 mm.

Biology. "An Gräsern" (Wagner & Franz, 1961). In German D.R. 2 generations (Schiemenz, 1964). In xerophilous and mesophilous biotopes. Hibernation in the egg stage (Schiemenz, 1969b). On *Achillea millefolium* (Allen, 1966). In 1948 I found three specimens on *Hyssopus officinalis* in Ög. Adults in June–September.

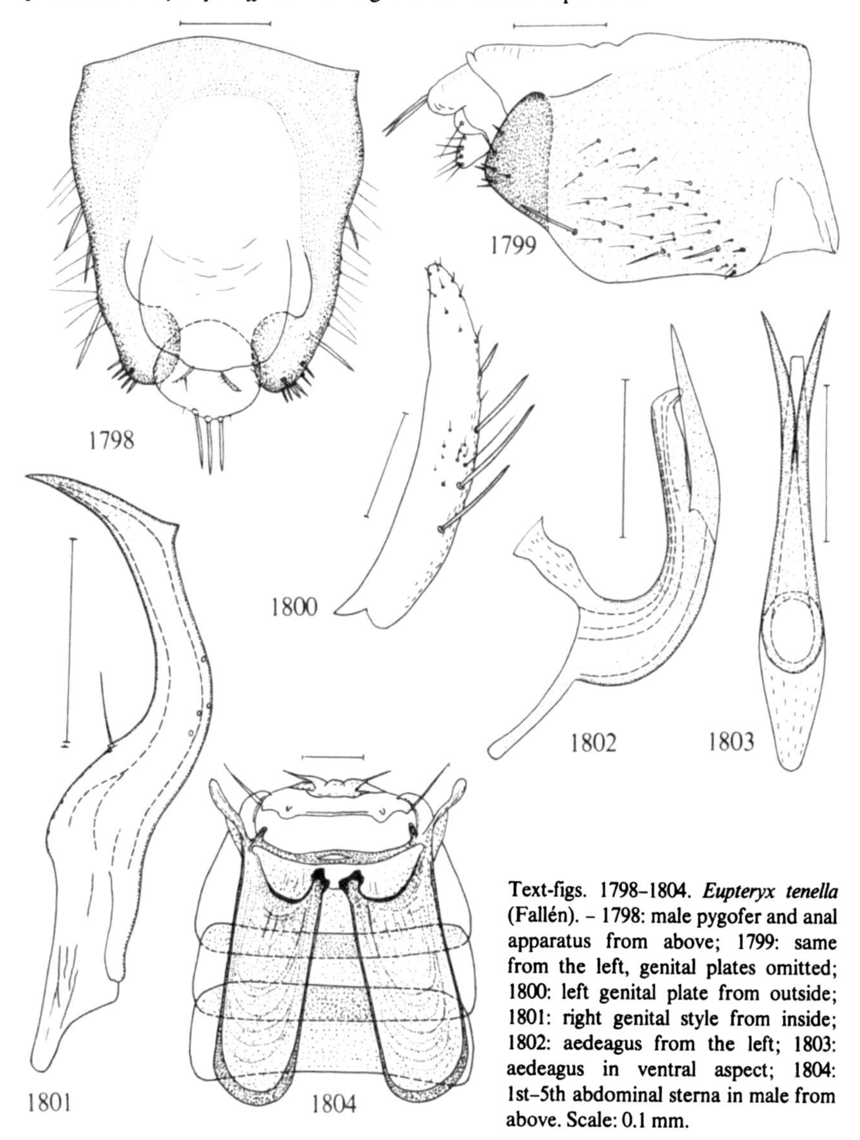

Text-figs. 1798–1804. *Eupteryx tenella* (Fallén). – 1798: male pygofer and anal apparatus from above; 1799: same from the left, genital plates omitted; 1800: left genital plate from outside; 1801: right genital style from inside; 1802: aedeagus from the left; 1803: aedeagus in ventral aspect; 1804: 1st–5th abdominal sterna in male from above. Scale: 0.1 mm.

Genus *Aguriahana* Distant, 1918

Aguriahana Distant, 1918: 105.
  Type-species: *Aguriahana metallica* Distant, 1918, by original designation.
Eupteroidea Young, 1952: 92.
  Type-species: *Typhlocyba stellulata* Burmeister, 1841, by original designation.
Asymmetropteryx Dlabola, 1958: 52.
  Type-species: *Typhlocyba pictilis* Stål, 1853, by original designation.
Wagneripteryx Dlabola, 1958: 53.
  Type-species: *Cicadula germari* Zetterstedt, 1838, by original designation.

Apical part of fore wing widened. Transverse vein connecting peripheric vein in hind wing with M reaching the latter distally of transverse vein between M and R. Basal part of male genital style short, tapering. Genital plate of male usually with a group of small pegs proximally of apex. In Denmark and Frnnoscania three species.

Key to species of *Aguriahana*

1 Fore body and fore wings unicolorous greyish green, yellowish green or olive-green without dark markings. On *Pinus* ... 254. *germari* (Zetterstedt)
– At least fore wings with dark markings ... 2
2 (1) Larger, 3.8–4.4 mm. Apex of fore wing with a shallow incision (Plate-fig. 120). On *Tilia, Prunus* etc. ... 256. *stellulata* (Burmeister)
– Smaller, 3.3–3.6 mm. Apex of fore wing rounded (Plate-fig. 121). On *Vaccinium myrtillus* and *Rhododendron* sp. ... 255. *pictilis* (Stål)

254. ***Aguriahana germari*** (Zetterstedt, 1838)
Plate-fig. 150, text-figs. 1270, 1805–1812.

Cicadula germari Zetterstedt, 1838: 301.

Elongate, shining, greenish yellow to greyish green to olive-coloured. Vertex frontally angular. Apical part of fore wing fumose. Hind wings fumose with dark veins. Dorsum and venter of abdomen greenish black with light segment borders, venter sometimes largely light. Male pygofer lobe as in Text-fig. 1805, genital plate as in Text-figs. 1806, 1807, genital style as in Text-fig.1808, aedeagus as in Text-figs. 1809, 1810, 1st–6th abdominal sterna in male as in Text-fig. 1811, 2nd abdominal tergum as in Text-fig.

Text-figs. 1805–1812. *Aguriahana germari* (Zetterstedt). – 1805: right pygofer lobe of male from outside; 1806: left genital plate from outside; 1807: apex of left genital plate from above; 1808: right genital style from above; 1809: aedeagus in ventral aspect; 1810: aedeagus from the right; 1811: 1st–6th abdominal sterna in male from above; 1812: 2nd abdominal tergum in male from the front. Scale: 0.1 mm.

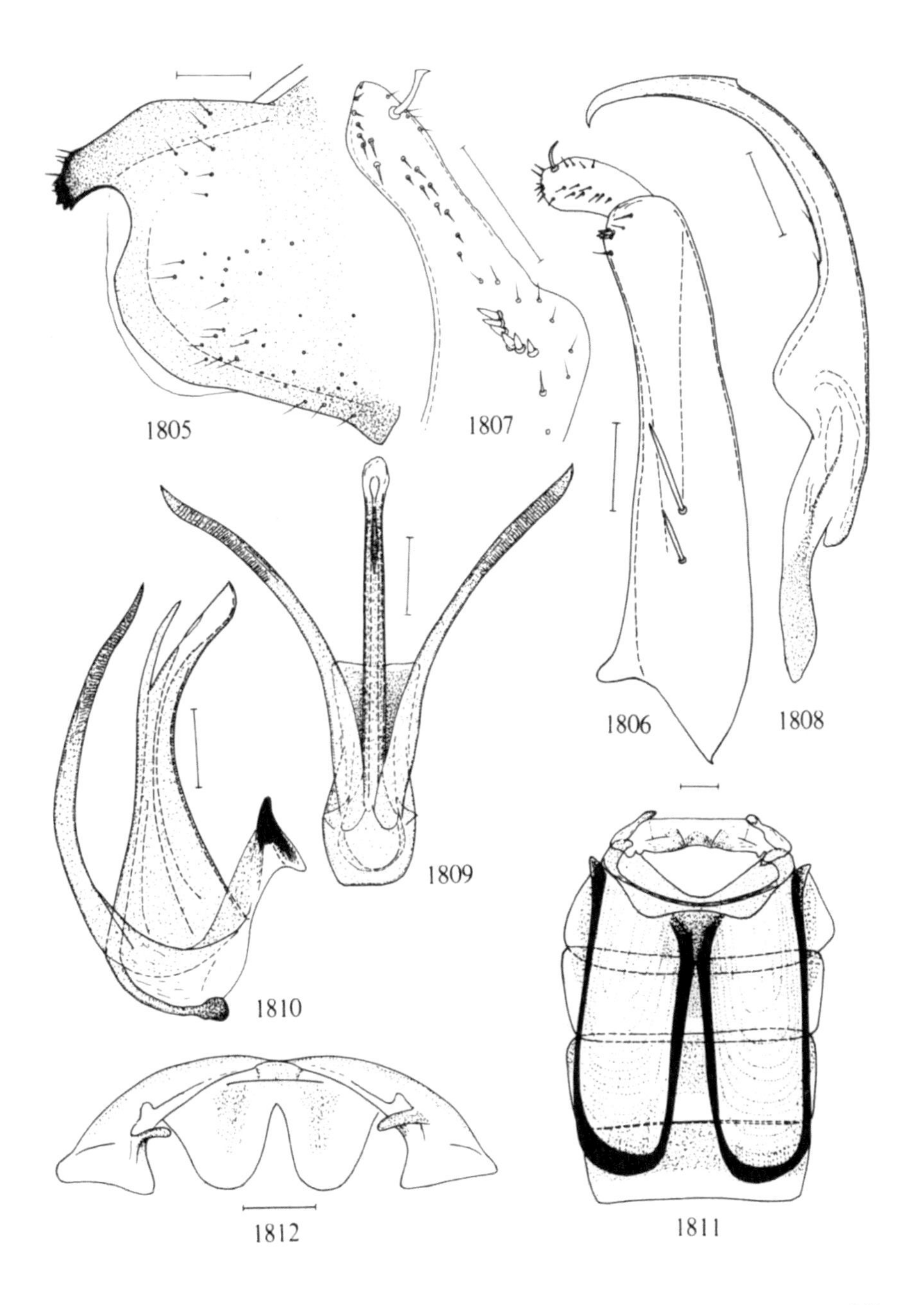
1805
1807
1806
1808
1809
1810
1811
1812

1812. Overall length 4.0–4.6 mm. – Last instar larva entirely pale brown, tergum with sparse raised dark spots, hairless, abdomen with very small lateral setae, setae on tibiae also short.

Distribution. Denmark: so far found in EJ, SJ, NEZ and B. Common and widespread in Sweden, Sk.–Nb., Lu.Lpm. – Widespread in Norway, found in AK, HEs, HEn, On, Bv, TEi, HOy, Nsi, Nnø, TRi. – Moderately common in southern and central East Fennoscandia, found as far to the north as in Sb, Kb and Ks. Also in Vib and Kr. – Widespread in Europe, also in Algeria, Altai Mts., Georgia, Kazakhstan, w. Siberia, Mongolia, Maritime Territory.

Biology. On *Pinus silvestris,* adults in June–October.

255. ***Aguriahana pictilis*** (Stål, 1853)
Plate-fig. 121, text-figs. 1813–1818.

Typhlocyba pictilis Stål, 1853: 176.

Elongate, white or dirty white, shining. Head frontally produced, rounded, medially considerably longer than near eyes, dorsally along fore border with or without a fine brownish line from eye to eye. Pronotum on each side behind eye usually with a longitudinal black-brown streak continuing on mesopleurum. Fore wing with fumose to blackish markings as in Plate-fig. 121. Genital segment in both sexes often partly brownish or blackish. Extreme apices of hind tibiae and hind tarsi often black. Male pygofer lobe as in Text-fig. 1813, genital plate as in Text-fig. 1814, genital style as in Text-fig. 1815, aedeagus as in Text-figs. 1816, 1817, 1st–5th abdominal sterna in male as in Text-fig. 1818. Overall length (♂♀) 3.3–3.6 mm. – Larvae occurring in two colour varieties: one entirely yellowish, another largely brownish black with legs yellow except hind tibiae and apices of hind femora. Head, pronotum, fore wing buds, abdomen and tibiae with many long erect setae, those on fore margin of head black, those on wing buds dirty yellow, abdominal setae white.

Distribution. Very rare in Denmark, only found in EJ: Silkeborg 4.–8.IX.1919 (Jensen-Haarup). – Scarce, locally abundant in central and northern Sweden, found in Boh., Sdm., Upl., Vstm., Dlr., Med., Jmt., Vb., Nb. – Norway: AK: Nesodden 5.VIII.-1946 (Holgersen); Bø: Ringerike, Krogkleven (O. M. Reuter), near Krøderen (teste Dworakowska, 1972). – Scarce in southern East Fennoscandia, found in Ab, N, St, Ta, Sa, Kb; Kr. – Netherlands, German F.R., Switzerland, Austria, Bohemia, Moravia, Poland, Estonia, Latvia, Lithuania, n. Russia, Altai Mts., Maritime Territory, Korean Peninsula.

Biology. In moist *Oxalis-Myrtillus* spruce woods (Linnavuori, 1952). On *Vaccinium myrtillus* (Sahlberg, 1871; Ossiannilsson, 1946c; Wagner & Franz, 1961). Food-plants: *Vaccinium myrtillus* L., *Rhododendron* sp (Dworakowska, 1972). Adults in July–September.

256. ***Aguriahana stellulata*** (Burmeister, 1841)
Plate-fig. 120, text-figs. 1819–1824.

Typhlocyba stellulata Burmeister, 1841: [3], pl.[13], fig. 1.

Body with sides arched, convex, greatest width somewhat frontally of middle. Pearl-white, shining, fore wings on distal half with fuscous and black markings, as in Plate-fig.

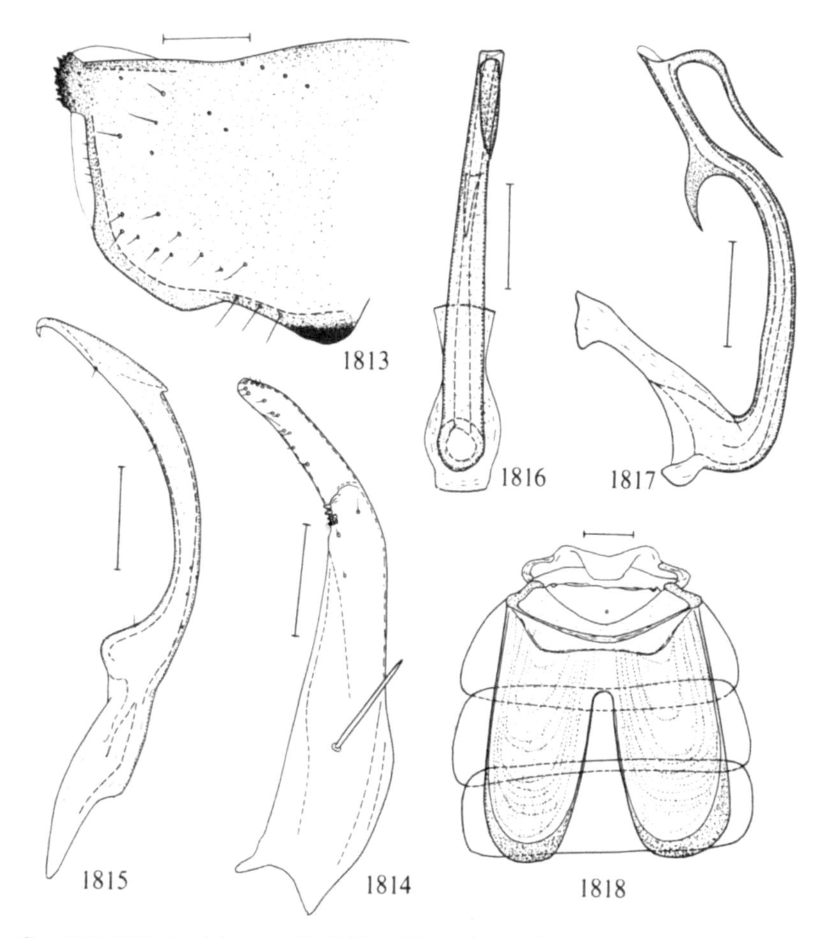

Text-figs. 1813–1818. *Aguriahana pictilis* (Stål). – 1813: right pygofer lobe of male from outside; 1814: left genital plate from outside; 1815: right genital style from above; 1816: aedeagus in ventral aspect; 1817: aedeagus from the left; 1818: 1st–5th abdominal sterna in male from above. Scale: 0.1 mm.

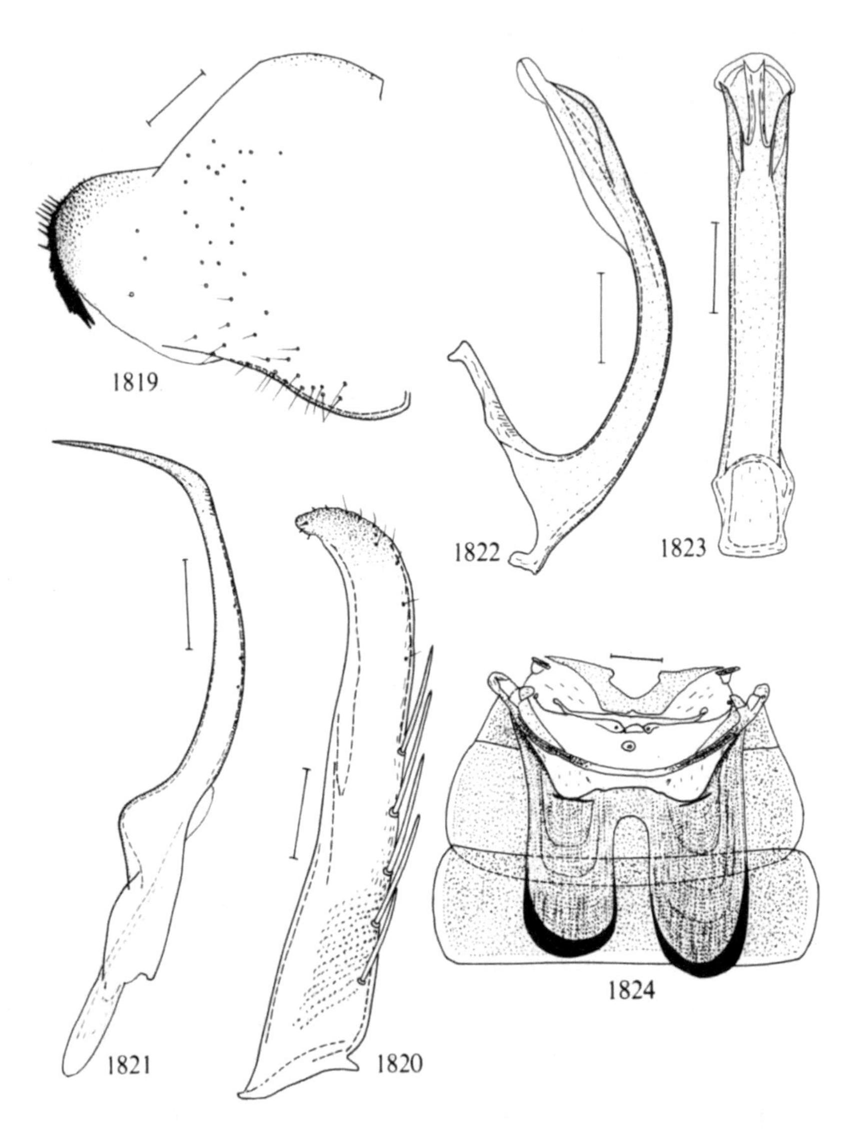

Text-figs. 1819–1824. *Aguriahana stellulata* (Burmeister). – 1819: right pygofer lobe of male from outside; 1820: left genital plate from outside; 1821: right genital style from above; 1822: aedeagus from the left; 1823: aedeagus in ventral aspect; 1824: 1st–4th abdominal sterna in male from above. Scale: 0.1 mm.

120. Male pygofer lobe as in Text-fig. 1819, genital plate as in Text-fig. 1820, genital style as in Text-fig. 1821, aedeagus as in Text-figs. 1822, 1823, 1st–4th abdominal sterna in male as in Text-fig. 1824. Overall length (♂♀) 3.8–4 mm. - Last instar larva pale yellow with areas of dark brown coloration, dorsally densely covered with setae much varying in length. Setae in brown areas may be dark.

Distribution. Rare in Denmark, only a few records from SZ and NEZ. - Not uncommon in southern Sweden, found in Sk., Sm., Sdm., Upl., Nrk., Vstm., Dlr. - Norway: HEs: Bröttum 4.IX.1972 (G. Söderman in litt.). - Scarce in southern East Fennoscandia, found in Ab, N, Ta, Tb; Kr. - England, France, Netherlands, German D.R. and F.R., Switzerland, Austria, Hungary, Bulgaria, Bohemia, Moravia, Slovakia, Romania, Poland, Latvia, Lithuania, Moldavia, m. Russia, Ukraine, Algeria, Tunisia, Mongolia, Maritime Territory, Japan; introduced into North America.

Biology. "An Birken und Kirschbäumen" (Kuntze, 1937). On *Tilia cordata* (Ossiannilsson, 1946c). Hibernation in the egg stage (Müller, 1957). On *Tilia* and *Prunus cerasus* (Wagner & Franz, 1961). On "Weichselkirsche" (Günthart, 1971). "Food plants: *Tilia cordata* Mill., *Tilia* sp., *Prunus cerasus* L., *P. avium* L., *P. insititia* Juslen, *Ulmus* sp., *Crataegus* sp., *Acer mono* Maxim".(Dworakowska, 1972). Adults in July–September.

Genus *Alnetoidia* Dlabola, 1958

Alnetoidia Dlabola, 1958: 55.
Type-species: *Cicadula alneti* Dahlbom, 1850, by original designation.

Elongate species, in habitus and colour resembling *Edwardsiana* Zachvatkin. Vertex rounded, medially 1 1/2 times as long as near eyes. Fore wing as in Text-fig. 1266. Pygofer with dorsal and ventral prolongation. Genital plates elongate, each near middle with 3–4 macrosetae in a row. Genital style short, basally broad, apical process slender, almost parallel-sided, straight or faintly curved. In Europe one species.

257. ***Alnetoidia alneti*** (Dahlbom, 1850)
Text-figs. 1266, 1271, 1825–1829.

Cicadula alneti Dahlbom, 1850: 181.
Typhlocyba coryli Tollin, 1851: 70, 72.
Zygina mali Edwards, 1915: 209.

Elongate, shining, whitish yellow or light yellow. Resembling e.g. *Fagocyba douglasi* (Edwards). Vertex frontally rounded. Fore wings light citrine or bright yellow usually also in apical part, only extreme apex of the latter fumose. (Fore wings whitish or whitish yellow in f. *coryli* (Tollin) and f. *mali* (Edwards)). Male pygofer lobe as in Text-fig. 1825, genital style as in Text-fig. 1826, aedeagus as in Text-figs. 1827, 1828, 2nd–5th abdominal sterna in male as in Text-fig. 1829. Overall length (♂♀) 3.9–4.3 mm.

(3.3–3.75 mm in f. *coryli*). – Last instar larvae uniform pale yellow. Head anteriorly with one pair of long setae, vertex with a pair of somewhat shorter setae near eyes, pronotum with two pairs of moderately long setae on sides. Setae on mesonotum and fore wing pads short. Abdominal terga 3–8 each with 1–2 pairs of long setae and some short marginal hairs. Tibial setae short on fore and middle legs, longer on hind legs.

Distribution. Common and widespread in Denmark and Sweden (Sk.–Nb.). – Widespread, apparently common in southern and central Norway, found in AK, HEn, Os, On, Bø, VE, TEi, VAy, Ry, Ri, HOy, HOi, SFi, STi. – Very common in southern and central East Fennoscandia, found in Al, Ab, N, Ta, Sa, Oa, Tb, Sb, Kb; Kr; f. (ssp.?) *coryli* recorded from Ta: Lammi 16.VIII.1949 (Linnavuori). – Widespread in Europe, also in Georgia, Maritime Territory and Japan. F. *coryli* had been recorded from Germany, France, England, Bohemia, Slovakia.

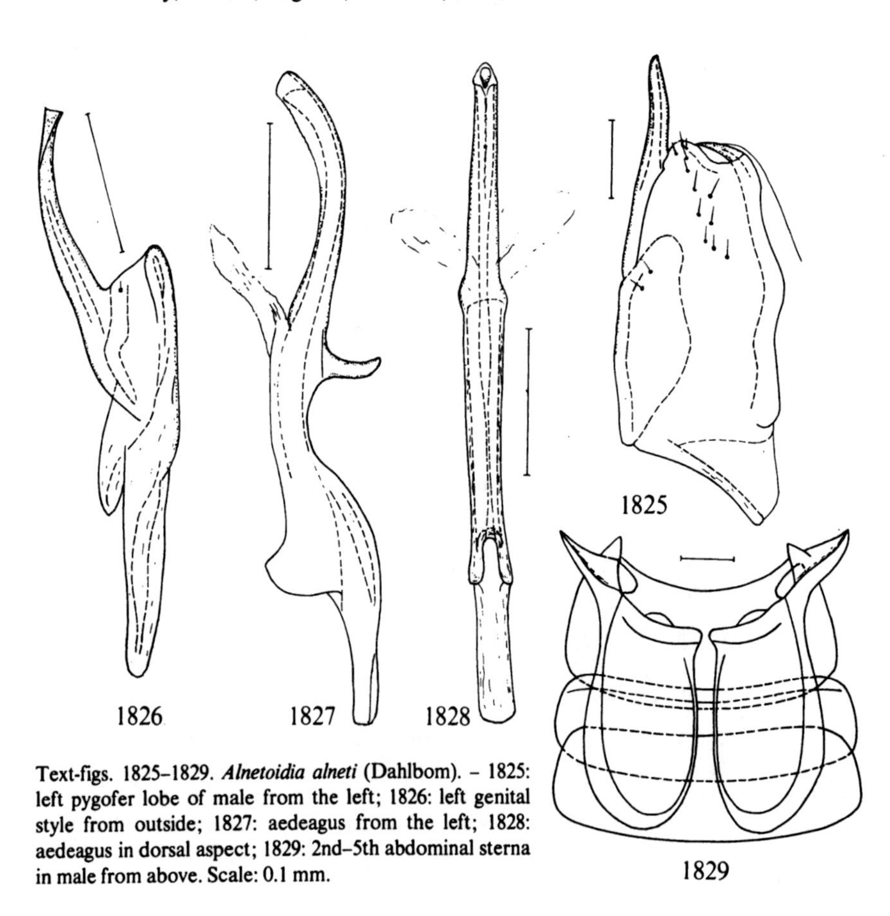

Text-figs. 1825–1829. *Alnetoidia alneti* (Dahlbom). – 1825: left pygofer lobe of male from the left; 1826: left genital style from outside; 1827: aedeagus from the left; 1828: aedeagus in dorsal aspect; 1829: 2nd–5th abdominal sterna in male from above. Scale: 0.1 mm.

Biology. On *Alnus glutinosa* and *Carpinus betulus* (Kuntze, 1937). "Die Imagines und Larven dieser Art verursachen Saugflecken auf Erlen, Ahorn und Cornus; nur die Imagines fanden wir auf Apfel, Pflaumen und Traubenkirschen, Sorbus und Linde" (Günthart, 1971). Hibernation takes place in the egg stage; only one generation in Holland (Everhuis, 1955, according to Günthart, l.c.). Polyphagous, reared on *Acer campestre, A. pseudoplatanus, Alnus, Betula, Carpinus, Corylus, Crataegus, Fagus, Quercus, Sorbus, Ulmus* (Claridge & Wilson, 1976). F. (ssp.?) *coryli* is associated with hazel. - Adults in June–September.

Note. I have no definite opinion on the taxonomic status of f. *coryli*. "Only very detailed behavioural and genetic studies will solve this problem" (Claridge & Wilson, 1976, p.249). I agree in this.

Genus *Hauptidia* Dworakowska, 1970

Hauptidia Dworakowska, 1970a: 620.
Type-species: *Typhlocyba distinguenda* Kirschbaum, 1868, by original designation.

"Head distinctly narrower than pronotum; pronotum broadened at hind part and face long and comparatively narrow. Pygophore side provided with an upper hook arising from it; a delicate, sclerotized ornamentation consisting of small spines at dorsal margin, at basal lower angle and at hind margin of pygophore side; short and slender setae grouped chiefly on areas covered by the mentioned sculpture; several short and rigid setae below and distad of upper pygophore appendage. Anal tube appendages short or absent. Subgenital plate rather short, pigmented on 2/3 of its length from tip and ornamented on this area with short rows of minute spines forming a scaly pattern; not numerous setae scattered on upper part of plate; marginal row of setae multiplicates at base of plate, setae becoming gradually a little shorter and thicker than at tip of plate; several macrosetae form a row parallel to marginal microsetae. Connective Y-shaped, with some sclerotization in middle part. Paramere with simple apical part like in *Asianidia* Zachv. and *Zyginidia* Hpt., sensory structures without any distinct setae. Praeatrium of penis very broad; rather short shaft provided with some appendages and ledges" (Dworakowska, 1970). In Fennoscandia one species.

258. ***Hauptidia distinguenda*** (Kirschbaum, 1868)
Plate-fig. 151, text-figs. 1830–1836.

Typhlocyba distinguenda Kirschbaum, 1868b: 183.
Zygina fasciaticollis Rey, 1891: 254.

Pale dirty yellow to whitish yellow, often with a greenish tinge. Anteclypeus entirely light brownish or with two brownish longitudinal bands. Frontoclypeus with two brownish longitudinal bands confluent on upper part. Vertex with two rounded black spots. Pronotum with a brown transverse band extending from side to side, often with an additional shorter transverse band more anteriorly. Scutellum with a black spot on each lateral corner, caudal apex often orange-coloured. Fore wing with clavus, cubital cell and subcostal cell strongly fumose, veins pale, light-bordered, apical part of fore wing less strongly fumose. Abdomen largely brownish black with light segment borders, male genital plates yellowish, apices dark. Male pygofer lobe as in Text-fig. 1830, genital plate as in Text-fig. 1831, genital style as in Text-fig. 1832, aedeagus as in Text-figs. 1833, 1834, 2nd and 3rd abdominal sterna in male as in Text-figs. 1835, 1836. Overall length (♂♀) 2.9–3.25 mm.

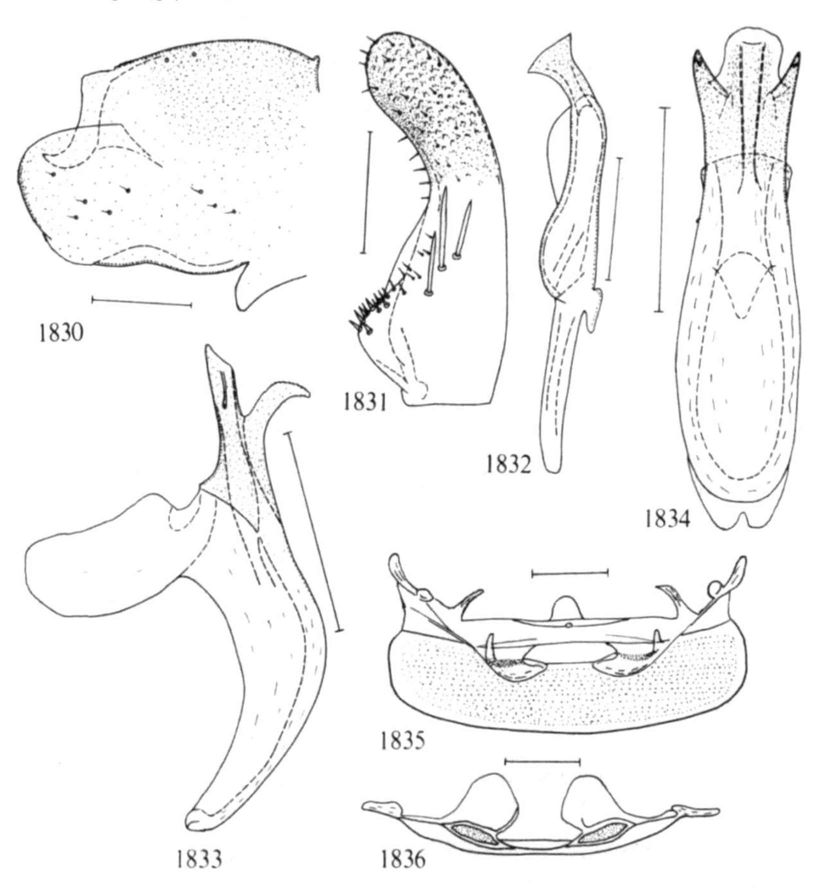

Text-figs. 1830–1836. *Hauptidia distinguenda* (Kirschbaum). – 1830: right pygofer lobe of male from outside, with right anal tooth on inside (broken line); 1831: left genital plate from outside; 1832: right genital style from above; 1833: aedeagus from the left; 1834: aedeagus in ventral aspect; 1835: 2nd and 3rd abdominal sterna in male from above; 1836: 2nd abdominal sternum in male in frontal aspect. Scale: 0.1 mm.

Distribution. Not found in Denmark, Norway and East Fennoscandia. - Scarce, locally abundant in southern Sweden, found in Sk., Brunnby, Kullen, near the lighthouse 24.3.1973 (R. Danielsson); Öl., Resmo 19.VII.1950 (Ossiannilsson); Upl.: Uppsala, Lurbo 1.VII.1950, Uppsala-Näs, Ytternäs 24.VI.1951, Rasbo, Hovgården 24.VIII.1952, Vaksala, Jälla 23.X.1952, Gryta, vicinity of Sjövik 8.VII.1971 (Ossiannilsson). - German F.R., France, Spain, Italy, Switzerland, Austria, Bulgaria, Romania, Moravia, Poland, Cyprus, Israel, Armenia.

Biology. On *Geranium Robertianum* Linné, adults in March–October.

Genus *Zyginidia* Haupt, 1929

Zyginidia Haupt, 1929b: 268.

Type-species: *Typhlocyba scutellaris* Herrich-Schäffer, 1838, by original designation.

Head slightly wider than pronotum. Lora long and narrow. Upper pygofer appendage usually forked. Appendages of anal tube long, slender. Submarginal row of setae on genital plate with a gap at middle dividing it into a basal and a subapical row, basal row consisting of thick and densely arranged pegs; macrosetae fairly large. Apex of genital style with a single apical broad and truncate extension, as in *Hauptidia*. Connective with central part semi-vitreous, lower central tooth sclerotized. Aedeagus usually laterally compressed, with paired basal appendages, shaft with a scaly surface sculpture near phallotreme. In Denmark and Fennoscandia two species.

Key to species of *Zyginidia*

1 Smaller, length 2.5–3 mm. Basal appendages of aedeagus little longer than shaft, which has no apical processes (Text-figs. 1840, 1841) .... 259. *pullula* (Boheman)

\- Larger, length about 3.4 mm. Basal appendages of aedeagus considerably longer than shaft, with a wrinkled surface structure; aedeagus near phallotreme with a pair of short processes (Text-figs. 1847, 1848) .... 260. *mocsaryi* (Horváth).

259. ***Zyginidia pullula*** (Boheman, 1845)

Plate-fig. 136, text-figs. 1837-1843.

Typhlocyba pullula Boheman, 1845b: 160.

Typhlocyba scutellaris J. Sahlberg, 1871: 182, nec Herrich-Schäffer, 1838.

Dirty yellow, not very elongate. Anteclypeus usually brownish, frontoclypeus along side margins with brownish muscle traces. Upper part or face largely yellowish brown, the dark colour continuing on vertex, apex of head and a pair of transverse diffuse spots on disk being brown or yellowish brown. Pronotum behind spots on vertex with a pair of similar spots (sometimes black), caudally of these a dark transverse band in front of the pale hind border. Scutellum with a black spot on each lateral corner and a third spot on caudal apex. Forewings greyish to greenish to yellow with indistinctly limited dark longitudinal bands in clavus and corium, apical part fumose. Thoracal venter black spotted, abdomen largely black, in male ventrally transversely striped. Anal tube and upper appendages of pygofer in male as in Text-fig. 1837, genital plate as in

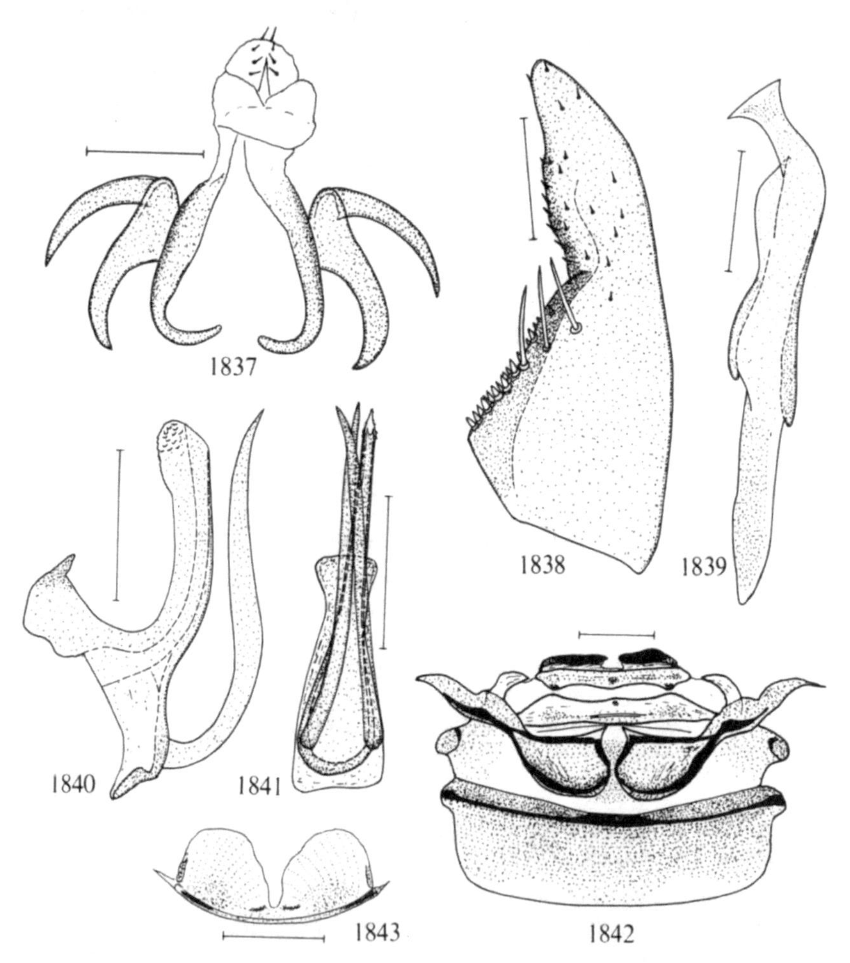

Text-figs. 1837–1843. *Zyginidia pullula* (Boheman). – 1837: male anal apparatus and upper pygofer appendages from behind (depressed under coverglass); 1838: left genital plate from outside; 1839: right genital style from above; 1840: aedeagus from the left; 1841: aedeagus in ventral aspect; 1842: 1st–4th abdominal sterna in male from above; 1843: 1st abdominal sternum in male from the front. Scale: 0.1 mm.

Text-fig. 1838, often apically infuscate, genital style as in Text-fig. 1839, aedeagus as in Text-figs. 1840, 1841, 1st–4th abdominal sterna in male as in text-figs. 1842, 1843. Overall length (♂♀) 2.5–3 mm.

Distribution. Scarce in Denmark, found only in EJ, NEZ and B. – Sporadic, locally

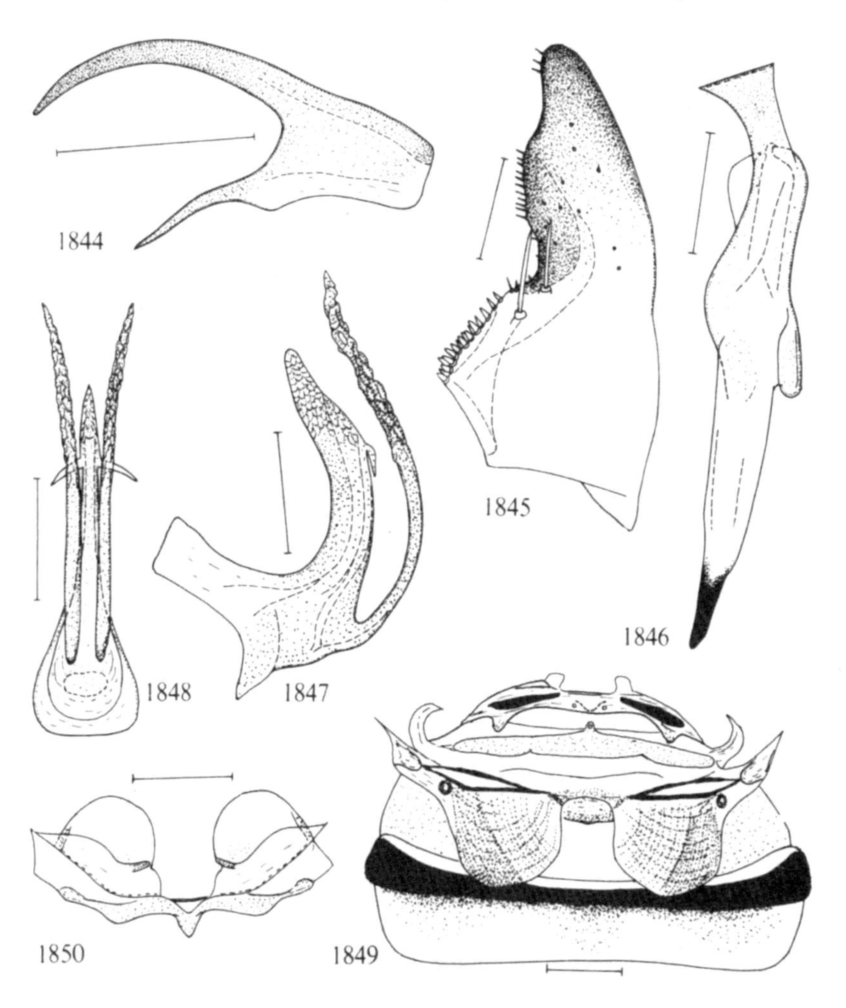

Text-figs. 1844–1850. *Zyginidia mocsaryi* (Horváth). – 1844: right upper pygofer appendage of male from the right; 1845: left genital plate from outside; 1846: right genital style from above; 1847: aedeagus from the left; 1848: aedeagus in ventral aspect; 1849: 1st–4th abdominal sterna in male from above; 1850: 1st abdominal sternum of male from the front. Scale: 0.1 mm.

common in southern Sweden, found in Öl., Gtl., Ög., Vg., Sdm., Upl., Nrk. - So far not found in Norway. - East Fennoscandia: only found in Al: Eckerö 19.-27.VII.1943, and in Al: Finström, by Håkan Lindberg. - German F.R., Spain, Italy, Hungary, Romania, Moravia, Slovakia, Yugoslavia, Albania, Bulgaria, Ukraine, Anatolia, Kazakhstan, Mongolia.

Biology. On dry meadows and steppes, adults in June-September.

260. ***Zyginidia mocsaryi*** (Horváth, 1910)
Text-figs. 1844-1850.

Erythroneura mocsaryi Horváth, 1910: 176.
Erythroneura silvicola Ossiannilsson, 1937a: 25.

Light yellow, resembling *pullula* but longer and comparatively more slender. Upper pygofer appendage as in Text-fig. 1844, genital plate as in Text-fig. 1845, genital style as in Text-fig. 1846, aedeagus as in Text-figs. 1847, 1848, 1st-4th abdominal sterna in male as in Text-figs. 1849, 1850. Overall length (♂♀) 3.4-3.6 mm.

Distribution. Not found in Denmark, Norway and East Fennoscandia. - Rare in Sweden, found in Bl.: Förkärla, Kvalmsö 4.VIII.1952 (Gyllensvärd); Gtl.: Roma 1.VIII.1935 (Ossiannilsson). - German D.R. and F.R., Switzerland, Austria, Slovakia, Romania, Poland, Estonia, Latvia, Ukraine.

Biology. Mainly in xerophilous, also in mesophilous biotopes; two generations in Central Europe, hibernation in the egg stage (Schiemenz, 1969b). On grasses in woods (Dworakowska, 1970c). In the Teucrio-Seslerietum (Müller, 1978). Lives on *Sesleria coerulea* (Vilbaste, 1974).

Genus *Zygina* Fieber, 1866

Zygina Fieber, 1866a: 509.
Type-species: *Typhlocyba nivea* Mulsant & Rey, 1855, by monotypy.
Flammigeroidia Dlabola, 1858: 56 (subgenus).
Type-species: *Cicada flammigera* Fourcroy, 1785, by original designation.
Hypericiella Dworakowska, 1970a: 559 (subgenus).
Type-species: *Typhlocyba hyperici* Herrich-Schäffer, 1836, by original designation.

Male pygofer without macrosetae, armed with one or two pairs of processes. Genital style with distinct preapical lobe and single truncate flat apical extension, without second apical extension (Text-figs. 1856, 1928). Aedeagus with or without appendages. Species often brightly coloured, but presence or absence of red colour in the *flammigera* group cannot be used as a character for separating species since this has been shown to depend on individual age (Günthart, 1979). Larvae (in species so far examined) with antennae longer than body. In Denmark and Fennoscandia two subgenera with in all eleven species.

Key to species of *Zygina*

1 Scutellum light, caudal apex black or dark brown. Vertex often with a black or dark brown spot shaped as an ink-bottle, i.e. with a broader caudal and a narrower frontal part (Plate-figs. 137, 138). Hind wing almost parallel-sided. Process of male pygofer forked (Text-fig. 1925). Basal submarginal setae of genital plate peg-like (Text-fig. 1926) (Subgenus *Hypericiella* Dworakowska) .... 271. *hyperici* (Herrich-Schäffer)

– Scutellum entirely light or entirely dark. Vertex without a black spot. Hind wing tapering towards apex. Process of male pygofer simple. Basal submarginal setae on genital plate thin (Subgenus *Zygina* Fieber) 2

2 (1) Tibiae dirty yellow. Small species, length at most 2.75 mm. On *Calluna* .... 270. *rubrovittata* (Lethierry)

– Tibiae whitish or whitish yellow. Not on *Calluna* 3

3 (2) Vertex comparatively long, ratio median length/width between eyes = 0.6–0.68 (Text-fig. 1860) .... 262. *angusta* Lethierry

– Vertex shorter, ratio below 0.6 4

4 (3) Hind tarsus of male long (about 1 mm), 1st segment ventrally with many setae considerably longer than diameter of tarsus (Text-fig. 1869). Hind tarsus of female about 2/3 mm in length, 1st segment ventrally with at least some setae longer than diameter of tarsus (Text-fig. 1871). Hind tibiae in both sexes with an "extra" ventral row of setae (Text-figs. 1868, 1870) .... 263. *rosea* (Flor)

– Ventral setae of 1st tarsal segment in both sexes shorter than diameter of tarsus (see e.g. Text-fig. 1852). No "extra" ventral row of setae on hind tibiae (Text-fig. 1851) 5

5 (4) Scutellum chocolate-brown or at least fumose, even in weakly pigmented specimens. Lateral margin of red longitudinal band in clavus, when present, exactly following claval suture (Plate-fig. 139) .... 261. *flammigera* (Fourcroy)

– Scutellum not fumose 6

6 (5) Setae on dorsum of anal tube in male arranged in two groups (Text-fig. 1899) .... 267. *suavis* Rey

– Setae on dorsum of male anal tube not arranged in two groups 7

7 (6) Hind tarsus of male half as long as hind tibia or longer, entirely black or fuscous (Text-fig. 1878). Clavus fumose, sometimes only faintly. Apodemes of 2nd abdominal sternum in male short (Text-fig. 1884) .... 264. *tiliae* (Fallén)

– Hind tarsus of male at most half as long as hind tibia, usually shorter. If hind tarsus of male is entirely infuscate, clavus is not fumose, and apodemes of 2nd abdominal sternum are long 8

8 (7) Apodemes of 2nd abdominal sternum in male long, usually extending considerately caudad of hind border of 3rd sternum (Text-fig. 1897) .... 266. *rosincola* (Cerutti)
– Apodemes of 2nd abdominal sternum in male shorter 9
9 (8) Apodemes of 2nd abdominal sternum in male large though smaller than in *rosincola* (Text-fig. 1909). Dorsal margin of shaft of aedeagus strongly curved (Text-fig. 1908) . 268. *schneideri* (Günthart)
– Apodemes of 2nd abdominal sternum short. Dorsal margin of shaft of aedeagus less strongly curved (Text-figs. 1889, 1914) 10
10 (9) Larger, 2.85–4.25 mm. On *Salix* sp. .... 265. *ordinaria* (Ribaut)
– Smaller, 2.4–2.6 mm (♂), 2.5–2.9 mm (♀). On *Salix repens* and *rosmarinifolia* .... 269. *salicina* Mitjaev.

261. ***Zygina (Zygina) flammigera*** (Fourcroy, 1785)
Plate-figs. 139, 152, text-figs. 1851–1859.

Cicada flammigera Fourcroy, 1785: 190.
Cicada blandula Rossi, 1792: 217.
Zygina pruni Edwards, 1924: 56.

Elongate, yellowish white, shining. Vertex frontally moderately produced, obtuse rounded. A caudally widening longitudinal band on vertex and pronotum, the entire scutellum, and 4th apical cell of fore wing fumose. Metanotum and clavus not fumose. Male hind tarsus with 3rd and distal half of 2nd segment black or fuscous (Text-fig. 1852). Dorsum of fore body and fore wings (in quite mature specimens) with carmine red markings somewhat varying in extension. Vertex with or without two red zigzag stripes close to each other. Red markings often present also on the face. Pronotum with a red middle stripe gradually widening caudad, often divided by a pale median line. Scutellum infuscate, usually chocolate brown. Fore wing with a broad red zigzag longitudinal band, proximal half situated in clavus, exteriorly exactly limited by claval suture, basal margin quite oblique forming a distinct angle with outer margin. Small red spots often present also outside this band, e. g. near the wax area. In immature specimens initially only fumose markings are present, later yellow markings appear ("*pruni*" Edwards), finally the colour changes into bright red. Also the appearance of black pigment in hind tarsi of male is connected with aging. Male pygofer lobe as in Text-fig. 1853, male anal tube as in Text-fig. 1854, genital plate as in Text-fig. 1855, genital style as in Text-fig. 1856, aedeagus as in Text-figs. 1857, 1858, 1st and 2nd abdominal sterna in male as in Text-fig. 1859. Overall length (♂♀) 3.2–3.4 mm.

Distribution. Common and widespread in Denmark as well as in southern and central Sweden, Sk.–Dlr. – Probably common also in southern Norway, but so far reliable records only from AK: Sem and Ås (Taksdal, Baeschlin), and VAy: Søgne (Holgersen). – Common in southern East Fennoscandia, recorded from Ab, N, Ta, Kb; Kr. – Widespread in Europe, also in Cyprus, Kazakhstan, Uzbekistan, Kirghizia, Mongolia, Maritime Territory; Nearctic region.

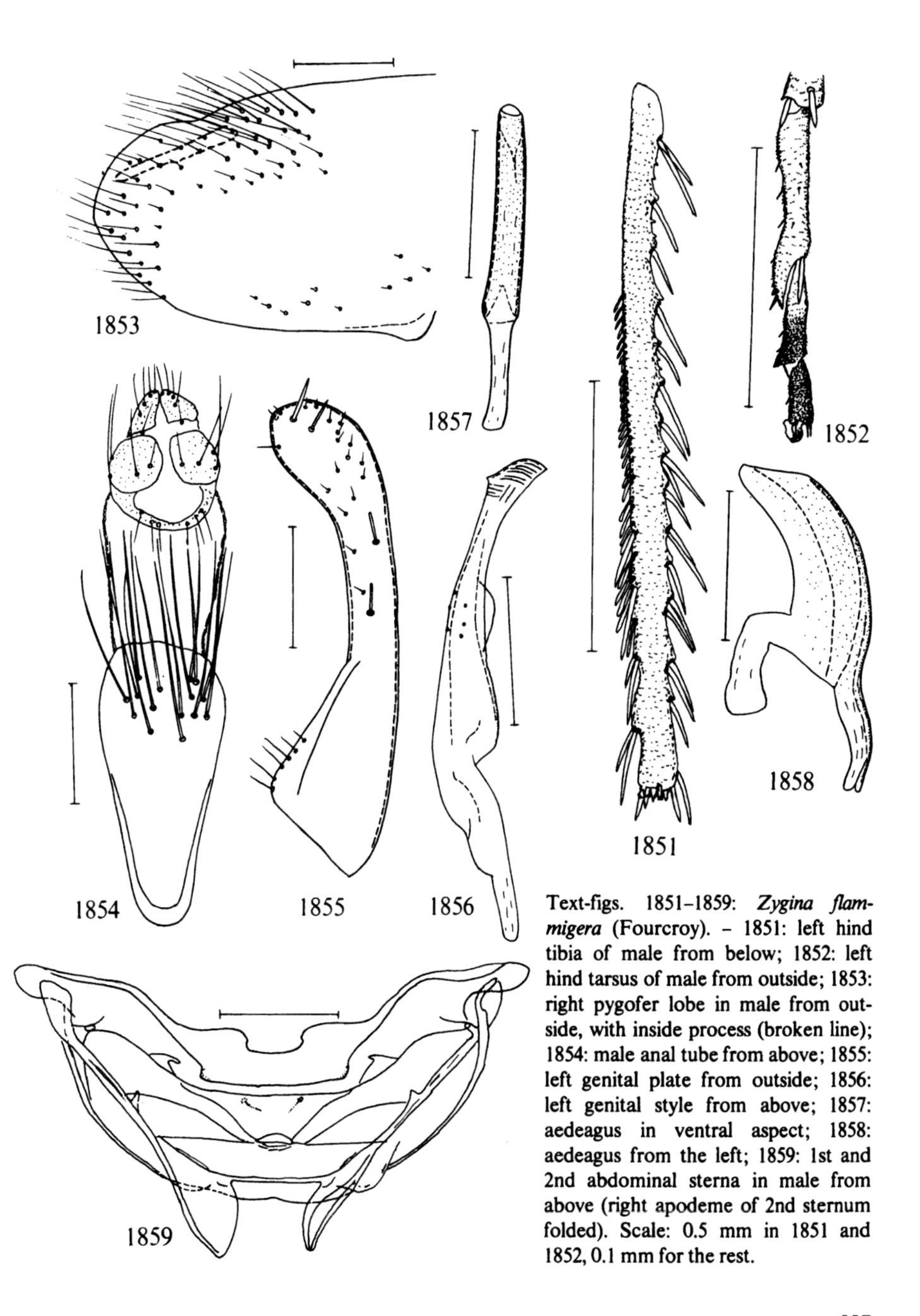

Text-figs. 1851–1859: *Zygina flammigera* (Fourcroy). – 1851: left hind tibia of male from below; 1852: left hind tarsus of male from outside; 1853: right pygofer lobe in male from outside, with inside process (broken line); 1854: male anal tube from above; 1855: left genital plate from outside; 1856: left genital style from above; 1857: aedeagus in ventral aspect; 1858: aedeagus from the left; 1859: 1st and 2nd abdominal sterna in male from above (right apodeme of 2nd sternum folded). Scale: 0.5 mm in 1851 and 1852, 0.1 mm for the rest.

Biology. Hibernates in the adult stage (Müller, 1957; Schwoerbel, 1957; Schiemenz, 1964). "Überwintert auf wintergrünen Holzgewächsen" (Wagner & Franz, 1961). "Polyphag an Laubholz" (Schiemenz, 1964). Collected on "Kirschen, Weichselkirsche, Apfel, Pflaume, Mespilus, Cornus, Buxus", reared on "Kirschen, Apfel, Pflaumen" (Günthart, 1971). "Charakteristisch ist, dass *E. flammigera* die schöne karminrote Zickzackfärbung auch bei den Zuchtbedingungen (Langtag und hohe Temperaturen) nicht verliert (im Gegensatz zu *E. rhamnicola, E.* n.sp. oder *Zygina rhamni*). Auch im Freiland findet man im Sommer gleich gut gefärbte Tiere wie im Herbst oder im Frühling nach der Ueberwinterung" (Günthart, l.c.). The "E. n. sp." alluded to is *Zygina schneideri* (Günthart). We collected adults in March–October.

262. ***Zygina (Zygina) angusta*** Lethierry, 1874
Text-figs. 1860–1867.

Zygina angusta Lethierry, 1874: 280.
Zygina tiliae moesta Ferrari, 1882: 158.
Zygina neglecta Edwards, 1914: 171.
Zygina neglecta rubrinervis Edwards, 1914: 171.

Resembling *flammigera,* vertex longer, frontally angulate or rounded (Text-fig. 1860). Scutellum entirely fumose or fumose with a narrow light median stripe. Metanotum brown. Clavus fumose except on basal angle, 4th apical cell in fore wing fumose. Often with a caudad diverging fumose band on pronotum, sometimes also on vertex. Red longitudinal band in clavus indistinctly limited, anterior margin not angular, not exactly adhering to claval suture. Hind tarsus of male shorter than half hind tibia, pigmentation as in *flammigera.* In f. *rubrinervis* Edwards the veins of corium are covered by red pigment; in f. *moesta* Ferrari red pigment is absent on dorsum and fore wings. Male pygofer lobe as in Text-fig. 1861, anal tube as in Text-fig. 1862, genital plate as in Text-fig. 1863, genital style as in Text-fig. 1864, aedeagus as in Text-fig. 1865, 1st–3rd abdominal sterna in male as in Text-figs. 1866, 1867. Overall length (♂♀) 2.75–3.15 mm.

Distribution. So far not found in Denmark. – Sweden: one female only found in Sk.: Brunnby, Kullen 10.V.1966 (Ossiannilsson). – Norway: one female captured in AAy: Risør 1.VI.1902 (Warloe). – Not found in East Fennoscandia. – France, England, Scotland, Wales, Belgium, Netherlands, Italy, Austria, Bulgaria, German D.R. and F.R., Hungary, Romania, Yugoslavia, Poland, s. Russia, Ukraine, Georgia, Azerbaijan, e. Siberia.

Biology. "Sur des arbres et arbustes divers (Chêne, Aulne, Prunier, Saules, Noisetier, Aubépine, Rosiers, Ronces, etc.). Elle se réfugie l'hiver dans les ronciers, sur le Fragon, le Buis et les arbres verts" (Ribaut, 1936b). "Wirtspflanze: Crataegus mon." (Günthart, 1974). The single Swedish specimen was captured on *Crataegus.*

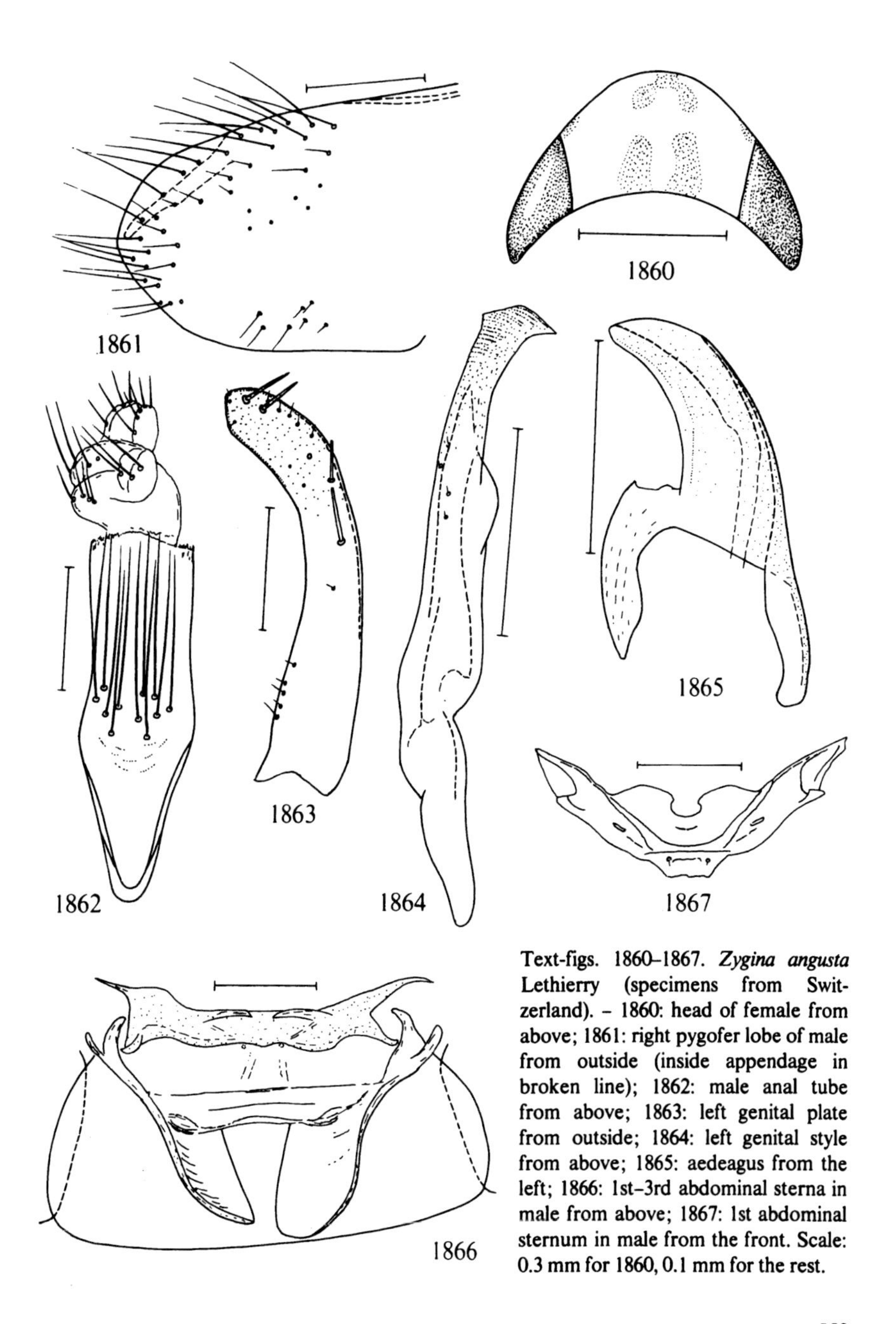

Text-figs. 1860–1867. *Zygina angusta* Lethierry (specimens from Switzerland). – 1860: head of female from above; 1861: right pygofer lobe of male from outside (inside appendage in broken line); 1862: male anal tube from above; 1863: left genital plate from outside; 1864: left genital style from above; 1865: aedeagus from the left; 1866: 1st–3rd abdominal sterna in male from above; 1867: 1st abdominal sternum in male from the front. Scale: 0.3 mm for 1860, 0.1 mm for the rest.

263. ***Zygina (Zygina) rosea*** (Flor, 1861)
Text-figs. 1868–1877.

? *Typhlocyba roseipennis* Tollin, 1851: 72.
Typhlocyba rosea Flor, 1861a: 403.

Differing from our remaining species of the present genus by the structure of hind tibiae and hind tarsi in both sexes. Head moderately produced, rounded. Head, pronotum, scutellum and clavus not fumose, metanotum partly brownish. Apical part of fore wing except 3rd cell faintly fumose. Red colour on dorsum and fore wings much varying in extension. Specimens collected during summer are usually entirely or almost entirely pale; in one specimen captured 26.III. distal 1/2 of M and Cu is red and broadly red-bordered, transverse veins red, head and pronotum without red pigment. On the other hand, clavus and corium may be more or less covered by red colour. Male pygofer lobe as in Text-fig. 1872, anal tube as in Text-fig. 1873, genital plate as in Text-fig. 1874, genital style as in Text-fig. 1875, aedeagus as in Text-fig. 1876, 1st and 2nd abdominal sterna as in Text-fig. 1877. Overall length (♂♀) 3.35–3.4 mm.

Distribution. Not found in Denmark. – Rare in southern and central Sweden, found in Hall.: Tönnersjö, Hilleshult 8.VIII.1934 (Ossiannilsson); Sm.: Vislanda 1.VI.1935 and Annerstad, Skeen 1.VI.1935 (Ossiannilsson); Ög.: Borensberg 18.VIII.1934 (Ossiannilsson); Upl.: Uppsala 12.IV.1936 (Orstadius), Uppsala, Norby 26.III.1950 (Ossian-

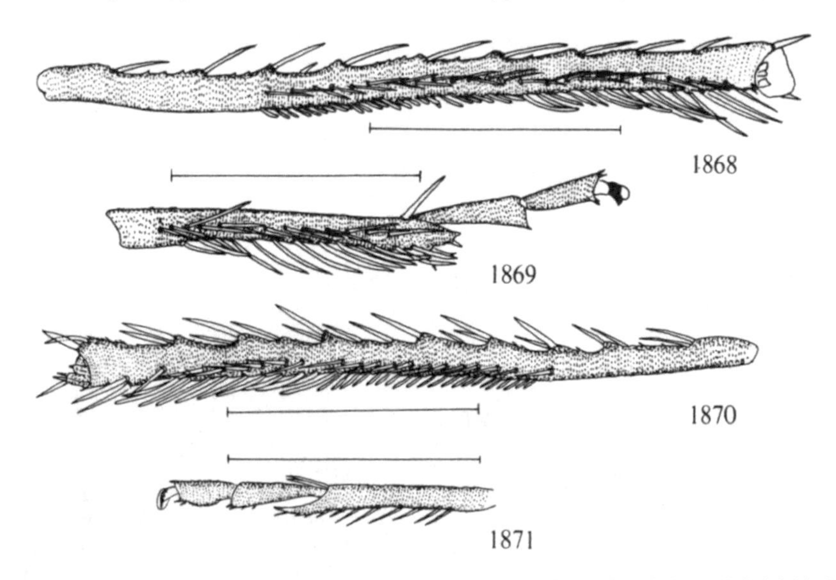

Text-figs. 1868–1871. *Zygina rosea* (Flor). – 1868: left hind tibia of male from below; 1869: left hind tarsus of male from outside; 1870: right hind tibia of female from below; 1871: right hind tarsus of female from outside. Scale: 0.5 mm.

nilsson); Dlr.: Sollerön 20.VIII.1918 (Tullgren), Rättvik, Glisstjärn 3.VI.1977 (T. Tjeder); Hls.: Edsbyn, Bäck 29.IX.1979 (Bo Henriksson). - Norway: Siebke (1874) recorded *rosea* from AK: Kongshavn and Baekkelaget, and from Bø: Ringerike, also from Valders and Romsdalen. - Rare in East Fennoscandia, recorded from Ab: Ispois (Reuter), Raisio and Ruissalo (Linnavuori); N: Esbo (Karvonen); Ta: Hattula

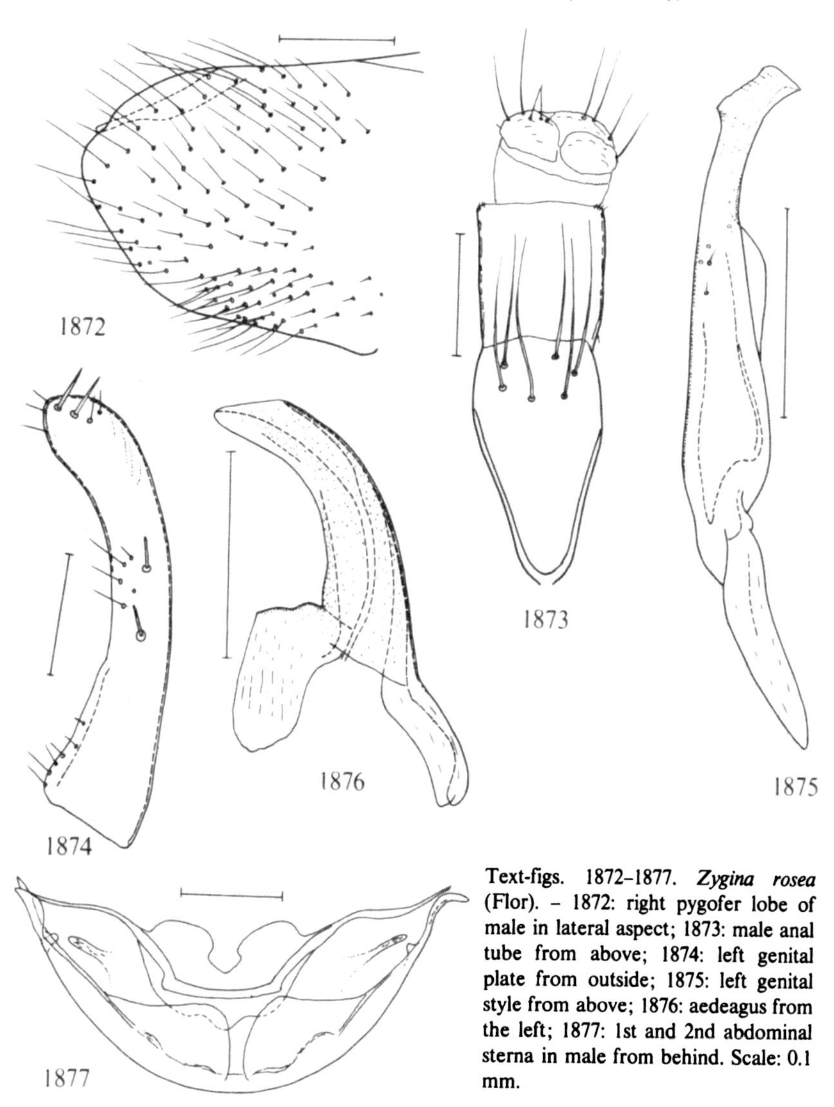

Text-figs. 1872-1877. *Zygina rosea* (Flor). - 1872: right pygofer lobe of male in lateral aspect; 1873: male anal tube from above; 1874: left genital plate from outside; 1875: left genital style from above; 1876: aedeagus from the left; 1877: 1st and 2nd abdominal sterna in male from behind. Scale: 0.1 mm.

(Nuorteva); Vib: Kivinebb (J. Sahlberg), Sakkola (Håkan Lindberg); Kr: Mjatusow at Swir (J. Sahlberg). – England, France, Belgium, Netherlands, German F.R., Switzerland, Italy, Austria, Hungary, Bohemia (?), Yugoslavia, Estonia, Latvia, m. Russia, Ukraine.

Biology. Flor (1861a) and Sahlberg (1871) collected *Z. rosea* on conifers, Linnavuori (1952a) and Nuorteva (1952a) on *Betula*. Apparently hibernation takes place in the adult stage.

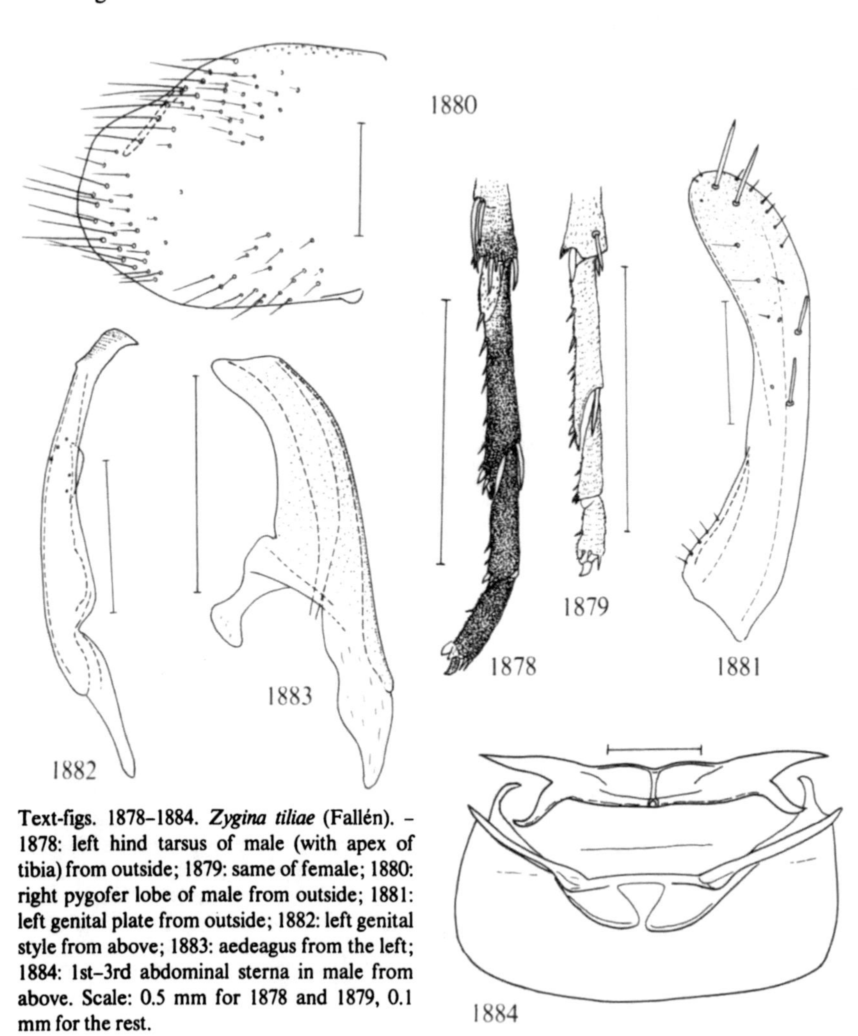

Text-figs. 1878–1884. *Zygina tiliae* (Fallén). – 1878: left hind tarsus of male (with apex of tibia) from outside; 1879: same of female; 1880: right pygofer lobe of male from outside; 1881: left genital plate from outside; 1882: left genital style from above; 1883: aedeagus from the left; 1884: 1st–3rd abdominal sterna in male from above. Scale: 0.5 mm for 1878 and 1879, 0.1 mm for the rest.

264. ***Zygina (Zygina) tiliae*** (Fallén, 1806)
Plate-fig. 153, text-figs. 1878–1884.

Cicada tiliae Fallén, 1806: 42.
Erythroneura tiliae peruncta Ribaut, 1936b: 47.

Resembling *flammigera*, scutellum entirely pale, or infuscate only on side margins. Metanotum brownish. Clavus, subcostal cell and most of apical part of fore wing fumose; vertex and pronotum often also with a fumose median stripe. Red band in clavus, when present, not adhering to claval suture. In f. *peruncta* Ribaut the entire clavus and almost the entire cubital cell are red spotted as well as the veins of corium. Hind tarsus of male usually more than half as long as hind tibia, in mature specimens entirely black (text-fig. 1878); hind tarsus of female (Text-fig. 1879) less than half length of hind tibia, pale (Text-fig. 1879). Male pygofer lobe as in Text-fig. 1880, genital plate as in Text-fig. 1881, genital style as in Text-fig. 1882, aedeagus as in Text-fig. 1883, 1st–3rd abdominal sterna in male as in Text-fig. 1884. Overall length (♂♀) 2.9–3.2 mm.

Distribution. Denmark: so far found in EJ, NEZ and B. - Common in southern and central Sweden, Sk.–Ång. - Norway: found in AK: Sem, Asker 18.IX.1974 and 19.VIII.1975 (Taksdal, Baeschlin); HEs: Løten 29.VIII.1961 (Holgersen); Ry: Fløjsvik, Høle 20.VII.1952 (Holgersen). - Scarce in East Fennoscandia, found in Al, Ab, N, St, Ta; Vib., Kr. - Widespread in Europe, also found in Georgia.

Biology. "Présentant les mêmes moeurs que *E. angusta*" (Ribaut, 1936b). Hibernation takes place in the adult stage, often on conifers. Adults collected in March–November.

265. ***Zygina (Zygina) ordinaria*** (Ribaut, 1936)
Text-figs. 1885–1890.

Erythroneura ordinaria Ribaut, 1936b: 47.
Erythroneura ordinaria pandellei Ribaut, 1936b: 47.
Erythroneura ordinaria variegata Ribaut, 1936b: 47.

Resembling *tiliae*, only distal 1/2 of male hind tarsus dark. Ratio length of hind tarsus/ length of hind tibiae in male = 0.41–0.46, in female = 0.38–0.41. Metanotum brownish. Veins in corium often blood-red and red-bordered. In f. *variegata* (Ribaut) the entire cubital cell and veins in corium are red mottled, clavus often entirely covered with red pigment. In f. *pandellei* (Ribaut) the fumose tinge is absent in clavus, replaced by strongly extended red colour. Male pygofer lobe as in Text-fig. 1885, male anal tube as in Text-fig. 1886, genital plate as in Text-fig. 1887, genital style as in Text-fig. 1888, aedeagus as in Text-fig. 1889, 1st–3rd abdominal sterna in male as in Text-fig. 1890. Overall length 2.85–4.25 mm.

Distribution. Not found in Denmark, nor in Norway. - Rare in Sweden, only found in Öl.: Högsrum, Halltorps hage 28–31.VIII.1976 (H. Andersson & R. Danielsson);

Plate-fig. 37. *Cicadetta montana* (Scop.) ♀, × 2.0.

Plate-fig. 38. *Lepyronia coleoptrata* (L.) ♀, × 5.7.

Plate-fig. 39. Same, last instar larva, × 5.7.

Plate-fig. 40. *Peuceptyelus coriaceus* (Fall.), × 6.7.

Plate-fig. 41. *Aphrophora alni* (Fall.), × 6.

Plate-fig. 42. Same, last instar larva, × 6.

Plate-fig. 43. *Aphrophora costalis* Mats., × 5.

Plate-fig. 44. Same, last instar larva, × 7.

Plate-fig. 45. *Neophilaenus campestris* (Fall.), × 8.5.

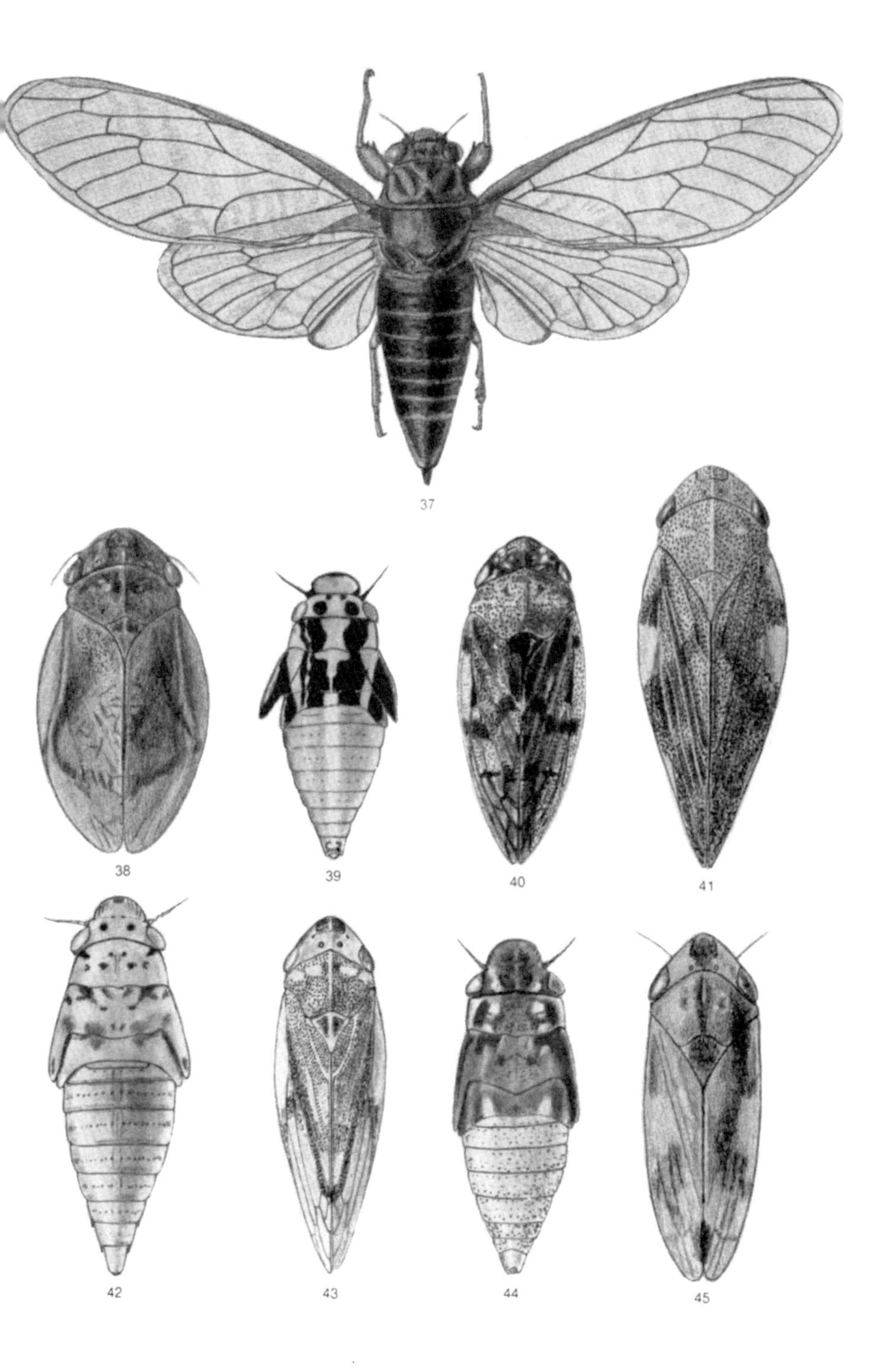

37
38
39
40
41
42
43
44
45

Plate-fig. 46. *Gargara genistae* (F.), × 9.

Plate-fig. 47. *Centrotus cornutus* (L.), × 7.

Plate-fig. 48. *Philaenus spumarius* (L.) f. typica, × 8.5.

Plate-fig. 49. *Ledra aurita* (L.), × 3.5.

Plate-fig. 50. *Macropsis infuscata* (J. Sahlb.), left fore wing, × 5.

Plate-fig. 51. *Macropsis fuscinervis* (Boh.), left fore wing, × 5.

Plate-fig. 52. *Macropsis cerea* (Germ.), left fore wing, × 5.

Plate-fig. 53. *Agallia brachyptera* (Boh.) ♂, × 16.

Plate-fig. 54. *Megophthalmus scanicus* (Fall.) ♂, × 17.

Plate-fig. 55. *Ulopa reticulata* (F.), × 13.

Plate-fig. 56. *Neophilaenus lineatus* (L.), × 8.5.

Plate-fig. 57. *Neophilaenus exclamationis* (Thunb.), × 10.

Plate-fig. 58. *Agallia venosa* (Fourcr.) ♂, × 14.

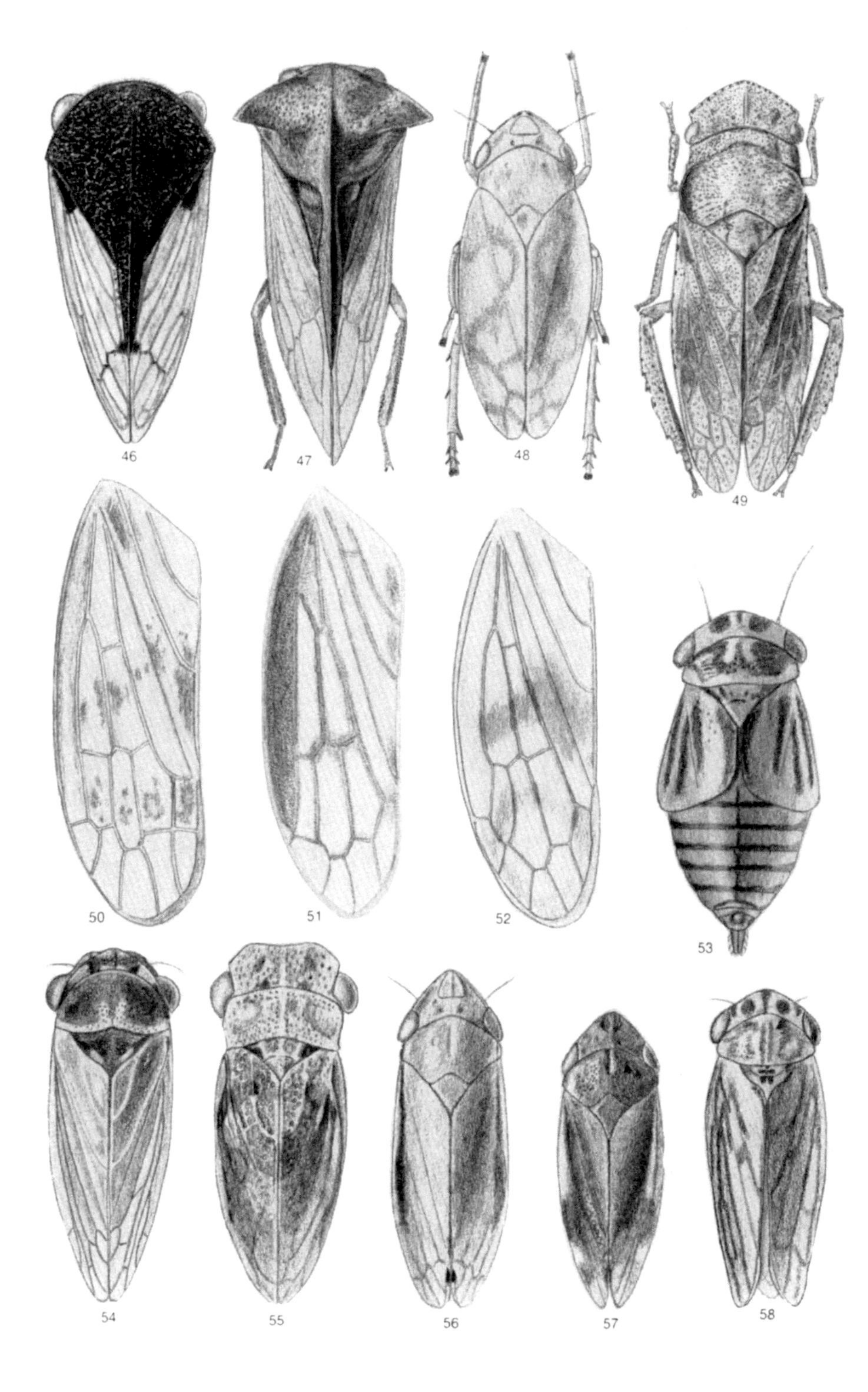
46
47
48
49
50
51
52
53
54
55
56
57
58

Plate-fig. 59. *Oncopsis flavicollis* (L.) var. *luteomaculata* W. Wagn. ♀, × 10.

Plate-fig. 60. *Oncopsis flavicollis* (L.) f. typica ♀, × 10.

Plate-fig. 61. *Pediopsis tiliae* (Germ.), × 10.

Plate-fig. 62. *Cicadella viridis* (L.) ♂, × 10.

Plate-fig. 63. Same ♀, × 8.5.

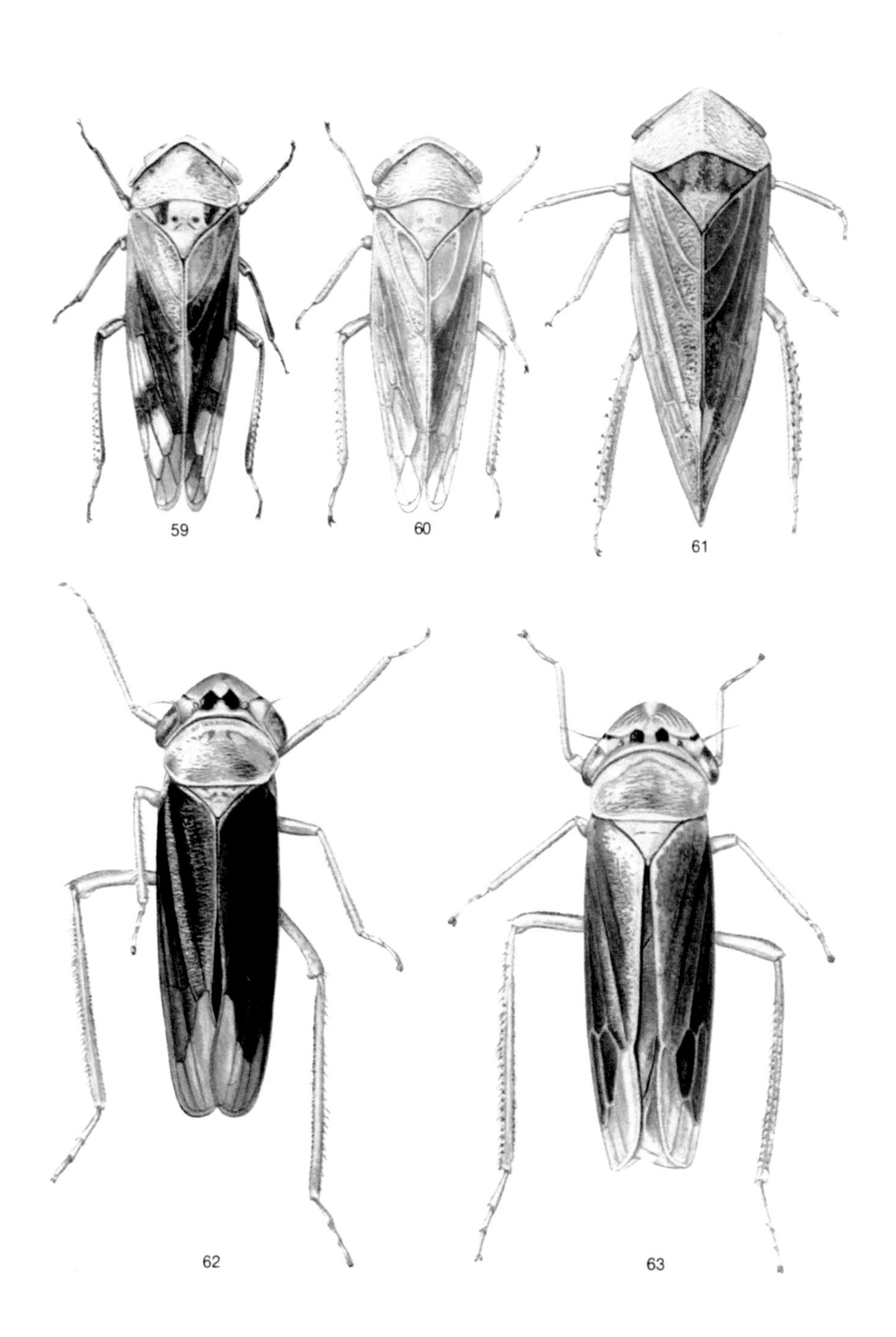

59
60
61
62
63

Plate-fig. 64. *Agallia brachyptera* (Boh.), last instar larva, × 22.

Plate-fig. 65. *Idiocerus lituratus* (Fall.) ♂, × 8.

Plate-fig. 66. *Idiocerus stigmaticalis* Lewis ♂, left fore wing, × 7.

Plate-fig. 67. *Stenidiocerus poecilus* (H.-S.), left fore wing, × 8.

Plate-fig. 68. *Metidiocerus elegans* (Flor), left fore wing, × 9.

Plate-fig. 69. *Idiocerus herrichii* Kbm. ♂, left fore wing, × 8.

Plate-fig. 70. *Tremulicerus tremulae* (Estl.), left fore wing, × 8.

Plate-fig. 71. *Aphrodes makarovi* Zachv. ♂, × 8.5.

Plate-fig. 72. *Planaphrodes bifasciata* (L.) ♂, × 8.5.

Plate-fig. 73. Same ♀, × 8.5.

Plate-fig. 74. *Planaphrodes trifasciata* (Fourcr.) ♂, × 8.5.

Plate-fig. 75. *Anoscopus albifrons* (L.) ♂, × 10.

Plate-fig. 76. *Anoscopus serratulae* (F.) ♂, × 9.

Plate-fig. 77. *Eupelix cuspidata* (F.) f. typica ♂, × 6.5.

Plate-fig. 78. *Eupelix cuspidata* (F.) f. *depressa* (F.) ♂, head from above, × 6.5.

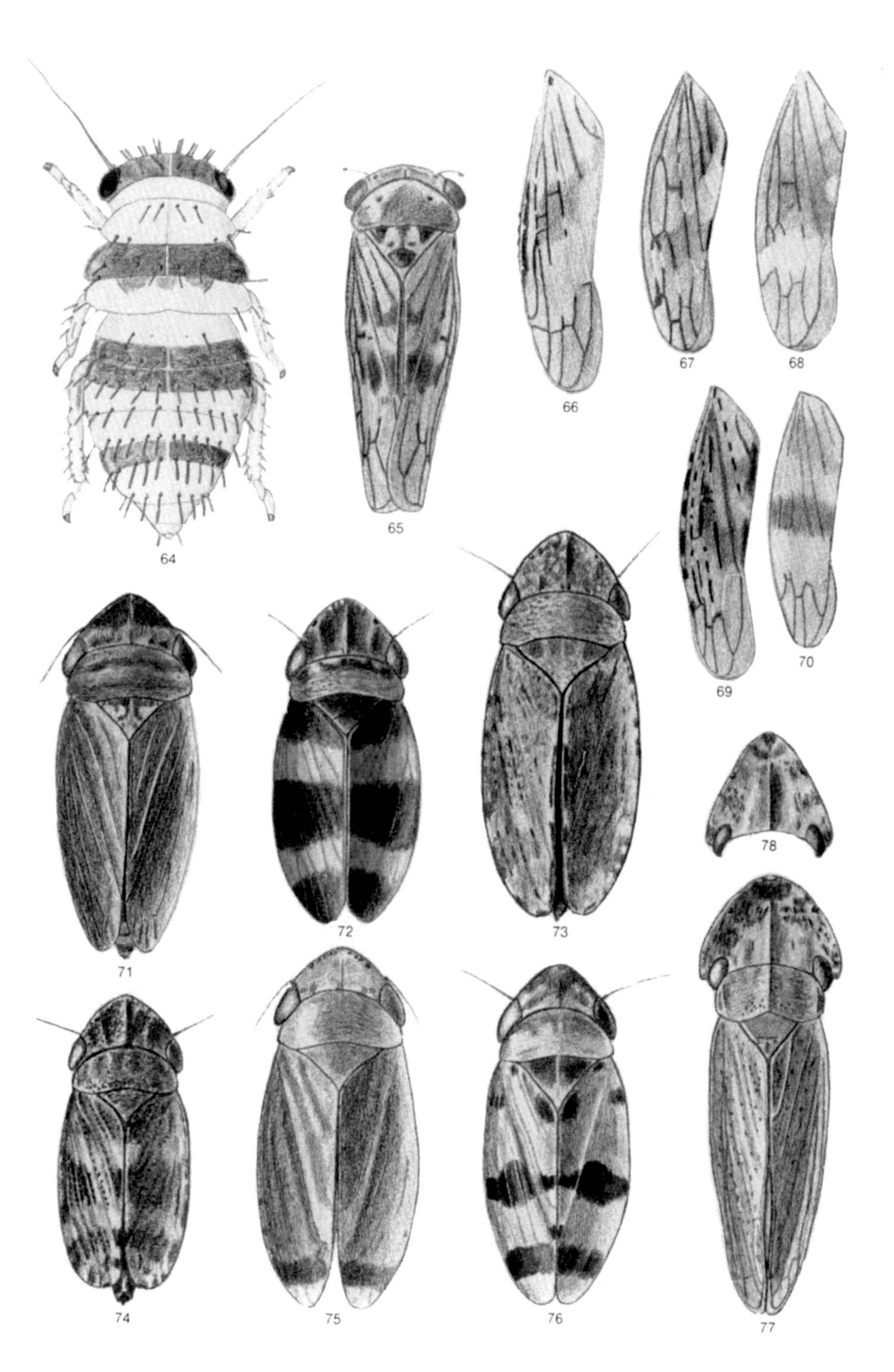

64
65
66
67
68
69
70
71
72
73
74
75
76
77
78

Plate-fig. 79. *Populicerus populi* (L.) ♂, × 10.

Plate-fig. 80. *Macropsis prasina* (Boh.) ♂, × 10.

Plate-fig. 81. *Iassus lanio* (L.) ♀, × 8.5.

Plate-fig. 82. *Evacanthus acuminatus* (F.) ♀, × 10.

Plate-fig. 83. *Evacanthus interruptus* (L.) ♂, × 10.

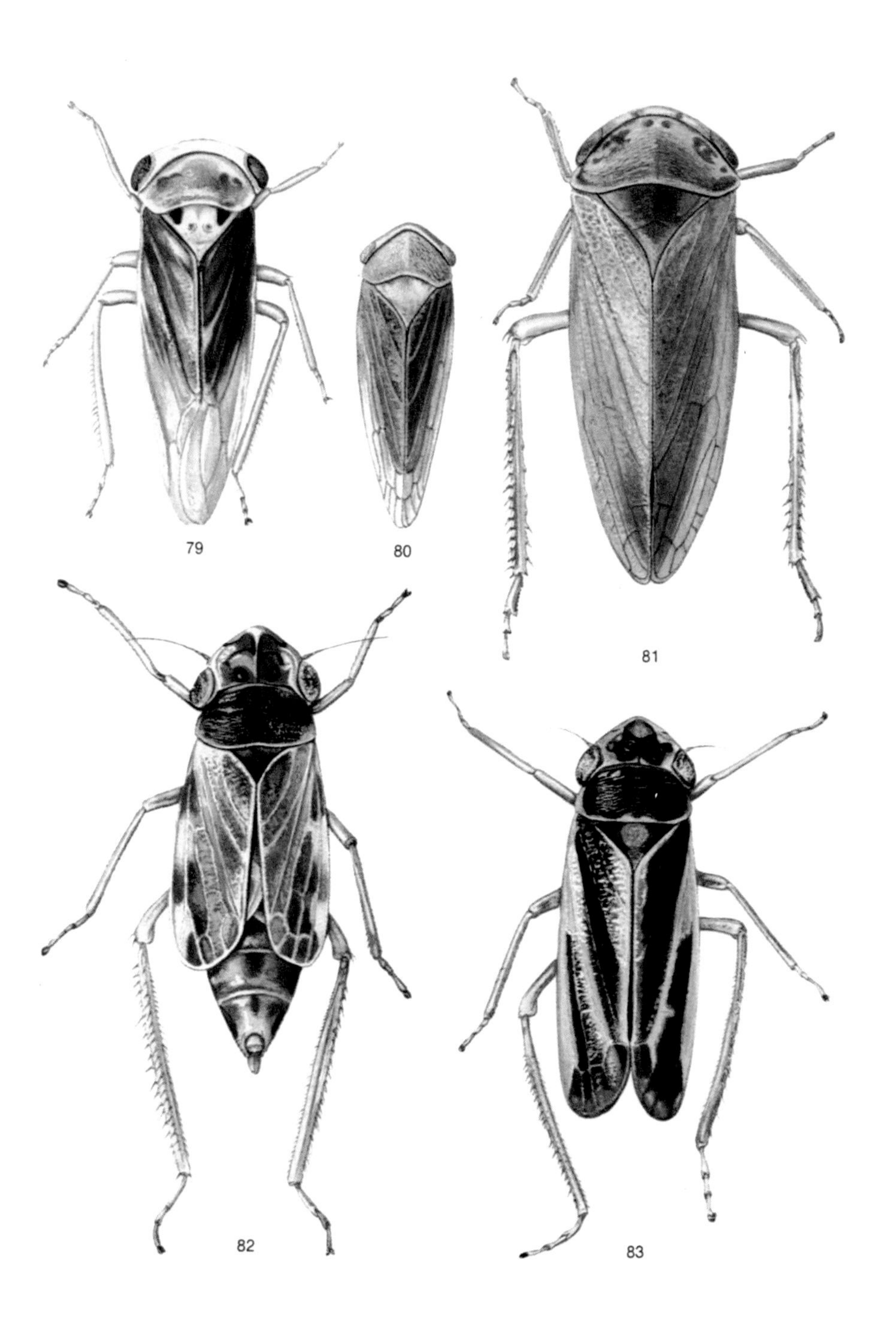
79
80
81
82
83

Plate-fig. 84. *Anoscopus flavostriatus* (Don.) ♂, × 15.

Plate-fig. 85. *Anoscopus histrionicus* (F.) ♂, × 12.5.

Plate-fig. 86. *Bathysmatophorus reuteri* J. Sahlb. ♀ f. brach., × 6.3.

Plate-fig. 87. Same ♂ f. macr., left fore wing, × 8.

Plate-fig. 88. *Stroggylocephalus agrestis* (Fall.) ♂, × 10.

Plate-fig. 89. *Empoasca smaragdula* (Fall.), × 13.

Plate-fig. 90. *Eupterycyba jucunda* (H.-S.), × 14.

Plate-fig. 91. *Ribautiana ulmi* (L.), × 14.

Plate-fig. 92. *Linnavuoriana decempunctata* (Fall.), × 14.

Plate-fig. 93. *Linnavuoriana sexmaculata* (Hardy), head and pronotum, × 14.

Plate-fig. 94. *Micantulina micantula* (Zett.), left fore wing, × 5.5.

Plate-fig. 95. *Ribautiana tenerrima* (H.-S.), left fore wing, × 5.5.

Plate-fig. 96. *Eurhadina pulchella* (Fall.), left fore wing, × 5.5.

Plate-fig. 97. *Eurhadina concinna* (Germ.), left fore wing, × 5.5.

Plate-fig. 98. *Eurhadina ribauti* W. Wagn. f. *umbrata* (Rib.), left fore wing, × 6.

Plate-fig. 99. *Eupteryx urticae* (F.), left fore wing, × 5.5.

Plate-fig. 100. *Eupteryx stachydearum* (Hardy), left fore wing, × 5.5.

Plate fig. 101. *Eupteryx thoulessi* Edw., left fore wing, × 5.5.

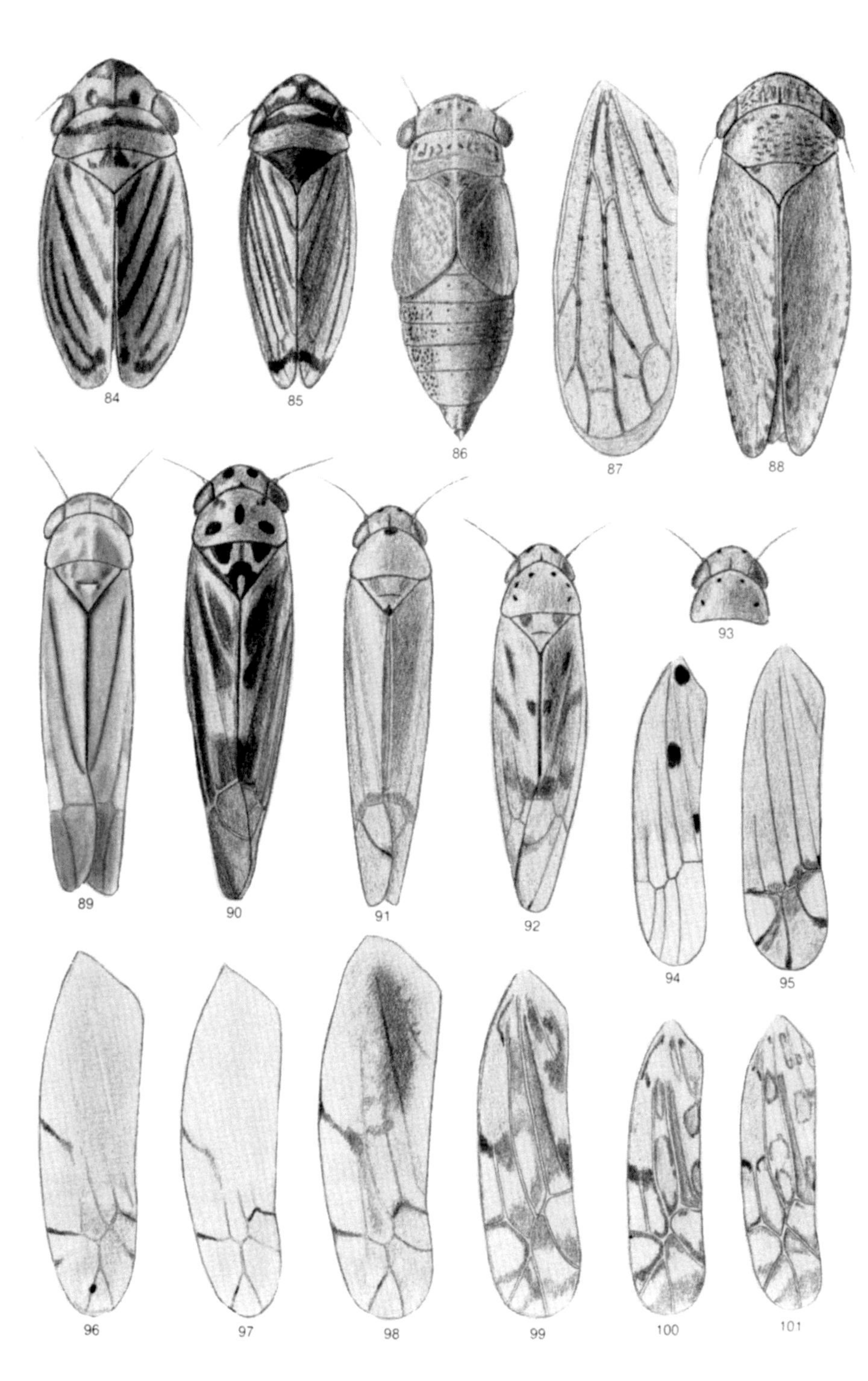
84
85
86
87
88
89
90
91
92
93
94
95
96
97
98
99
100
101

Plate-fig. 102. *Alebra albostriella* (Fall.) ♂, × 12.5.

Plate-fig. 103. *Alebra albostriella* (Fall.) ♀ f. typica, × 12.5.

Plate-fig. 104. *Alebra albostriella* (Fall.) f. *discicollis* (H.-S.) ♀, × 12.5.

Plate-fig. 105. *Alebra wahlbergi* (Boh.) ♀, × 12.5.

Plate-fig. 106. *Notus flavipennis* (Zett.) ♀, × 12.5.

Plate-fig. 107. *Empoasca rufescens* (Mel.) ♀, × 12.5.

Plate-fig. 108. *Empoasca vitis* (Göthe) ♀, × 12.5.

Plate-fig. 109. *Empoasca apicalis* (Flor) ♀, × 12.5.

Plate-fig. 110. *Erythria aureola* (Fall.) ♂, × 12.5.

Plate-fig. 111. *Chlorita viridula* (Fall.) ♀, × 12.5.

Plate-fig. 112. *Fagocyba cruenta* (H.-S.), × 12.5.

Plate-fig. 113. *Edwardsiana rosae* (L.) ♂, × 12.5.

Plate-fig. 114. *Edwardsiana gratiosa* (Boh.) ♀, × 12.5.

Plate-fig. 115. *Edwardsiana geometrica* (Schrnk.) ♀, × 12.5.

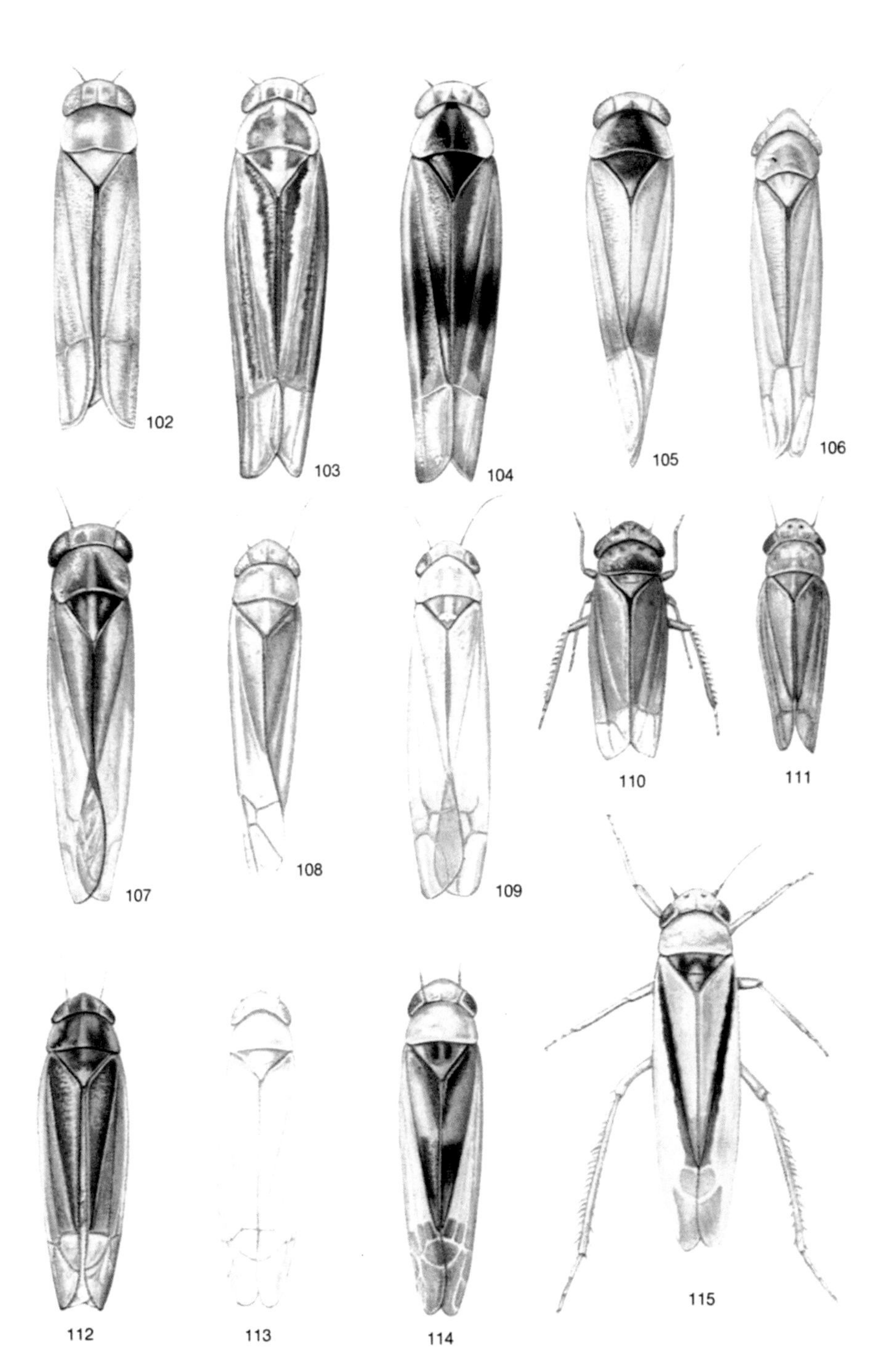
102
103
104
105
106
107
108
109
110
111
112
113
114
115

Plate-fig. 116. *Eupteryx tenella* (Fall.) × 16.

Plate-fig. 117. *Eupteryx artemissiae* (Kbm.), × 17.

Plate-fig. 118. *Eupteryx signatipennis* (Boh.), × 16.

Plate-fig. 119. *Eupteryx urticae* (F.), × 18.

Plate-fig. 120. *Aguriahana stellulata* (Burm.), left fore wing, × 13.

Plate-fig. 121. *Aguriahana pictilis* (Stål), left fore wing, × 13.

Plate-fig. 122. *Eupteryx thoulessi* Edw. ♀, face, × 43.

Plate-fig. 123. *Eupteryx stachydearum* (Hdy.), face, × 43.

Plate-fig. 124. *Eupteryx cyclops* Mats., face, × 43.

Plate-fig. 125. Same, head from above, × 43.

Plate-fig. 126. *Eupteryx cyclops* Mats., f. *trilobata* Rib., head from above, × 43.

Plate-fig. 127. *Eupteryx calcarata* Oss., head from above, × 43.

Plate-fig. 128. Same, another specimen, head from above, × 43.

Plate-fig. 129. *Eupteryx stachydearum* (Hdy.), head from above, × 43.

Plate-fig. 130. *Eupteryx thoulessi* Edw., head from above, × 43.

Plate-fig. 131. *Eupteryx aurata* (L.) ♀, left fore wing, × 16.

Plate-fig. 132. *Eupteryx atropunctata* (Gze.), left fore wing, × 16.

Plate-fig. 133. *Eupteryx vittata* (L.), left fore wing, × 17.

Plate-fig. 134. *Eupteryx notata* Curt., left fore wing, × 17.

Plate-fig. 135. *Arboridia parvula* (Boh.), × 17.

Plate-fig. 136. *Zyginidia pullula* (Boh.), fore body from above, × 19.

Plate-fig. 137. *Zygina hyperici* (H.-S.) ♀, fore body from above, × 19.

Plate-fig. 138. Same ♀, another specimen, × 19.

Plate-fig. 139. *Zygina flammigera* (Fourcr.), × 19.

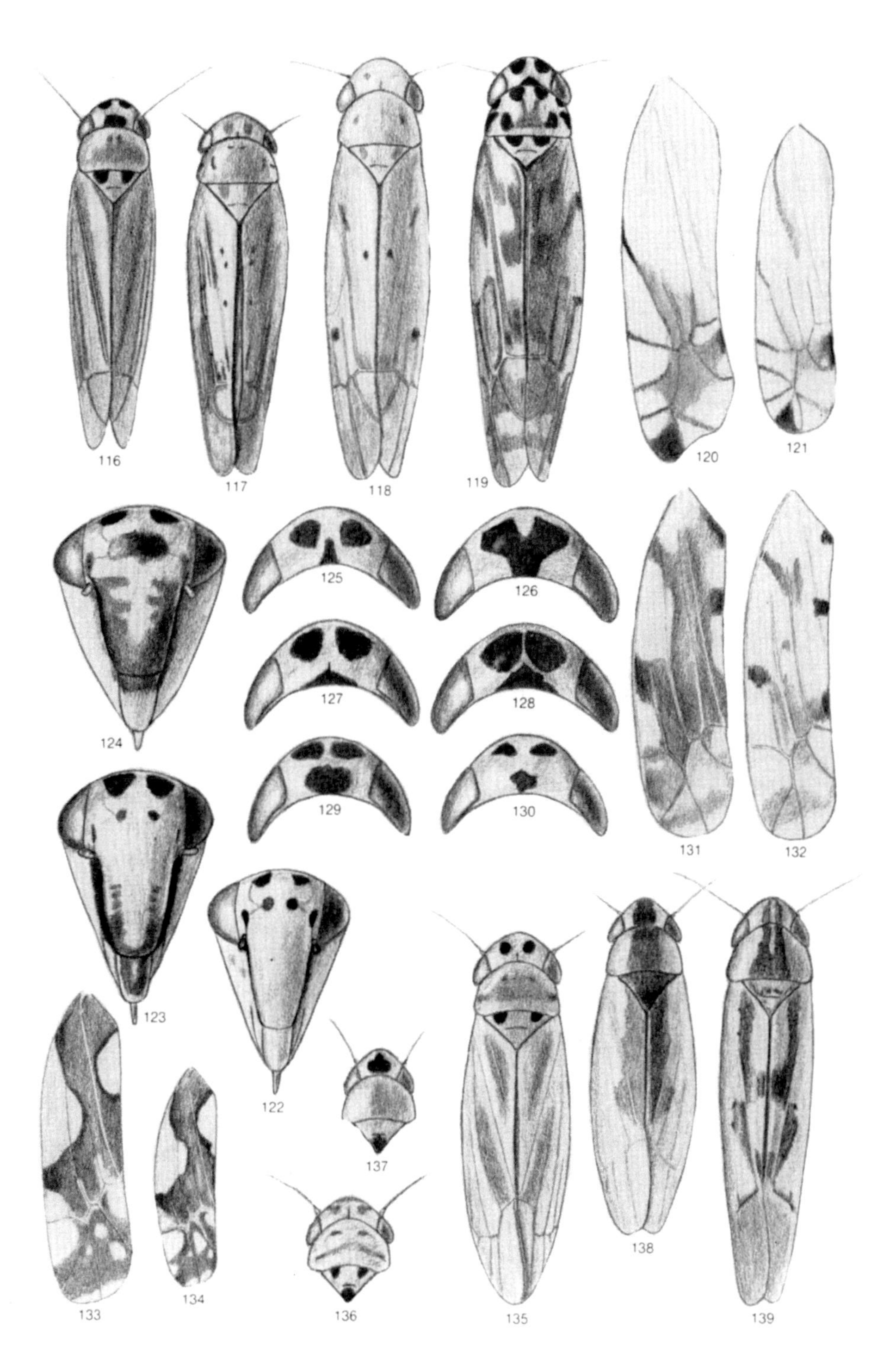
116
117
118
119
120
121
124
125
126
127
128
129
130
131
132
123
122
137
136
133
134
135
138
139

Plate-fig. 140. *Typhlocyba bifasciata* Boh. ♀, × 12.5.

Plate-fig. 141. *Typhlocyba quercus* (F.) ♀, × 12.5.

Plate-fig. 142. *Eurhadina pulchella* (Fall.) f. *thoracica* (Fieb.) ♀, × 12.5.

Plate-fig. 143. *Eurhadina pulchella* f. *ornatipennis* (Curt.) ♀, × 12.5.

Plate-fig. 144. *Eupteryx aurata* (L.) ♀, × 12.5.

Plate-fig. 145. *Eupteryx atropunctata* (Gze.) ♀, × 12.5.

Plate-fig. 146. *Eupteryx cyclops* Mats. ♀, × 12.5.

Plate-fig. 147. *Eupteryx stachydearum* (Hdy.) ♀, × 12.5.

Plate-fig. 148. *Eupteryx thoulessi* Edw. ♀, × 12.5.

Plate-fig. 149. *Eupteryx vittata* (L.) ♀, × 12.5.

Plate-fig. 150. *Aguriahana germari* (Zett.) ♀, × 12.5.

Plate-fig. 151. *Hauptidia distinguenda* (Kbm.) ♀, × 12.5.

Plate-fig. 152. *Zygina flammigera* (Fourcr.) ♂, × 12.5.

Plate-fig. 153. *Zygina tiliae* (Fall.) ♂, × 13.3.

Plate-fig. 154. *Zygina rosincola* (Cer.) ♂, × 12.

Plate-fig. 155. *Zygina rubrovittata* (Leth.) ♀, × 13.8.

Plate-fig. 156. *Zygina hyperici* (H.-S.) ♀, × 13.5.

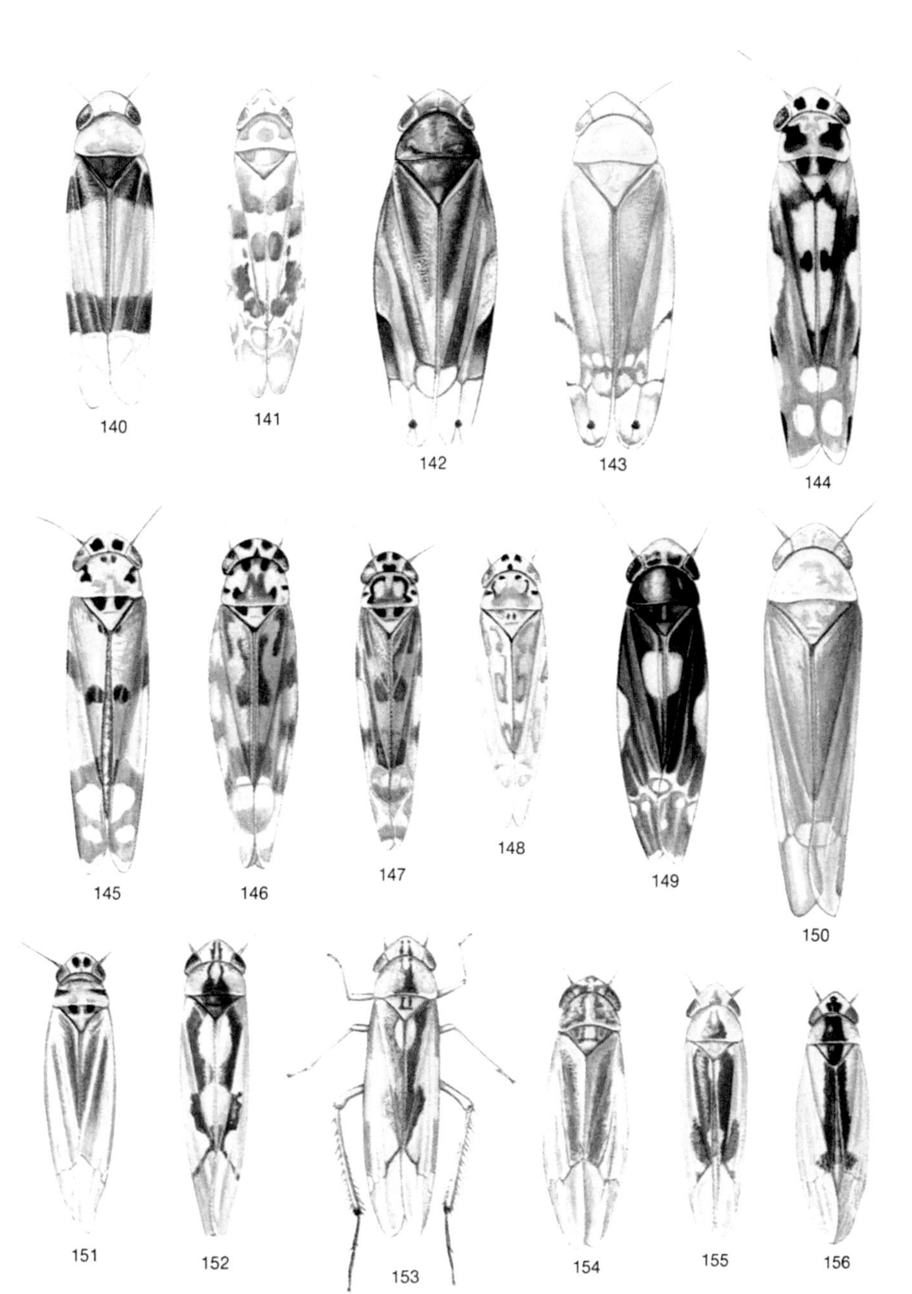
140
141
142
143
144
145
146
147
148
149
150
151
152
153
154
155
156

Upl.: Stockholm, Experimentalfältet 22.X.1939 (Ossiannilsson), Uppsala, Åsen 17.X.-1972 (Ossiannilsson). - East Fennoscandia: recorded from Ab, Ta, Sa, but these records have not been checked after the dividing of *ordinaria* by Günthart (1974). - France, Switzerland.

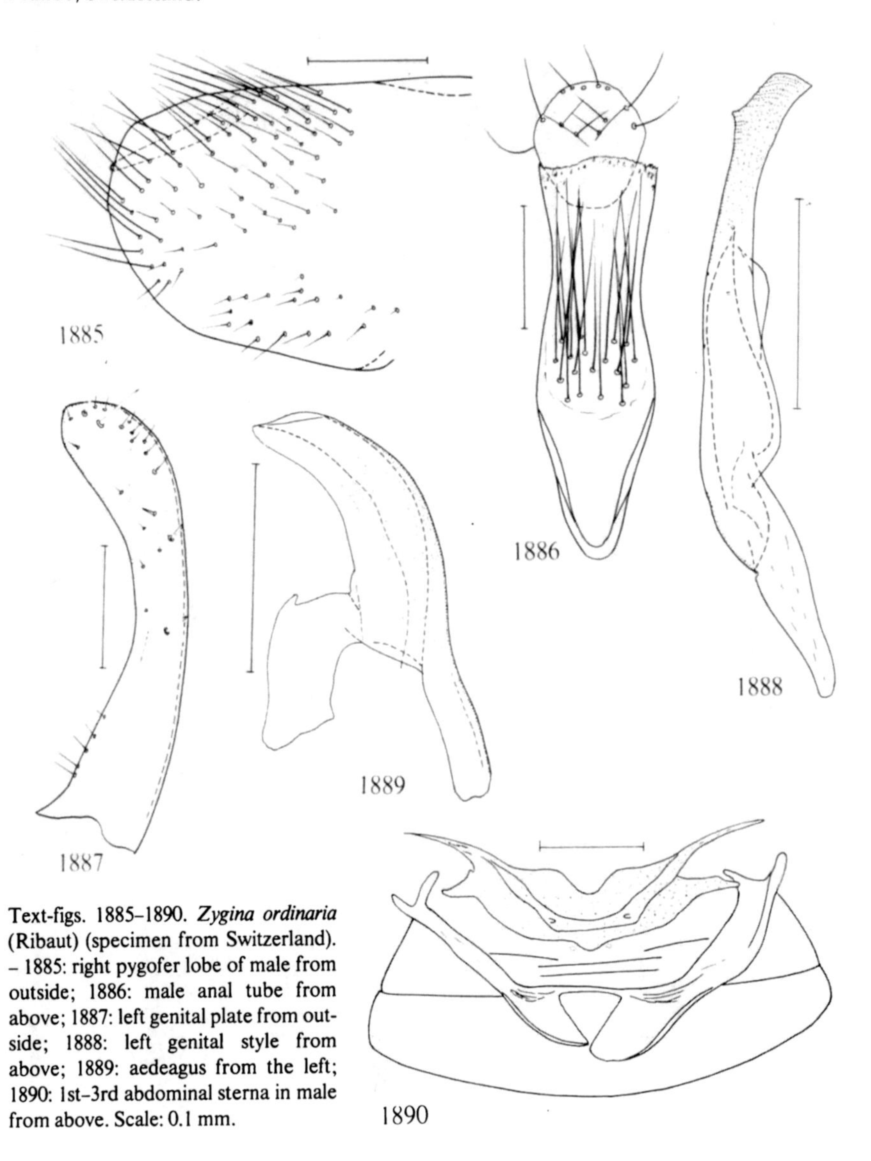

Text-figs. 1885–1890. *Zygina ordinaria* (Ribaut) (specimen from Switzerland). - 1885: right pygofer lobe of male from outside; 1886: male anal tube from above; 1887: left genital plate from outside; 1888: left genital style from above; 1889: aedeagus from the left; 1890: 1st-3rd abdominal sterna in male from above. Scale: 0.1 mm.

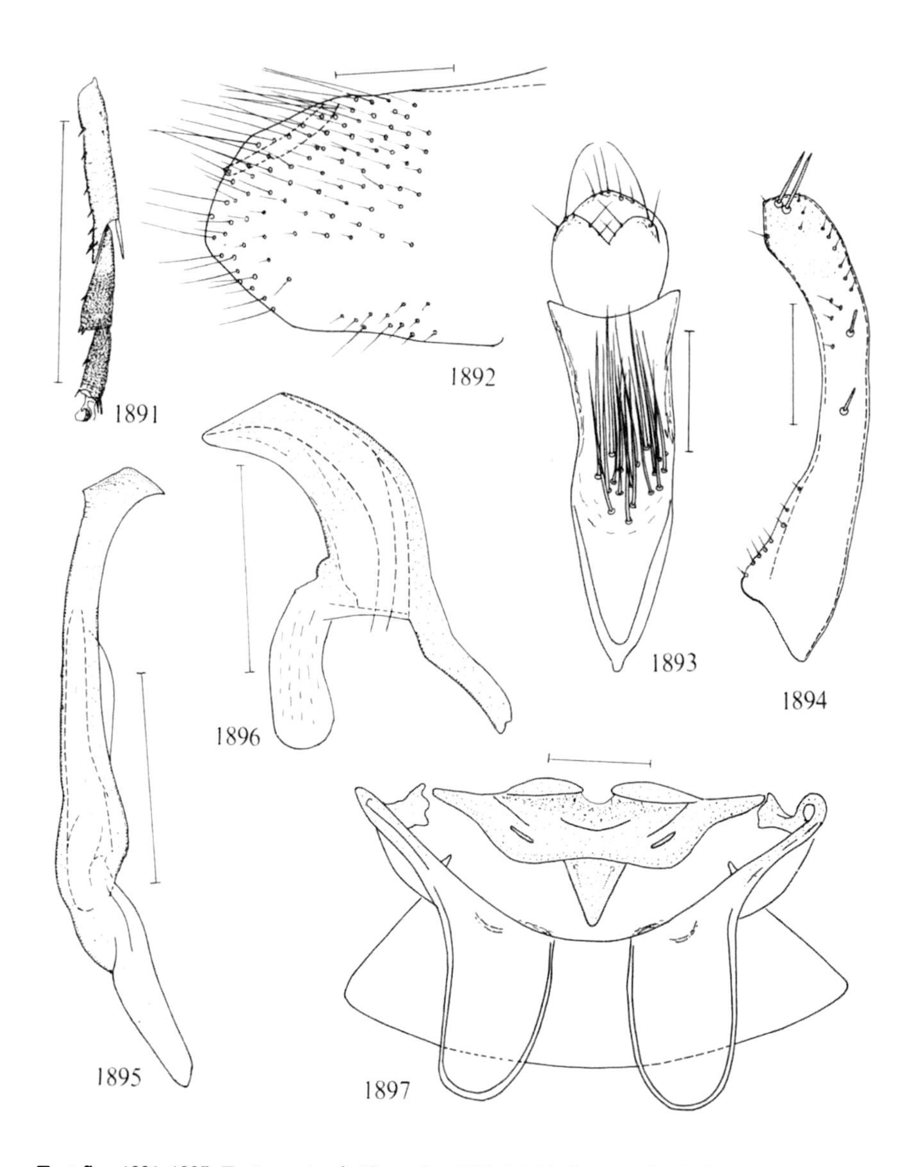

Text-figs. 1891–1897. *Zygina rosincola* (Cerutti). – 1891: left hind tarsus of male from outside; 1892: right pygofer lobe of male from outside; 1893: male anal tube from above; 1894: left genital plate from outside; 1895: left genital style from above; 1896: aedeagus from the left; 1897: 1st–3rd abdominal sterna in male from above. Scale: 0.5 mm for 1891, 0.1 mm for the rest.

Biology. Probably on *Salix* spp. (Ribaut, 1936b); reared on *Salix* sp. with narrow leaves (Günthart, in litt.). Univoltine (Günthart, 1974).

266. ***Zygina (Zygina) rosincola*** (Cerutti, 1939)
Plate-fig. 154, text-figs. 1891–1897.

Erythroneura rosincola Cerutti, 1939: 84.

Resembling *ordinaria.* Male hind tarsus as in Text-fig. 1891, male pygofer lobe as in Text-fig. 1892, male anal tube as in Text-fig. 1893, genital plate as in Text-fig. 1894, genital style as in Text-fig. 1895, aedeagus as in Text-fig. 1896, 1st–3rd abdominal sterna as in Text-fig. 1897. Overall length (♂) 2.75–2.9 mm.

Distribution. Denmark: so far found in EJ only: Skæring 2.X.1965 (Trolle), and Århus 26.IX.1966 (Trolle). – Apparently not uncommon in southern Sweden, so far found in Öl.: Högsrum, Halltorps hage 3.–6.VIII.1976, 28.–31.VIII.1976 (H. Andersson & R. Danielsson); Sm.: Torpa 14.VIII.1934 (Ossiannilsson); Ög.: Landeryd 5.VII.1934, and Rystad, Fröstad 25.V.1934 (Ossiannilsson); Upl.: Solna 24.IX.1940, 25.XI.1941, 10.IV.1942 (Ossiannilsson), Uppsala, Åsen 17.X.1972, Nysätra near the church 17.X.-1972 (Ossiannilsson). – Norway: AK: Sem, Asker 18.IX.1974 (Taksdal). – Not recorded from East Fennoscandia. – Switzerland.

Biology. Host-plant: *Rosa,* also found on *Crataegus* (Günthart, 1974). Taksdal collected his two males on *Pyrus communis.* Univoltine (Günthart, l.c.). I found specimens on *Picea abies* 25.XI. and 10.IV, so hibernation appears to take place in the adult stage.

267. ***Zygina (Zygina) suavis*** Rey, 1891
Text-figs. 1898–1903.

Zygina blandula suavis Rey, 1891: 225.
Zygina rhamnicola Horváth, 1903b: 556.
Erythroneura inconstans Ribaut, 1936b: 48.

Resembling *tiliae,* fore body often without red markings, or pronotum with two parallel red longitudinal stripes or with one parallel-side red longitudinal band. Red longitudinal band of fore wing not reaching to scutellar border of clavus, exterior margin only roughly following claval suture. Extended red markings often present also in corium. Clavus, scutellum and pronotum not fumose, metanotum more or less infuscate. Apical part of fore wing partly faintly fumose. Hind tarsi of male entirely or only partly infuscate, sometimes entirely pale. Ratio length of hind tarsus/length of hind tibia in male = 0.44–0.50, in female = 0.41–0.46. Male pygofer lobe as in Text-fig. 1898, male anal tube as in Text-fig. 1899, genital plate as in Text-fig. 1900, genital style as in Text-fig. 1901, aedeagus as in Text-fig. 1902, 1st–3rd abdominal sterna in male as in Text-fig. 1903. Overall length (♂♀) 2.9–3.05 mm.

Distribution. So far not found in Denmark, nor in Norway. - Scarce in Sweden, found in Sk.: Ängelholm 4.X.1939 (Ossiannilsson); Ög.: Borensberg 18.VIII.1934 (Ossiannilsson); Upl.: Solna 20.X.1942 (Ossiannilsson); Hls.: Edsbyn, Sötmyran 18.IX.-1977 (Bo Henriksson). - East Fennoscandia: recorded from *Rhamnus frangula* in Ab: Åbo, Ispoinen, 1948 and 1949, by Linnavuori (1950). I have not seen these specimens but the host-plant note makes the identification probable. - England, France, German

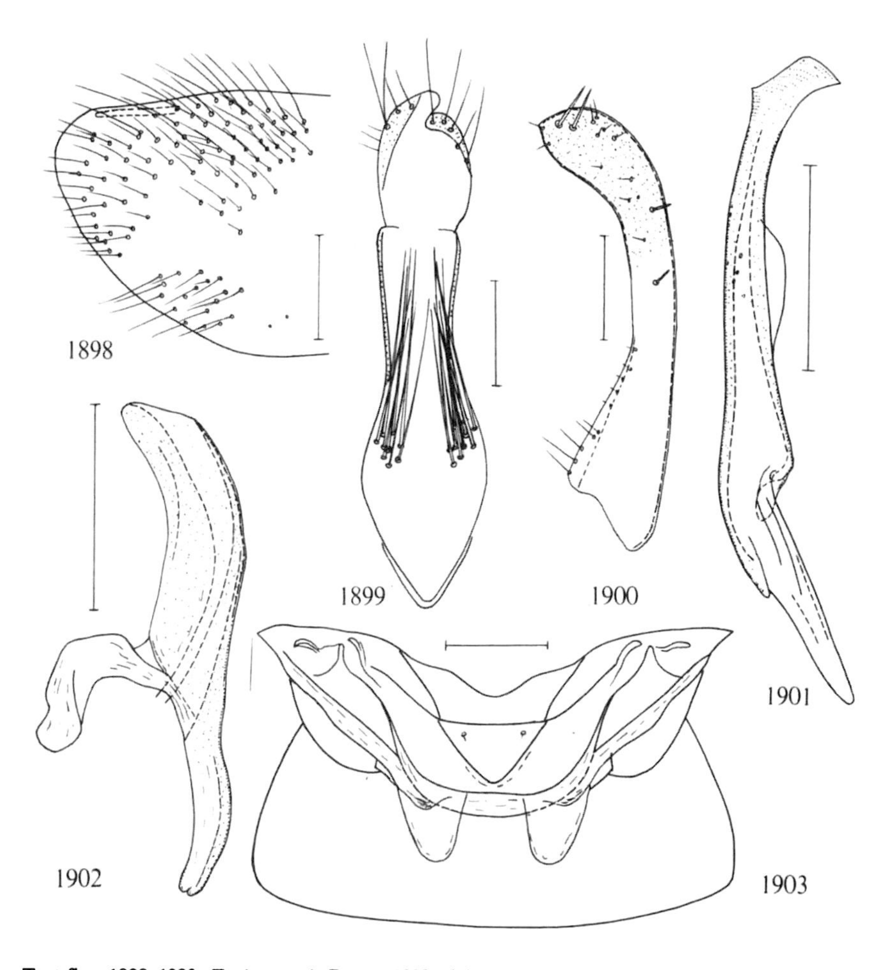

Text-figs. 1898–1903. *Zygina suavis* Rey. - 1898: right pygofer lobe of male from outside; 1899: male anal tube from above; 1900: left genital plate from outside; 1901: left genital style from above; 1902: aedeagus from the left; 1903: 1st–3rd abdominal sterna in male from above. Scale: 0.1 mm.

D.R. and F.R., Switzerland, Austria, Italy, Bohemia; further distribution uncertain since all records published before 1974 (Günthart) must be revised.

Biology. On *Rhamnus frangula* and *cathartica* (Vidano, 1959). At least two generations p.a. (in Switzerland) (Günthart, 1974).

268. ***Zygina (Zygina) schneideri*** (Günthart, 1974)
Text-figs. 1904–1909.

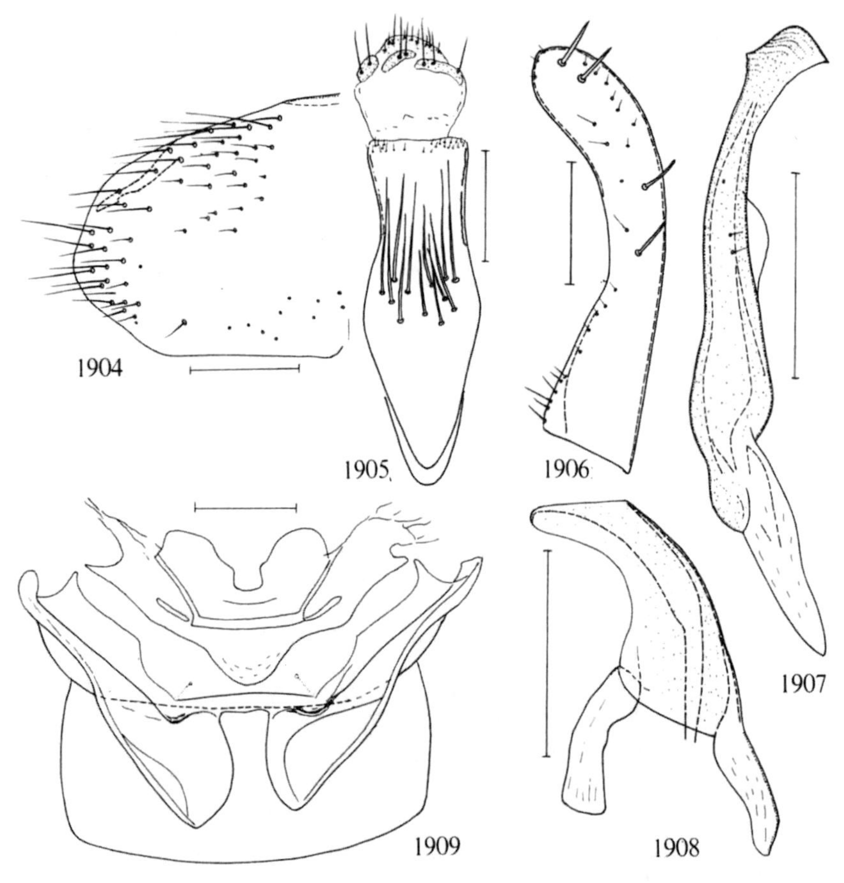

Text-figs. 1904–1909. *Zygina schneideri* (Günthart). – 1904: right pygofer lobe of male from outside; 1905: male anal tube from above; 1906: left genital plate from outside; 1907: left genital style from above; 1908: aedeagus from the left; 1909: 1st–3rd abdominal sterna in male from above (apodemes laterally folded). Scale: 0.1 mm.

Erythroneura (Flammigeroidia) schneideri Günthart, 1974: 24.

As *Z. suavis,* differing primarily by the structure of male anal tube (Text-fig. 1905) and by the size of the apodemes of 2nd abdominal sternum in the male (Text-fig. 1909). Last

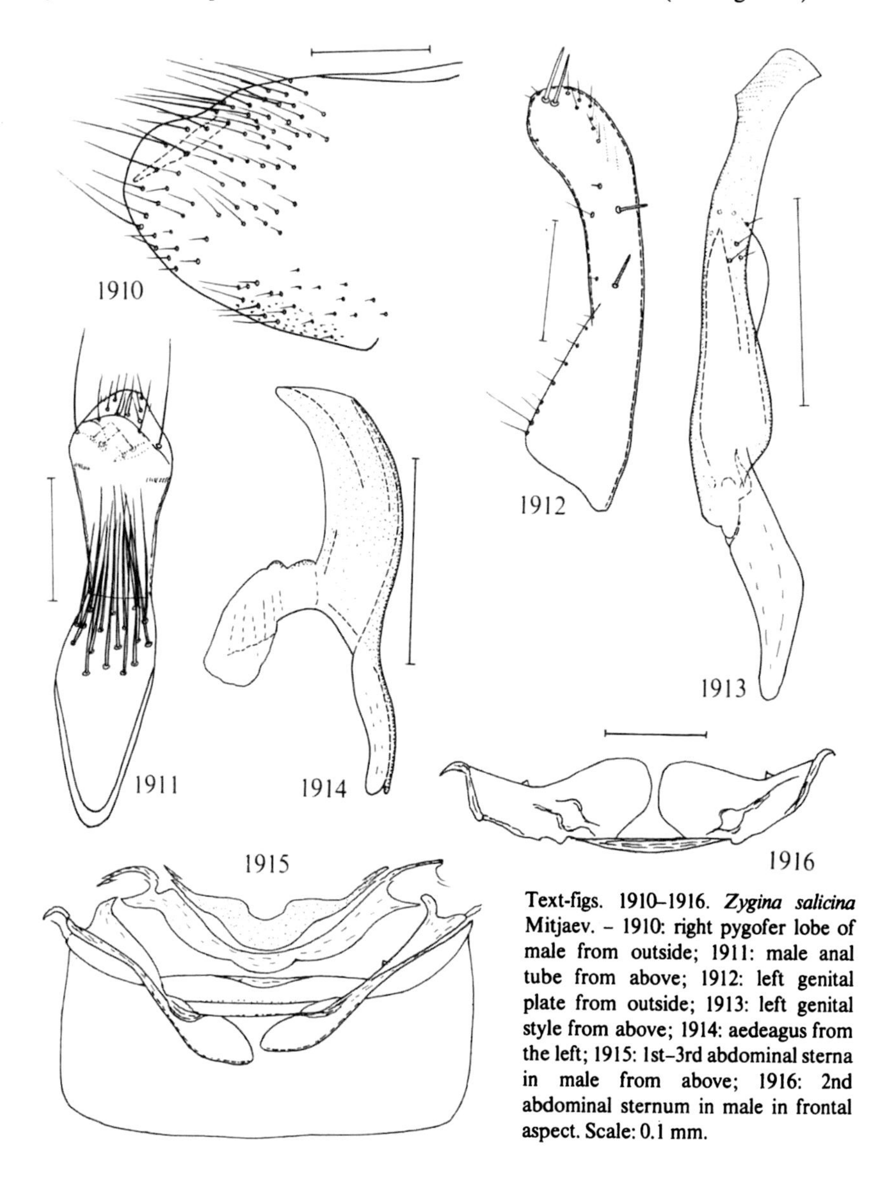

Text-figs. 1910–1916. *Zygina salicina* Mitjaev. – 1910: right pygofer lobe of male from outside; 1911: male anal tube from above; 1912: left genital plate from outside; 1913: left genital style from above; 1914: aedeagus from the left; 1915: 1st–3rd abdominal sterna in male from above; 1916: 2nd abdominal sternum in male in frontal aspect. Scale: 0.1 mm.

hind tarsal segment and distal half of 2nd hind tarsal segment in male black. Male pygofer lobe as in Text-fig. 1904, genital plate as in Text-fig. 1906, genital style as in Text-fig. 1907, aedeagus as in Text-fig. 1908. Measurements (after Günthart, 1974): total length of male: mean 2.86 mm (2.80–2.96), of female: mean 2.91 (2.80–3.00); vertex in male, length 0.18 mm, width 0.29 mm, ratio l/w = 0.60; vertex in female, length 0.17 mm, width 0.30 mm, ratio l/w = 0.57; ratio length of hind tarsus/length of hind tibia in male = 0.44 (0.41–0.50).

Distribution. Not recorded from Denmark. – Sweden: so far found by the present author in Sk.: Brunnby, Kullen 26.VIII.1964; Ög.: Viby 27.VIII.1933, Rystad, Fröstad 3.XI.1974; Upl.: Experimentalfältet 22.X.1939. – Found in Norway, AK: Ås and Asker 4.VI.1969, 17.X.1973, 19.VI., 6.VIII, 7.IX.1974, 29.IX.1975 by Taksdal and Baeschlin. – So far not recorded from East Fennoscandia. – Switzerland.

Biology. Collected on *Prunus avium*, reared on *Malus, Sorbus, Prunus* spp. At least 2 generations p.a. In summer conditions (long-day light) present in pale form only, during winter (short-day) developing bright orange-red colour (Günthart, 1974). I found *Zygina schneideri* on *Ulmus* and *Picea abies;* Taksdal and Baeschlin collected it from *Pyrus communis, Pyrus malus,* plum trees, *Fragaria* cult.

269. ***Zygina (Zygina) salicina*** Mitjaev, 1975
Text-figs. 1910–1916.

Zygina salicina Mitjaev, 1975: 580.

Resembling *Z. ordinaria* (Ribaut), distinctly smaller (♂ 2.4–2.6 mm, ♀ 2.5–2.9 mm, according to Mitjaev, 1975). Male pygofer lobe as in Text-fig. 1910, anal tube in male as in Text-fig. 1911, genital plate as in Text-fig. 1912, genital style as in Text-fig. 1913, aedeagus as in Text-fig. 1914, 1st–3rd abdominal sterna in male as in Text-figs. 1915, 1916.

Distribution. Denmark: found in NEJ: Svinkløv in May, 1966, and in October, 1967, by L. Trolle. – Not recorded from Sweden, Norway and East Fennoscandia. – Kazakhstan.

Biology. On *Salix rosmarinifolia* (Mitjaev) and *Salix repens* (Trolle).

270. ***Zygina (Zygina) rubrovittata*** (Lethierry, 1869)
Plate-fig. 155, text-figs. 1917–1923.

Typhlocyba rubrovittata Lethierry, 1869: 363.
Typhlocyba (Zygina) ericetorum J. Sahlberg, 1871: 185.

Resembling *tiliae* but much smaller, main colour of body and fore wings more yellowish. Vertex frontally more produced and more angular in female than in male. Fore wings fumose, especially in their apical part. Hind tarsi of male entirely black, apical part of male hind tibia often also black. Males often without red markings on fore body, females usually with a red longitudinal band on head and pronotum. Fore

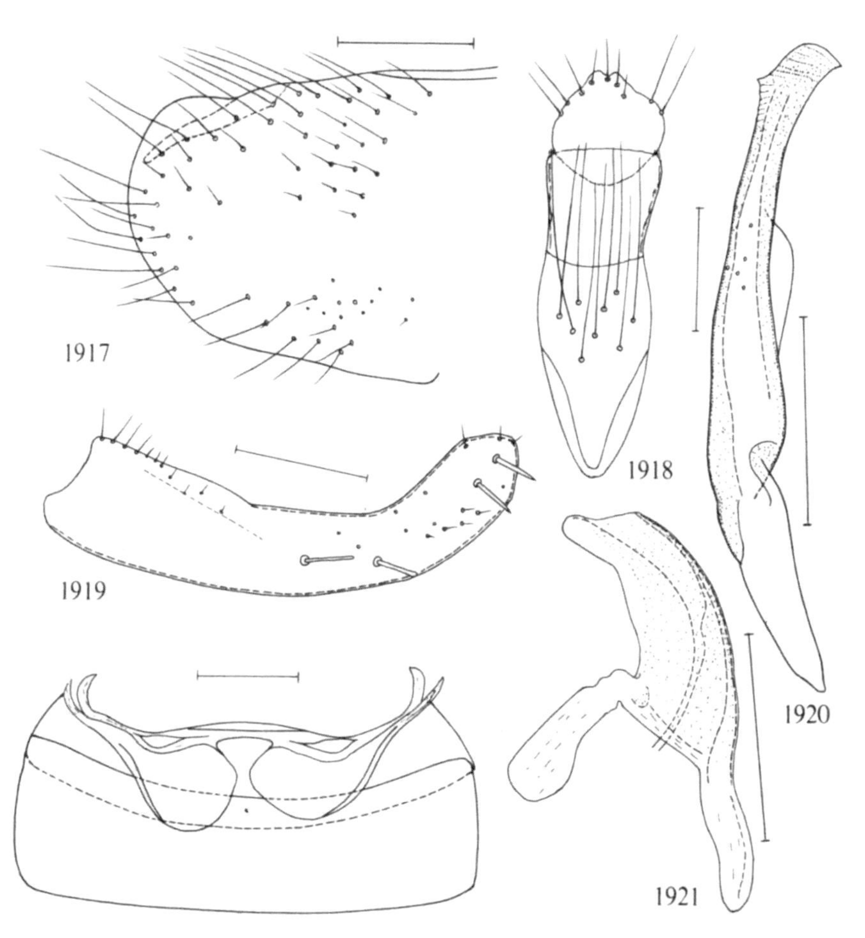

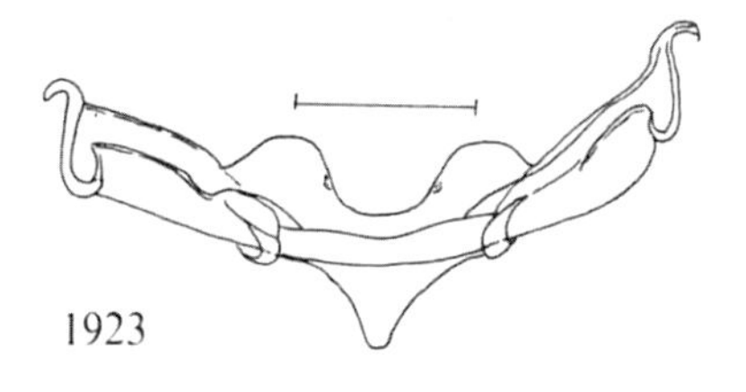

Text-figs. 1917–1923. *Zygina rubrovittata* (Lethierry). – 1917: right pygofer lobe of male from outside; 1918: male anal tube from above; 1919: left genital plate from outside; 1920: left genital style from above; 1921: aedeagus from the left; 1922: 2nd–4th abdominal sterna in male from above; 1923: 1st abdominal sternum in male in frontal aspect. Scale: 0.1 mm.

wings usually with a more or less distinctly marked red longitudinal band. Male pygofer lobe as in Text-fig. 1917, male anal tube as in Text-fig. 1918, genital plate as in Text-fig. 1919, genital style as in Text-fig. 1920, aedeagus as in Text-fig. 1921, 1st–4th abdominal sterna in male as in Text-figs. 1922, 1923. Overall length (♂♀) 2.3–2.75 mm.

Distribution. Not very common and abundant in Denmark, found in EJ, WJ, NEJ and B. – Common in southern and central Sweden, Sk.–Dlr., Ång., Vb. – Norway: found in Bø: Drammen (Warloe); TEi: Øvrebø, Bø (Holgersen); AAy: Lillesand (Holgersen); VAy: Lyngdal (Holgersen). – Common in southern and central East Fennoscandia, found in Al, Ab–Ta, Tb, Sb, Kb; Vib, Kr. – Widespread in Europe.

Biology. On *Calluna vulgaris*. Adults in July–September.

271. ***Zygina (Hypericiella) hyperici*** (Herrich-Schäffer, 1836)
Plate-figs. 137, 138, 156, text-figs. 1924–1930.

Typhlocyba hyperici Herrich-Schäffer, 1836: 4.
Typhlocyba coronula Boheman, 1845b: 160.
Typhlocyba placidula Stål, 1853: 176.

Moderately elongate, faintly shining, vertex frontally rounded. Male light yellow, usually only caudal apex of scutellum, metanotum and center of abdominal dorsum black, apical part of fore wing fumose. Vertex in female with a black, often red-bordered spot shaped as in Plate-fig. 137; pronotum in female often with a broad longitudinal band frontally black, caudally dark red; scutellum largely black, each fore wing with a sanguine longitudinal band along the black commissural border; dorsum of abdomen largely black, apex of saw-case dark. Dark markings on upper side often strongly reduced also in females, the *black* markings on vertex and scutellar apex being more constant than the rest. Dorsum of male anal tube without long setae. Male pygofer lobe as in Text-fig. 1924, process of pygofer lobe as in Text-fig. 1925, genital plate as in Text-fig. 1926, genital style as in Text-fig. 1927, aedeagus as in Text-figs. 1928, 1929, 1st–3rd abdominal sterna in male as in Text-fig. 1930. Overall length 2.45–2.6 mm. – Last instar larvae yellowish white with a still lighter median longitudinal line on dorsum, without dark markings. Dorsal setae absent except on 9th abdominal segment where some short hairs are present. Tibial setae short and weak. Distal half of antennal flagellum blackish.

Distribution. Denmark: fairly common in Jytland, otherwise scarce. – Locally common in southern Sweden, often abundant, found Sk.–Upl., Nrk. – Norway: found in AK: Bækkelaget and Ormøen (Siebke), and in AK: Asker 30.VI.1969 (Taksdal). – Scarce and sporadic in southern East Fennoscandia, found in Ab, N, Ta, Sa. – Widespread in Europe, also in Kazakhstan, Uzbekistan, Kirghizia.

Biology. On *Hypericum perforatum*. In mesophilous and especially in xerophilous biotopes (Schiemenz, 1969b). In southern Sweden (vicinity of Stockholm) 2 generations. Adults in June–September.

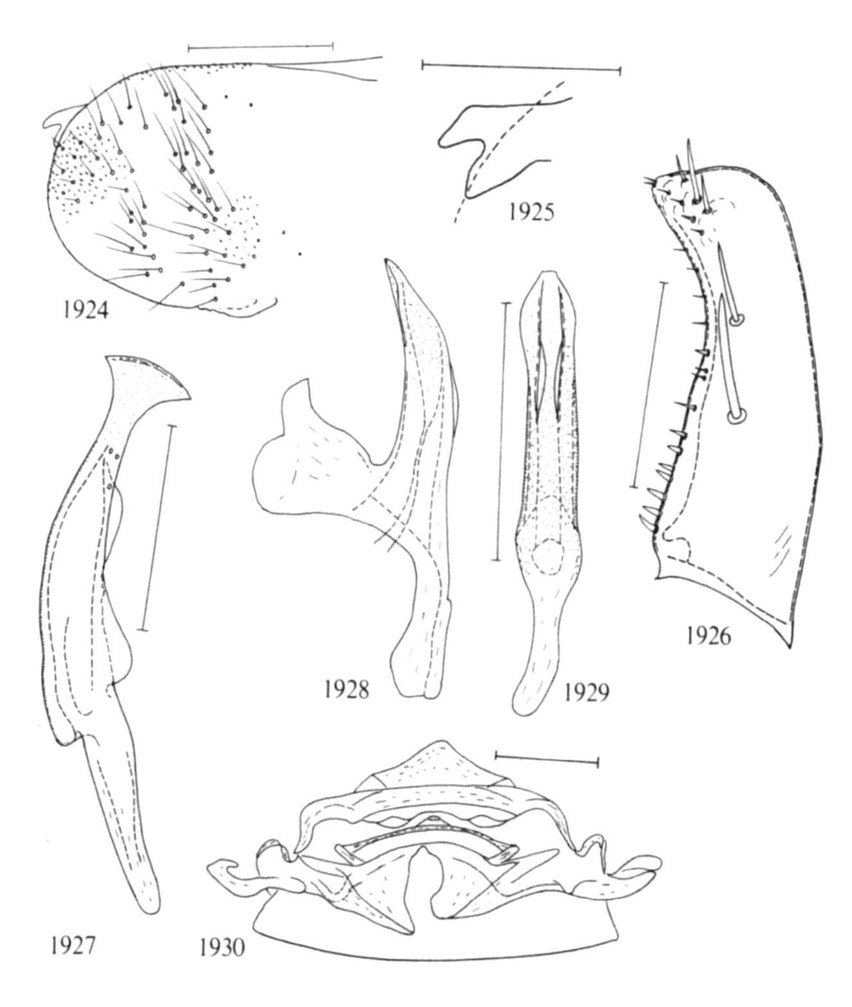

Text-figs. 1924–1930. *Zygina hyperici* (Herrich-Schäffer). – 1924: right pygofer lobe of male from outside; 1925: process of right pygofer lobe of male from outside; 1926: left genital plate from outside; 1927: left genital style from above; 1928: aedeagus from the left; 1929: aedeagus in ventral aspect; 1930: 1st–3rd abdominal sterna in male from above. Scale: 0.05 mm. for 1925, 0.1 mm for the rest.

Genus *Arboridia* Zachvatkin, 1946

Zyginidia, subgenus *Arboridia* Zachvatkin, 1946: 153.
Type-species: *Typhlocyba parvula* Boheman, 1845, by original designation.

Body moderately elongate. Fore wings less than four times as logn as broad. Frontoclypeus in lateral aspect very convex. Face less than twice as long as broad between eyes. Pygofer of male with a pair of simple or forked upper appendages. Genital plate with a small number of macrosetae on basal half. Genital style apically unsymmetrical, with a "second eptension" (Text-fig. 1933). Aedeagus with a pair of processes arising from socle. In Denmark and Fennoscandia one species.

272. ***Arboridia parvula*** (Boheman, 1845)
Plate-fig. 135, text-figs. 1931–1936.

Typhlocyba parvula Boheman, 1845b: 161.
Erythroneura disjuncta Ribaut, 1931c: 404 (nec Mc Atee, 1920).

Dirty yellow, fore body faintly shining, fore wings strongly shining. Anteclypeus black. Frontoclypeus in male usually narrowly, in female broadly black-edged. Vertex with two rounded black spots. Pronotum with a broad brownish transverse band often occupying the major part of its surface. Along fore border of this band there are some small black spots. Scutellum with a black spot on each lateral corner. Fore wing with two brownish longitudinal bands, one in clavus and another on both sides of Cu. Apical part of fore wing fumose. Thoracal venter black spotted, abdomen entirely or largely black. Pygofer process in male as in Text-fig. 1931, genital plate as in Text-fig. 1932, genital style as in Text-fig. 1933, aedeagus as in Text-figs. 1934, 1935, 1st–3rd abdominal sterna in male as in Text-fig. 1936. Overall length (♂♀) 2.5–3 mm. – Last instar larva brown mottled, legs light. Head a little wider than pronotum, frontally rounded. Length of antennae about 3/7 of body length. Setae of dorsum and fore margin of head stump, apically abruptly cut off. Fore margin of head with 4 moderately long macrosetae and a few shorter setae. Vertex with one long macroseta near each eye. Pronotum with two pairs of long setae near fore border, two pairs of long or somewhat shorter setae near hind border, and a few shorter setae. Mesonotum with 3–4 pairs of long setae, each fore wing bud with 6 long macrosetae. Metanotum with one pair of macrosetae, hind wing buds each with 3 macrosetae. Abdomen dorsally with 4 longitudinal rows of macrosetae, most terga with two pairs. Tibial setae short.

Distribution. Rare in Denmark, only a few records from EJ and LFM. – Not uncommon in southern Sweden, found in Sk., Öl., Ög., Vg., Dlsl., Upl. – Norway: found in HEs: Eidskog 11.VII.1974 (teste Ossiannilsson, 1977). – Rare and sporadic in East Fennoscandia, found in Ab: Raisio (Linnavuori); Ta: Lammi 3.VIII.1948 (Linnavuori); also in Oa (Raatikainen & Vasarainen, 1973); Kr: Jalguba (J. Sahlberg). – Widespread in Europe, also in Tunisia, Georgia, Azerbaijan, Kazakhstan, Tadzhikistan, m. Siberia, Maritime Territory.

Biology. On *Filipendula ulmaria* (Ossiannilsson, 1946c). "On *Rubus chamaemorus* in the Kareva bog. It lives quite near the bog surface. It hibernates as the adult, and imagos are found throughout the summer" (Linnavuori, 1952a). Hibernation takes place in the adult stage (Müller, 1957). Two generations (Schiemenz, 1969b). "Wirtspflanze: *Rubus idaeus*" (Günthart, 1974). We found adults in May–October.

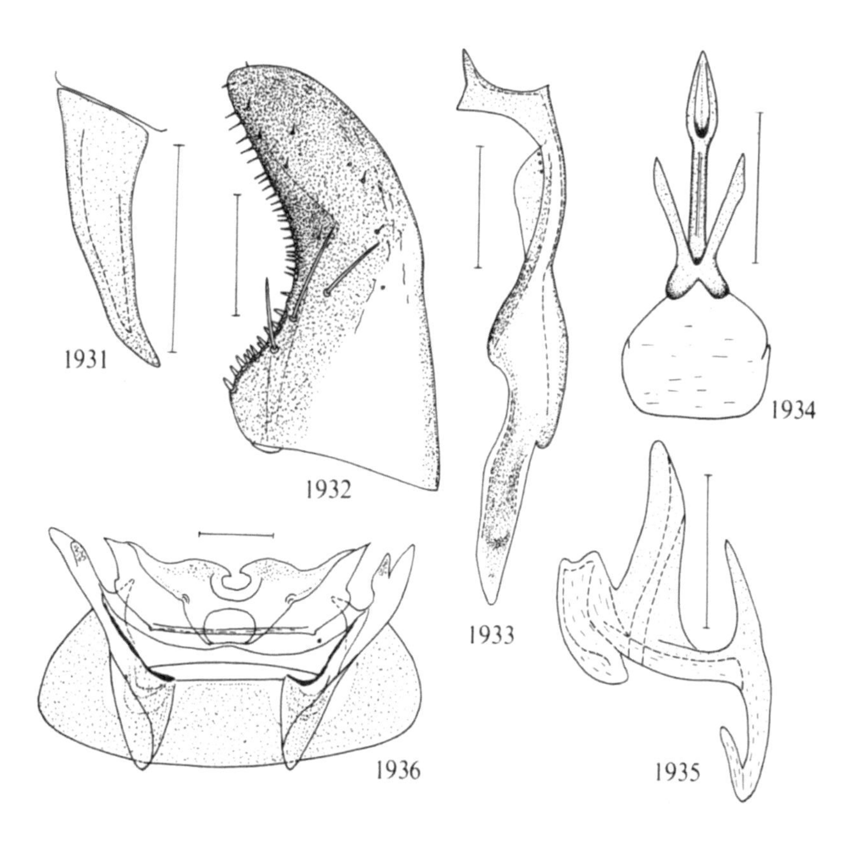

Text-figs. 1931–1936. *Arboridia parvula* (Boheman). – 1931: right upper pygofer process in male from outside (as seen through pygofer wall); 1932: left genital plate from outside; 1933: right genital style from inside; 1934: aedeagus in postero-ventral aspect; 1935: aedeagus from the left; 1936: 1st–3rd abdominal sterna in male from above. Scale: 0.1 mm.

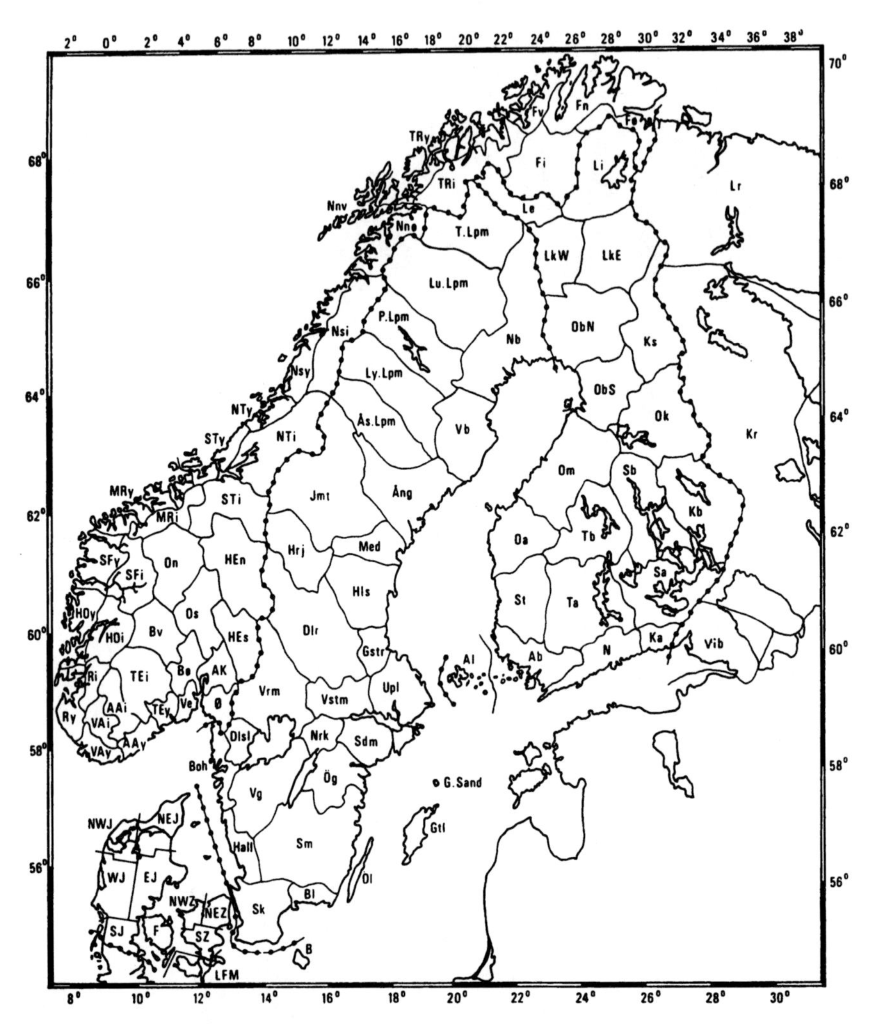

List of abbreviations for the provinces used throughout the text, on the map and in the following tables.

DENMARK

SJ	South Jutland	LFM	Lolland, Falster, Møn
EJ	East Jutland	SZ	South Zealand
WJ	West Jutland	NWZ	North West Zealand
NWJ	North West Jutland	NEZ	North East Zealand
NEJ	North East Jutland	B	Bornholm
F	Funen		

SWEDEN

Sk.	Skåne	Vrm.	Värmland
Bl.	Blekinge	Dlr.	Dalarna
Hall.	Halland	Gstr.	Gästrikland
Sm.	Småland	Hls.	Hälsingland
Öl.	Öland	Med.	Medelpad
Gtl.	Gotland	Hrj.	Härjedalen
G. Sand.	Gotska Sandön	Jmt.	Jämtland
Ög.	Östergötland	Ång.	Ångermanland
Vg.	Västergötland	Vb.	Västerbotten
Boh.	Bohuslän	Nb.	Norrbotten
Dlsl.	Dalsland	Ås. Lpm.	Åsele Lappmark
Nrk.	Närke	Ly. Lpm.	Lycksele Lappmark
Sdm.	Södermanland	P. Lpm.	Pite Lappmark
Upl.	Uppland	Lu. Lpm.	Lule Lappmark
Vstm.	Västmanland	T. Lpm.	Torne Lappmark

NORWAY

Ø	Østfold	HO	Hordaland
AK	Akershus	SF	Sogn og Fjordane
HE	Hedmark	MR	Møre og Romsdal
O	Opland	ST	Sør-Trøndelag
B	Buskerud	NT	Nord-Trøndelag
VE	Vestfold	Ns	southern Nordland
TE	Telemark	Nn	northern Nordland
AA	Aust-Agder	TR	Troms
VA	Vest-Agder	F	Finnmark
R	Rogaland		

n northern   s southern   ø eastern   v western   y outer   i inner

FINLAND

Al	Alandia	Kb	Karelia borealis
Ab	Regio aboensis	Om	Ostrobottnia media
N	Nylandia	Ok	Ostrobottnia kajanensis
Ka	Karelia australis	ObS	Ostrobottnia borealis, S part
St	Satakunta	ObN	Ostrobottnia borealis, N part
Ta	Tavastia australis	Ks	Kuusamo
Sa	Savonia australis	LkW	Lapponia kemensis, W part
Oa	Ostrobottnia australis	LkE	Lapponia kemensis, E part
Tb	Tavastia borealis	Li	Lapponia inarensis
Sb	Savonia borealis	Le	Lapponia enontekiensis

USSR

Vib Regio Viburgensis   Kr Karelia rossica   Lr Lapponia rossica

Printed in the United States
By Bookmasters